Nonlinear Optimization and Applications

Nonlinear Optimization and Applications

Edited by

G. Di Pillo

University of Rome "La Sapienza"
Rome, Italy

and

F. Giannessi

University of Pisa
Pisa, Italy

Plenum Press • New York and London

Library of Congress Cataloging-in-Publication Data

On file

Proceedings of the International School of Mathematics "G. Stampacchia" 21st Workshop on Nonlinear Optimization and Applications, held June 13 – 21, 1995, in Erice, Italy

ISBN 0-306-45316-9

A Division of Plenum Publishing Corporation
233 Spring Street, New York, N. Y. 10013

10 9 8 7 6 5 4 3 2 1

Printed in the United States of America

PREFACE

This volume contains the edited texts of the lectures presented at the workshop on *Nonlinear Optimization: Theory and Applications*, held in Erice at the "G. Stampacchia" School of Mathematics of the "E. Majorana" International Centre for Scientific Culture June 13–21, 1995.

The meeting was conceived to review and discuss recent advances and promising research trends concerning theory, algorithms, and innovative applications in the field of Nonlinear Optimization. This is a field of mathematics which is providing viable tools in engineering, in economics and in other applied sciences, and which is giving a great contribution also in the solution of the more practiced linear optimization problems.

The meeting was attended by approximately 70 people from 18 countries. Besides the lectures, several formal and informal discussions took place. The result was a broad exposure providing a wide and deep understanding of the present research achievements in the field.

We wish to express our appreciation for the active contributions of all the participants in the meeting. Our gratitude is due to the Ettore Majorana Center in Erice, which offered its facilities and stimulating environment: its staff was certainly instrumental for the success of the meeting. Our gratitude is also due to Francisco Facchinei and Massino Roma for the time spent in the organization of the workshop, and to Giuliana Cai for the careful typesetting of this volume. We are indebted to the Italian National Research Council, and in particular to its Committees on Engineering Sciences, on Mathematical Sciences, and on Technological Researches for their financial support. Finally, we address our thanks to Plenum Publishing Corp. for having offered to publish this volume and for their continuing cooperation.

Gianni Di Pillo (Roma)
Franco Giannessi (Pisa)

CONTENTS

TOWARDS A DISCRETE NEWTON METHOD WITH MEMORY FOR LARGE-SCALE OPTIMIZATION

Byrd Richard H.[1], Nocedal Jorge[2] and Zhu Ciyou[2]

[1] Computer Science Department
University of Colorado at Boulder
Boulder Colorado 80309

[2] Department of Electrical Engineering and Computer Science
Northwestern University
Evanston Il 60208
e-mail: nocedal@eecs.nwu.edu

Abstract. A new method for solving large nonlinear optimization problems is outlined. It attempts to combine the best properties of the discrete-truncated Newton method and the limited memory BFGS method, to produce an algorithm that is both economical and capable of handling ill-conditioned problems. The key idea is to use the curvature information generated during the computation of the discrete Newton step to improve the limited memory BFGS approximations. The numerical performance of the new method is studied using a family of functions whose nonlinearity and condition number can be controlled.

Key words. Discrete Newton method, variable metric method, truncated Newton method, large scale optimization, nonlinear optimization, limited memory method.

1. INTRODUCTION

In some important classes of large unconstrained optimization problems, the objective function f and its gradient g are available, but the Hessian matrix cannot be computed or stored. Problems of this type are often solved by the discrete-truncated Newton method or the limited memory BFGS method, but both methods have significant drawbacks that sometimes lead to unacceptable computing times. The discrete Newton method can require many gradient evaluations per iteration, whereas the limited memory method can be very slow on ill-conditioned problems. In order to remedy these limitations we propose an algorithm that attempts to combine, in a novel way, the best properties of each method. The purpose of this paper is to outline our approach and test it in a controlled setting.

[1] This author was supported by NSF grant CCR-9101795, ARO grant DAAL 03-91-G-0151, and AFOSR grant AFOSR-90-0109.

[2] These authors were supported by National Science Foundation Grants CCR-940081 and ASC-9213149, and by Department of Energy Grant DE-FG02-87ER25047-A004.

It is convenient to begin our discussion by reviewing the truncated Newton method. Each iteration of this type of method generates a search direction d_k that approximately solves the Newton equations

$$\nabla^2 f(x_k) d_k = -g(x_k). \tag{1.1}$$

For large problems it is not cost effective to solve the system (1.1) exactly since this may require excessive computing time, and an approximation may result in a sufficiently good search direction. Therefore an inner iterative method such as the linear conjugate method [5] is used to solve the system (1.1) approximately. The accuracy of the solution of (1.1) can be controlled by the norm of the residual,

$$r_k \equiv \nabla^2 f(x_k) d_k + g(x_k),$$

which is forced to decrease as the iterates converge to a solution [1],[15]. An alternative is to terminate the inner iteration when sufficient decrease in the quadratic model

$$\frac{1}{2} d_k^T \nabla^2 f(x_k) d_k + g(x_k)^T d_k$$

has been obtained [11].

When the Hessian $\nabla^2 f(x_k)$ is not positive definite, the inner conjugate gradient iteration may generate a direction v of negative curvature, i.e. a vector v such that $v^T \nabla^2 f(x_k) v < 0$. If this occurs, the inner iteration is terminated and the last estimate obtained prior to the generation of the negative curvature direction is taken as the search direction d_k. Because the inner conjugate gradient iteration is truncated (when either the desired accuracy is reached, or negative curvature is detected), this method is called the *truncated Newton method.*

A modification of the truncated Newton method allows us to solve problems where the Hessian $\nabla^2 f(x_k)$ is not available. It is based on the observation that the conjugate gradient method only needs the product of $\nabla^2 f(x_k)$ with certain displacement vectors v – and does not require the Hessian matrix itself. These products can be approximated by finite differences,

$$\nabla^2 f(x_k) v \approx \frac{g(x_k + \epsilon v) - g(x_k)}{\epsilon}, \tag{1.2}$$

where ϵ is a small differencing parameter. Since each iteration of the conjugate gradient method performs one product $\nabla^2 f(x_k) v$, it requires a new evaluation of the gradient g of the objective function. A method that uses (1.2) is called a *discrete Newton method* [13],[10]. The combination of these two ideas gives the discrete-truncated Newton method, which has been implemented, for example, in the codes of Nash [9] and Schlick and Fogelson [14].

After computing the search direction d_k, we define the new iterate by

$$x_{k+1} = x_k + \alpha_k d_k, \tag{1.3}$$

where α_k is a step length parameter. In this paper we will assume that α_k satisfies the strong Wolfe conditions (cf. [2] or [4])

$$f(x_k + \alpha_k d_k) \le f(x_k) + c_1 \alpha_k g(x_k)^T d_k \tag{1.4}$$

$$|g(x_k + \alpha_k d_k)^T d_k| \le c_2 |g(x_k)^T d_k| \tag{1.5}$$

where $0 < c_1 < c_2 < 1$ are constants.

We should note that the product $\nabla^2 f(x_k)v$ can be computed by automatic differentiation techniques [6] instead of the finite difference (1.2). Automatic differentiation has the important advantage of being accurate. However it is at least as expensive as finite differences in terms of computing time. The algorithms presented below can use the automatic differentiation technique, but for concreteness the details of the discussion will be framed in terms of the discrete version (1.2) of the method.

The accuracy with which (1.1) is solved has a marked effect on the behavior of the discrete-truncated Newton method: if only one iteration of the inner conjugate gradient method is performed, the method reduces to steepest descent, whereas high accuracy results in an approximation to Newton's method.

The discrete-truncated Newton method can be very useful for solving large problems but suffers from two drawbacks. It is not easy to guess the accuracy with which the Newton equations (1.1) are to be solved, so as to obtain a good rate of convergence and at the same time avoid an unnecessarily large number of gradient evaluations. If the accuracy in the solution of (1.1) is too low, convergence will be very slow, but if the inner convergence test is too stringent, an excessive number of gradient evaluations will be made during the inner conjugate gradient iteration. In each of these extreme cases the discrete-truncated Newton method is likely to be less efficient than the limited memory BFGS method described in the next section. The second deficiency is that, in comparison with a quasi-Newton method, discrete-truncated Newton spend a great amount of functional information into each major iteration, but this information is not used in subsequent iterations. In contrast, a quasi-Newton method always carries the information gathered about the Hessian of objective function from one step to the next.

How can we improve the cost-effectiveness of the discrete-truncated Newton method? An idea that comes to mind is to save the curvature information generated by the inner conjugate gradient iteration at the iterate x_k and use it at the next iterate x_{k+1}. This can be done in various ways.

Our approach consists of viewing the inner cycle of conjugate gradient iterations from the point of view of variable metric methods. We note that the gradient differences (1.2) contain useful information about the curvature of the objective function f which can be saved in the form of a limited memory quasi-Newton matrix. Once this information has been stored, it may be unnecessary to perform a discrete Newton step at the next iteration since a limited memory step using this iteration matrix may produce good progress towards the solution. Before we can explain this idea in more detail, we need to digress and review the main features of the limited memory BFGS method.

2. LIMITED MEMORY BFGS

The limited memory BFGS method (L-BFGS) is an adaptation, to large problems, of the standard BFGS method. Like the discrete-truncated Newton method, it does not require knowledge about the Hessian matrix, but uses only function and gradient information. The iteration of the L-BFGS method is of the form (1.3), where α_k is a steplength satisfying the Wolfe conditions (1.4)-(1.5), and where the search direction d_k is of the form

$$d_k = -H_k g(x_k). \tag{2.1}$$

The iteration matrix H_k, which is not formed explicitly, is based on the BFGS updating formula. In the standard BFGS method the inverse Hessian approximation H_k is updated at every iteration by means of the formula

$$H_{k+1} = V_k^T H_k V_k + \rho_k s_k s_k^T, \tag{2.2}$$

where

$$\rho_k = 1/y_k^T s_k, \qquad V_k = I - \rho_k y_k s_k^T, \tag{2.3}$$

and

$$s_k = x_{k+1} - x_k, \qquad y_k = g(x_{k+1}) - g(x_k).$$

We say that the matrix H_{k+1} is obtained by updating H_k using the pair $\{s_k, y_k\}$.

The $n \times n$ BFGS matrices H_k will generally be dense, so that storing and manipulating them is prohibitive when the number of variables is large. To circumvent this problem, the limited memory BFGS method does not form these matrices but only stores a certain number, say m, of pairs $\{s_k, y_k\}$ that define them implicitly through the BFGS update formula (2.2)-(2.3). The parameter m therefore determines how much memory is used by the algorithm. The product $H_k g(x_k)$ is obtained by performing a sequence of inner products involving $g(x_k)$ and these m pairs $\{s_k, y_k\}$. After computing the new iterate, we save the most recent correction pair $\{s_k, y_k\}$ – unless the storage is full, in which case we delete the oldest pair from the set $\{s_i, y_i\}$ to make room for the newest one, $\{s_k, y_k\}$. In summary, limited memory BFGS algorithm always keeps the m most recent pairs $\{s_i, y_i\}$ to define the iteration matrix. This approach is suitable for large problems because it has been observed in practice that small values of m (say $m \in [3, 20]$) very often give satisfactory results [7], [3].

Let us now describe the updating process in more detail. Suppose that the current iterate is x_k and that we have stored the m pairs $\{s_i, y_i\}$, $i = k-m, ..., k-1$. We first define the "basic matrix" $H_k^{(0)} = \gamma_{k-1} I$ where

$$\gamma_{k-1} = \frac{s_{k-1}^T y_{k-1}}{y_{k-1}^T y_{k-1}}. \tag{2.4}$$

We then update $H_k^{(0)}$ m times using the BFGS formula (2.2)-(2.3) and the m pairs $\{s_i, y_i\}$, $i = k-m, ..., k-1$. From (2.2) we see that H_k can be written as

$$\begin{aligned} H_k &= \left(V_{k-1}^T \cdots V_{k-m}^T\right) H_k^{(0)} \left(V_{k-m} \cdots V_{k-1}\right) \\ &+ \rho_{k-m} \left(V_{k-1}^T \cdots V_{k-m+1}^T\right) s_{k-m} s_{k-m}^T \left(V_{k-m+1} \cdots V_{k-1}\right) \\ &+ \rho_{k-m+1} \left(V_{k-1}^T \cdots V_{k-m+2}^T\right) s_{k-m+1} s_{k-m+1}^T \left(V_{k-m+2} \cdots V_{k-1}\right) \\ &+ \vdots \\ &+ \rho_{k-1} s_{k-1} s_{k-1}^T. \end{aligned} \tag{2.5}$$

A recursive formula described in [12] takes advantage of the symmetry of this expression to compute the product $H_k g(x_k)$ efficiently.

The numerical performance of the limited memory method L-BFGS is often very good in terms of total computing time. However the method suffers from two major drawbacks: it is not rapidly convergent, and on ill-conditioned problems it can require an excessive number of function and gradient evaluations. In contrast, the discrete-truncated Newton method can be designed to give fast convergence, and to cope with ill-conditioned problems; its drawback is that it often requires too many gradient evaluations in (1.2).

3. THE NEW APPROACH

Our aim is to design an algorithm that is as economical as L-BFGS in terms of gradient evaluations, but is efficient on ill-conditioned problems. There are two ways of viewing our goal. One of them is to try to improve the L-BFGS method by interleaving discrete-truncated Newton steps and using the curvature information generated during the inner conjugate gradient iteration to improve the L-BFGS matrix H_k. The second view is to note that the discrete Newton method does not possess memory: the gradient differences used at an iterate are not used at the next one. Therefore the idea is to save the information from these gradient differences in the form of a limited memory matrix, and avoid performing discrete-truncated Newton steps during the next few iterations.

Let us first introduce some notation that will help us explain the idea of saving information from the conjugate gradient cycle into a limited memory matrix.

Let $\{x_l\}$ be the iterates generated by the optimization algorithm. Suppose that at the iterate x_k we perform a discrete-truncated Newton step, i.e. we approximately solve (1.1) by the conjugate gradient (CG) method, using the finite difference (1.2). Let us write the CG iteration at x_k as

$$z^{i+1} = z^i + \sigma^i v^i, \tag{3.1}$$

where σ^i is the steplength and where the directions $\{v^i\}$ are conjugate with respect to the Hessian matrix $\nabla^2 f(x_k)$; see e.g. [5]. The initial guess in the CG iteration is set to be $z^1 = 0$. Suppose that p iterations of the CG method are performed; then the search direction of the optimization algorithm at x_k is given by $d_k = z^p$. To perform the multiplication $\nabla^2 f(x_k)v^i$, which is needed at every CG iteration, we compute the finite difference

$$\frac{g(x_k + \epsilon v^i) - g(x_k)}{\epsilon}.$$

The inner CG iteration (3.1) is terminated when

$$r^i \equiv \nabla^2 f(x_k) z^i + g(x_k) \tag{3.2}$$

satisfies $\|r^i\| \leq c g(x_k)$, for some constant $c > 0$.

Let us also define

$$s_k^i = \epsilon v^i, \qquad y_k^i = g(x_k + \epsilon v^i) - g(x_k), \quad i = 1, \ldots, p. \tag{3.3}$$

Since the CG iteration is terminated if negative curvature is detected, these vectors satisfy the curvature condition

$$(s_k^i)^T y_k^i > 0,$$

and can be used to define a limited memory BFGS matrix H_k as described in the previous section. We conjecture that the information saved in this manner can be quite useful because the conjugate gradient iteration will explore the function along directions that would normally not be seen by the limited memory method.

The new algorithm could proceed as follows. At the starting point x_0 we perform a discrete-truncated Newton step leading to x_1, and construct a limited memory matrix H_1 using the vectors (3.3) generated during the inner conjugate gradient iteration. The matrix H_1 will also incorporate the pair

$$s_0 = x_1 - x_0, \qquad y_0 = g_1 - g_0 \tag{3.4}$$

corresponding the total step. Thus H_1 is obtained by updating a multiple of the identity matrix $p+1$ times. We now proceed with the limited memory method generating iterates $x_2, x_3, \ldots$, and updating the limited memory matrices $H_2, H_3, \ldots$, at each iteration using the information $(s_1, y_1), (s_2, y_2), \ldots$, from the total steps.

At some iterate x_t it is judged that a discrete-truncated Newton step should be computed again. We then approximately solve the linear system

$$\nabla^2 f(x_t) d_t = -g(x_t) \tag{3.5}$$

using the conjugate gradient method, and store the new inner vectors (3.3). This produces the search direction d_t leading to the new iterate x_{t+1}. At this point we remove from the limited memory matrix H_t the information obtained from the previous cycle of conjugate gradient iterations and replace it with the latest set of vectors (3.3) generated by the inner CG iterations used to solve (3.5). Thus the matrix H_{t+1} is obtained by updating a multiple of the identity matrix using the most recent inner steps (3.3) plus (possibly) a few previous outer pairs $(s_{t-j}, y_{t-j}), \ldots, (s_t, y_t)$.

We can think that the storage is divided in two: one part contains the pairs $\{s^i, y^i\}$ generated during the inner CG iteration, and the other contains pairs $\{s_k, y_k\}$ corresponding to the outer (or total) steps of the algorithm. The information corresponding to the outer steps is continuously updated as in the standard limited memory method: the latest correction vector replaces the oldest one. The information from the inner CG iteration could be kept intact until a new truncated Newton step is performed, and then completely refreshed, or it could be lumped in with the outer iteration pairs and refreshed in the same last-in-first-out fashion.

In summary the algorithm combines discrete-truncated Newton and limited memory steps. But the algorithm does not simply alternate them. The key idea is to preserve the information generated during the inner CG iteration to improve the quality of the limited memory matrix used in subsequent steps.

3.1. Variations

Even though we have outlined our approach in some detail, many variations of it are possible and some details of implementation have to be made more precise. We now list and discuss some of these points.

(i) We need to develop a criterion to determine when to trigger a new discrete-truncated Newton step. This could be done with a regular frequency – say 1 discrete-truncated Newton step for every 10 limited memory steps – but an automatic criterion based on the observed behavior of the objective function would be more appropriate.

(ii) As in any truncated Newton method, the termination test on the inner CG iteration needs to be chosen carefully. But in our approach we also need to determine if all, or only some, of the pairs generated during the inner CG iteration should be stored.

(iii) The limited memory correction pairs available at an iterate where a discrete-truncated Newton step is to be performed could be used as a preconditioner for the CG iteration. This poses the additional question of what types of pairs should be used in the preconditioner: only the inner pairs $\{s^i, y^i\}$, only the outer pairs $\{s_k, y_k\}$, or both? A simple form of preconditioning of this type is used by Nash [10].

(iv) How long the two blocks information, from the inner and outer iterations, are to be kept in memory must be determined, as discussed above. Also, the *order* in which limited memory updating is performed has not been specified. The limited memory matrix H_{k+1} could be obtained by updating a multiple of the identity matrix using first the inner information $\{s^i, y^i\}$, followed by the most recent outer pairs (s_k, y_k). Or the order could be reversed, updating first with the outer information and then with the inner information. (In the limited memory method L-BFGS the criterion is simple: the oldest pairs are used first.)

(v) This brings up the choice of the scaling parameter γ_k that initializes limited memory updating; see (2.4). It could be based on inner or outer information pairs.

Since there appear to be too many algorithmic choices, many of which could have an important impact on performance, we will not attempt to experiment with different combinations. Instead we would like to test the main idea underlying in this approach: *is it beneficial to save the inner pairs* $\{s^i, y^i\}$ *in the limited memory matrix?* To try to answer this question, we will perform a set of controlled experiments.

4. NUMERICAL INVESTIGATION

We will test three algorithms. The first one is the limited memory BFGS method (L-BFGS) as described in [7]. It is used mainly as a benchmark. The other two algorithms combine discrete-truncated Newton and limited memory steps in a regular manner: a discrete Newton step is performed at iterations $6, 16, 26, \ldots$, and all other steps are limited memory BFGS steps. The inner CG iteration that computes the discrete Newton step is terminated either when the residual (3.2) satisfies $\|r^i\|_2 \leq 10^{-2}$, or when the total number of CG iterations is 20. (Since all the test functions will be strongly convex, there is no need to include a test that terminates the iteration when negative curvature is detected). The difference between these two methods lies in the type of information they save, as we now discuss in detail.

DINEMO. (Discrete Newton method with Memory). The first iterations are identical to those of the L-BFGS method. When we reach the first point x_k at which a discrete-truncated Newton step is to be performed, we clear the storage by discarding all the correction pairs saved in the limited memory matrix. We then compute the discrete-truncated Newton step and save all the pairs $\{s^i, y^i\}$ $i = 1, \ldots, p$ generated during the inner CG iteration, as well as the pair (s_k, y_k) corresponding to the outer step. At the new iterate x_{k+1} we construct a limited memory BFGS matrix H_{k+1} as follows: (i) the scaling parameter (2.4) is defined by the outer pair (s_k, y_k); (ii) we update the matrix $\gamma_k I$ $p+1$ times using the inner pairs $\{s^i, y^i\}$ $i = 1, \ldots, p$, and the outer pair (s_k, y_k) (in that order). We then use H_{k+1} to perform a limited memory BFGS step. We continue generating limited memory steps – and adding the new correction pairs to storage – until a new discrete-truncated Newton step is to be performed.

The maximum number of limited memory corrections stored is $m = 29$. Since the maximum number of inner CG steps is $maxcg = 20$, and since a discrete-truncated Newton step is performed at every 10 iterations, limited memory corrections are only discarded prior to taking a discrete-truncated Newton step.

ALTERNATE. In this method the limited memory and discrete-truncated Newton steps are alternated, but the information from the inner CG iteration is *not saved.* This method is identical to DINEMO, except for the construction of the limited memory matrix H_{k+1} immediately following a discrete-truncated Newton step. In ALTERNATE the scaling parameter (2.4) is defined by the outer pair (s_k, y_k), and H_{k+1} is obtained by updating $\gamma_k I$ once using the pair (s_k, y_k).

We will measure the effect of saving information from the inner CG cycle by comparing DINEMO, which saves this information, with ALTERNATE, which does not. The three methods use the same line search. It is performed by the routine of Moré and Thuente [8] with parameters $c_1 = 10^{-4}$ and $c_2 = 0.9$ in (1.4)-(1.5).

We chose the following quartic objective function in $n = 100$ variables to perform our tests,

$$\min \frac{1}{2}(x-1)^T D(x-1) + \frac{\sigma}{4}\left((x-1)^T B(x-1)\right)^2 + 1,$$

where D is a positive definite diagonal matrix, σ is a parameter that controls the deviation from quadratic, and

$$B = U^T U, \quad \text{with} \quad U = \begin{bmatrix} 1 & \cdots & 1 \\ & \ddots & \vdots \\ & & 1 \end{bmatrix}.$$

The starting point was chosen as $(-1)^i \times 50$ for $i = 1, \ldots, 100$.

In the first experiment, the matrix D was chosen as

$$D = \text{diag}\left[(1+\epsilon)^{-50}, (1+\epsilon)^{-49}, \ldots, (1+\epsilon)^{49}\right], \tag{4.1}$$

where $\epsilon > 0$ is a variable parameter that determines the condition number of D. In our tests we used the values $\epsilon = 0$, $\epsilon = 0.05$, and $\epsilon = 0.09$ which give rise, respectively, to the condition numbers of $1, 125$, and 5073 in D. We also tried several values for the parameter σ to observe the behavior of the methods on non-quadratic functions.

Since we are interested in highlighting the differences between the three methods, we used a very stringent stopping test. The runs were terminated when both

$$f(x_k) \leq 1 + 10^{-14} \tag{4.2}$$

and

$$g(x_k)^T g(x_k) \leq 10^{-14} \tag{4.3}$$

(note that the optimal objective function value is 1). On some runs, only one of the stopping tests was satisfied, and this is indicated by a * in the tables below, but this is of no particular importance since all the runs achieved essentially the same accuracy. Our code always evaluates the function and gradient simultaneously, and the tables below give the total number of function/gradient evaluations.

The results of the first experiment are given in Table 1. When $\epsilon = \sigma = 0$ the objective function is a quadratic with unit Hessian, and would be minimized using only 1 function/gradient evaluation if an exact line search were used. However our code always tries the unit steplength and accepts it if it satisfies the Wolfe conditions (1.4)-(1.5). Because of this, 6 function/gradient evaluations were required by each of the three methods in this case.

	ϵ/σ			
Algorithm	0/0	0/.06	0/.12	0/.18
L-BFGS	6	131	138	151
DINEMO	6	110	115	115
ALTERNATE	6	133	136	148
	ϵ/σ			
Algorithm	.05/0	.05/.06	.05/.12	.05/.18
L-BFGS	134	208	211	218
DINEMO	153	212	211	210
ALTERNATE	154	246	248	274
	ϵ/σ			
Algorithm	.09/0	.09/.06	.09/.12	.09/.18
L-BFGS	683	607	607	600
DINEMO	899	922	*740	926
ALTERNATE	1084	*936	*811	868

Table 1. Number of function/gradient evaluations when D is given by (4.1). The symbol * indicates that only the stopping test (4.3) was met.

Let us analyze the results of Table 1. When the problem is very well conditioned ($\epsilon = 0$) the new algorithm (DINEMO) performs quite well. However, as ϵ increases, its performance relative to L-BFGS deteriorates. DINEMO does well compared with ALTERNATE (the method that does not save the information generated by the inner CG iteration), except for large values of ϵ and σ. We may be tempted to draw the following conclusion from this test: saving the information from the inner conjugate gradient cycle accelerates the iteration, but for ill-conditioned problems the cost incurred does not pay off – since L-BFGS ends up being the winner when $\epsilon = .09$. It turns out, however, that the condition number of the problem is not the determining factor in the efficiency of the methods, but that the distribution of eigenvalues plays a crucial role, as we will show below. Note that when D is given by (4.1), its eigenvalues are more or less evenly spread, and that this situation is very disadvantageous to the (unpreconditioned) inner conjugate gradient iteration, since the problem has 100 distinct eigenvalues and we only allow 20 conjugate gradient iterations.

In the second experiment we alter D so that its eigenvalue distribution changes but its condition number remains the same. We let the 5 smallest and 6 largest elements be as in (4.1), but make all the other elements 1. Thus, if the new diagonal matrix is denoted by $\hat{D} = \text{diag}(\hat{d}_1, ..., \hat{d}_n)$ we have

$$\begin{aligned} \hat{d}_i &= d_i \quad i = 1, ..., 5 \\ \hat{d}_i &= 1 \quad i = 6, ..., 94 \\ \hat{d}_i &= d_i \quad i = 95, ..., 100, \end{aligned} \tag{4.4}$$

where $D = \text{diag}(d_1, ..., d_n)$ is defined by (4.1). The results are given in Table 2 and are strikingly different. We have omitted the results for $\epsilon = 0$, since they are the same as in Table 1. All three methods required substantially fewer function evaluations, but the new method (DINEMO) performs consistently better than the other two. It is interesting to note that DINEMO has a clear advantage of ALTERNATE, indicating that the strategy of saving inner CG information is useful.

	ϵ/σ			
Algorithm	.05/0	.05/.06	.05/.12	.05/.18
L-BFGS	56	172	177	182
DINEMO	43	139	142	166
ALTERNATE	44	198	206	215
	ϵ/σ			
Algorithm	.09/0	.09/.06	.09/.12	.09/.18
L-BFGS	96	291	291	288
DINEMO	52	175	180	180
ALTERNATE	72	256	253	248

Table 2. Number of function/gradient evaluations when D is given by (4.4).

The matrix $\hat{D}$ given by (4.4) has only 12 distinct eigenvalues, and since the inner CG cycle is allowed to perform 20 steps it will completely solve the Newton equations (1.1) when $\sigma = 0$. Therefore this case may be too simple.

In the third experiment we alter D further so that the inner CG iteration is not able to solve the Newton equations. We leave the 5 smallest and 6 largest eigenvalues as in (4.4) but now split the eigenvalue of 1 into 89 eigenvalues which are contained in the interval $[0.6, 9.4]$. The new diagonal matrix, which we denote by $\bar{D}$ is given by

$$\begin{aligned} \bar{d}_i &= d_i \quad i = 1, ..., 5 \\ \bar{d}_i &= i/10 \quad i = 6, ..., 94 \\ \bar{d}_i &= d_i \quad i = 95, ..., 100. \end{aligned}$$

The eigenvalues of $\bar{D}$ are still clustered into three groups, but the middle cluster has a relatively wide spread of eigenvalues which prevents the CG iteration from "seeing" the cluster of smallest eigenvalues. Note that the condition numbers of $D, \hat{D}$ and $\bar{D}$ remain the same. The results of the third experiment are given in Table 3.

	ϵ/σ			
Algorithm	.05/0	.05/.06	.05/.12	.05/.18
L-BFGS	102	194	190	191
DINEMO	98	178	208	208
ALTERNATE	123	218	219	247
	ϵ/σ			
Algorithm	.09/0	.09/.06	.09/.12	.09/.18
L-BFGS	264	415	359	354
DINEMO	222	333	274	331
ALTERNATE	364	405	*444	435

Table 3. Number of function/gradient evaluations when D is given by (4.1). The symbol * indicates that only the stopping test (4.3) was met.

Table 3 shows that when D is well conditioned ($\epsilon = .05$, which corresponds to a condition number of 125) L-BFGS and DINEMO are comparable. This is in contrast to Table 2 where DINEMO had a clear advantage in this case. Thus collecting incomplete information during the inner CG iteration is not as beneficial for the new method DINEMO. On the other hand, when the condition number of D is large ($\epsilon = .09$, corresponding to a condition numer of 5073) DINEMO performs better than L-BFGS. Since from Table 1 we know that DINEMO does very well when $\epsilon = 0$, we see that, overall, it performs quite well in this test. We note once more that DINEMO outperforms ALTERNATE by a wide margin, indicating again that the inner CG information is useful.

In the tables we have reported only the number of function/gradient evaluations since this is the key measure of performance. But we should mention that DINEMO and ALTERNATE required a substantially smaller number of iterations than L-BFGS.

We performed several other experiments. In one of them the 25 smallest and 26 largest eigenvalues of the diagonal matrix were as in (4.1), and all the other eigenvalues were 1. In that test DINEMO clearly outperformed the two other methods for all values of ϵ. We also repeated the first experiment in which D is given by (4.1), using various stopping criteria for the inner conjugate gradient iteration. We observed that the more information is collected in the inner cycle, the better DINEMO performed.

We have drawn the following conclusions from these experiments. Collecting information from the inner CG cycle and saving it in the form of a limited memory matrix is certainly beneficial since DINEMO clearly outperformed ALTERNATE in the great majority of the tests. The benefits of saving this information depend on the eigenvalue structure of the Hessian. The more information is captured during the inner CG iteration, the more competitive DINEMO is with respect to L-BFGS. Not only is DINEMO faster in terms of number of outer iterations, but its cost effectiveness in terms of function/gradient evaluations increases by capturing more eigenvalue information. We believe that a fully developed version of DINEMO is likely to become a powerful general-purpose optimization algorithm. However to achieve this goal, all the implementation questions listed in §3.1 need to be investigated. We intend to do so in the near future.

REFERENCES

[1] R.S. Dembo, S.C. Eisenstat and T. Steihaug, "Inexact Newton methods". *SIAM J. Numer. Anal.* 19, pp. 400–408, 1982.

[2] J. E. Dennis, Jr. and R. B. Schnabel, *"Numerical Methods for Unconstrained Optimization and Nonlinear Equations"*. Englewood Cliffs, N.J., Prentice-Hall, 1983.

[3] J.C. Gilbert and C. Lemaréchal, "Some numerical experiments with variable storage quasi-Newton algorithms." *Mathematical Programming* 45, pp. 407–436, 1989.

[4] P. E. Gill, W Murray and M. H. Wright, *"Practical Optimization"*. London, Academic Press, 1981.

[5] G.H. Golub and C.F. Van Loan, *"Matrix Computations"*. (Second Edition), The John Hopkins University Press, Baltimore and London, 1989.

[6] A. Griewank, "On automatic differentiation." In *"Mathematical Programming"* (M. Iri and K. Tanabe, eds.), Kluwer Academic Publishers, pp. 83–107, Tokyo, 1989.

[7] D. C. Liu and J. Nocedal, "On the limited memory BFGS method for large scale optimization methods." *Mathematical Programming* 45 pp. 503-528, 1989.

[8] J. J. Moré and D.J. Thuente, "Line search algorithms with guaranteed sufficient decrease." *ACM Transactions on Mathematical Software* 20, no. 3, pp. 286-307, 1994.

[9] S.G. Nash, "User's guide for TN/TNBC: FORTRAN routines for nonlinear optimization". Report 397, Mathematical Sciences Dept., The Johns Hopkins University, 1984.

[10] S.G. Nash, "Preconditioning of truncated-Newton methods." *SIAM Journal on Scientific and Statistical Computing* 6, pp. 599–616, 1985.

[11] S.G. Nash and J. Nocedal, "A Numerical Study of the Limited Memory BFGS Method and the Truncated-Newton Method for Large Scale Optimization." *SIAM Journal on Optimization* 1, no. 3, pp. 358-372, 1991.

[12] J. Nocedal, "Updating quasi-Newton matrices with limited storage." *Mathematics of Computation* 35, pp. 773-782, 1980.

[13] D.P. O'Leary, "A discrete Newton algorithm for minimizing a function of many variables." *Mathematical Programming* 23, pp. 20–33, 1982.

[14] T. Schlick and A. Fogelson, "TNPACK - A truncated Newton package for large-scale problems: I. Algorithms and usage." *ACM Transactions on Mathematical Software* 18, no. 1, pp. 46-70, 1992.

[15] T. Steihaug (1983), "The conjugate gradient method and trust regions in large scale optimization." *SIAM J. Num. Anal.* 20, pp. 626–637, 1983.

ON REGULARITY FOR GENERALIZED SYSTEMS AND APPLICATIONS

Castellani Marco[1], Mastroeni Giandomenico[2] and Pappalardo Massimo[3]

Dipartimento di Matematica,

Via Buonarroti, 2 – 56127 Pisa.

e–mail: [1]castella@dm.unipi.it

e–mail: [2]giando@karush.dm.unipi.it

e–mail: [3]pappalar@dm.unipi.it

Abstract. In the field of constrained optimization the nonvacuity or the boundedness of the generalized Lagrange multiplier set is guaranteed under some regularity conditions (or constraint qualification; the difference in the terminology consisting of whether or not the condition involves the objective function). This type of analysis is now-a-days well stated also for nondifferentiable optimization. Moreover, the great development of these topics has been enforced with the recent results that establish strict relationships between regularity conditions (as well as metric regularity) and calmness, exact penalization and stability. The nature of all these conditions is of analytical type. On the other hand, a new approach has been recently proposed for establishing regularity conditions. It mainly exploits geometrical tools and takes into account that regularity conditions for optimality can be expressed as geometrical conditions for certain types of separation or more generally they are conditions which guarantee the impossibility of a system. This paper aims to give a characterization of regularity conditions for generalized systems and to apply it to the study of optimality conditions.

Key Words. Regularity conditions, image space, Lagrange multipliers, generalized systems.

1. INTRODUCTION

We aim to study conditions for a generalized system to be impossible. The systems

Nonlinear Optimization and Applications
Edited by G. Di Pillo and F. Giannessi, Plenum Press, New York, 1996

we will consider are of the following type:

$$\begin{cases} f(x) > 0 \\ g(x) \in C \\ x \in \mathbb{R}^n \end{cases} \tag{S}$$

where $f : \mathbb{R}^n \to \mathbb{R}$, $g : \mathbb{R}^n \to \mathbb{R}^m$ and $C \subseteq \mathbb{R}^m$ is a nonempty and convex cone with apex at the origin.

It is well known that many important mathematical questions, like optimality for constrained extremum problems as well as the study of solutions to Variational Inequalities and problems with inclusion constraints, can be reduced to the impossibility of a system.

Following the approach introduced in [5] and developed in [3,4], we will analyse such a topic by means of separation arguments in the image space associated to (S).

The impossibility of (S) can be established by proving that two suitable subsets of the image space lie in two disjoint level sets of a linear functional.

The existence and regularity of such a functional will be the core of the sequel. To this aim, first of all, we will introduce the concept of regular separation analogously to the classical case for generalized multipliers [6, 9, 12] and then the new concept of total regularity [2] of the linear separation. When the system (S) represents an optimality condition for a constrained extremum problem, these regularity conditions becomes regularity conditions for the Kuhn–Tucker–type optimality conditions.

Further applications to optimization problems will be considered in the last section expecially as regards generalized first–order optimality conditions.

We recall the main definitions and notations that will be used in the sequel. $B(\bar{x}, \delta)$ is the ball of center $\bar{x}$ and radius δ. If $x \in \mathbb{R}^n$, $x \geq 0$ iff $x_i \geq 0$, for each $i = 1, \dots, n$; $\mathbb{R}^n_+ := \{x \in \mathbb{R}^n : x \geq 0\}$; $\mathbb{R}^n_- := \{x \in \mathbb{R}^n : x \leq 0\}$; if $A, B \subseteq \mathbb{R}^n$, $A \times B$ denotes the Cartesian product and

$$\begin{aligned} A + B :=& \{x \in \mathbb{R}^n : \exists a \in A,\ \exists b \in B :\ a + b = x\}, \\ A - B :=& \{x \in \mathbb{R}^n : \exists a \in A,\ \exists b \in B :\ a - b = x\}. \end{aligned}$$

Let $M \subseteq \mathbb{R}^n$, $\operatorname{cl} M$, $\operatorname{conv} M$ and $\operatorname{int} M$ denote the closure, the convex hull and the interior of M, respectively. Moreover

$$\operatorname{cone} M := \{y \in \mathbb{R}^n : y = \lambda x,\ \lambda \geq 0,\ x \in M\}.$$

If C is a cone of $\mathbb{R}^n$ with apex at the origin,

$$C^* := \{y \in \mathbb{R}^m : \langle x, y \rangle \geq 0,\ \forall x \in C\}$$

is the positive polar cone of C; C is said pointed iff $C \cap (-C) = \{0\}$. When $C \subseteq \mathbb{R}^m$ is a convex cone, the function $g : \mathbb{R}^n \to \mathbb{R}^m$ is said a *C–convex function* iff

$$g((1 - \alpha)x + \alpha y) - (1 - \alpha)g(x) - \alpha g(y) \in C, \qquad \forall \alpha \in [0, 1],\ \forall x, y \in \mathbb{R}^n.$$

In particular when $C = \mathbb{R}^m_-$, g is convex.

We recall the main definitions concerning the image space associated to the system (S).

Definition 1.1.

i) The set

$$\mathcal{K} := \{(u,v) \in \mathbb{R} \times \mathbb{R}^m : \ u = f(x), \ v = g(x), \ x \in \mathbb{R}^n\}$$

is the *image* associated to the system (S).

ii) Let

$$\mathcal{H} := \{(u,v) \in \mathbb{R} \times \mathbb{R}^m : \ u > 0, \ v \in C\},$$

the set

$$\mathcal{E} := \mathcal{K} - \operatorname{cl} \mathcal{H}$$

is the *extended image* associated to (S). The space where $\mathcal{K}$, $\mathcal{H}$ and $\mathcal{E}$ run is called *image space.*

It is immediate to see that the system (S) is impossible iff $\mathcal{K} \cap \mathcal{H} = \emptyset$. The next theorem states, in the image space, a result which holds under an assumption which will be analysed in detail in the next section.

Theorem 1.1. *If*

$$\mathcal{H} + \operatorname{cl} \mathcal{H} = \mathcal{H}, \tag{1.1}$$

then the system (S) is impossible iff $\mathcal{E} \cap \mathcal{H} = \emptyset$.

Proof. "Only if". It is enough to note that

$$\mathcal{E} - \mathcal{H} = (\mathcal{K} - \operatorname{cl} \mathcal{H}) - \mathcal{H} = \mathcal{K} - (\operatorname{cl} \mathcal{H} + \mathcal{H}) = \mathcal{K} - \mathcal{H}.$$

"If". It follows from the inclusion $\mathcal{K} \subseteq \mathcal{E}$. □

Remark 1.1. Theorem 1.1 shows that, under the assumption (1.1), the impossibility of (S) is equivalent to the disjunction between $\mathcal{E}$ and $\mathcal{H}$. In many cases $\mathcal{E}$ turns out to have better properties than $\mathcal{K}$ and therefore it is easier to analyse the disjunction between $\mathcal{E}$ and $\mathcal{K}$ with respect to that between $\mathcal{K}$ and $\mathcal{H}$ [5]. Moreover (1.1) is satisfied in the most important cases which arise from the applications.

In the next section the concept of regularity will be connected to separation (Sect. 3), optimality of order 0 (Sect. 4) and of order 1 (Sect. 5).

2. ABOUT THE CONDITION $\mathcal{H} + \operatorname{cl} \mathcal{H} = \mathcal{H}$

In this section $\mathcal{H} \subseteq \mathbb{R}^n$ denotes a generic cone with apex at the origin and not necessarily that used elsewhere in this paper.

Necessary conditions and sufficient conditions for (1.1) to hold are established. First of all we show some examples for which (1.1) does not hold.

Example 2.1.

1) $\mathcal{H} = \{(x,y) \in \mathbb{R}^2 : x > 0, y = 0\} \cup \{(x,y) \in \mathbb{R}^2 : x = 0, y > 0\}$; $(1,0) \in \mathcal{H}$, $(0,1) \in \operatorname{cl}\mathcal{H}$ but $(1,1) \notin \mathcal{H} + \operatorname{cl}\mathcal{H}$.
2) $\mathcal{H} = \{(x,y) \in \mathbb{R}^2 : x \neq 0, y = 0\}$; $(0,0) \in \mathcal{H} + \operatorname{cl}\mathcal{H}$ but $(0,0) \notin \mathcal{H}$.
3) $\mathcal{H} = \{x \in \mathbb{R}^3 : x > 0\} \cup \{x \in \mathbb{R}^3 : x_1 > 0, x_2 = x_3 = 0\}$; $(1,0,0) \in \mathcal{H}$, $(1,0,1) \in \operatorname{cl}\mathcal{H}$ but $(2,0,1) \notin \mathcal{H}$.

The following result holds.

Theorem 2.1. *Suppose that $\mathcal{H} \neq \operatorname{cl}\mathcal{H}$. If (1.1) holds, then we have:*

i) *$\mathcal{H}$ is convex;*
ii) $\mathcal{H} \cap (-\operatorname{cl}\mathcal{H}) = \emptyset$.

Proof.

i) Suppose, ab absurdo, that there exist $a, b \in \mathcal{H}$ such that $a + b \notin \mathcal{H}$. This contradicts the hypothesis because $b \in \operatorname{cl}\mathcal{H}$.
ii) We observe that $0 \notin \mathcal{H}$. Ab absurdo, if $\mathcal{H} \cap (-\operatorname{cl}\mathcal{H}) \neq \emptyset$, then there exists $x \in \mathcal{H}$ such that $-x \in \operatorname{cl}\mathcal{H}$ which implies $0 \in \mathcal{H}$. □

Observe that the case $\mathcal{H} = \operatorname{cl}\mathcal{H}$ is of no interest since (1.1) trivially holds.

Remark 2.1. From Theorem 2.1 it follows that, if $\mathcal{H} \neq \operatorname{cl}\mathcal{H}$, then $\mathcal{H}$ must be pointed in order to have (1.1). Anyway, this condition is not sufficient (see Example 3).

We now need the following result.

Lemma 2.1. *Let the cone $\mathcal{H}$ be convex and have nonempty interior. Then*

$$\operatorname{int}\mathcal{H} + \operatorname{cl}\mathcal{H} = \operatorname{int}\mathcal{H}.$$

Proof. The inclusion $\supseteq$ is immediate since $0 \in \operatorname{cl}\mathcal{H}$. Let now $h_1 \in \operatorname{int}\mathcal{H}$ and $h_2 \in \operatorname{cl}\mathcal{H}$. There exists $\delta > 0$ such that $[h_1 + B(0,\delta)] \subseteq \mathcal{H}$; this and the convexity of $\mathcal{H}$ imply

$$[h_1 + B(0,\delta)] + h_2 \subseteq \operatorname{cl}\mathcal{H}.$$

Therefore $[(h_1 + h_2) + B(0,\delta)] \subseteq \operatorname{cl}\mathcal{H}$, that is $(h_1 + h_2) \in \operatorname{int}(\operatorname{cl}\mathcal{H}) = \operatorname{int}\mathcal{H}$. □

Theorem 2.2. *Let $\mathcal{H}$ be a convex cone. If one of the following conditions*

i) *$\mathcal{H}$ is closed,*
ii) *$\mathcal{H}$ is open with a nonempty interior,*
iii) $\mathcal{H} = \operatorname{cl}\mathcal{H} \setminus \{0\}$

is verified, then (1.1) holds.

Proof.

i) The inclusion $\supseteq$ is immediate since $0 \in \operatorname{cl}\mathcal{H}$. The opposite inclusion holds because $\mathcal{H} = \operatorname{cl}\mathcal{H}$ and the convexity of $\operatorname{cl}\mathcal{H}$ imply $\operatorname{cl}\mathcal{H} + \operatorname{cl}\mathcal{H} = \operatorname{cl}\mathcal{H}$.

ii) Since by assumption $\mathcal{H} = \operatorname{int}\mathcal{H}$, the thesis follows from Lemma 2.1.

iii) We have to prove that

$$(\operatorname{cl}\mathcal{H} \setminus \{0\}) + \operatorname{cl}\mathcal{H} = \operatorname{cl}\mathcal{H} \setminus \{0\}.$$

Since $\operatorname{cl}\mathcal{H} \setminus \{0\}$ is a convex cone and $(\operatorname{cl}\mathcal{H} \setminus \{0\}) + \{0\} = \operatorname{cl}\mathcal{H} \setminus \{0\}$ we obtain

$$(\operatorname{cl}\mathcal{H} \setminus \{0\}) + (\operatorname{cl}\mathcal{H} \setminus \{0\}) \subseteq \operatorname{cl}\mathcal{H} \setminus \{0\}$$

The opposite inclusion follows observing that $0 \in \operatorname{cl}\mathcal{H}$. □

Remark 2.2. When $0 \in \mathcal{H}$ then (1.1) holds iff $\mathcal{H}$ is closed.

Remark 2.3. We now return to the abovementioned examples. Example 1 shows a case where $\mathcal{H} = \operatorname{cl}\mathcal{H} \setminus \{0\}$, $\mathcal{H}$ is pointed but (1.1) does not hold. Example 2 shows a case where $\mathcal{H} = \operatorname{cl}\mathcal{H} \setminus \{0\}$, $\operatorname{cl}\mathcal{H}$ is not pointed but (1.1) does not hold. Example 3 shows a case where $\mathcal{H}$ is convex, $\operatorname{cl}\mathcal{H}$ is pointed, $0 \notin \mathcal{H}$ but (1.1) does not hold.

Remark 2.4. Suppose that $\mathcal{H} = \mathcal{H}_1 \times \mathcal{H}_2$ with $\mathcal{H}_1$ and $\mathcal{H}_2$ satisfying (1.1). Then $\mathcal{H}$ fulfils (1.1). In fact, it is easy to see that

$$\mathcal{H} + \operatorname{cl}\mathcal{H} = (\mathcal{H}_1 \times \mathcal{H}_2) + \operatorname{cl}(\mathcal{H}_1 \times \mathcal{H}_2) = (\mathcal{H}_1 + \operatorname{cl}\mathcal{H}_1) \times (\mathcal{H}_2 + \operatorname{cl}\mathcal{H}_2) = \mathcal{H}_1 \times \mathcal{H}_2 = \mathcal{H}.$$

3. LINEAR SEPARATION

We study the disjunction of the sets $\mathcal{E}$ and $\mathcal{H}$ by showing that they lie in two disjoint level sets of a suitable functional. In this work we will focus our attention on the simplest kind of separating functional: the linear one.

Definition 3.1. We say that $\mathcal{E}$ and $\mathcal{H}$ are *linearly separated* iff there exists a hyperplane whose equation is

$$\pi_{\theta,\lambda}(u,v) := \theta u + \langle \lambda, v \rangle = 0$$

and such that

$$\pi_{\theta,\lambda}(u,v) \geq 0, \quad \forall (u,v) \in \mathcal{H} \tag{3.1}$$

$$\pi_{\theta,\lambda}(u,v) \leq 0, \quad \forall (u,v) \in \mathcal{E}. \tag{3.2}$$

Remark 3.1. Condition (3.1) implies $(\theta,\lambda) \in \mathcal{H}^* = \mathbb{R}_+ \times C^*$.

In the sequel the separation (when achieved) will be always *proper*. Moreover, the existence of a separating hyperplane does not ensure, by itself, the disjunction of sets $\mathcal{E}$ and $\mathcal{H}$. In fact, it is easily seen that, if $\theta = 0$, we may have $\mathcal{H} \cap \mathcal{E} \neq \emptyset$ since the halfline $\{(u,v) \in \mathbb{R} \times \mathbb{R}^m : u > 0,\ v = 0\}$ is included both in $\mathcal{H}$ and in the separating hyperplane, $\mathcal{E}$ being included in its nonpositive level set. Therefore, in order to achieve the disjunction between $\mathcal{H}$ and $\mathcal{E}$, it is necessary to introduce further assumptions which we will refer to as regularity conditions.

In particular, following [2], we will make distinction between the concepts of "regularity" and "total regularity" for the linear separation: the regularity guarentees the existence of a separating hyperplane $\pi_{\theta,\lambda}(u,v) = 0$ with $\theta > 0$, while total regularity ensures that every separating hyperplane $\pi_{\theta,\lambda}(u,v) = 0$ has a positive θ coefficient. In other words, the regularity does not prevent us from the possibility of having an irregular separation, which cannot happen under the hypothesis of total regularity. More precisely, consider the following sets

$$M^0 = \{\lambda \in C^* \backslash \{0\} : \pi_{0,\lambda} \text{ fulfils (3.1) and (3.2)}\},$$
$$M^1 = \{\lambda \in C^* : \pi_{1,\lambda} \text{ fulfils (3.1) and (3.2)}\}.$$

Remark 3.2. $\mathcal{E}$ and $\mathcal{H}$ admit a linear separation iff $M^0 \cup M^1 \neq \emptyset$.

Remark 3.3. We stress the fact that the difference between M^0 and M^1 is made by the first multiplier θ being zero or positive. Analogous definitions can be given for any multipliers $\lambda_1, \ldots, \lambda_m$.

If $M^1 = \emptyset$ and $M^0 \neq \emptyset$, then the separation is irregular as it happens in the following example:

Example 3.1. Consider the system

$$\begin{cases} -x > 0 \\ -x^2 \geq 0. \end{cases}$$

In this case the only separating hyperplane for the sets $\mathcal{E}$ and $\mathcal{H}$ is that of equation $v = 0$.

Definition 3.2. We say that $\mathcal{E}$ and $\mathcal{H}$ admit a *regular linear separation* iff $M^1 \neq \emptyset$; if, furthermore, $M^0 = \emptyset$, then the separation is said *totally regular* [2].

Remark 3.4. The existence of a regular linear separation guarantees that $\mathcal{E} \cap \mathcal{H} = \emptyset$ and then the impossibility of the system (S).

The next result characterizes the existence of a regular linear separation. To this aim, we introduce the following set

$$\hat{\mathcal{H}} := \{(u,v) \in \mathbb{R} \times \mathbb{R}^m : u > 0,\ v \in C \cup \{0\}\},$$

and prove the following result.

Theorem 3.1.

$$M^1 \neq \emptyset \qquad \Leftrightarrow \qquad \text{cl cone conv } \mathcal{E} \cap \hat{\mathcal{H}} = \emptyset.$$

Proof. $\Rightarrow$ For each $\lambda \in M^1$ we consider the set

$$S_\lambda := \{(u,v) \in \mathbb{R} \times \mathbb{R}^m : u + \langle \lambda, v \rangle \leq 0\}.$$

It is immediate to observe that

$$\text{cl cone conv } \mathcal{E} \subset S_\lambda.$$

Recalling that $\lambda \in C^*$ we have that

$$u + \langle \lambda, v \rangle \geq u > 0, \quad \forall (u,v) \in \hat{\mathcal{H}}.$$

$\Leftarrow$ Given $\bar{u} > 0$, there exists $\rho > 0$ such that

$$\text{cl cone conv } \mathcal{E} \cap B((\bar{u},0),\rho) = \emptyset.$$

Then

$$\text{cl cone conv } \mathcal{E} \cap \text{ cone } B((\bar{u},0),\rho) = \{(0,0)\}.$$

Since the two previous sets are closed and convex cones, there exists a hyperplane whose equation be $\theta u + \langle \lambda, v \rangle = 0$, which separates them, we have to show that $\theta \neq 0$. For each $v \in \mathbb{R}^m$

$$\left(\bar{u}, \rho \frac{v}{\|v\|}\right) \in \text{ cone } B((\bar{u},0),\rho);$$

therefore $\theta = 0$ would imply

$$\langle \lambda, v \rangle \geq 0, \quad \forall v \in \mathbb{R}^m,$$

that is $\lambda \equiv 0$, which contradicts the existence of the hyperplane. □

Theorem 3.1 generalizes a result stated in [5]. It is possible to give a further characterization of the condition given in the previous theorem; in fact the set $\hat{\mathcal{H}}$ can be replaced with the subset

$$\hat{\mathcal{H}}_u := \{(u,v) \in \mathbb{R} \times \mathbb{R}^m : u > 0, \; v = 0\}.$$

To this aim let us state some preliminary results.

Lemma 3.1. *Let $A \subseteq \mathbb{R}^{m+1}$. We have*

$$\text{conv}\,(A - \text{cl}\,\mathcal{H}) - \mathcal{H} \subseteq \text{conv}\,(A - \text{cl}\,\mathcal{H}).$$

Proof. Let $x \in \text{conv}\,(A - \text{cl}\,\mathcal{H})$ and $h \in \mathcal{H}$; then there exist $a_1, \ldots, a_{m+2} \in A$, $h_1, \ldots, h_{m+2} \in \text{cl}\,\mathcal{H}$ and $\lambda_1, \ldots, \lambda_{m+2} \geq 0$ with $\sum_{i=1}^{m+2} \lambda_i = 1$, such that

$$x = \sum_{i=1}^{m+2} \lambda_i(a_i - h_i).$$

Therefore

$$x - h = \sum_{i=1}^{m+2} \lambda_i(a_i - h_i) - \sum_{i=1}^{m+2} \lambda_i h = \sum_{i=1}^{m+2} \lambda_i(a_i - h_i - h).$$

Since $\text{cl}\,\mathcal{H}$ is a convex cone then $h + h_i \in \text{cl}\,\mathcal{H}$, for each i, that is $x - h \in \text{conv}\,(A - \text{cl}\,\mathcal{H})$.

□

Lemma 3.2. *Let $A \subseteq \mathbb{R}^{m+1}$. We have*

$$\text{cl cone conv}\,(A - \text{cl}\,\mathcal{H}) - \mathcal{H} \subseteq \text{cl cone conv}\,(A - \text{cl}\,\mathcal{H}).$$

Proof. Let $x \in \text{cl cone conv}\,(A - \text{cl}\,\mathcal{H})$ and $h \in \mathcal{H}$. Therefore there exist $\{t_k\}_{k\in\mathbb{N}} \subseteq \mathbb{R}_+ \setminus \{0\}$ and $\{x_k\}_{k\in\mathbb{N}} \subseteq \text{conv}\,(A - \text{cl}\,\mathcal{H})$ such that $t_k x_k \to x$. Let $h_k = \frac{1}{t_k} h \in \mathcal{H}$; by Lemma 3.1 we have

$$x_k - h_k \in \text{conv}\,(A - \text{cl}\,\mathcal{H}),$$

which implies that

$$t_k x_k - h = t_k(x_k - h_k) \in \text{cone conv}\,(a - \text{cl}\,\mathcal{H}).$$

Since $t_k x_k \to x$ we have

$$x - h \in \text{cl cone conv}\,(A - \text{cl}\,\mathcal{H}).$$

□

We are now in the position to state the announced result.

Theorem 3.2. *Let $A \subseteq \mathbb{R}^{m+1}$. We have*

$$\text{cl cone conv}\,(A - \text{cl}\,\mathcal{H}) \cap \hat{\mathcal{H}} = \emptyset$$

iff

$$\text{cl cone conv}\,(A - \text{cl}\,\mathcal{H}) \cap \hat{\mathcal{H}}_u = \emptyset. \tag{3.3}$$

Proof. The "only if" part is immediate. Since $\hat{\mathcal{H}} = \mathcal{H} \cup \mathcal{H}_u$, to prove the "if" part it is enough to show that (3.3) implies

$$\text{cl cone conv}\,(A - \text{cl}\,\mathcal{H}) \cap \mathcal{H} = \emptyset.$$

Ab absurdo, let $(\bar{u}, \bar{v}) \in \text{cl cone conv}\,(A - \text{cl}\,\mathcal{H}) \cap \mathcal{H}$; since $\bar{u} > 0$ and $\bar{v} \in C$ we have

$$\left(\frac{\bar{u}}{2}, \bar{v}\right) \in \mathcal{H}.$$

Applying Lemma 3.2 we obtain $(\frac{\bar{u}}{2}, 0) = (\bar{u}, \bar{v}) - (\frac{\bar{u}}{2}, \bar{v}) \in \text{cl cone conv}\,(A - \text{cl}\,\mathcal{H}) \cap \mathcal{H}_u$ which contradicts (3.3).

□

Theorem 3.2 generalizes an analogous result stated in [4]. Moreover we observe that (3.3) is the *image regularity condition* when $A = \mathcal{K}$ and $\mathcal{E}$ is convex [3]. In [2] we give a characterization of total regularity in the image space. In that case we are also able to give the interpretation of the condition in the space $\mathbb{R}^n$. In particular we show that if $M^0 \cup M^1 \neq \emptyset$, then $M^0 = \emptyset$ iff

$$\operatorname{conv} \mathcal{E} \cap (\mathbb{R} \times \operatorname{int} C) \neq \emptyset. \tag{3.4}$$

In the space where x runs (i.e. $\mathbb{R}^n$), condition (3.4) is equivalent to affirm that there exist $x_i \in \mathbb{R}^n$ and $\lambda_i \geq 0$ with $i = 1 \ldots m$ such that $\sum_{i=1}^{m+2} \lambda_i = 1$ and

$$\sum_{i=1}^{m+2} \lambda_i g(x_i) \in \operatorname{int} C. \tag{3.5}$$

We observe that, when g is a C–convex function, (3.5) is equivalent to the well–known Slater condition [7, 8]

$$\exists \bar{x} \in \mathbb{R}^n : \qquad g(\bar{x}) \in \operatorname{int} C.$$

4. APPLICATIONS TO OPTIMIZATION PROBLEMS

We now want to analyse the particular case where the impossibility of (S) represents an optimality condition for a constrained extremum problem of the kind

$$\begin{cases} \min \varphi(x) \\ g(x) \in C \end{cases} \tag{P}$$

where $\varphi : \mathbb{R}^n \to \mathbb{R}$ and $C \subseteq \mathbb{R}^m$ is a closed and convex cone with apex at the origin. Obviously a feasible point $\bar{x}$ is a (global) minimum point for (P) iff the system

$$\begin{cases} \varphi(\bar{x}) - \varphi(x) > 0 \\ g(x) \in C \end{cases} \tag{S$_P$}$$

is impossible.

We show that the existence of a linear separation between the sets $\mathcal{E}$ and $\mathcal{H}$ associated to (S_P) is equivalent to the existence of a John saddle–point of the generalized Lagrangean function associated to (P), namely

$$L(\theta, \lambda, x) := \theta\varphi(x) - \langle \lambda, g(x) \rangle \ .$$

Definition 4.1. A point $(\bar{\theta}, \bar{\lambda}, \bar{x}) \in \mathbb{R}_+ \times C^* \times \mathbb{R}^n$ with $(\bar{\theta}, \bar{\lambda}) \neq (0, 0)$, such that

$$L(\bar{\theta}, \lambda, \bar{x}) \leq L(\bar{\theta}, \bar{\lambda}, \bar{x}) \leq L(\bar{\theta}, \bar{\lambda}, x), \quad \forall x \in \mathbb{R}^n, \quad \forall \lambda \in C^*$$

is called a *John saddle–point* of $L(\bar{\theta}, \lambda, x)$ on $\mathbb{R}^n \times C^*$.

The following theorem is a generalization of an analogous one stated in [4].

Theorem 4.1. *$\pi_{\bar\theta,\bar\lambda}$ is a separating hyperplane for the sets $\mathcal{E}$ and $\mathcal{H}$ iff $(\bar\theta,\bar\lambda,\bar x)$ is a John saddle-point for $L(\bar\theta,\lambda,x)$ on $\mathbb{R}^n \times C^*$.*

Proof. "Only if". Since $\pi_{\bar\theta,\bar\lambda}$ separates $\mathcal{E}$ and $\mathcal{H}$ the following inequality holds

$$\bar\theta u + \langle\bar\lambda, v\rangle \le 0, \quad \forall (u,v)\in\mathcal{E}$$

where, in this case,

$$\mathcal{E} = \{(u,v)\in\mathbb{R}\times\mathbb{R}^m : u \le \varphi(\bar x)-\varphi(x),\ v = g(x)-c,\ c\in C,\ x\in\mathbb{R}^n\}.$$

The previous inequality is equivalent to

$$\bar\theta(\varphi(\bar x)-\varphi(x)) + \langle\bar\lambda, g(\bar x)\rangle \le 0, \quad \forall x\in\mathbb{R}^n\,. \tag{4.1}$$

(4.1) computed for $x=\bar x$ and the feasibility of $\bar x$ imply

$$\langle\bar\lambda, g(\bar x)\rangle = 0.$$

Therefore (4.1) is equivalent to

$$\bar\theta\varphi(\bar x) - \langle\bar\lambda, g(\bar x)\rangle \le \bar\theta\varphi(x) - \langle\bar\lambda, g(x)\rangle, \quad \forall x\in\mathbb{R}^n$$

that is

$$L(\bar\theta,\bar\lambda,\bar x) \le L(\bar\theta,\bar\lambda,x), \quad \forall x\in\mathbb{R}^n\,.$$

Moreover, for each $\lambda\in C^*$

$$\bar\theta\varphi(\bar x) - \langle\lambda, g(\bar x)\rangle \le \bar\theta\varphi(\bar x) - \langle\bar\lambda, g(\bar x)\rangle$$

that is

$$L(\bar\theta,\lambda,\bar x) \le L(\bar\theta,\bar\lambda,\bar x), \quad \forall\lambda\in C^*.$$

"If". The inequality

$$\bar\theta\varphi(\bar x) - \langle\lambda, g(\bar x)\rangle \le \bar\theta\varphi(\bar x) - \langle\bar\lambda, g(\bar x)\rangle, \quad \forall\lambda\in C^*$$

computed for $\lambda = 0\in C^*$ gives $-\langle\bar\lambda, g(\bar x)\rangle \ge 0$. Since $\bar x$ is feasible and $\bar\lambda\in C^*$ we have that $\langle\bar\lambda, g(\bar x)\rangle \ge 0$, therefore we have

$$\langle\bar\lambda, g(\bar x)\rangle = 0.$$

This implies that the inequality

$$L(\bar\theta,\bar\lambda,\bar x) \le L(\bar\theta,\bar\lambda,x), \quad \forall x\in\mathbb{R}^n$$

is equivalent to (4.1), so that $\pi_{\bar\theta,\bar\lambda}$ is a separating hyperplane for $\mathcal{E}$ and $\mathcal{H}$. □

We recall that if $(\bar{\theta}, \bar{\lambda}, \bar{x})$ is a John saddle–point then it is a solution, in the differentiable problems, of the Fritz–John necessary optimality conditions. Therefore any statement that ensures the regularity of the linear separation can be considered as a regularity condition for the problem (P) in view of the Karush–Kuhn–Tucker conditions.

5. FIRST ORDER OPTIMALITY CONDITIONS

In this section we consider the problem (P) in the particular case where $C = \mathbb{R}^m_-$. We analyse first order optimality conditions of (P), that can be formulated by means of the impossibility of a suitable system. To do this we will denote respectively by $D^-\varphi(\bar{x}; y)$ and $D^+\varphi(\bar{x}; y)$ the lower and the upper Dini derivative of the function φ at $\bar{x}$ in the direction y; let $I := \{1, \ldots, m\}$ and, for a given $\bar{x} \in \mathbb{R}^n$, $I(\bar{x}) := \{i \in I : g_i(\bar{x}) = 0\}$. For the sake of completeness, we give the following result.

Theorem 5.1. *Let $\bar{x}$ be a local optimal solution for (P). If $g_i(x)$ are upper semicontinuous for each $i \in I \backslash I(\bar{x})$; then the following system is impossible:*

$$\begin{cases} D^-\varphi(\bar{x}; y) < 0 \\ D^+ g_i(\bar{x}; y) < 0, \quad i \in I(\bar{x}). \end{cases} \tag{5.1}$$

Proof. Ab absurdo there exists $\bar{y} \in \mathbb{R}^n$ that solves (5.1); then there exists $\{t_k\}_{k \in \mathbb{N}} \downarrow 0$ such that definitively

$$\frac{\varphi(\bar{x} + t_k \bar{y}) - \varphi(\bar{x})}{t_k} < 0$$

and therefore there exists $k_0 \in \mathbb{N}$ such that, for each $k > k_0$, we have $\varphi(\bar{x} + t_k \bar{y}) < \varphi(\bar{x})$. Moreover, definitively, also

$$\frac{g_i(\bar{x} + t_k \bar{y}) - g_i(\bar{x})}{t_k} < 0$$

and then for each $i \in I(\bar{x})$ there exists $k_i \in \mathbb{N}$ such that for every $k > k_i$, we have

$$g_i(\bar{x} + t_k \bar{y}) < g_i(\bar{x}) = 0.$$

Finally $\bar{x} + t_k \bar{y} \to \bar{x}$ and, taking into account the upper semicontinuity of g_i, for each $i \in I \backslash I(\bar{x})$, there exists $k_i \in \mathbb{N}$ such that for each $k > k_i$, we have

$$g_i(\bar{x} + t_k \bar{y}) < g_i(\bar{x}) - \frac{g_i(\bar{x})}{2} < 0.$$

So, for each $k \geq \max\limits_{i \in I \cup \{0\}} k_i$, defining $x_k = \bar{x} + t_k \bar{y}$ we observe that x_k is feasible, $x_k \to \bar{x}$ and $\varphi(x_k) < \varphi(\bar{x})$, which contradicts the assumption. □

We develop the analysis of the impossibility of system (5.1) using the results obtained in Section 3. We observe that in general a linear separation for the image of system (5.1) does not exist. To ensure the existence of a linear separation we substitute the Dini directional derivatives with a convex approximate or we suppose that they are convex by themselves.

Remark 5.1. A classical approach [10,11] to substitute a Dini derivative with its upper sublinear approximate is to consider a function $\Psi : \mathbb{R}^n \to]-\infty, +\infty]$ which is sublinear and such that

$$\Psi(y) \geq D^{\pm}\varphi(\bar{x}; y), \quad \forall y \in \mathbb{R}^n .$$

Moreover we recall, for example, that if φ is a Lipschitz function then the Clarke derivative is an upper convex approximate of the upper Dini derivative.

It is necessary to consider upper approximates (i.e. functions greater than Dini derivative) because in this case the system

$$\begin{cases} \varphi^k(\bar{x}; y) < 0 \\ g_i^{k_i}(\bar{x}; y) < 0, \quad i \in I(\bar{x}) \end{cases} \tag{5.2}$$

is impossible in a local optimal solution $\bar{x}$, if φ^k is an upper (convex) approximate of $D_-\varphi$ and $g_i^{k_i}$ are upper (convex) approximates of D_+g_i. In this situation we have

$$\mathcal{E} = \{(u, v) \in \mathbb{R} \times \mathbb{R}^s : u \leq -\varphi^k(\bar{x}; y), \; v_i \geq g_i^{k_i}(\bar{x}; y), \quad i = 1, \ldots, s, \quad y \in \mathbb{R}^n\},$$
$$\mathcal{H} = \{(u, v) \in \mathbb{R} \times \mathbb{R}^s : u > 0, \; v < 0\}$$

and $s = |I(\bar{x})|$, i.e. the cardinality of the set $I(\bar{x})$.

Since $\mathcal{E}$ is convex, for the convexity of φ^k and $g_i^{k_i}$, the impossibility of system (5.2) is equivalent to the disjunction between $\mathcal{E}$ and $\mathcal{H}$. So $\mathcal{E}$ and $\mathcal{H}$ admit a linear separation, that is there exists a nonzero vector $(\theta, \lambda) \in (\mathbb{R}_+ \times \mathbb{R}^s_-)^* = \mathbb{R}_+ \times \mathbb{R}^s_-$ such that

$$\theta(-\varphi^k(\bar{x}; y)) + \sum_{i \in I(\bar{x})} \lambda_i g_i^{k_i}(\bar{x}; y) \leq 0, \quad \forall y \in \mathbb{R}^n . \tag{5.3}$$

Due to Theorem 3.1 and to the sublinearity of φ^k and $g_i^{k_i}$, $\mathcal{E}$ is a closed and convex cone; therefore we can suppose $\theta > 0$ in (5.3) iff

$$\mathcal{E} \cap \hat{\mathcal{H}} = \emptyset \tag{5.4}$$

where, in this case, $\hat{\mathcal{H}} = \{(u, v) \in \mathbb{R} \times \mathbb{R}^s : u > 0, \; v \leq 0\}$.

It is interesting to give an interpretation of condition (5.4) in terms of the problem (P). To this aim we recall this definition [1]

Definition 5.1. Let $\bar{x} \in \mathbb{R}^n$, φ be a real valued function and $\varphi^k(\bar{x}; y)$ an upper sublinear approximation of a Dini derivative.

$$D^<_{\varphi^k}(\bar{x}) := \{y \in \mathbb{R}^n : \varphi^k(\bar{x}; y) < 0\}$$

is the *cone of strong k–descent directions* of φ at $\bar{x}$;

$$D^{\leq}_{\varphi^k}(\bar{x}) := \{y \in \mathbb{R}^n : \varphi^k(\bar{x}; y) \leq 0\}$$

is the *cone of weak k–descent directions* of φ at $\bar{x}$.

In this case we have

$$M^1 = \{\lambda \in \mathbb{R}^s_- : (1, \lambda) \ \text{ fulfils (5.3)}\}$$

omitting the dependence on $\bar{x}$ for the sake of simplicity and we can prove the following theorem.

Theorem 5.2. *Let $\bar{x}$ be a local optimal solution for (P). The following statements are equivalent:*

1) $M^1 \neq \emptyset$;

2) $D^{<}_{\varphi^k}(\bar{x}) \cap \left(\bigcap_{i \in I(\bar{x})} D^{\leq}_{g_i^{k_i}}(\bar{x}) \right) = \emptyset$;

3) *0 is a global optimal solution of*

$$\begin{cases} \min \varphi^k(\bar{x}; y) \\ g_i^{k_i}(\bar{x}; y) \leq 0, \quad i \in I(\bar{x}). \end{cases} \tag{5.5}$$

Proof. From Theorem 3.1 we have

$$M^1 \neq \emptyset \ \Leftrightarrow \text{cl cone conv } \mathcal{E} \cap \hat{\mathcal{H}} = \emptyset \ \Leftrightarrow \ \mathcal{E} \cap \hat{\mathcal{H}} = \emptyset.$$

1) $\Rightarrow$ 2) is immediate.

2) $\Rightarrow$ 1) Ab absurdo, assume that $M^1 = \emptyset$, therefore $\mathcal{E} \cap \hat{\mathcal{H}} \neq \emptyset$. This means that there exists $\bar{y} \in \mathbb{R}^n$ such that

$$\begin{cases} \varphi^k(\bar{x}; y) < 0 \\ g_i^{k_i}(\bar{x}; y) \leq 0, \quad i \in I(\bar{x}) \end{cases}$$

which contradicts 2).

2) $\Leftrightarrow$ 3) follows from Definition 5.1. □

Remark 5.2. Problem (5.5) is a first order approximation of (P); when φ and g are differentiable, problem (5.5) is the linearized of (P).

REFERENCES

[1] M. Castellani, "Refinements on solvability results for sublinear functions". Pubbl. 3.205(772), Dip. di Matematica, Univ. di Pisa, 1994.

[2] M. Castellani, G. Mastroeni and M. Pappalardo, "Separation of sets, Lagrange multipliers and totally regular extremum problems". Forthcoming.

[3] P.H. Dien, G. Mastroeni, M. Pappalardo and P.H. Quang, "Regularity conditions for constrained extremum problems via image space". *J. Optim. Theory Appl.*, 81, pp. 19–37, 1994.

[4] P.H. Dien, G. Mastroeni, M. Pappalardo and P.H. Quang, "Regularity conditions for constrained extremum problems via image space: the linear case". In Lecture Notes in Econom. and Math. Systems, 405, Komlosi–Rapcsack–Schaible eds., Springer–Verlag, 1994.

[5] F. Giannessi, "Theorems of the Alternative and Optimality Conditions". *J. Optim. Theory Appl.*, 42, pp. 331–365, 1984.

[6] A. Jourani, "On constraint qualifications and Lagrange multipliers in nondifferentiable programming problems", *J. Optim. Theory Appl.*, 81, pp. 533–548, 1944.

[7] O.L. Mangasarian, "Nonlinear Programming". New York Academic Press, 1969.

[8] O.L. Mangasarian and S. Fromovitz, "The Fritz–John necessary optimality condition in the presence of equality and inequality constraints". *J. Math. Anal. Appl.*, 7, pp. 37–47, 1967.

[9] R.R. Merkovsky and D.E. Ward, "General constraint qualifications in nondifferentiable programming", *Math. Programming*, 47, pp. 389–405, 1990.

[10] D.E. Ward, "Convex directional derivatives in Optimization". In Lecture Notes in Econom. and Math. Systems, 345, Cambini–Castagnoli–Martein–Mazzoleni–Schaible eds., Springer–Verlag, 1990.

[11] D.E. Ward, "Chain rules for nonsmooth functions". *J. Math. Anal. Appl.*, 158, pp. 519–538, 1991.

[12] J. Zowe and S. Kurcyusz, "Regularity and stability for the mathematical programming problem in Banach space". *Appl. Math. Optim.*, 5, pp. 49–62, 1979.

AN ALGORITHM USING QUADRATIC INTERPOLATION FOR UNCONSTRAINED DERIVATIVE FREE OPTIMIZATION

Conn Andrew R.[1] and Toint Philippe L.[2]

[1] IBM T.J. Watson Research Center,
P.O.Box 218, Yorktown Heights, NY 10598, USA.
[2] Department of Mathematics,
Facultés Universitaires ND de la Paix,
61, rue de Bruxelles, B-5000 Namur, Belgium, EU.

Abstract. This paper explores the use of multivariate interpolation techniques in the context of methods for unconstrained optimization that do not require derivative of the objective function. A new algorithm is proposed that uses quadratic models in a trust region framework. The algorithm is constructed to require few evaluations of the objective function and is designed to be relatively insensitive to noise in the objective function values. Its performance is analyzed on a set of 20 examples, both with and without noise.

Key words. Nonlinear optimization, trust regions, derivative free algorithms.

1. INTRODUCTION

We are concerned, in this paper, with the problem of minimizing an objective function whose value is determined by measuring some quantity in the real world. This measure may be of a physical nature (for instance, the depth of a certain layer in geophysical exploration) or be related to other contexts. We will focus on the case where there are no constraints on the problem variables. The generalization to simple bounds, in particular, is quite straightforward, as indicated at the end of the paper. Moreover, since the proposed method is derivative free, one might want to handle constrained problems using an exact penalty function.

Three important features characterize these types of problems. Firstly, the cost of obtaining a function value, that is of performing the measure for particular values of the problem variables, is typically very high. This calls for optimization techniques that make optimal use of all such evaluations, possibly at the expense of more extensive linear algebra calculations within the algorithm itself. The second important feature is that the nature of the function evaluation or some other reasons, prevents the computation of any associated derivatives (gradient or Hessian), a serious drawback for many optimization methods. Finally, the considered measure is usually subject to error itself, introducing some "noise" on the objective evaluation, which puts additional requirements on the minimization's robustness.

Note that these problem features may make the calculation of derivatives by finite differences unattractive. Indeed, the additional function evaluations required in the differencing, by this technique, may be very costly and, most importantly, finite differencing can be unreliable in the presence of noise if no specific action is taken to adapt the differentiation step size to the noise level. Since automatic differentiation (see Griewank [22], for example) is not applicable to a "physical" measurement procedure, we thus may be forced to consider algorithms that do not approximate objective function derivatives for a given value of the problem variables.

By extension, we will also consider in this paper unconstrained optimization problems whose objective function is the result of a complex and costly numerical procedure (such as, for example, in the analysis of the in flight vibration of a helicopter rotor), possibly involving some considerable noise (due, for instance, to truncation or approximation in the calculation defining the objective). At variance with the framework described above, automatic differentiation may often be applied to such cases, but the computed derivatives then include differentiation of the noise itself, making the calculated gradients of questionable value to measure the local slope. Furthermore, automatic differentiation is not applicable when the source code for evaluating the objective function is unavailable. Finally, it may not always be straightforward to use, as is for example the case in fluid dynamics calculations where, according to Burns [7], it may generate unwanted dependence on discretization parameters or on the introduction of artificial viscosity.

Derivative free optimization methods have a long history and we refer the reader to Dixon [16], Himmelblau [24] or Polyak [34] for extensive discussion and references. These methods come in essentially five different classes. The first class contains the algorithms which use finite-difference approximations of the objective function's derivatives in the context of a gradient based method, such as nonlinear conjugate gradients or quasi-Newton methods (see, for instance, Stewart [46], Dennis and Schnabel [14], Gill *et al.* [19] and Gill *et al.* [20]). The methods in the second class are often referred to as "pattern search" methods, because they are based on the exploration of the variables' space using a well specified geometric pattern, typically a simplex. They were investigated by Spendley *et al.* [44], Hooke and Jeeves [25] and Nelder and Mead [31], the algorithm proposed by the latter still being one of the most popular minimization technique in use today. More recent developments of pattern search methods include proposals by Torczon [50], Dennis and Torczon [15], Buckley [6] and Elster and Neumaier [18]. The approaches of the third type are based instead on random sampling and were developed by Box [3], Brooks [5] and Kelly and Wheeling [26], to cite a few. The methods of the fourth class are based, as for many methods using derivatives, on the use of successive one-dimensional minimizations (line searches) along selected directions. These directions may be chosen amongst the set of coordinate basis vectors, as in Elkin [17], Ortega and Rheinboldt [32] or Lucidi and Sciandrone [27], with possible reorientation of the basis as described in Rosenbrock [40] and Swann [47], or on sets of mutually conjugate directions, as proposed by Powell [35] and later developed by Brent [4]. Finally, the algorithms of the fifth class are based on the progressive building and updating of a model of the objective function, as proposed by Powell in [38] for linear models and in [39] for quadratic ones. There is also a related class of "global modelling" methods, that uses Design of Experiments (DOE) interpolation models. For instance, in a problem with ten variables one may determine 50 suitably chosen function values (perhaps by using optimal designs, see for example Owen [33]) for determining an initial model that satisfies a maximum likelihood estimator (MLE) criteria and one may use an additional 50 evaluations to refine the model. Details are given in Booker [2], Mitchell *et al.* [28], Morris *et al.* [30] and Sacks *et al.* [41].

The approach developed below belongs to the "model building and updating" class and will be closely related to the proposal of Powell [39]. Following ideas expressed in this reference, we will consider a trust region framework where the objective function's model is built by multivariate (quadratic) interpolation. At variance with Powell's proposal, we will however insist on the ability of our algorithm to take long steps and also to progress as early as possible with every available function evaluation.

The purpose of this paper is to present the current state of the authors' ideas in what is likely to be a longer term project. It is organized as follows: Sect. 2 introduces the problem, notation and the algorithm, while the results of preliminary numerical experience are discussed in Sect. 3. Some conclusions and perspectives are outlined in Sect. 4.

2. ALGORITHMIC CONCEPTS

We consider the problem of finding a vector $x \in \mathbb{R}^n$, a solution of the unconstrained nonlinear program

$$\min_{x \in \mathbb{R}^n} f(x), \tag{2.1}$$

where $f(\cdot)$ is a twice differentiable function from $\mathbb{R}^n$ into $\mathbb{R}$. Although the derivatives of $f(\cdot)$ may exist, we assume that they cannot be calculated. We also assume that, although $f(x)$ can be evaluated at any x, the cost of such an evaluation is high compared to that of solving square dense systems of linear equations in n variables. The notation $\langle x, y\rangle$ will be used throughout to denote the Euclidean inner product of x and y and $\|x\|$ will denote the Euclidean norm of x.

The algorithm proposed in this note belongs to the class of "trust-region" methods. Such algorithms are iterative and build, around the current iterate, a model of the true objective function which is cheaper to evaluate and easier to minimize than the objective function itself. This model is assumed to represent the objective function well in a so-called trust region, typically a ball centered at the current iterate, x_c say. The radius of this ball, traditionally denoted by Δ, is called the trust region radius and indicates how far the model is thought to represent the objective function well. A new trial point is then computed, which minimizes or sufficiently reduces the model *within the trust region* and the true objective function is evaluated at this point. If the achieved objective function reduction is sufficient compared to the reduction predicted by the model, the trial point is accepted as the new iterate and the trust region possibly enlarged. On the other hand, if the achieved reduction is poor compared to the predicted one, the current iterate is typically unchanged [1] and the trust region is reduced. This process is then repeated until convergence (hopefully) occurs.

The Quadratic Model and How to Improve It

One of the main ingredients of a trust region algorithm is thus the choice of an adequate objective function model. We will here follow a well established tradition in

[1]This, of course, does not prevent the algorithm recording the best point found so far, and returning to this point at the end of the calculation. We note that there also exist trust region methods in which any point lower than the current one is accepted as the next iterate (see Powell [37], for instance).

choosing a quadratic model of the form

$$m(x_c + s) = f(x_c) + \langle g, s \rangle + \frac{1}{2}\langle s, Hs \rangle, \tag{2.2}$$

where g is a vector of $\mathbb{R}^n$ and where H is a square symmetric matrix of dimension n. However, we will depart from many trust-region algorithms in that g and H will not be determined by the (possibly approximate) first and second derivatives of $f(\cdot)$, but rather by imposing that the model (2.2) interpolates function values at past points, that is we will impose that

$$m(x) = f(x) \tag{2.3}$$

for each vector x in a set I such that $f(x)$ is known for all $x \in I$. Note that this interpolation technique is also used by Powell in [38] and [39]. Note also that the cardinality of I must be equal to

$$p = \frac{1}{2}(n+1)(n+2) \tag{2.4}$$

to ensure that the quadratic model is entirely determined by the equations (2.3). However, if $n > 1$, this last condition is not sufficient to guarantee the existence of an interpolant. It is indeed well-known (see De Boor and Ron [12] or Sauer and Xu [42], for instance) that the points in I must also satisfy some geometric constraints: for instance, six points on a line do not determine a two dimensional quadratic. When the geometry of the points in I is such that the interpolant exists, we follow Sauer and Xu [42] and say that I is *poised*. If we choose a basis $\{\phi_i(\cdot)\}_{i=1}^p$ of the linear space of n-dimensional quadratics, $I = \{x_1, \ldots, x_p\}$ is poised when

$$\delta(I) = \det \begin{pmatrix} \phi_1(x_1) & \cdots & \phi_1(x_p) \\ \vdots & & \vdots \\ \phi_p(x_1) & \cdots & \phi_p(x_p) \end{pmatrix} \tag{2.5}$$

is non-zero. Of course, the quality of the model (2.2) as an approximation of the objective function around x_c will be dependent on the geometry of the considered interpolation points, and thus on the value of $|\delta(I)|$. Following Powell [39], we will say that this geometry, and hence the model, is *good* (with respect to x_c and the radius Δ) when all the points in I are no further away from x_c than 2Δ and when the value of $|\delta(I)|$ cannot be doubled by adjusting one of the points of I to an alternative value within distance Δ from x_c.

In derivative based trust-region methods, the radius Δ is decreased whenever sufficient decrease in the objective function is not obtained at the computed trial point (the iteration is then said to be *unsuccessful*). This technique aims at improving the model within the trust region, since Taylor's theorem indicates that the derivative based model (2.2) better fits $f(\cdot)$ in a smaller neighborhood of x_c. However, this improvement is not an immediate consequence of reducing Δ in our case, since our model is based on interpolating function values rather than derivatives. In order to ensure progress of the algorithm away from stationary points, we therefore have to explicitly improve the interpolation model at unsuccessful iterations, either by discarding interpolation points that are too far away from x_c or by improving the geometry of I. Note that both these actions imply that I is modified, which usually means that a new point, x_+, such that $\|x_+ - x_c\| \leq \Delta$, and its associated objective function value $f(x_+)$ must be computed. If we wish to make the geometry of I as good as possible, we therefore need a measure of the improvement obtained by replacing some past point

$x_- \in I$ by x_+, possibly comparing several choices for x_-. We consider two cases. First, if $\|x_- - x_c\| \le \Delta$, a suitable measure is $|\delta(I)|$, and we therefore wish to compute the factor by which $|\delta(I)|$ is multiplied when x_- is replaced by x_+. Remarkably, this factor is independent of the basis $\{\phi_i\}$ and is equal to $|L(x_+, x_-)|$, where $L(\cdot, x_-)$ is the Lagrange interpolation function whose value is one at x_- and at all other points of I is zero[2]. This very nice result was pointed out by Powell in [39]. Hence, if $\|x_- - x_c\| \le \Delta$, it makes sense to replace x_- by

$$x_+ = \arg \max_{\|x - x_c\| \le \Delta} |L(x, x_-)|. \tag{2.6}$$

On the other hand, if $\|x_- - x_c\| > \Delta$, it is important to take this inequality into account when choosing a suitable replacement x_+. One possible method, again suggested by Powell [39], is to compare x_+ not with x_- directly, but rather with the best point on the segment joining x_c to x_- limited to the ball of radius Δ around x_c. This "scaled down" version of x_- is the vector that maximizes $|L(x_c + td_-, x_i)|$ for $t \in [0, \Delta]$, where $d_- = (x_- - x_c)/\|x_- - x_c\|$. Hence, x_+ may be chosen in this case as

$$x_+ = \arg \max_{\|x - x_c\| \le \Delta} S(x, x_-), \tag{2.7}$$

where

$$S(x, x_-) = \frac{|L(x, x_-)|}{\min[1, \max_{t \in [0,\Delta]} |L(x_c + td_-, x_-)|]}. \tag{2.8}$$

The minimum in the denominator of (2.8) guarantees that the scaled down version of x_-, namely $\arg\max_{t \in [0,\Delta]} |L(x_c + td_-, x_-)|$, is treated exactly as any other point within distance Δ from x_c (that is according to (2.6)). This feature of $S(\cdot, \cdot)$ and the definition of the Lagrange interpolation function imply that

$$|L(x, x_-)| = S(x, x_-) \text{ whenever } \|x_- - x_c\| \le \Delta, \tag{2.9}$$

and $S(\cdot, \cdot)$ may thus be used instead of $|L(\cdot, \cdot)|$ in (2.6), making the distinction on $\|x_- - x_c\|$ unnecessary. Note that (2.9), the identity $L(x_-, x_-) = 1$ and the definition (2.8) imply that, for each x_- and each x_+ determined from x_- by (2.7), one has that

$$S(x_+, x_-) \ge 1. \tag{2.10}$$

Note also that the Lagrange interpolation function $L(\cdot, \cdot)$ is a quadratic determined by function value interpolation, and therefore only exists, together with $S(\cdot, \cdot)$, if I is poised.

A special situation occurs in this respect in the first iterations of the algorithm. Due to the assumed high cost of a function evaluation, we may wish to define a model of the type (2.2) as soon as a few objective values have been computed. This means that, although we have a set of points poised for quadratic interpolation, the objective function value may not be known for each of them. As above, we denote by I the set of points for which the objective value is known, and we denote by J the set of remaining points, where the objective value is still unknown. Thus I may contain less than p points, although the set $I \cup J$ is poised for quadratic interpolation. The Lagrange interpolation function $L(\cdot, \cdot)$ and $S(\cdot, \cdot)$ are thus well defined, but the model (2.2) is no longer fully specified by the interpolation conditions (2.3). Two solutions are then possible to determine a suitable model.

[2]Note that $L(\cdot, \cdot)$ thus depends on all points in I, and not just on its two explicit arguments.

- The first is to take out the remaining degrees of freedom in the model by imposing a variational criterion. We have considered here computing the model of minimal Frobenius norm, that is the model for which $\|g\|^2 + \|H\|_F^2$ is minimal, which still satisfies the interpolation conditions (2.3). Note that this choice depends on the scaling of the problem's variables.

- The second solution is to build an sub-quadratic model, that is a model in which not all degrees of freedom of a full quadratic model are exploited but which does interpolate all the data points available. Such models can be obtained by the multivariate interpolation algorithms proposed by Sauer and Xu [42] and Sauer [43], for example. The actual form of the model is determined, in this approach, by the number and geometry of the available interpolation points and by the polynomial basis used to span the linear space of multivariate quadratics. We refer the reader to Sauer and Xu [42] for further details.

Therefore computing a suitable model when $|I| < p$ is possible. Of course, if such a model needs to be improved, we will bias our procedure to reduce the cardinality of J, therefore enriching the information on which the model is based. This also means that, for some iterations, the Lagrangian interpolation function $L(\cdot,\cdot)$, and thus the function $S(\cdot,\cdot)$, now depend on the set of points in $I \cup J$ rather than I alone.

We finally note that, when the model must be improved after an unsuccessful iteration, it may happen that points where the objective function has been previously computed are no further away than Δ from x_c and yet do not belong anymore to I. It is then sometimes possible to improve the geometry of the current set $I \cup J$ by replacing one point in $I \cup J$ by such a previous point, and this procedure can be repeated as long as improvement of the model's geometry is obtained, without requiring new evaluations of the objective function.

We are now ready to describe the algorithm we propose to use to improve the quadratic model after an unsuccessful iteration. In this description and later in the paper, we denote by

$$M = \{x \in \mathbb{R}^n \mid f(x) \text{ is known } \} \tag{2.11}$$

the set of points where the objective function has been evaluated so far.

A1: Geometry improvement

The sets I, J and M, the radius Δ and the current point x_c are given.

Step 1: attempt to reuse past points that are close to x_c.

For each point $x_i \in M \setminus I$ such that $\|x_i - x_c\| \leq \Delta$,

- determine which of the current interpolation points in $(I \cup J) \setminus \{x_c\}$ can be exchanged with x_i to maximally improve the interpolation set geometry, that is compute

$$x_- = \arg \max_{x_j \in (I \cup J) \setminus \{x_c\}} S(x_i, x_j); \tag{2.12}$$

- perform the exchange if the improvement is sufficient, that is, if $S(x_i, x_-) \geq 2$. In this case, redefine

$$I = (I \setminus \{x_-\}) \cup \{x_i\} \tag{2.13}$$

if $x_- \in I$, or

$$I = I \cup \{x_i\} \text{ and } J = J \setminus \{x_-\} \tag{2.14}$$

otherwise.

If at least one exchange (2.13) or (2.14) has been performed, successfully terminate Algorithm A1.

Step 2: attempt to replace a point of J distant from x_c.
Determine the point in J that is furthest from x_c, that is

$$x_- = \arg\max_{x_i \in J} \|x_i - x_c\|. \tag{2.15}$$

If $\|x_- - x_c\| > 2\Delta$, find a better point closer to x_c to replace x_-, that is compute

$$x_+ = \arg\max_{\|x - x_c\| \le \Delta} S(x, x_-), \tag{2.16}$$

calculate $f(x_+)$, set $M = M \cup \{x_+\}$, $I = I \cup \{x_+\}$, $J = J \setminus \{x_-\}$ and successfully terminate Algorithm A1.

Step 3: attempt to replace a point of I distant from x_c.
Determine the point in I that is furthest from x_c, that is

$$x_- = \arg\max_{x_i \in I} \|x_i - x_c\|. \tag{2.17}$$

If $\|x_- - x_c\| > 2\Delta$, find a better point close to x_c to replace x_-, that is compute

$$x_+ = \arg\max_{\|x - x_c\| \le \Delta} S(x, x_-), \tag{2.18}$$

calculate $f(x_+)$, set $M = M \cup \{x_+\}$, perform the exchange $I = (I \setminus \{x_-\}) \cup \{x_+\}$ and successfully terminate Algorithm A1.

Step 4: attempt to replace a point of J close to x_c.
If J is not empty, find the point in J whose replacement maximally improves the interpolation set geometry, that is compute

$$x_- = \arg\max_{x_i \in J} [\arg\max_{\|x - x_c\| \le \Delta} S(x, x_i)] \tag{2.19}$$

and let x_+ be the x that realizes the inner maximum in (2.19). Then calculate $f(x_+)$, set $M = M \cup \{x_+\}$, $I = I \cup \{x_+\}$, $J = J \setminus \{x_-\}$ and successfully terminate Algorithm A1.

Step 5: attempt to replace a point of I close to x_c.
Find the point in $I \setminus \{x_c\}$ whose replacement maximally improves the interpolation set geometry, that is compute

$$x_- = \arg\max_{x_i \in I \setminus \{x_c\}} [\arg\max_{\|x - x_c\| \le \Delta} S(x, x_i)] \tag{2.20}$$

and let x_+ be the x that realizes the inner maximum in (2.20). Then, if $S(x_+, x_-) \ge 2$, calculate $f(x_+)$, set $M = M \cup \{x_+\}$, perform the exchange $I = (I \setminus \{x_-\}) \cup \{x_+\}$ and successfully terminate Algorithm A1.

Step 6: geometry deemed satisfactory.
No significant geometry improvement can be identified with the current Δ: unsuccessfully terminate Algorithm A1.

end of A1

In this algorithm, we have consistently used the function $S(\cdot,\cdot)$ to measure the geometrical improvement of the interpolation set. We have also attempted to reduce J by considering the elimination of points of this set first (in Steps 2 and 4). If Step 6 is reached, this means that the model's geometry is good (in the sense defined above), given the current value of the trust-region radius, and thus that further improvement around x_c will only be obtained by reducing this radius, forcing the interpolation points to be closer to x_c if necessary. Note that the geometry of the interpolation set cannot deteriorate in Steps 2 to 4, because of (2.10).

2.1. The Trust-Region Step

After examining how the model is built and how it can be improved if necessary, a suitable step is computed using this model by applying the standard trust-region technology: the step s is calculated that attempts to minimize the model (2.2) within the ball centered at x_c and of radius Δ, that is

$$s = \arg \min_{\|s\| \leq \Delta} m(x_c + s). \tag{2.21}$$

This calculation can be exact (see Hebden [23], Moré [29] or Dennis and Schnabel [14] for instance) or approximate (see Powell [36], Dennis and Mei [13], Steihaug [45], Toint [49] or Conn *et al.* [10] for examples). As we take the view that an objective function evaluation is very costly, we opt for the first choice. The algorithm we currently use for solving (2.21) is somewhat involved because it covers the case where H has negative eigenvalues (with the possible occurrence of the "hard case" where g is orthogonal to the eigenvector of H corresponding to its smallest eigenvalue), and will not be detailed here. Note also that, as indicated by Powell [39], the complexity of solving (2.21) is identical to that of maximizing $S(x, x_-)$ in a ball of radius Δ, a calculation that occurs in Algorithm A1.

Once s is known, we then (as is standard) compute the ratio of achieved vs. predicted reduction

$$\rho = \frac{f(x_c) - f(x_c + s)}{m(x_c) - m(x_c + s)}. \tag{2.22}$$

However, instead of immediately proceeding to update the current iterate and trust-region radius, as would be traditional, we first examine if we cannot afford a possibly much longer step in the case where the model fits the true objective well enough. The motivation of these "long steps" or "jumps" is to use the current information to progress as much as possible using the current information (again keeping in mind the high cost of evaluating the objective). More precisely, we first test if $\rho \in [0.9, 1.1]$, indicating an excellent ratio of achieved to predicted reduction. In this case, we examine in succession all the past points $x_i \in M \setminus I$, determine, for each of these points, the ratio

$$\rho_i = \frac{f(x_i) - f(x_c)}{m(x_i) - m(x_c)}, \tag{2.23}$$

and compute the maximal distance $\|x_i - x_c\|$ for all x_i such that the ratio (2.23) is only slightly worse than (2.22), that is $\rho_i \in [0.85, 1.15]$. Let Θ denote this maximum distance. If it is much larger than Δ, the model (2.2) is thus likely to be valid in a much larger region than that in which the step s has been computed. A larger trial step, or jump, d may then be computed as

$$d = \arg \min_{\|s\| \leq \Theta} m(x_c + s), \tag{2.24}$$

with a chance of success very comparable to that of the original s. We may then decide to use the step d instead of the shorter s (redefining $s = d$) whenever sufficient progress is made, that is when

$$\rho_d = \frac{f(x_c) - f(x_c + d)}{m(x_c) - m(x_c + d)} \geq 0.05 \text{ and } f(x_c + d) < \min(f(x_c), f(x_c + s)). \quad (2.25)$$

The mechanism of these jumps provides the possibility of very rapid progress when the model is adequate. This is for instance the case when the objective function is itself quadratic in the region of interest, exactly or approximately.

More generally, once a new step s has been determined and the objective function evaluated at $x_c + s$, one then has to decide if this trial point should replace the current iterate x_c. As is usual in trust-region methods, the new point is accepted if the ratio of achieved vs. predicted reduction is larger than some small constant (0.05 in our implementation). If this is the case, one naturally includes the new point in the interpolation set I, simultaneously dropping the point in $I \cup J$ whose replacement by $x_c + s$ is most beneficial for the interpolation set geometry, whenever dropping is necessary (for example if $|I| = p$). The point $x_c + s$ may also be included in I even if descent is not obtained, if the interpolation set geometry is improved substantially by dropping another point of $(I \cup J) \setminus \{x_c\}$. However, both these possibilities may fail, in which case the iteration is declared unsuccessful, x_c is kept unchanged and an attempt is made to improve the geometry underlying the model using Algorithm A1. This latter action may also be advisable when the achieved reduction is small, say when $\rho \leq 0.15$, even if the iteration is successful.

Finally, the trust-region radius must be updated. Strictly speaking, one might consider decreasing Δ only when all other methods for improving the geometry fail (that is when Algorithm A1 terminates unsuccessfully at Step 6) and the proposed trust-region step fails, but this makes the reduction of Δ much too slow, as it typically takes of the order of p iterations to reach this stage. Instead, we propose to reduce Δ at all unsuccessful iterations, i.e. geometric and trust-region iterations. Similarly, Δ may be increased at every iteration that is clearly successful (that is when $\rho \geq 0.75$, say). A new iteration may then begin.

A last practical concern is to define an adequate stopping criterion. Two cases may occur in our algorithm. The first is when the objective function can no longer be significantly improved (as a consequence of the noise, for example), even if the trust-region radius is small, say below a tolerance ϵ_Δ, say. The second case is when a point x is found such that $f(x) - f_\ell$ is below a small tolerance ϵ_f, where f_ℓ is a lower bound on the objective function value. Such a bound is indeed often known by the user: for instance, $f_\ell = 0$ is a trivial choice for least-squares calculations. It can of course be set to minus infinity if no information is available.

2.2. The Complete Algorithm

After outlining the mechanism of the algorithm, we are now in position to formally state it in full detail. In the description that follows, we denote by η_r and η_a the relative and absolute noise levels on the evaluation of $f(x)$, given x.

General Algorithm

The values for x_1, ϵ_Δ, f_ℓ, ϵ_f, $r > 1$, Δ_1, η_a and η_r are given.

Step 0: initialization.

Define $\Delta = \Delta_1$ and set $M = I = \{x_1, x_2\}$ where x_2 is a random point different from x_1 such that $\|x_2 - x_1\| \leq \Delta$. Compute $f(x_1)$ and $f(x_2)$ and set $x_c = x_1$. Define p according to (2.4) and $J = \{x_3, \dots x_p\}$ such that $I \cup J$ is poised.

Step 1: start a new iteration.
Select $x_t = \arg\min_{x \in I} f(x)$. If

$$f(x_t) < f(x_c) - 0.1\epsilon_\Delta \Delta, \tag{2.26}$$

set $x_c = x_t$.

Step 2: convergence test.
If

$$\Delta < \epsilon_\Delta \quad \text{or} \quad f(x_t) < f_\ell + \epsilon_f, \tag{2.27}$$

stop and return x_t as an approximate solution.

Step 3: compute a quadratic model, if possible.

variant 1: If $|I| = p$, attempt to construct a quadratic model of the form

$$m(x_c + s) = f(x_c) + g^T s + \frac{1}{2} s^T H s \tag{2.28}$$

by defining the vector g and the symmetric matrix H such that

$$m(x) = f(x) \quad \text{for each } x \in I. \tag{2.29}$$

Else, that is if $|I| < p$, attempt to construct a quadratic model of the form (2.28) by defining the vector g and the symmetric matrix H as the solution of the problem

$$\min_{g,H} \|g\|^2 + \|H\|_F^2 \tag{2.30}$$

such that (2.29) holds. In both cases, if the model cannot be safely computed, attempt to improve the model using Algorithm A1 and go to Step 7.

variant 2: Construct a (possibly sub-) quadratic model such that

$$m(x) = f(x) \tag{2.31}$$

for each x in a subset of I which is poised and of maximum cardinality.

Step 4: compute a short step.
Solve the trust region problem

$$\min_{\|s\| \leq \Delta} m(x_c + s) \tag{2.32}$$

for the step s. If the predicted model reduction is too small, that is if

$$m(x_c) - m(x_c + s) < \max\left[\epsilon_\Delta \Delta, \frac{\eta_a}{2}, \frac{\eta_r}{2}|f(x_c + s)|\right], \tag{2.33}$$

then attempt to improve the model's geometry using Algorithm A1 and go to Step 7. Otherwise set $M = M \cup \{x_c + s\}$, compute $f(x_c + s)$ and the ratio (2.22).

Step 5: attempt a long step.

If the model fits the objective well at $x_c + s$, that is if $\rho \in [0.9, 1.1]$, compute

$$\Theta = \min(1000\Delta, \max \|x_i - x_c\|), \tag{2.34}$$

where the maximum is taken on all $x_i \in M \setminus I$ such that

$$1.15(m(x_i) - m(x_c)) \geq f(x_i) - f(x_c) \geq 0.85(m(x_i) - m(x_c)) \tag{2.35}$$

and

$$\|x_i - x_c\| \geq \Delta. \tag{2.36}$$

If $\Theta > (1 + r)\|s\|$, compute a long step d by solving

$$\min_{\|d\| \leq \Theta} m(x_c + d). \tag{2.37}$$

If $\|d\| > (1 + r)\|s\|$, then set $M = M \cup \{x_c + d\}$ and evaluate $f(x_c + d)$. In this latter case and if (2.25) holds, redefine $s = d$ and $\rho = \rho_d$.

Step 6: possibly add the new point to the interpolation set.

If sufficient descent is obtained, that is if $\rho \geq 0.05$, compute the best points in I and J that can be replaced by $x_c + s$, that is determine

$$x_I = \arg\max_{x \in I} S(x_c + s, x) \quad \text{and} \quad x_J = \arg\max_{x \in J} S(x_c + s, x) \tag{2.38}$$

and define

$$S_I = S(x_c + s, x_I) \quad \text{and} \quad S_J = S(x_c + s, x_J). \tag{2.39}$$

Then set

$$\begin{array}{ll} I = (I \setminus \{x_I\}) \cup \{x_c + s\} & \text{if } S_I > 2S_J, \\ I = I \cup \{x_c + s\} \text{ and } J = J \setminus \{x_J\} & \text{if } S_I \leq 2S_J. \end{array} \tag{2.40}$$

Else, determine the best points in $I \setminus \{x_c\}$ and J that can be replaced by $x_c + s$, that is

$$x_I = \arg\max_{x \in I \setminus \{x_c\}} S(x_c + s, x) \quad \text{and} \quad x_J = \arg\max_{x \in J} S(x_c + s, x) \tag{2.41}$$

and define S_I and S_J according to (2.39). If the corresponding geometry of the interpolation set is improved, that is if $\max(S_I, S_J) > 1$, set

$$\begin{array}{ll} I = (I \setminus \{x_I\}) \cup \{x_c + s\} & \text{if } S_I > S_J, \\ I = I \cup \{x_c + s\} \text{ and } J = J \setminus \{x_J\} & \text{if } S_I \leq S_J. \end{array} \tag{2.42}$$

If both the preceding conditions fail or if $\rho < 0.15$, attempt to improve the model's geometry using Algorithm A1.

Step 7: update the trust region radius.

• If the step s is successful, that is if $\rho \geq 0.75$, then set

$$\Delta = \min\left[r\Delta, \max(\Delta, r\|s\|)\right]. \tag{2.43}$$

• Else, if $\rho < 0.05$ and $\Delta > \epsilon_\Delta r$ and if the model has been computed at Step 3 without running into conditioning problems, set

$$\Delta = \max(\epsilon_\Delta r, \frac{\Delta}{r}). \tag{2.44}$$

Then, if Algorithm A1 has failed to improve the geometry with the previous value of Δ, apply it again with the updated value (2.44).

• If both the preceding conditions fail and Algorithm A1 has failed to improve the model's geometry with the current value of Δ, set

$$\Delta = \frac{\Delta}{r}. \tag{2.45}$$

• In all cases, go to Step 1.

end of General Algorithm

We now need to comment on some features introduced in the algorithm, but not discussed earlier.

1. We note that, for the algorithm to be well-defined, one needs the set $I \cup J$ to remain poised throughout the calculation. Indeed, this property is required for the Lagrange interpolation function $L(\cdot,\cdot)$, and hence for the function $S(\cdot,\cdot)$, to exist. We also note that, at every step of the algorithm, *except in (2.40)*, care is taken that introducing a new point in the set $I \cup J$ does not deteriorate the interpolation set geometry (since a new point x_+ is only introduced if $S(x_+, x_-) \geq 1$ for some $x_- \in I \cup J$), and hence that the poised nature of $I \cup J$ is maintained. However, no such test is performed when sufficient descent is obtained, in which case (2.40) forces the introduction of $x_c + s$ in I. For the algorithm to be well-defined, we therefore need to prove that the determinant $\delta(I \cup J)$ defined in (2.5) remains nonzero, when this exchange is performed.

   If we define

   $$\phi(x) = (\phi_1(x), \ldots, \phi_p(x))^T \tag{2.46}$$

   for any $x \in \mathbb{R}^n$, the fact that $\delta(I \cup J) \neq 0$ before the exchange implies that the columns of this determinant form a basis of $\mathbb{R}^p$, and hence that

   $$\phi(x_c + s) = \sum_{x_j \in (I \cup J)} \alpha_j \phi(x_j) \tag{2.47}$$

   for some coefficients α_j. Because some quadratic polynomials do not vanish at $x_c + s$, $\phi(x_c + s)$ is not the zero vector and there must thus exists at least one j such that $\alpha_j \neq 0$. If $x_j \in I$, we may then perform the first exchange of (2.40) with $x_I = x_j$ and obtain that

   $$\delta((I \setminus \{x_j\}) \cup \{x_c + s\} \cup J) = \alpha_j \delta(I \cup J) \neq 0. \tag{2.48}$$

   (We obtain the same result if $x_j \in J$, performing then the second exchange of (2.40) with $x_J = x_j$.) This implies that S_I or S_J is nonzero and thus that the interpolation set remains poised after (2.40), as desired.

   In theory, the above analysis guarantees that the model (2.2) and the Lagrange interpolation functions are always well-defined in the course of the computation. In practice however, it may happen that, although theoretically nonzero, $\delta(I \cup J)$ becomes rather small. In this case, the model based on all available interpolation points is numerically difficult to compute. The second of our algorithmic variants then automatically reduces the interpolation set to provide a well-defined model based on fewer points. The details of this procedure are once more described by Sauer and Yuan [42]. The first variant instead forces improvement of the geometry of the interpolation set until this problem disappears: this is the purpose of the test at the end of the section of Step 3 related to this variant.

2. Another difficulty hidden in our formulation is the choice of the set J at Step 0. We currently randomly generate points in a ball centered at x_1 and of radius Δ_1 until the set $I \cup J$ is poised. We then repeatedly apply Algorithm A1 until the geometry (as measured by $S(\cdot,\cdot)$) cannot be improved by a factor higher than 1.1. Other techniques are obviously possible and may be preferable. In particular, it is possible to let the user specify both x_2 and the points in J as well as x_1, or to let the points in $I \cup J$ to be composed of vertices of a nondegenerate simplex and the mid-points of its edges.

3. As for Algorithm A1, we have biased the formulation of Step 6 to encourage I to contain p points (thus yielding $J = \emptyset$) as soon as possible.

4. The parameter r in Step 7 gives the amount by which the trust-region radius is decreased on unsuccessful iterations, or increased on successful ones. We currently use $r = 1.5$. Observe that Step 7 does not automatically reduce Δ when this reduction would produce a value below the threshold ϵ_Δ, which would in turn cause immediate termination at the next iteration. Instead, attempts are made to improve the quadratic model (2.2) as much as possible with the current Δ. As a result, the radius is decreased (and the algorithm terminated) only when no improvement in function value appears possible within a region of radius ϵ_Δ.

   On the other hand, this termination mechanism may be judged too slow, as it typically requires of the order of p iterations with

$$\Delta \in [\epsilon_\Delta, r\epsilon_\Delta]. \tag{2.49}$$

   When the dimension increases, the algorithm may thus take a significant number of iterations in order to realize that a minimum has been found. One possible remedy is to also fit, for values of Δ satisfying (2.49), a linear model (whose geometry is good) around the current iterate. Building such a linear model typically requires of the order of n iterations only, which is much lower than p when n grows. The calculation may then be terminated whenever this linear model does not yield any further significant descent in a region of radius ϵ_Δ around the current iterate x_c.

5. The mechanism of Step 7 follows the traditional framework for trust-region methods, and may be refined in several ways. For instance, one might increase the value of r if several iterations of the same nature (successful or unsuccessful) occur in sequence. One may also replace (2.43) by

$$\Delta = \max\left(\Delta_{\max}, \min\left[r\Delta, \max(\Delta, r\|s\|)\right]\right) \tag{2.50}$$

   in order to impose an upper bound on the step length, a sometimes wise modification in the context of finite precision arithmetic.

6. Observe finally that the predicted reduction is, in (2.33), compared to the noise on the objective function value: a predicted reduction that is comparable to this noise is not considered as significant. This natural comparison is in accordance with the recommendations of Conn *et al.* [11].

3. PRELIMINARY NUMERICAL EXPERIENCE

We now report on some very preliminary numerical experience using the proposed algorithm. The two variants (using minimum norm and sub-quadratic models, respectively) have been programmed in MATLAB 4.1.1 [48] and tested on a set of 20 small dimensional unconstrained problems from the CUTE collection (see Bongartz *et al.* [1]). All calculations were performed on an IBM RS6000 workstation, with the algorithms parameters given by

$$\epsilon_\Delta = 10^{-3}, \quad \epsilon_f = 10^{-5}, \quad f_\ell = -\infty \text{ and } \Delta_1 = 1. \tag{3.1}$$

The model computation of Step 3 in the algorithm was judged ill-conditioned when the condition number of the associated Vandermonde system exceeds $10^4/\epsilon_M$, where ϵ_M is the machine precision (for variant 1 when $|I| < p$) or when the smallest pivot in the interpolation scheme (see Sauer [43] for a precise definition) was smaller than 10^{-8} (for all other cases). Successful termination was assumed for both algorithms when the stopping criterion (2.27) were satisfied, while failure was declared if an arithmetic error occurred in the calculation.

For comparison, we also ran LANCELOT (see Conn *et al.* [10]) using finite differences and the symmetric rank-one quasi-Newton update, all other algorithmic options being set to their default values. Because we have assumed that the cost of computing the objective function value dominates all other algorithmic costs, we will only discuss the results obtained in terms of number of objective function evaluations. Furthermore, since both variants of the new method choose the initial points in $I \cup J$ randomly, we only report, for these variants, the average number of function evaluations taken over 10 runs, this average being then rounded to the nearest integer.

We first examine the case where the objective function can be evaluated without noise, in which case we may set

$$\eta_a = \eta_r = 0. \tag{3.2}$$

Table 1 reports the number of objective function evaluations required for convergence for each of the 20 problems. The first column of this table gives the name of the considered problem, as specified in the CUTE collection, the second column indicates its number of variables, and columns three to five give the number of functions evaluations for the two variants of our algorithm (the column heading "min.-norm" referring to the variant using minimum Frobenius norm models, and "sub-Q" referring to the variant using sub-quadratic ones), and for LANCELOT[3], respectively.

A few tentative conclusions may be drawn from the results presented in this table.

1. The number of function evaluations required to minimize the test functions remains moderate.

2. The variant using the minimum Frobenius norm models and that using sub-quadratic models do not appear to behave very differently on our examples, although we note a slight advantage for the first.

[3]Because LANCELOT reports the number of function iterations (each of which evaluates the objective function once) and the number of gradient evaluations separately, the total number of function evaluations was estimated according to the formula $1 +$ number of iterations $+$ n(number of gradient evaluations $+ 1$), where the first 1 takes the initial objective function evaluation into account, and where number of gradient evaluations has been increased by one to reflect that the package uses a mix of forward and central differences to estimate the gradients.

problem's name	n	min.-norm	sub-Q	LANCELOT
AKIVA	2	64	57	41
BARD	2	51	65	43
BEALE	2	38	43	39
BOX3	3	36	40	35
BRKMCC	2	18	21	14
BROWNDEN	4	95	96	failed
CLIFF	2	28	33	83
CRAGGLVY	4	117	136	80
CUBE	2	144	123	89
DENSCHNA	2	28	30	20
GULF	3	213	237	143
HAIRY	2	51	53	363
HATFLDE	3	67	74	60
HELIX	3	93	91	76
KOWOSB	4	74	90	114
PFIT1LS	3	166	failed	219
ROSENBR	2	106	93	90
SCHMVETT	3	49	53	35
SISSER	2	24	26	35
SNAIL	2	321	326	778

Table .1: Number of function evaluations in the absence of noise

3. In the absence of noise, both variants have an efficiency which is comparable to that of the finite-difference version of LANCELOT. They even seem to outperform this package for the more ill-conditioned or more nonlinear cases (CLIFF, HAIRY, PFIT1LS (for the variant 1 only) and SNAIL).

The details of the runs also show that the long steps significantly improve the performance of both variants.

We now turn to the case where noise is present in the objective function and give, in Table 2, the number of function evaluations required for algorithm termination for two different noise levels. We have chosen the absolute and relative noise levels to be equal to 0.0001 and 0.01. We note that, as expected, noise typically prevents the algorithm finding the true minimum of the test functions. Moreover, there is no guarantee that the algorithm will stop at a point whose function value differs from the true noise-free minimum value by an amount comparable to the noise level, as it may indeed stop at any point at which the slope is of the order of the noise. As before, the number shown are averages over 10 runs, rounded to the nearest integer. The first number in each entry of columns three and four corresponds to a noise level of 0.0001 and the second to a level of 0.01. In all cases, η_a and η_r were chosen equal and identical to the considered noise level.

We also indicate in Table 2, the number of function evaluations required by a trust-region code in which the finite-difference step size was chosen as a function of the noise level, according to the recommendations of Dennis and Schnabel [14], pages 97–99. This code relies on LANCELOT subroutines, but uses the machinery of this package in a *context for which it was not originally designed.* The resulting program however represents, in the authors' view, a reasonable yard-stick for measuring the performance of the new methods versus a more traditional technique. When results for this code are shown within square brackets, this indicates that the final objective function value was substantially above the optimal noise-free value, even taking the noise into account. For instance, the code did not succeed to reduce the objective function beyond its value at the starting point for problem SCHMVETT and noise level 0.01. These cases may be considered as practical failures, although they cannot formally be interpreted as such, because one of the specified stopping criteria[4] was met.

problem's name	n	min.-norm	sub-Q	finite-differences trust-region code
AKIVA	2	57/26	48/31	[23]/[15]
BARD	2	39/26	48/33	44/40
BEALE	2	35/23	40/26	49/68
BOX3	3	30/21	33/28	47/120
BRKMCC	2	17/16	20/16	13/18
BROWNDEN	4	89/73	90/76	54/54
CLIFF	2	25/20	30/26	86/86
CRAGGLVY	4	89/60	105/67	85/90
CUBE	2	130/60	112/48	77/158
DENSCHNA	2	26/21	28/24	20/23
GULF	3	141/25	168/33	1791/198
HAIRY	2	48/40	48/42	[200]/[18]
HATFLDE	3	54/37	55/38	86/110
HELIX	3	89/75	87/75	80/92
KOWOSB	4	42/23	63/48	81/138
PFIT1LS	3	134/23	failed	203/126
ROSENBR	2	102/65	88/64	95/85
SCHMVETT	3	44/22	47/34	24/[7]
SISSER	2	24/20	24/20	35/53
SNAIL	2	327/13	315/18	[207]/[36]

Table .2: Number of function evaluations for absolute and relative noise levels 0.0001/0.01

These results allow us to draw three more conclusions.

1. The comparable performance of both variants is confirmed in the presence of noise. It also seems to be significantly higher than that of the finite-differences based code.

2. Both variants seem substantially more robust that the finite-differences based code.

3. The effort (expressed in number of function evaluations) decreases when the noise increases when one of the two new algorithms is used. This is to be expected

[4]In order to provide meaningful comparison with the other algorithms, successful termination was assumed when either the approximate gradient was smaller than 10^{-5} or the trust region smaller than 10^{-3}. The initial trust-region radius was set to one and the total number of function evaluations was estimated as for LANCELOT without noise.

because more severe noise causes the minimization to stop sooner. However, we do not always observe this phenomenon for the finite-differences based code.

Furthermore, a closer look at the detailed results show that, although not guaranteed (as already mentioned), the best function value reported by the new algorithm is very often within the range $[f_*, f_*(1+\eta_r)+\eta_a]$, where f_* is the minimum objective value for the noiseless problem.

These preliminary conclusions should of course be taken with caution, given the very limited amount of testing reported here. They are however encouraging.

4. CONCLUSIONS AND PERSPECTIVES

We have presented the current stage of development of a derivative free algorithm for unconstrained optimization. This algorithm is based on the trust-region paradigm and allows large steps to be taken when appropriate. Furthermore, the choice of models in the first iterations of the algorithm permits a substantial reduction of the objective function at the outset of the calculation. The algorithm is acceptably efficient and its performance is comparable, in the absence of noise, to that of good finite-difference quasi-Newton methods. Moreover, its behaviour also seems satisfactory in the presence of noise.

The authors realize that much additional development is necessary before the ideas presented here can result in a final algorithm and corresponding software. In particular, the following directions of investigation appear to be of interest and are the subject of ongoing work.

- If multivariable interpolation allows for sub-quadratic models, it also allows for models that are polynomials of degree higher than two. The use of these models suggests a number of interesting questions concerning their solution and the handling of the associated geometry.

- The amount of linear algebra involved at each iteration of the current version of the algorithm is relatively high, in accordance with the view that this cost is dwarfed by that of evaluating the objective function. However, there are cases where this workload may become excessive when the problem's dimension increases and the cost of evaluating the objective function is not very high. Variants of the algorithm requiring less linear algebra work per iteration are thus of interest.

  Furthermore, our current implementation of the variant using sub-quadratic models uses the Newton-type interpolation method of Sauer [43], but yet bases its choice of a new interpolation point on (2.7)–(2.8), which uses the Lagrange interpolation functions. This is somewhat inefficient and should be modified to use the Newton interpolation functions throughout.

- Applying the ideas developed in this paper in the context of large-scale problems is also possible, in particular by the using the partially separable structure (see Griewank and Toint [21], Conn *et al.* [9] or Conn *et al.* [10], for instance) which is very often associated with large problems. One can indeed use multivariate interpolation techniques to build a model for each of the element functions in a partially separable decomposition, provided such a decomposition is available.

But this obviously raises a number of questions on how to maintain a suitable overall model and how to handle the geometry at the level of each element.

- The inclusion of constraints in our framework is also a very interesting development. A first stage is of course to adapt our technique in order to handle bound constrained problems, which can be done, for instance, by suitably modifying the trust region definition and using the ℓ_∞-norm for determining its shape (see Conn *et al.* [8] or Conn *et al.* [10]). But we also wish to handle more general constraints. As mentioned in the introduction, this can be done in several ways, including augmented Lagrangian techniques or exact penalty functions.

- The development of a proper convergence theory covering the final version of our algorithm is highly desirable.

- Besides the case where the evaluation of the objective function involves some noise whose level cannot be controlled, another interesting case is that of objective functions for which the accuracy of evaluation may be specified in advance, with the understanding that a more accurate evaluation may be (possibly substantially) more costly. It is thus of interest to adapt the algorithm presented here in such a way that it specifies, for each evaluation, a degree of accuracy that is sufficient for the algorithm to proceed efficiently while keeping the evaluation cost as low as possible.

- Finally, it is clear that considerable further numerical experience is needed to really assess the techniques discussed above, both from the reliability and efficiency point of view.

As implied by these comments and perspectives, the ideas presented here are thus but a preliminary second step in the development of a robust and efficient algorithm for optimization problems in which derivatives are unavailable. The first step was Powell's work in [38] and [39]. Research in this domain remains challenging and, in the authors' experience, clearly meets a strong and explicit need in several application areas.

ACKNOWLEDGEMENT

The authors are grateful to K. Mints and M. Ferris for their interest in this research and their useful comments.

The research of the first author was supported in part by the Advanced Research Projects Agency of the Department of Defense and was monitored by the Air Force Office of Scientific Research under Contract No F49620-91-C-0079. The United States Government is authorized to reproduce and distribute reprints for governmental purposes notwithstanding any copyright notation hereon.

REFERENCES

[1] I. Bongartz, A. R. Conn, N.I.M. Gould, and Ph. L. Toint. "CUTE: Constrained and Unconstrained Testing Environment". *ACM Transactions on Mathematical Software*, 21(1):123–160, 1995.

[2] A. J. Booker. "DOE for computer output. Technical Report BCSTECH-94-052". Boeing Computer Services, 1994.

[3] M. J. Box. "A comparison of several current optimization methods, and the use of transformations in constrained problems". *Computer Journal*, 9, 1966.

[4] R. P. Brent. *"Algorithms for Minimization Without Derivatives"*. Prentice-Hall, Engelwood Cliffs, USA, 1973.

[5] S. H. Brooks. "A discussion of random methods for seeking maxima". *Journal of Operations Research*, 6, 1958.

[6] A. G. Buckley. "A derivative-free algorithm for parallel and sequential optimization". presentation at the NATO ASI on Algorithms for Continuous Optimization, Il Ciocco, 1994.

[7] J. Burns. "The sensitivity equation approach to optimal control". Presentation at the IMA Workshop on Large-Scale Optimization, Minneapolis, 1995.

[8] A. R. Conn, N. I. M. Gould, and Ph. L. Toint. "Global convergence of a class of trust region algorithms for optimization with simple bounds". *SIAM Journal on Numerical Analysis*, 25:433–460, 1988. See also same journal 26:764–767, 1989.

[9] A. R. Conn, N. I. M. Gould, and Ph. L. Toint. "An introduction to the structure of large scale nonlinear optimization problems and the LANCELOT project". In R. Glowinski and A. Lichnewsky, editors, *Computing Methods in Applied Sciences and Engineering*, pages 42–51, Philadelphia, USA, 1990. SIAM.

[10] A. R. Conn, N. I. M. Gould, and Ph. L. Toint. *"LANCELOT: a Fortran package for large-scale nonlinear optimization (Release A)"*. Number 17 in Springer Series in Computational Mathematics. Springer Verlag, Heidelberg, Berlin, New York, 1992.

[11] A. R. Conn, Nick Gould, A. Sartenaer, and Ph. L. Toint. "Global convergence of a class of trust region algorithms for optimization using inexact projections on convex constraints". *SIAM Journal on Optimization*, 3(1):164–221, 1993.

[12] C. De Boor and A. Ron. "Computational aspects of polynomial interpolation in several variables". *Mathematics of Computation*, 58(198):705–727, 1992.

[13] J. E. Dennis and H. H. W. Mei. "Two new unconstrained optimization algorithms which use function and gradient values". *Journal of Optimization Theory and Applications*, 28(4):453–482, 1979.

[14] J. E. Dennis and R. B. Schnabel. *"Numerical methods for unconstrained optimization and nonlinear equations"*. Prentice-Hall, Englewood Cliffs, USA, 1983.

[15] J. E. Dennis and V. Torczon. "Direct search methods on parallel machines". *SIAM Journal on Optimization*, 1(4):448–474, 1991.

[16] L. C. W. Dixon. *"Nonlinear Optimisation"*. The English Universities Press Ltd, London, 1972.

[17] R. Elkin. *"Convergence Theorems for Gauss-Seidel and Other Minimization Algorithms"*. PhD thesis, University of Maryland, College Park, 1968.

[18] C. Elster and A. Neumaier. "A grid algorithm for bound constrained optimization of noisy functions". *IMA Journal of Numerical Analysis*, 15:585–608, 1995.

[19] P. E. Gill, W. Murray, and M. H. Wright. *"Practical Optimization"*. Academic Press, London and New York, 1981.

[20] P. E. Gill, W. Murray, M. A. Saunders, and M. Wright. "Computing forward-difference intervals for numerical optimization". *SIAM Journal on Scientific and Statistical Computing*, 4:310–321, 1983.

[21] A. Griewank and Ph. L. Toint. "On the unconstrained optimization of partially separable functions." In M. J. D. Powell, editor, *Nonlinear Optimization 1981*, pages 301–312, London and New York, 1982. Academic Press.

[22] A. Griewank. "On automatic differentiation". In M. Iri and K. Tanabe, editors, *Mathematical Programming: recent developments and applications*, pages 83–108, Dordrecht, NL, 1989. Kluwer Academic Publishers.

[23] M. D. Hebden. "An algorithm for minimization using exact second derivatives". Technical Report T.P. 515, AERE Harwell Laboratory, Harwell, UK, 1973.

[24] D. M. Himmelblau. *"Applied Nonlinear Programming"*. McGraw-Hill, New-York, 1972.

[25] R. Hooke and T. A. Jeeves. "Direct search solution of numerical and statistical problems". *Journal of the ACM*, 8:212–229, 1961.

[26] R. J. Kelly and R. F. Wheeling. "A digital computer program for optimizing nonlinear functions". Technical report, Mobil Oil Corp., Research Dept., Princeton, New Jersey, 1962.

[27] S. Lucidi and M. Sciandrone. "A coordinate descent method without derivatives". Technical Report 10-95 (in preparation), University of Rome "La Sapienza", Rome, 1995.

[28] T. J. Mitchell, J. Sacks, W. J. Welch, and H. P. Wynn. "Design and analysis of computer experiments". *Statistical Science*, 4(4):409–435, 1989.

[29] J. J. Moré. "The Levenberg-Marquardt algorithm: implementation and theory". In G. A. Watson, editor, *Proceedings Dundee 1977*, Berlin, 1978. Springer Verlag. Lecture Notes in Mathematics.

[30] M. Morris, C. Currin, T. J. Mitchell, and D. Ylvisaker. "Bayesian prediction of deterministic functions, with applications to the design and analysis of computer experiments". *Journal of the American Statistical Association*, 86(416):953–963, 1991.

[31] J. A. Nelder and R. Mead. "A simplex method for function minimization". *Computer Journal*, 7:308–313, 1965.

[32] J. M. Ortega and W. C. Rheinboldt. *"Iterative solution of nonlinear equations in several variables"*. Academic Press, New York, 1970.

[33] A. B. Owen. "Orthogonal arrays for computer experiments, integration and visualization". *Statistica Sinica*, 2:439–452, 1992.

[34] B. Polyak. *"Introduction to Optimization"*. Optimization Software Inc., New York, 1987.

[35] M. J. D. Powell. "An efficient method for finding the minimum of a function of several variables without calculating derivatives". *Computer Journal*, 17:155–162, 1964.

[36] M. J. D. Powell. "A new algorithm for unconstrained optimization". In J. B. Rosen, O. L. Mangasarian, and K. Ritter, editors, *Nonlinear Programming*, New York, 1970. Academic Press.

[37] M. J. D. Powell. "On the global convergence of trust region algorithms for unconstrained minimization". *Mathematical Programming*, 29(3):297–303, 1984.

[38] M. J. D. Powell. "A direct search optimization method that models the objective and constraint functions by linear interpolation". In *Advances in Optimization and Numerical Analysis, Proceedings of the Sixth Workshop on Optimization and Numerical Analysis, Oaxaca, Mexico*, volume 275, pages 51–67, Dordrecht, NL, 1994. Kluwer Academic Publishers.

[39] M. J. D. Powell. "A direct search optimization method that models the objective by quadratic interpolation". Presentation at the 5th Stockholm Optimization Days, 1994.

[40] H. H. Rosenbrock. "An automatic method for finding the greatest or least value of a function". *Computer Journal*, 3:175–184, 1960.

[41] J. Sacks, H. P. Wynn, T. J. Mitchell, W. J. Welch, R. J. Buck, and M. Morris. "Screening, predicting and computer experiments". *Technometrics*, 34(1):15–25, 1992.

[42] Th. Sauer and Yuan Xu. "On multivariate Lagrange interpolation". *Mathematics of Computation*, 64:1147–1170, 1995.

[43] Th. Sauer. "Computational aspects of multivariate polynomial interpolation". *Advances in Computational Mathematics*, 3:219–238, 1995.

[44] W. Spendley, G. R. Hext, and F. R. Himsworth. "Sequential application of simplex designs in optimization and evolutionary operation". *Technometrics*, 4, 1962.

[45] T. Steihaug. "The conjugate gradient method and trust regions in large scale optimization". *SIAM Journal on Numerical Analysis*, 20(3):626–637, 1983.

[46] G. W. Stewart. "A modification of Davidon's minimization method to accept difference approximations of derivatives". *Journal of the ACM*, 14, 1967.

[47] W. H. Swann. "Report on the development of a new direct search method of optimization". Technical Report Research Note 64/3, I.C.I., Central Instruments Laboratory, 1964.

[48] The Mathworks Inc. "Matlab reference guide". The Mathworks Inc., 1992.

[49] Ph. L. Toint. "Towards an efficient sparsity exploiting Newton method for minimization". In I. S. Duff, editor, *Sparse Matrices and Their Uses*, pages 57–88, London, 1981. Academic Press.

[50] V. Torczon. "On the convergence of the multidirectional search algorithm". *SIAM Journal on Optimization*, 1(1):123–145, 1991.

MASSIVELY PARALLEL SOLUTION OF LARGE SCALE NETWORK FLOW PROBLEMS[1]

De Leone Renato[1], Merelli Emanuela[2] and Buti Sara[1]

[1] Dipartimento di Matematica e Fisica
Università di Camerino
Camerino, Italy
e-mail: deleone@camcic.unicam.it

[2] Diploma Universitario in Informatica
Università di Camerino
Camerino, Italy
e-mail: merelli@camcic.unicam.it

Abstract. Two massively parallel algorithms for large scale linear and convex quadratic network flow problems are proposed and studied. The methods are based on the alternating step method for monotropic programming. The original network flow problem is decomposed in simple subproblems involving only few variables for which solution in closed form exists. Computational results obtained on the CRAY T3D show that the methods hold the promise of solving extremely large size problems.

Key words. Parallel Computing, Network Flow Problems, Decomposition Methods.

1. INTRODUCTION

One of the aim of recent research in mathematical programming is to determine algorithms that will efficiently use the available massively parallel computers. Decomposition provides a general basic scheme for constructing parallel algorithms. The original problem is divided in smaller independent subproblems that can be concurrently solved; at the end of the parallel phase some sort of coordination is needed to determine if the optimal solution has been obtained or to construct a new different splitting of the problem. The main contribution of the paper is the definition of two new algorithms that can be efficiently implemented on massively parallel architecture for convex separable network flow problems. We derive both algorithms in a very simple way from the same general alternating step scheme.

In Sect.2, we will describe the Alternating Direction Method for solving convex optimization problems with linear constraints. From this very general scheme, various fine–grain and coarse–grain parallel algorithms have been derived [4,5,8,3,2] for various classes of problems. In Sect.3 we present two different fine–grain massively parallel algorithms for network flow problems with convex separable objective function. These algorithms were implemented on the CRAY T3D parallel machines and

[1]This work was supported by CNR Grant n. 95.00732.CT01 and by a grant from CINECA

the implementation issues are discussed in Sect.4. Finally computational results are shown in Sect.5.

We describe now some of the general notations that will be used in this paper. We let $\mathbb{R}^n$ to denote the real n–dimensional space. For a vector $x \in \mathbb{R}^n$ and for a $m \times n$ real matrix A, we denote the transposes by x^T and A^T. Subscripts will be used for components of vectors, while superscript will denote iterates in a sequence. The symbol $\|x\|$ will denote the Euclidean norm of the vector x, $\|x\| = (x^T x)^{\frac{1}{2}}$. For a function $g : \mathbb{R}^n \to \mathbb{R}$, the gradient of g will be denoted by ∇g.

2. THE ALTERNATING DIRECTION (ADI) METHOD

In this section we introduce the Alternating Direction Method (ADI) for minimizing the sum of two convex functions subject to linear constraints:

$$\begin{array}{ll} \underset{r,s}{\text{minimize}} & F(r) + G(s) \\ \text{subject to} & Mr + Ns = q \end{array} \tag{2.1}$$

The Augmented Lagrangian function associated with the above problem is given by

$$L_\lambda(r,s,p) := F(r) + G(s) + p^T(Mr + Ns - q) + \frac{\lambda}{2}\|Mr + Ns - q\|_2^2 \tag{2.2}$$

where λ is a positive scalar. A single iteration of the Alternating Direction Method consists of minimizing the Augmented Lagrangian function (2.2) with respect to r with s and p fixed, then minimize with respect to s with r and p fixed and finally modify the multiplier p by taking a step along the steepest ascent direction with r and s fixed. More specifically, starting from an arbitrarily chosen initial point (r^0, s^0, p^0), the new point $(r^{t+1}, s^{t+1}, p^{t+1})$ is obtained from the previous iterate (r^t, s^t, p^t) as follows:

$$r^{t+1} \in \underset{r}{\operatorname{argmin}}\, L_\lambda(r, s^t, p^t) \tag{2.3a}$$

$$s^{t+1} \in \underset{s}{\operatorname{argmin}}\, L_\lambda(r^{t+1}, s, p^t) \tag{2.3b}$$

$$p^{t+1} = p^t + \lambda \nabla_p L_\lambda(r^{t+1}, s^{t+1}, p^t). \tag{2.3c}$$

Sufficient conditions for the convergence of the ADI method have been derived by using the theory of maximal monotone operators [10,4,6] or the saddle–point theory of Lagrangian functions [7]. Self–contained proofs of convergence of the Alternating Direction Method can be found in [1] and [8] where a complete bibliography can also be found. In particular, it is possible to show [8, Theorem 2.3.1] that if the functions F and G are extended–real valued, proper, convex functions, if problem (2.1) has a lagrangean saddle–point and if subproblems (2.3) are solvable, then the sequence $\{F(r^t) + G(s^t)\}$ converges to the optimal value of (2.1). Moreover, the sequence $\{p^t\}$ converges to the optimal dual multiplier for problem (2.1). In addition, if the matrices M and N have full column rank [8, Corollary 2.5.3] then the sequence $\{r^t, s^t\}$ converges to an optimal primal point of (2.1).

Our interest in this method is mainly motivated by the possibility of deriving from this general scheme parallel fine–grain and coarse–grain algorithms for special classes of problems. In [8] and [2] various coarse–grain parallel algorithms were derived for convex block–angular problems. By introducing new variables a new equivalent formulation of the original problem was obtained and functions $F(.)$ and $G(.)$ were defined in such a way that the first minimization problem in the Alternating Direction Method (step

(2.3a)) naturally decomposes in independent subproblems. The coupling constraints were taken into account only in the second step and closed form solution for (2.3b) was obtained. This is the coordination step and communication among the processors is required. The linear constraints in (2.1) reflect the equality of newly introduced variables and original variables of the problem.

The three splitting methods derived in [8] are coarse–grain parallel algorithms requiring at each step the solution (in parallel) of non–trivial subproblems and a very simple coordination phase. The number of subproblems that can be solved in parallel is equal to the number of blocks in the original problem. In the next section we will derive (still from the general Alternating Direction Method) two fine–grain massively parallel algorithms for network flow problems with convex separable objective function. Again the minimization in step (2.3a) of the Alternating Direction Method, will naturally decompose in independent subproblems. However, the number of subproblems here is equal to the number of nodes and arcs in the network problem and solutions in closed form can be obtained for the linear and quadratic case. Therefore, the proposed algorithm can be successfully implemented on massively parallel machines.

3. MASSIVELY PARALLEL ADI ALGORITHMS FOR NETWORK FLOW PROBLEM

Given a directed graph $\mathcal{G}(\mathcal{N}, \mathcal{A})$ with m nodes and n arcs, the minimum–cost network flow problem that we consider here is:

$$\begin{array}{ll} \text{minimize} & \sum_{(i,j)\in\mathcal{A}} f_{ij}(x_{ij}) \\ \text{subject to} & \sum_{j\in\delta_+(i)} x_{ij} - \sum_{j\in\delta_-(i)} x_{ji} = b_i, \quad i \in \mathcal{N} \\ & l_{ij} \le x_{ij} \le u_{ij}, \quad (i,j) \in \mathcal{A} \end{array} \tag{3.1}$$

where for each node i, $\delta_+(i)$ is the subset of arcs leaving node i and $\delta_-(i)$ is the subset of arcs entering node i. We are mostly interested in the convex quadratic case

$$f_{ij}(a) := \tfrac{1}{2} d_{ij} a^2 + c_{ij} a$$

although many of the results apply to general convex differentiable functions.

In the sequel we will derive from the general scheme discussed in Sect.2 two different algorithms suitable for massively parallel computation. The main difference between the two algorithms is in the fact that the first algorithm does not necessarily produce iterates that satisfy the simple bounds on the variables.

3.1. The algorithm MASSNET 1

For the network flow problem (3.1) each variable x_{ij} participates in four distinct constraints corresponding to the flow conservation constraints at the source node, the flow conservation constraint at the target node, the lower and the upper bound constraints. In this first algorithm we replace the single flow variable x_{ij} with four variables $x1_{ij}, x2_{ij}, x3_{ij}$ and $x4_{ij}$. Note that each new variable is present is one and only one constraint and in the *coupling* constraints imposing the equality of the four values for the flow on the arc. For the generic arc (i,j), the variable $x1_{ij}$ is associated with the flow conservation constraint at the source node i, the variable $x2_{ij}$ is associated with the flow conservation constraint at the target node j, while $x3_{ij}$ and $x4_{ij}$ are associated with the simple flow bounds.

The equivalent formulation of problem (3.1) we consider is:

$$\begin{array}{lll} \text{minimize} & \sum_{(i,j)\in\mathcal{A}} f_{1ij}(x_{1ij}) + f_{2ij}(x_{2ij}) + f_{3ij}(x_{3ij}) + f_{4ij}(x_{4ij}) & \\ \text{subject to} & \sum_{j\in\delta_+(i)} x_{1ij} - \sum_{j\in\delta_-(i)} x_{2ji} = b_i, \quad i\in\mathcal{N} & \\ & \left.\begin{array}{r} l_{ij} \le x_{3ij} \\ x_{4ij} \le u_{ij} \\ x_{1ij} = x_{2ij} = x_{3ij} = x_{4ij} \end{array}\right\} (i,j)\in\mathcal{A} & \end{array} \tag{3.2}$$

where

$$f_{1ij}(a) + f_{2ij}(a) + f_{3ij}(a) + f_{4ij}(a) = f_{ij}(a), \quad \forall a \in \mathbb{R} \text{ and } \forall (i,j)\in\mathcal{A}.$$

Define now

$$r := \begin{bmatrix} r_{1ij} \\ r_{2ij} \\ r_{3ij} \\ r_{4ij} \end{bmatrix} \text{ and } s := \begin{bmatrix} s_{1ij} \\ s_{2ij} \\ s_{3ij} \\ s_{4ij} \end{bmatrix}$$

and

$$F(r) := \begin{cases} \sum_{(i,j)\in\mathcal{A}} \sum_{l=1}^{4} f_{lij}(r_{lij}) & \text{if } \sum_{j\in\delta_+(i)} r_{1ij} - \sum_{j\in\delta_-(i)} r_{2ji} = b_i, \quad i\in\mathcal{N} \\ & \quad r_{3ij} \ge l_{ij}, \quad r_{4ij} \le u_{ij}, \quad (i,j)\in\mathcal{A} \\ +\infty & \text{otherwise,} \end{cases} \tag{3.3}$$

$$G(s) := \begin{cases} 0 & \text{if } s_{1ij} = s_{2ij} = s_{3ij} = s_{4ij}, \quad (i,j)\in\mathcal{A} \\ +\infty & \text{otherwise.} \end{cases} \tag{3.4}$$

Problem (3.2) is now equivalent to

$$\begin{array}{ll} \underset{r,s}{\text{minimize}} & F(r) + G(s) \\ \text{subject to} & r - s = 0. \end{array} \tag{3.5}$$

Note that at the end of each iteration, the values of s_1, s_2, s_3 and s_4 are identical. For simplicity we will indicate this value by x. Moreover the minimization problem in step (2.3a) decomposes in $m + 2n$ simpler independent subproblems that can be solved in a parallel way.

Algorithm MASSNET 1

- Start with arbitrary values for the flow vector x^0 and the multipliers $p^0 = [p_1{}^0, p_2{}^0, p_3{}^0, p_4{}^0]^T$. Set $t = 0$
- Having x^t and $p^t = [p_1{}^t, p_2{}^t, p_3{}^t, p_4{}^t]^T$ compute x^{t+1} and $p^{t+1} = [p_1{}^{t+1}, p_2{}^{t+1}, p_3{}^{t+1}, p_4{}^{t+1}]^T$ as follows:

For each node $i \in \mathcal{N}$ solve (in parallel)

$$\begin{array}{ll} \underset{r}{\text{minimize}} & \sum_{j\in\delta_+(i)} \left[f_{1ij}(r_{1ij}) + p_{1ij}^t\left(r_{1ij} - x_{ij}^t\right) + \frac{\lambda}{2}\left(r_{1ij} - x_{ij}^t\right)^2 \right] + \\ & \sum_{j\in\delta_-(i)} \left[f_{2ji}(r_{2ji}) + p_{2ji}^t\left(r_{2ji} - x_{ji}^t\right) + \frac{\lambda}{2}\left(r_{2ji} - x_{ji}^t\right)^2 \right] \\ \text{subject to} & \sum_{j\in\delta_+(i)} r_{1ij} - \sum_{j\in\delta_-(i)} r_{2ji} = b_i \end{array} \tag{3.6a}$$

to obtain new values for $r_1{}^{t+1}$ and $r_2{}^{t+1}$. For each arc $(i,j) \in \mathcal{A}$ solve (in parallel) the problems:

$$\begin{array}{ll} \underset{r}{\text{minimize}} & f_{3ij}(r_{3ij}) + p_{3ij}^t\left(r_{3ij} - x_{ij}^t\right) + \frac{\lambda}{2}\left(r_{3ij} - x_{ij}^t\right)^2 \\ \text{subject to} & r_{3ij} \geq l_{ij} \end{array} \tag{3.6b}$$

and

$$\begin{array}{ll} \underset{r}{\text{minimize}} & f_{4ij}(r_{4ij}) + p_{4ij}^t\left(r_{4ij} - x_{ij}^t\right) + \frac{\lambda}{2}\left(r_{4ij} - x_{ij}^t\right)^2 \\ \text{subject to} & r_{4ij} \leq u_{ij} \end{array} \tag{3.6c}$$

to obtain new values for $r_3{}^{t+1}$ and $r_4{}^{t+1}$.

Determine x^{t+1} by solving:

$$\underset{x}{\text{minimize}} \sum_{l=1}^{4} \left[p_l{}^{tT}\left(r_l{}^{t+1} - x\right) + \frac{\lambda}{2}\left\|r_l{}^{t+1} - x\right\|_2^2 \right] \tag{3.7}$$

Finally,

$$p_l{}^{t+1} = p_l{}^t + \lambda\left(r_l{}^{t+1} - x^{t+1}\right) \tag{3.8}$$

for $l = 1, \ldots, 4$.

Note that if the initial multiplier p^0 is chosen in such a way that

$$p_1{}^0 + p_2{}^0 + p_3{}^0 + p_4{}^0 = 0$$

the same property holds for all iterates p^t and the solution of problem (3.7) is simply obtained averaging the four values for the flow on each arc as obtained in (3.6a), (3.6b), and (3.6c):

$$x^{t+1} = \frac{1}{4}\left(r_1{}^{t+1} + r_2{}^{t+1} + r_3{}^{t+1} + r_4{}^{t+1}\right).$$

Moreover, when the functions $f_{lij}(.)$ are convex quadratic, solution in closed form exists for problems (3.6a), (3.6b), and (3.6c). The most expansive part of the computation is in solving (3.6a) in parallel for each node. A quadratic problem with as many variables as entering and leaving arcs and with a single constraint but not upper or lower bounds on the flow must be solved.

For the generic node $i \in \mathcal{N}$ we determine the optimal value of the multiplier associated with the single constraint in (3.6a) and the current flow for the arcs leaving node i and for the arcs entering the node. The dual optimal multiplier is proportional to the violation of the conservation constraint for a flow vector that is the current flow vector x^t modified by taking into account the linear term in $f_1(.)$ or $f_2(.)$ and the current multiplier p^t.

3.2. The Algorithm MASSNET 2

The main feature of the previous approach is that we do not attempt to enforce the bound constraints, although it would be quite simple to do so. From a computational point of view it seems more reasonable to satisfy the flow bounds at each iteration. The second algorithm we propose, tries to overcome this difficulty.

Instead of replacing the single flow variable x_{ij} with four new variables, we only introduce two new variables x_{1ij} and x_{2ij} corresponding to the flow conservation constraints at source and target node.

The equivalent formulation of problem (3.1) we consider is:

$$\begin{array}{lll} \text{minimize} & \sum\limits_{(i,j)\in\mathcal{A}} f1_{ij}(x1_{ij}) + f2_{ij}(x2_{ij}) & \\ \text{subject to} & \sum\limits_{j\in\delta_+(i)} x1_{ij} - \sum\limits_{j\in\delta_-(i)} x2_{ji} = b_i, \quad i\in\mathcal{N} & \\ & l_{ij} \le x1_{ij} = x2_{ij} \le u_{ij}, \quad (i,j)\in\mathcal{A} & \end{array} \tag{3.9}$$

where now

$$f1_{ij}(a) + f2_{ij}(a) = f_{ij}(a), \qquad \forall a \in \mathbb{R} \text{ and } \forall (i,j)\in\mathcal{A}.$$

Define now

$$r := \begin{bmatrix} r1_{ij} \\ r2_{ij} \end{bmatrix} \text{ and } s := \begin{bmatrix} s1_{ij} \\ s2_{ij} \end{bmatrix}$$

and

$$F(r) := \begin{cases} \sum\limits_{(i,j)\in\mathcal{A}} f1_{ij}(r1_{ij}) + f2_{ij}(r2_{ij}) & \text{if } \sum\limits_{j\in\delta_+(i)} r1_{ij} - \sum\limits_{j\in\delta_-(i)} r2_{ji} = b_i, \quad i\in\mathcal{N} \\ +\infty & \text{otherwise,} \end{cases} \tag{3.10}$$

$$G(s) := \begin{cases} 0 & \text{if } l_{ij} \le s1_{ij} = s2_{ij} \le u_{ij}, \quad (i,j)\in\mathcal{A} \\ +\infty & \text{otherwise.} \end{cases} \tag{3.11}$$

Problem (3.9) is now equivalent to

$$\begin{array}{ll} \underset{r,s}{\text{minimize}} & F(r) + G(s) \\ \text{subject to} & r - s = 0. \end{array} \tag{3.12}$$

Note that at the end of each iteration, the values of $s1$ and $s2$ are identical. For simplicity we will indicate this value by x.

Algorithm MASSNET 2

- Start with arbitrary values for the flow vector x^0 and the multipliers $p^0 = [p1^0, p2^0]^T$. Set $t = 0$
- Having x^t and $p^t = [p1^t, p2^t]^T$ compute x^{t+1} and $p^{t+1} = [p1^{t+1}, p2^{t+1}]^T$ as follows:

  For each node $i \in \mathcal{N}$ solve

$$\begin{array}{ll} \underset{r}{\text{minimize}} & \sum\limits_{j\in\delta_+(i)} \left[f1_{ij}(r1_{ij}) + p1^t_{ij}\left(r1_{ij} - x^t_{ij}\right) + \frac{\lambda}{2}\left(r1_{ij} - x^t_{ij}\right)^2 \right] + \\ & \sum\limits_{j\in\delta_-(i)} \left[f2_{ji}(r2_{ji}) + p2^t_{ji}\left(r2_{ji} - x^t_{ji}\right) + \frac{\lambda}{2}\left(r2_{ji} - x^t_{ji}\right)^2 \right] \\ \text{subject to} & \sum\limits_{j\in\delta_+(i)} r1_{ij} - \sum\limits_{j\in\delta_-(i)} r2_{ji} = b_i. \end{array} \tag{3.13}$$

  Determine x^{t+1} by solving:

$$\begin{array}{ll} \underset{x}{\text{minimize}} & \sum\limits_{l=1}^{2} \left[p_l^{tT}\left(r_l^{t+1} - x\right) + \frac{\lambda}{2}\left\| r_l^{t+1} - x \right\|_2^2 \right] \\ \text{subject to} & l \le x \le u. \end{array} \tag{3.14}$$

Finally,

$$p_1^{t+1} = p_1^t + \lambda \left(r_1^{t+1} - x^{t+1} \right) \tag{3.15}$$
$$p_2^{t+1} = p_2^t + \lambda \left(r_2^{t+1} - x^{t+1} \right) \tag{3.16}$$
$$\tag{3.17}$$

for $l = 1, \ldots, 4$.

Again, the solution of problem (3.14) is trivial to obtain being (for each arc) the average value of the flows obtained by solving problem (3.13), modified by the associate multipliers $p_{1}{}^t_{ij}$ and $p_{2}{}^t_{ij}$ and projected in the bounds.

As for the previous algorithm, the most expensive part consists of finding the solution for problem (3.13) that is essentially the same as (3.6a) and therefore the same technique can be adopted.

4. IMPLEMENTATION ISSUES

In this section we describe the implementation of the algorithms presented in the previous section on the CRAY T3D parallel machine.

The CRAY T3D is a distributed memory machine with up to 2048 processors capable of up to 300 GFlops peak performance. Each node in the CRAY T3D contains 2 processing elements (DEC α 21064 chip with peak performance of 150 MFlops) and 8 Mw of DRAM. Each node is directly connected to six neighborhood nodes and a maximum of 300 MByte/sec transfer rate in the six directions can be achieved.

The machine supports the message passing style of programming (via PVM) as well as the data parallel, global address and work sharing paradigm. We used this second style and we wrote our program using the MPP Fortran data parallel programming model, that includes many of the features of the High Performance Fortran such as array operations and WHERE statements. Since the user has explicit control over the execution of single PEs, the SPMD (Single–Program–Multiple–Data) is fully supported. In this programming model we distinguish between data objects (scalar or arrays) that are private to a task and those that are shared among all tasks. Shared data are accessible to all task and (if the object is an array) is distributed across multiple PEs.

For both the algorithms proposed here the most computationally expensive part is solving step (3.6a) or step (3.13) and, in particular, determining the multiplier associated with the single constraint of these subproblems. Therefore we used an implementation scheme that tries to minimize the communication and computation in this step of the algorithm. As in Zenios and Lasken[11] and Eckstein[5] that proposed and implemented on the Connection Machine CM2 a similar algorithm, we associate to each arc two processing elements, one corresponding to the source node and one to the target node for the arc. At the expense of duplicating information, we obtain that data for arcs incident to a specific node are now grouped together. The main advantage is that a segmented scan operation can be used to compute (in parallel and for all nodes) the violation of the flow conservation constraint and, therefore, the dual variable needed in step (3.6a) or step (3.13). The segmented scan is a parallel prefix reduction function generally available on parallel machine. We refer the interested reader to [5] for a more complete discussion of this operation. Unfortunately in the current release of the MPP Fortran for CRAY T3D, this operation is not available and therefore we had to simulate it using much more expensive statements. Future release of the compiler will support this reduction operation.

Finally results for larger network problems with strictly convex quadratic costs with quadratic term equal to 0.5 and 1.0 are reported in Table 5.

5. COMPUTATIONAL RESULTS

In view of the lack of the segmented scan operation in the current release of the FORTRAN compiler, we were unable to solve large network flow problems on the CRAY T3D. We concentrate on the performance of the algorithms for different values of the penalty parameter λ for linear and strictly convex quadratic problems.

All algorithms were implemented on the 64–PEs CRAY T3D available at CINECA, Bologna, Italy. The test problems have been generated by a modified version of NETGEN [9].

Table 1 reports the solution error after 10,000 iterations for the MASSNET 1 and MASSNET 2 algorithms and for the alternating step method of Eckstein [5] for a network flow problem with linear costs. The parameter λ is fixed to the value shown on the first column of the table. We observe that for larger values of λ the proposed algorithm outperforms the method proposed by Eckstein while the situation is reversed when small values of the parameter λ are used.

Much better convergence result can be achieved using an adaptive scheme for the parameter λ as shown in Table 2. We increase the value of the parameter λ up to 10.0 starting from the value in the first column of Table 1. For the case when a solution error of 10^{-7} is achieved, we report in parenthesis the number of iterations. The performance of the two proposed algorithms appears to be very similar. The best result has been obtained with MASSNET 1 where a high accuracy of 10^{-7} is achieved in 8332 iterations.

Table 3 reports the number od iterations needed to achieve a solution error of 10^{-7} for the same problem as in Table 1 but with strictly convex costs with quadratic term all equal to 0.5 and 1.0.

Finally results for larger network problems with strictly convex quadratic costs with quadratic term equal to 0.5 and 1.0 are reported in Table 4 and Table 5. For the quadratic cases we do not have direct comparisons with the algorithm proposed by Eckstein. As expected, fewer iterations are needed to achieve high accuracy for the strictly convex case. Also, the MASSNET 2 algorithm outperforms MASSNET 1 in some cases by a factor of 2. The final table (Table 5) shows that as few as 200 iterations are needed to achieve convergence for problems with thousands of nodes and arcs. This limited number of iterations (compared to the linear case) opens the possibility of using proximal point approaches for efficiently solve the network flow problem with linear objective function.

6. CONCLUSIONS

Decomposition methods provide a basic scheme for constructing parallel algorithms. Coarse–grain and massively parallel fine–grain algorithms can be derived from the general Alternating Direction Method. Although much more computational experience is needed, splitting methods for special classes of problems hold the promise of solving extremely large problem instances. While convergence of the proposed algorithms is

λ	MASSNET 1	MASSNET 2	Eckstein
0.05	0.263E+00	0.255E+00	0.255E+00
0.1	0.221E+00	0.200E+00	0.182E+00
0.5	0.268E-02	0.255E+00	0.816E-03
1.0	0.799E-03	0.396E-03	0.229E-04
5.0	0.928E-07*	0.338E-03	0.283E-03

Table 1: Solution error after 10,000 iterations for varying starting values of λ. Linear case, number of nodes = 128, number of arcs = 512. (* Solution obtained after 8971 iterations.)

λ	MASSNET 1	MASSNET 2
0.05	0.385E-05	0.290E-03
0.1	0.160E-04	0.224E-03
0.5	0.406E-05	0.172E-03
1.0	0.192E-04	0.562E-07(8898)
5.0	0.990E-07(8332)	0.935E-07(8993)

Table 2: Solution error after 10,000 iterations for varying starting values of λ with adaptive updating scheme of λ. Linear case, number of nodes =128, number of arcs = 512

λ	MASSNET 1		MASSNET 2	
	$d_{ij} = 0.5$	$d_{ij} = 1.0$	$d_{ij} = 0.5$	$d_{ij} = 1.0$
0.05	2255	1605	1314	1191
0.1	2185	1453	1163	1036
0.5	1914	1143	897	675
1.0	1851	1038	799	551
5.0	2214	1108	1061	513

Table 3: Number of iterations needed to achieve a solution error of 10^{-7} for varying starting values of λ. Quadratic case, number of nodes = 128, number of arcs = 512

λ	MASSNET 1		MASSNET 2	
	$d_{ij} = 0.5$	$d_{ij} = 1.0$	$d_{ij} = 0.5$	$d_{ij} = 1.0$
0.05	3049	1995	2020	1497
0.1	2938	1928	1870	1348
0.5	2703	1618	1574	1037
1.0	2530	1531	1459	935
5.0	2604	1385	1324	739

Table 4: Number of iterations needed to achieve a solution error of 10^{-7} for varying starting values of λ. Quadratic case, number of nodes = 512, number of arcs = 1024

λ	MASSNET 1		MASSNET 2	
	$d_{ij} = 0.5$	$d_{ij} = 1.0$	$d_{ij} = 0.5$	$d_{ij} = 1.0$
0.05	1164	1168	1130	1076
0.1	1005	1018	976	923
0.5	666	671	627	577
1.0	528	528	485	438
5.0	255	229	210	188

Table 5: Number of iterations needed to achieve a solution error of 10^{-1} for varying starting values of λ. Quadratic case, number of nodes = 1024, number of arcs = 4096

guarantee for convex (but not necessarily strictly convex) separable objective functions, computational results show fewer iterations are required for the strictly convex case. Therefore, approaches based on proximal point ideas could be used for the convex case.

ACKNOWLEDGMENTS

CRAY T3D computer time was kindly offered by the CINECA Consortium, Bologna, Italy.

REFERENCES

[1] D. Bertsekas and J. Tsitsiklis, "*Parallel and Distributed Computation.*" Prentice Hall, Englewood Cliffs, New Jersey, 1989.

[2] R. De Leone, R. Meyer and S. Kontogiogis, "Alternating direction splittinngs for block–angular parallel optimization". Technical Report 1217, University of Wisconsin, Computer Sciences Dept., February 1994. Submitted to Journal of Optimization Theory and Applications.

[3] R. De Leone, R. Meyer, S. Kontogiogis, A. Zakarian and G. Zakeri, "Coordination in coarse–grain decomposition". SIAM Journal on Optimization, 4(4):777–793, November 1994.

[4] J. Eckstein, "Splitting methods for monotone operators with applications to parallel optimization". Massachussetts Institute of Technology, Department of Civil Engineering, 1989.

[5] J. Eckstein , "The alternating step method for monotropic programming on the Connection Machine CM–2". ORSA Journal on Computing, 5(1):293–318, 1993.

[6] J. Eckstein and B.P. Bertsekas, "On the Douglas–Rachford splitting method and the proximal point method for maximal monotone operators". Mathematical Programming, Series A, 55(3):293–318, 1992.

[7] R. Glowinski and P. Le Tallec, "Augmented Lagrangian and Operator-Splitting Methods in Nonlinear Mechanics". Society for Industrial and Applied Mathematics, 1989.

[8] S. Kontogiorgis, "Alternating Direction Methods for the Parallel Solution of Large-scale Block-structured Optimization Problems". University of Wisconsin – Madison, Department of Computer Sciences, 1994.

[9] D. Klingman, A. Napier and J. Stutz, "NETGEN-A program for Generating Large-Scale (Un)capacitated Assignment, Transportation, and Minimum Cost Network Problems". Management Science, 20:814–822, 1974.

[10] P.L. Lions and B. Mercier, "Splitting Algorithms for the Sum of Two Nonlinear Operators". SIAM Journal on Numerical Analusis, 16:964–979, 1979.

[11] S.A. Zenios and R. A. Lasken, "Nonlinear Network Optimization on a Massively Parallel Connection Machine". Annals of Operation Research, 14:147–165, 1988.

CORRECTION THEOREMS FOR NONSMOOTH SYSTEMS

Dem'yanov Vladimir F.

Applied Mathematics Department,
St. Petersburg State University,
Staryi Peterhof, 198904 Russia
e-mail: demyanov@robot.apmath.lgu.spb.su

Abstract. In 1974 H. Halkin formulated and solved the correction function problem in the case the functions involved are differentiable but not necessarily continuously differentiable. In this paper some generalizations of the Halkin correction theorem are stated for quasidifferentiable functions. A generalization of the Kakutani fixed-point theorem is used. Uniqueness and continuity conditions are formulated.

Key words. Correction theorem, fixed point, mean-value theorem, quasidifferentiable functions, Nonsmooth Analysis.

1. INTRODUCTION

In 1974 H. Halkin formulated and solved the so-called correction function problem [8] in a slightly nonsmooth setting (the functions were assumed to be differentiable but not necessarily continuously differentiable). In a few words his approach can be described as follows. Let a differentiable function $f : \mathbb{R}^n \to \mathbb{R}$ and points x and $x + \Delta$ be given. The mean-value theorem says that there exists a point $\bar{x} = x + \theta\Delta$ where $\theta = \theta(x, \Delta) \in (0, 1)$ such that

$$f(x + \Delta) - f(x) = (f'(x + \theta\Delta), \Delta). \tag{1.1}$$

H. Halkin proposed to find a "correction value" $\eta(\Delta)$ such that

$$f(x + \Delta + \eta(\Delta)) - f(x) = (f'(x), \Delta). \tag{1.2}$$

(Here, as usual, f' is the gradient of f and (a, b) stands for the scalar product of vectors a and b).

A very important property of the correction function $\eta(\Delta)$ (if it exists) is the fact that in many cases

$$\frac{\|\eta(\Delta)\|}{\|\Delta\|} \xrightarrow[\|\Delta\| \to 0]{} 0$$

while in (1.1) we only have $\theta(x, \Delta) \in (0,1)$.

A "justification" of the necessity to study the correction function problems (if one needs it) and some applications are discussed in [8].

In the present paper we formulate several correction function problems and solve them in the case of quasidifferentiable functions and systems of quasidifferentiable functions. The approach suggested allows to formulate and to solve the correction problems of the k–th order. To prove the results we need to employ the generalized Kakutani Theorem and the Xia mean–value theorem for quasidifferentiable functions. In Section 2 some necessary preliminary results are stated, in Section 3 the smooth case is discussed and in Section 4 the quasidifferentiable case is studied.

2. PRELIMINARIES

2.1. A Generalized Kakutani Theorem.

Let $K \subset \mathbb{R}^n$ be a given set, $F : K \to 2^{\mathbb{R}^N}$ be a multivalued mapping defined on K. Here 2^A is the set of all nonempty subsets of a set A. A point $x^* \in K$ is called a *fixed–point* of the mapping F if $x^* \in F(x^*)$.

Theorem 2.1 (Kakutani). Let K be a convex compact set of $\mathbb{R}^n$, a mapping $F : K \to 2^K$ be upper semicontinuous (u.s.c.) on K. If for each $x \in K$ the set $F(x) \subset K$ is convex and compact then there exists a point $x^* \in K$ such that $x^* \in F(x^*)$ (i.e. x^* is a fixed point of the mapping F on K).

This theorem was proved by S. Kakutani [10]. Another proof (using H. Nikaido's idea [13]) can be found in [11]. We need a slight generalization of this theorem (see [6]).

Let $K \subset \mathbb{R}^n$ be a convex compact set, mappings $G : K \to 2^{\mathbb{R}^{n\times n}}$ and $r : K \to 2^{\mathbb{R}^n}$ be u.s.c. Here $\mathbb{R}^{n\times n}$ is the family of all real $(n \times n)$–matrices.
Consider the mapping $\mathcal{Z} : K \to 2^{\mathbb{R}^{n\times n}} \times 2^{\mathbb{R}^n}$ where $\mathcal{Z} = [G, r]$. Assume that for every $x \in K$ the set $\mathcal{Z}(x) = [G(x), r(x)]$ is convex and compact in the space $\mathbb{R}^{n\times n} \times \mathbb{R}^n$ and that the sets $\mathcal{Z}(x)$ and $r(x)$ are totally bounded on K.

Due to the assumed upper–semicontinuity of the mappings G and r, the mapping $\mathcal{Z}$ is also n.s.c.

Let the mapping $\phi : K \to 2^{\mathbb{R}^n}$ be defined by

$$\phi(x) = G^{-1}(x)r(x) = \{A^{-1}h | A \in G(x),\ h \in r(x)\} \tag{2.1}$$

where

$$G^{-1}(x) = \{B = A^{-1} | A \in G(x)\}. \tag{2.2}$$

The mapping ϕ is defined only for those $x \in K$ where

$$\det A \neq 0 \quad \forall A \in G(x).$$

Assume that

$$|\det A| \geq \beta > 0 \quad \forall A \in \bar{G} \tag{2.3}$$

where

$$\bar{G} = \text{co}\,\{G(x)|x \in K\}.$$

Put

$$R = \text{co}\,\{r(x)|x \in K\},$$

$$\bar{\phi} = \bar{G}^{-1}R = \{A^{-1}h|A \in \bar{G},\ h \in R\}. \tag{2.4}$$

The problem is to find conditions for the existence of a fixed point of the mapping ϕ. The mapping ϕ is not convex–valued therefore it is not possible to apply Theorem 2.1 directly.

Theorem 2.2. If (2.3) holds and

$$\bar{\phi} \subset K \tag{2.5}$$

then the mapping ϕ defined by (1.1) and (1.2) has a fixed point on K i.e. there exists a point $x^* \in K$ such that

$$x^* \in \phi(x^8) \tag{2.6}$$

or, what is just the same, there exist $A_0 \in G(x^*)$ and $h_0 \in r(x^*)$ such that

$$x^* = A_0^{-1}h_0. \tag{2.7}$$

2.2. A Mean–value theorem for quasidifferentiable functions

Let a function f be defined and real–valued on an open set $S \subset \mathbb{R}^n$. We say that the function f is quasidifferentiable at a point $x \in S$ if it is directionally differentiable at x and there exist convex compact sets $\underline{\partial} f(x)$ and $\overline{\partial} f(x) \subset \mathbb{R}^n$ such that

$$f'(x, g) = \max_{v \in \underline{\partial} f(x)} (v, g) + \min_{w \in \overline{\partial} f(x)} (w, g) \quad \forall g \in \mathbb{R}^n \tag{2.8}$$

where $f'(x, g)$ stands for the directional derivative of f at x. Properties of quasidifferentiable functions can be found in (4.6).

Theorem 2.3 (Mean–value Theorem). Let f be quasidifferentiable on an open convex set S, $x \in S$, $x + \Delta \in S$. Then there exist $\tau \in (0, 1)$, $v \in \underline{\partial} f(x + \tau\Delta)$, $w \in \overline{\partial} f(x + \tau\Delta)$ such that

$$f(x + \Delta) - f(x) = (v + w, \Delta). \tag{2.9}$$

This theorem was proved by Z.–Q. Xia [17].

2.3. Addition of convex sets.

Let $A \subset \mathbb{R}^n$, $B \subset \mathbb{R}^n$ be convex compact sets. We shall use the following sums of these sets: $A + B$, $A \dot{+} B$, $A \ddot{+} B$.

$A + B$ is the well-known Minkowski sum:

$$A + B = \{v = a + b | a \in A, b \in B\}.$$

To define other sums let us construct the following functions. Take $g \in \mathbb{R}^n$ and put

$$\begin{aligned} A(g) &= \arg\max\{(v,g)|v \in A\} = \{v \in A|(v,g) = \max_{\bar{v}\in A} |(\bar{v},g)\}, \\ B(g) &= \arg\min\{(w,g)|w \in B\} = \{w \in B|(w,g) = \min_{\bar{w}\in B} |(\bar{w},g)\}, \\ C(g) &= A(g) + B(g). \end{aligned}$$

It is clear that

$$A(\lambda g) = A(g),\ B(\lambda g) = B(g),\ C(\lambda g) = C(g) \quad \forall \lambda > 0,$$

$$A(0_n) = A,\ B(0_n) = B,\ C(0_n) = A + B.$$

Here $0_n = (0, \ldots, 0) \in \mathbb{R}^n$.

It is known from Convex Analysis [14] that for almost every $g \in \mathbb{R}^n$ the set $A(g)$ is a singleton and for almost every $g \in \mathbb{R}^n$ the set $B(g)$ is a singleton, hence, for almost every $g \in \mathbb{R}^n$ the set $C(g)$ is a singleton as well.

Now let us put

$$\begin{aligned} A \dot{+} B &= \operatorname{cl} \operatorname{co} \{C(g) | \|g\| = 1,\ C(g) \text{ is a singleton}\}, \\ A \ddot{+} B &= \operatorname{cl} \operatorname{co} \{C(g) | \|g\| = 1\}. \end{aligned}$$

The sum $A \dot{+} B$ was introduced in [3], the sum $A \ddot{+} B$ was introduced by A.M. Rubinov. These both sums were studied in [15,16]. The sum $A \dot{+} B$ was effectively used in [12].

3. THE PROBLEM FORMULATION. THE SMOOTH CASE

3.1. Correction of a function.

Let a function $f : \mathbb{R}^n \to \mathbb{R}$ be continuously differentiable on an open set $S \subset \mathbb{R}^n$, $x \in \mathbb{R}^n$ be fixed. Then the following expansion holds:

$$f(x + \Delta) = f(x) + (f'(x), \Delta) + o(\Delta) \tag{3.1}$$

where

$$\frac{o(\Delta)}{\|\Delta\|} \xrightarrow[\|\Delta\| \to 0]{} 0. \tag{3.2}$$

Here $f'(x)$ is the gradient of f at x (a column vector), (a,b) is the scalar product of a and b.

P1 (The correction problem of the first type):

Find a vector function $\eta_1(\Delta)$ such that

$$f(x + \Delta + \eta_1(\Delta)) = f(x) + (f'(x), \Delta), \tag{3.3}$$

$$\frac{\eta_1(\Delta)}{\|\Delta\|} \xrightarrow[\|\Delta\|\to 0]{} 0_n. \tag{3.4}$$

P2 (The correction problem of the second type):

Find a vector function $\eta_2(\Delta)$ such that

$$f(x + \Delta) = f(x) + (f'(x), \Delta + \eta_2(\Delta)), \tag{3.5}$$

$$\frac{\eta_2(\Delta)}{\|\Delta\|} \xrightarrow[\|\Delta\|\to 0]{} 0_n. \tag{3.6}$$

P3 (The correction problem of the third type):

Find a vector function $\eta_3(\Delta)$ such that

$$f(x + \Delta) = f(x + \eta_3(\Delta)) + (f'(x + \eta_3(\Delta)), \Delta), \tag{3.7}$$

$$\frac{\eta_3(\Delta)}{\|\Delta\|} \xrightarrow[\|\Delta\|\to 0]{} 0_n. \tag{3.8}$$

If there exists a function $\eta_i(\Delta)$ $(i \in 1:3)$ then we say that the correction problem of the i–th type has a solution and then the function $\eta_i(\Delta)$ is called a correction function (or just a correction) of the i–th type of the function f in a neighbourhood of the point x.

Our aim is to find sufficient conditions under which a correction function exists and is unique. Note that the problem P1 was formulated and solved for systems of smooth functions by H. Halkin in [8].

The following results can easily be proved.

Lemma 3.1. If $f'(x) \neq 0_n$ then for any $\epsilon > 0$ there exist $\delta_1 > 0$, $\delta_2 > 0$ and vector functions $\eta_1(\Delta), \eta_2(\Delta)$ defined on S_{δ_1} and S_{δ_2} respectively such that

$$f(x + \Delta + \eta_1(\Delta)) = f(x) + (f'(x), \Delta) \quad \forall \Delta \in S_{\delta_1},$$

$$f(x + \Delta) = f(x) + (f'(x), \Delta + \eta_2(\Delta)) \quad \forall \Delta \in S_{\delta_2},$$

$$\|\eta_i(\Delta)\| \leq \epsilon \|\Delta\| \quad \forall \Delta \in S_{\delta_i}, \quad \forall i \in 1:2.$$

If, in addition, f is twice continuously differentiable then for any $\epsilon > 0$ there exist δ_3 and a vector function $\eta_3(\Delta)$ defined on S_{δ_3} such that

$$f(x + \Delta) = f(x + \eta_3(\Delta)) + (f'(x + \eta_3(\Delta)), \Delta) \quad \forall \Delta \in S_{\delta_3},$$

$$\|\Delta_3(\Delta)\| \leq \epsilon \|\Delta\| \quad \forall \Delta \in S_{\delta_3}.$$

Here, as usual, $S_\delta = \{\Delta \in \mathbb{R}^n \mid \|\Delta\| \leq \delta\}$.

Remark 3.1. Consider the one–dimensional case. Assume that $x \in \mathbb{R}$, f is twice continuously differentiable. Put $\Delta = \alpha g$ where $g \in \mathbb{R}$ is fixed, $\alpha \in \mathbb{R}$. If $f'(x) \neq 0$ then it is easy to show that there exist correction functions of all three types of the form

$$\eta_i(\Delta) = \eta_i(\alpha g) = \alpha^2 q_i(\alpha)$$

where

$$q_1(\alpha) \xrightarrow[\alpha\downarrow 0]{} -\frac{1}{2}\frac{f''(x)g^2}{f'(x)} = -A, \; q_2(\alpha) \xrightarrow[\alpha\downarrow 0]{} A, \; q_3(\alpha) \xrightarrow[\alpha\downarrow 0]{} A.$$

Something similar can be obtained in the n–dimensional case.

3.2. Correction of a system.

Let $f : \mathbb{R}^n \to \mathbb{R}^n$ (where $m \leq n$, $f = (f_1, \ldots, f_m)$) be a smooth vector function on an open set $S \subset \mathbb{R}^n$.

PS1 (The correction problem of the first type for a system): Find a vector–function $\eta_1(\Delta)$ such that

$$f_i(x + \Delta + \eta_1(\Delta)) = f_i(x) + (f_i'(x), \Delta) \quad \forall i \in 1 : m,$$

$$\frac{\eta_1(\Delta)}{\|\Delta\|} \xrightarrow[\|\Delta\|\to 0]{} 0_n.$$

PS2 (The correction problem of the second type for a system): Find a vector–function $\eta_2(\Delta)$ such that

$$f_i(x + \Delta) = f_i(x) + (f_i'(x), \Delta + \eta_2(\Delta)) \quad \forall i \in 1 : m,$$

$$\frac{\eta_2(\Delta)}{\|\Delta\|} \xrightarrow[\|\Delta\|\to 0]{} 0_n$$

PS3 (The correction problem of the third type for a system): Find a vector–function $\eta_3(\Delta)$ such that

$$f_i(x + \Delta) = f_i(x + \eta_3(\Delta)) + (f_i'(x + \eta_3(\Delta)), \Delta) \quad \forall i \in 1 : m,$$

$$\frac{\eta_3(\Delta)}{\|\Delta\|} \xrightarrow[\|\Delta\|\to 0]{} 0_n.$$

Put

$$f'(x) = \begin{pmatrix} (f_1'(x))^T \\ \vdots \\ (f_m'(x))^T \end{pmatrix}.$$

This is an $(m \times n)$–matrix.

The following lemma can easily be checked.

Lemma 3.2 If the rank of the matrix $f'(x)$ is m then for any $\epsilon > 0$ there exist $\delta_1 > 0$, $\delta_2 > 0$ and vector functions $\eta_1(\Delta)$, $\eta_2(\Delta)$ defined on S_{δ_1} and S_{δ_2} respectively such that

$$f(x + \Delta + \eta_1(\Delta)) = f(x) + f'(x)\Delta \quad \forall \Delta \in S_{\delta_1},$$
$$f(x + \Delta) = f(x) + f'(x)(\Delta + \eta_2(\Delta)) \quad \forall \Delta \in S_{\delta_2},$$
$$\|\eta_i(\Delta)\| \leq \epsilon\|\Delta\| \quad \forall \Delta \in S_{\delta_i}, \quad \forall i \in 1:2.$$

If, in addition, all the functions f_i ($i \in 1:m$) are twice continuously differentiable on S then for any $\epsilon > 0$ there exist $\delta_3 > 0$ and a vector function $\eta_3(\Delta)$ defined on S_{δ_3} such that

$$f(x + \Delta) = f(x + \eta_3(\Delta)) + f'(x + \eta_3(\Delta))\Delta \quad \forall \Delta \in S_{\delta_3},$$

$$\|\eta_3(\Delta)\| \leq \epsilon\|\Delta\| \quad \forall \Delta \in S_{\delta_3}.$$

Remark 3.1. Like in [8] it is possible to formulate the problem of finding a correction function of the order k.

4. CORRECTION OF A SYSTEM OF QUASIDIFFERENTIABLE FUNCTIONS

1. Let functions $f_i : \mathbb{R}^n \to \mathbb{R}$ ($i \in 1 : n$) be quasidifferentiable on an open set $S \subset \mathbb{R}^n$, $x \in S$. Then

$$f_i(x + \Delta) = f_i(x) + \varphi_i(x, \Delta) + o_i(\Delta) \quad \forall i \in 1 : n \tag{4.1}$$

where

$$\varphi_i(x, \Delta) = \max_{v \in \underline{\partial} f_i(x)} (v, \Delta) + \min_{w \in \bar{\partial} f_i(x)} (w, \Delta), \tag{4.2}$$

$$\frac{o_i(\alpha\Delta)}{\alpha} \xrightarrow[\alpha \downarrow 0]{} 0 \quad \forall \Delta \in \mathbb{R}^n, \tag{4.3}$$

$\mathcal{D}f_i(x) = [\underline{\partial} f_i(x), \bar{\partial} f_i(x)] \subset \mathbb{R}^n \times \mathbb{R}^n$ is a quasidifferential of the function f_i at the point x.

Assume that instead of (4.3) a more restrictive condition is satisfied:

$$\frac{o_i(\Delta)}{\|\Delta\|} \xrightarrow[\|\Delta\| \to 0]{} 0 \quad \forall i \in 1 : n. \tag{4.4}$$

PQDS1 (The correction problem of the first type for a system of quasidifferentiable functions): Find a vector function $\eta_1(\Delta)$ such that

$$f_i(x + \Delta + \eta_1(\Delta)) = f_i(x) + \varphi_i(x, \Delta) \quad \forall i \in 1 : n, \tag{4.5}$$

$$\frac{\eta_1(\Delta)}{\|\Delta\|} \xrightarrow[\|\Delta\| \to 0]{} 0_n. \tag{4.6}$$

PQDS2 (The correction problem of the second type for a system of quasidifferentiable functions): Find a vector function $\eta_2(\Delta)$ such that

$$f_i(x+\Delta) = f_i(x) + \varphi_i(x, \Delta + \eta_2(\Delta)) \quad \forall i \in 1:n, \tag{4.7}$$

$$\frac{\eta_2(\Delta)}{\|\Delta\|} \xrightarrow[\|\Delta\|\to 0]{} 0_n. \tag{4.8}$$

Remark 4.1. Since the quasidifferential mapping is, in general, discontinuous, it makes no sense to consider the correction problem of the third type.

Remark 4.2. It is clear from definition (4.2) that the functions φ_i are continuous in Δ (but not in x due to the discontinuity of the quasidifferential mapping).

Put

$$\mathcal{Z}(x) = \left\{ A = \begin{pmatrix} a_1^T \\ \vdots \\ a_n^T \end{pmatrix} \middle| a_i \in \underline{\partial} f_i(x) + \bar{\partial} f_i(x) \quad \forall i \in 1:n \right\}. \tag{4.9}$$

Theorem 4.1. If the mappings $\mathcal{Z}_i = \underline{\partial} f_i + \bar{\partial} f_i$ are bounded and upper semicontinuous (u.s.c.) in a neighbourhood of a point x and, besides,

$$|\det A| \geq \beta > 0 \quad \forall A \in \mathcal{Z}(x) \tag{4.10}$$

then for any $\epsilon > 0$ there exist $\delta_1 > 0$ and a vector function $\eta_1(\Delta)$ defined on S_{δ_1} such that

$$f_i(x + \Delta + \eta_1(\Delta)) = f_i(x) + \varphi_i(x, \Delta) \quad \forall \Delta \in S_{\delta_1} \quad \forall i \in 1:n, \tag{4.11}$$

$$\|\eta_1(\Delta)\| \leq \epsilon \|\Delta\| \quad \forall \Delta \in S_{\delta_1}. \tag{4.12}$$

For Δ sufficiently small $\eta_1(\Delta)$ is unique and continuous.

Proof. By the mean–value theorem for quasidifferentiable functions (Theorem 2.3)

$$f_i(x + \Delta + \eta) = f_i(x + \Delta) + (a_i, \eta) \tag{4.13}$$

where

$$a_i = v_i + w_i, \ v_i \in \underline{\partial} f_i(x + \Delta + \theta_i \eta), \tag{4.14}$$

$$w_i \in \bar{\partial} f_i(x + \Delta + \theta_i \eta), \quad \theta_i = \theta_i(\Delta, \eta) \in (0, 1).$$

It follows from (4.1) and (4.13) that

$$f_i(x + \Delta + \eta) - f_i(x) - \varphi_i(x, \Delta) = (a_i, \eta) + o_i(\Delta) \quad \forall i \in 1:n. \tag{4.15}$$

Put $\eta = \|\Delta\| \zeta$ (4.15) yields

$$\frac{1}{\|\Delta\|}[f_i(x + \Delta + \eta) - f_i(x) - \varphi_i(x, \Delta)] = (a_i, \zeta) + r_i(\Delta) \quad \forall i \in 1:x \tag{4.16}$$

where

$$r_i(\Delta) = \frac{o_i(\Delta)}{\|\Delta\|} \underset{\|\Delta\|\to 0}{\longrightarrow} 0. \tag{4.17}$$

By $\mathcal{A}_i(\eta)$ let us denote the set of all a_i's satisfying (4.13) and (4.14). Due to the upper semicontinuity of the mappings $\mathcal{Z}_i$ the mappings $\mathcal{A}_i$ are also u.s.c. (they are convex–valued as well). Put

$$\mathcal{A}(\eta) = \left\{ A = \begin{pmatrix} a_1^T \\ \vdots \\ a_n^T \end{pmatrix} \middle| a_i \in \mathcal{A}_i(\eta) \quad \forall i \in 1:n \right\}. \tag{4.18}$$

For sufficiently small Δ and bounded ζ the mapping $\mathcal{A}$ is u.s.c. and convex–valued. For sufficiently small Δ and $\eta = \|\Delta\|\zeta$ we get

$$|\det A| \geq \frac{1}{2}\beta \quad \forall A \in \mathcal{A}(\eta).$$

By the generalized Kakutani theorem (Theorem 2.2) there exists a fixed point of the mapping $\phi(\zeta) = -\mathcal{A}^{-1}(\|\Delta\|\zeta)r(\Delta)$ where $\mathcal{A}^{-1} = \{A^{-1} | A \in \mathcal{A}\}, (r(\Delta))^T = (r_1(\Delta), \ldots, r_n(\Delta))$ i.e. there exists $\zeta(\Delta)$ such that

$$\zeta(\Delta) \in -\mathcal{A}^{-1}(\|\Delta\|\zeta))r(\Delta).$$

It means that there exists $A \in \mathcal{A}(\|\Delta\|\zeta(\Delta))$ such that

$$\zeta(\Delta) = -A^{-1}r(\Delta). \tag{4.19}$$

(4.19) implies

$$f_i(x + \Delta + \|\Delta\|\zeta(\Delta)) - f_i(x) - \varphi_i(x, \Delta) = 0 \quad \forall i \in 1:n.$$

Hence, $\eta_1(\Delta) = \|\Delta\|\zeta(\Delta)$ satisfies (4.11), and (4.17) and (4.19) yield (4.6), which is equivalent to (4.12).

Now let us prove that for Δ sufficiently small the function $\eta_1(\Delta)$ is unique. Assume the opposite. Then there exist sequences $\{\Delta_k\}$, $\{z_{1k}\}$, $\{z_{2k}\}$ such that

$$\Delta_k \to 0_n, \quad \frac{z_{1k}}{\|\Delta_k\|} \to 0_n, \quad \frac{z_{2k}}{\|\Delta_k\|} \to 0_n, \quad z_{1k} \neq z_{2k} \quad \forall k, \tag{4.20}$$

$$f_i(x + \Delta_k + z_{1k}) = f_i(x + \Delta_k + z_{2k}) = f_i(x) + \varphi_i(x, \Delta_k) \quad \forall i \in 1:n, \quad \forall k.$$

By the mean–value theorem (Theorem 2.3) there exist $\theta_{ik} \in (0,1)$, $v_{ik} \in \underline{\partial} f_i(x_{ik})$, $w_{ik} \in \bar{\partial} f_i(x_{ik})$ where $x_k = x + \Delta_k + z_{1k} + \theta_{ik}(z_{2k} - z_{1k})$ such that

$$(v_{ik} + w_{ik}, z_{2k} - z_{1k}) = 0. \tag{4.21}$$

Put

$$g_k = \frac{z_{2k} - z_{1k}}{\|z_{2k} - z_{1k}\|}$$

(due to (4.20) $\|z_{2k} - z_{1k}\| \neq 0$). Since $\|g_k\| = 1$ then without loss of generality one can assume that

$$g_k \xrightarrow[k\to\infty]{} g_0 \;, \quad \|g_0\| = 1. \tag{4.22}$$

Since the mappings $\mathcal{Z}_i$ are bounded and u.s.c. then without loss of generality let us assume that

$$a_{ik} = v_{ik} + w_{ik} \xrightarrow[k\to\infty]{} a_i = v_i + w_i$$

where

$$v_i \in \underline{\partial} f_i(x), \;\; w_i \in \bar{\partial} f_i(x).$$

The equality (4.21) implies

$$(a_i, g_0) = 0 \quad \forall i \in 1:n. \tag{4.23}$$

It follows from (4.23) that $(A, g_0) = 0_n$ where $A = \begin{pmatrix} a_1^T \\ \vdots \\ a_n^T \end{pmatrix}$. Due to (4.22) it is possible only if $\det A = 0$ which contradicts (4.10). This contradiction proves the uniqueness.

Let Δ be a point where the function η_1 is uniquely defined. Let us show that η_1 is continuous at this point. Assume the contrary. Then there exists a sequence $\{\Delta_k\}$ such that

$$\Delta_k \to \Delta, \;\; \eta_1(\Delta_n) \not\to \eta_1(\Delta).$$

Without loss of generality assume that

$$\eta_1(\Delta_k) \to \bar{\eta} \neq \eta_1(\Delta). \tag{4.24}$$

By the definition of η_1

$$f_i(x + \Delta_k + \eta_1(\Delta_k)) = f_i(x) + \varphi_i(x, \Delta_k) \quad \forall i \in 1:n, \; \forall k.$$

The continuity of the functions f_i and φ_i in Δ yields

$$f_i(x + \Delta + \bar{\eta}) = f_i(x) + \varphi_i(x, \Delta_k) \quad \forall i \in 1:n. \tag{4.25}$$

The relations (4.24) and (4.25) contradict the uniqueness of $\eta_1(\Delta)$. □

Theorem 4.2. If (4.10) holds then for any $\epsilon > 0$ there exist $\delta_2 > 0$ and a vector function $\eta_2(\Delta)$ defined on S_{δ_2} such that

$$f_i(x + \Delta) = f_i(x) + \varphi_i(x, \Delta + \eta_2(\Delta)) \quad \forall \Delta \in S_{\delta_2}, \quad \forall i \in 1:n, \tag{4.26}$$

$$\|\eta_2(\Delta)\| \le \epsilon \|\Delta\| \quad \forall \Delta \in S_{\delta_2}. \tag{4.27}$$

Proof. By the mean–value theorem (Theorem 2.3)

$$\varphi_i(x, \Delta + \eta) = \varphi_i(x, \Delta) + (a_i, \eta) \tag{4.28}$$

where

$$a_i = v_i + w_i,\ v_i \in \underline{\partial}\varphi_{i\Delta}(x, \Delta + \theta_{i\eta}),\ w_i \in \bar{\partial}\varphi_{i\Delta}(x, \Delta + \theta_i\eta), \tag{4.29}$$

$$\theta_i = \theta_i(\Delta, \eta) \in (0, 1).$$

Note that (see (4.2))

$$\underline{\partial}\varphi_{i\Delta}(x, z) = \{v \in \underline{\partial}f_i(x) | (v, z) = \psi_{1i}(z)\} \subset \underline{\partial}f_i(x); \tag{4.30}$$

$$\bar{\partial}\varphi_{i\Delta}(x, z) = \{w \in \bar{\partial}f_i(x) | (w, z) = \psi_{2i}(z)\} \subset \bar{\partial}f_i(x);$$

$$\psi_{1i}(z) = \max_{v \in \underline{\partial}f_i(x)} (v, z),\ \psi_{2i}(z) = \min_{w \in \bar{\partial}f_i(x)} (w, z).$$

It follows from (4.1) and (4.28) that

$$f_i(x + \Delta) - f_i(x) - \varphi_i(x, \Delta + \eta) = -(a_i, \eta) + o_i(\Delta) \quad \forall i \in 1 : n. \tag{4.31}$$

Putting $\eta = \|\Delta\|\zeta$ we get from (4.31)

$$\frac{1}{\|\Delta\|}[f_i(x + \Delta) - f_i(x) - \varphi_i(x, \Delta + \eta)] = -(a_i, \zeta) + r_i(\Delta) \quad \forall i \in 1 : n \tag{4.32}$$

where

$$r_i(\Delta) = \frac{o_i(\Delta)}{\|\Delta\|} \xrightarrow[\|\Delta\| \to 0]{} 0. \tag{4.33}$$

Let $\mathcal{A}_i(\eta)$ be the set of all a_i's satisfying (4.28)–(4.29). It is clear from (4.30) that the mappings $\mathcal{A}_i$ are u.s.c. and convex–valued.
Like in the proof of Theorem 4.1 let

$$\mathcal{A}(\eta) = \left\{ A = \begin{pmatrix} a_1^T \\ \vdots \\ a_n^T \end{pmatrix} \middle| a_i \in \mathcal{A}_i(\eta) \quad \forall i \in 1 : n \right\}. \tag{4.34}$$

The mapping $\mathcal{A}$ is u.s.c. and convex–valued. Due to (4.10)

$$|\det A| \geq \beta > 0 \quad \forall A \in \mathcal{A}(\eta).$$

Put

$$r(\Delta) = (r_1(\Delta), \ldots, r_n(\Delta)).$$

By the generalized Kakutani theorem it follows from (4.33) that for Δ sufficiently small there exists a fixed point of the mapping

$$\phi(\zeta) = \mathcal{A}^{-1}(\|\Delta\|\zeta)r(\Delta)$$

where

$$\mathcal{A}^{-1} = \{A^{-1} | A \in \mathcal{A}\}.$$

Hence, there exists $\zeta(\Delta)$ such that

$$\zeta(\Delta) \in \mathcal{A}^{-1}(\|\Delta\|\zeta(\Delta))r(\Delta).$$

It means that one can find $A(\Delta) \in \mathcal{A}(\|\Delta\|\varsigma(\Delta))$ such that

$$\varsigma(\Delta) = (A(\Delta))^{-1} r(\Delta). \tag{4.35}$$

(4.35) and (4.32) yield

$$f_i(x+\Delta) - f_i(x) - \varphi_i(x, \Delta + \|\Delta\|\varsigma(\Delta)) = 0 \quad \forall i \in 1:n. \tag{4.36}$$

Relations (4.36) hold for Δ sufficiently small. It follows from (4.33), (4.35) and the boundedness of $\mathcal{D}f_i(x)$ that for any $\epsilon > 0$ one can find $\delta_2 > 0$ such that

$$\|\varsigma(\Delta)\| \leq \epsilon \quad \forall \Delta \in S_{\delta_2}. \tag{4.37}$$

The relations (4.36) and (4.37) imply that the relations (4.26) and (4.27) hold for the vector $\eta_2(\Delta) = \|\Delta\|\varsigma(\Delta)$. □

Remark 4.3. It is also possible to prove that for Δ sufficiently small the function $\eta_2(\Delta)$ is unique and continuous.

2. Let functions f_i ($i \in 1:n$) be defined and continuous on an open set $S \subset \mathbb{R}^n$ and quasidifferentiable at a point $x \in S$.

Put

$$\mathcal{Z}(x) = \left\{ A = \begin{pmatrix} a_1^T \\ \vdots \\ a_n^T \end{pmatrix} \middle| a_i \in \underline{\partial} f_i(x) \dot{+} \bar{\partial} f_i(x) \quad \forall i \in 1:n \right\}. \tag{4.38}$$

We formulate without proof the following propositions.

Theorem 4.3. If

$$|\det A| \geq \beta > 0 \quad \forall A \in \mathcal{Z}(x) \tag{4.39}$$

then for any $\epsilon > 0$ there exist δ_1 and a vector function $\eta_1(\Delta)$ defined on S_{δ_1} such that (4.11) and (4.12) hold.

Theorem 4.4. If condition (4.39) holds then the conclusion of Theorem 4.2 is also valid.

Remark 4.4. Note that if condition (4.39) holds (with $\mathcal{Z}$ defined by (4.38)) then Theorems 4.3 and 4.4 guarantee only that there exist corrections of the first and second order while Theorems 4.1 and 4.2 provided also their uniqueness and continuity. But condition (4.39) is less restrictive than the assumptions imposed in Theorems 4.1 and 4.2.

Remark 4.5. In the case of continuously codifferentiable functions (see the definition in [6,7]) it is possible to state and to solve the correction problem of the third type.

Remark 4.6. The case where the number of functions f_i is $m < n$ can also be treated in a similar way.

Remark 4.7. One can expect that correction problems can be solved in the case of arbitrary Lipschitz functions by making use of the Clarke subdifferential [2] and different implicit function theorems and mean–value theorems [1,2,9].

ACKNOWLEDGEMENTS

The author is greatly indebted to Dr. Kuntz and Dr. S. Scholtes who drew the author's attention to the paper [8] by H. Halkin and to professors F. Giannessi, D. Pallaschke, P.D. Panagiotopoulos for their interesting and useful discussions on the subject.

REFERENCES

[1] J.–P. Aubin and H. Frankowska, "Set valued Analysis". Birkhäuser, Basel, 1990.

[2] F. Clarke, "Optimization and Nonsmooth Analysis". J. Wiley–Interscience, New York, N.Y., 1983.

[3] V.F. Demyanov, "On a relation between the Clarke subdifferential and quasidifferential". Vestnik Leningrad Univ. Math. 13, pp. 183–189, (English Transl.), 1981.

[4] V.F. Demyanov and L.V. Vasiliev, "Nondifferentiable Optimization". Springer–Optimization Software, New York, 1985.

[5] V.F. Demyanov and A.M. Rubinov, "Quasidifferential Calculus". Optimization Software Inc., New York, 1986.

[6] V.F. Demyanov and A.M. Rubinov, "Constructive Nonsmooth Analysis". Peter Lang, Frankfurt a/M, Berlin 1995.

[7] F. Giannessi (Ed.), "Nonsmooth Optimization. Methods and Applications". Gordon & Breach Science Publishers, Singapore, 1992.

[8] H. Halkin, "Implicit Functions and Optimization Problems without continuous differentiability of the data". SIAM Journal on Control 12 (2), pp. 229–236, 1974.

[9] J.–B. Hiriart–Urruty, "Mean–value theorems in Nonsmooth Analysis". Numerical Funct. Anal. and Optimization 2, pp. 1–30, 1980.

[10] S. Kakutani, "A generalization of Brower's fixed point theorem", Duke Math. J., Vol. 8, pp. 457–459, 1941.

[11] L.V. Kantorovich and G.P. Akilov, "Functional Analysis". Pergamon Press, Oxford, 1982.

[12] L. Kuntz, "Implicit Functions and Regular Points in Quasidifferentiable Optimization". In: W. Oettli and D. Pallaschke, eds., "Advances in Operations Research", Springer, Berlin, 1992.

[13] H. Nikaido, "Convex structures and Economic Theory". Academic Press, New York, 1968.

[14] R.T. Rockafellar, "Convex Analysis". Princeton University Press, Princeton, N.J. 1970.

[15] A.M. Rubinov, "Differences of convex compact sets and their applications in Nonsmooth Analysis". See [7], pp. 366–378.

[16] A.M. Rubinov and I.S. Akhundov, "Difference of compact sets in the sense of Demyanov and its application to Nonsmooth Analysis". Optimization, Vol. 23, pp. 179–188, 1992.

[17] Z.-Q. Xia, "On mean-value theorems in Quasidifferential Calculus", J. Math. Research and Exposition, Vol. 4, pp. 681–684, (Dalian, D.I.T.) 1987.

SUCCESSIVE PROJECTION METHODS FOR THE SOLUTION OF OVERDETERMINED NONLINEAR SYSTEMS

Diniz-Ehrhardt Maria A.[1] and Martínez José Mario[2]

Department of Applied Mathematics, IMECC-UNICAMP
University of Campinas, CP 6065, 13081-970
Campinas SP, Brazil
[1] e-mail: cheti@ime.unicamp.br
[2] e-mail: martinez@ccvax.unicamp.br

Abstract. We analyze a generalization of the classical Kaczmarz method for overdetermined nonlinear systems of equations with a convex constraint, where the feasible region is, in general, empty. We prove a local convergence theorem to fixed points of the algorithmic mapping. We defined a stopping rule for ill-conditioned problems, based on the behavior of the increment norm $\|x^{k+1} - x^k\|$. We show numerical experiments.

Key words. Nonlinear systems, projection methods, ill-posed problems, stopping criteria, local convergence.

1. INTRODUCTION

In this paper we consider overdetermined nonlinear systems of equations

$$F(x) = 0 \tag{1.1}$$

where $F : \mathbb{R}^n \to \mathbb{R}^m, m \geq n$, that arise from discretization of ill-posed problems. See [1], [2], etc. In these cases, the finite-dimensional systems are very ill-conditioned and the "exact" least-squares solution has no physical meaning in terms of the original problem. Regularization techniques are usually invoked in order to obtain meaningful approximations to the "true solution". See [3], [4], [5], [6], [1], [2].

In [7] we analyzed a straightforward nonlinear generalization of the classical Cimmino's method for linear systems (see [8], [9]), we defined a new class of approximate solutions for (1.1) based on the fixed points of the algorithmic mapping and we observed that these fixed points seemed to provide better solutions for ill-posed problems than least-squares solutions. In [10] we modified the nonlinear Cimmino method incorporating an additional projection on a nonempty convex set Ω that represents a priori information on the solution of the real problem. We observed that, on some ill-posed problems, this improvement is essential for obtaining reasonable results.

In [11], R. J. Santos reported that a very large class of iterative *linear* methods exhibit regularization properties, in the sense that the best approximations to the real solution of the problem appear for moderate values of the iteration index, that can be detected using a generalized cross-validation test. He also showed how his criterion can be applied to image reconstruction problems, where the Algebraic Reconstruction Technique (ART) (see [12]) is used as linear solver. ART is one of the generalizations of Kaczmarz's method

[13] for solving linear systems of equations. This method solves a linear system $Ax = b$ by means of cyclic projections on the hyperplanes defined by each equation of the system. Many generalizations of this method to the convex feasibility problem and to compatible nonlinear equations can be found in the literature (see [14], [15], [16], [17], [18], [19], [20].)

In this paper, we consider the generalization of Kaczmarz's method to feasibility problems defined by an overdetermined system of nonlinear equations and a convex set. So, the mathematical problem that motivates the algorithm is to find $x \in \mathbb{R}^n$ such that

$$F(x) = 0 \text{ and } x \in \Omega. \tag{1.2}$$

A solution of (1.2) is necessarily a fixed point of the algorithmic mapping defined in Section 2. However, in many cases, fixed points of the algorithmic mapping are not solutions of (1.2). Moreover, in most interesting cases, (1.2) has no solutions and, many times, exact solutions are not interesting in practice due to intrinsic ill-posedness of the problem.

We prove a new local convergence theorem for this method that, as in the case of the Cimmino method, implies the classical local convergence theorem of the Gauss-Newton method.

From a practical point of view, it is essential to define suitable stopping criteria for methods as the one presented here, especially if we deal with very ill-conditioned problems. In this paper, we introduce a stopping criterion based on the behavior of the increment $\|x^{k+1} - x^k\|$ as a function of k. This criterion is related with the so called "L-curve" criterion [21]. We exhibit numerical experiments that illustrate how the new method with the introduced stopping rule behaves in practice.

Throughout this paper, we use the following notations: $\|\cdot\|$ is the 2-norm of vectors or matrices; $A^\dagger$ is the Moore-Penrose pseudoinverse of a matrix; $\{e_1 \ldots, e_q\}$ is the canonical basis of $\mathbb{R}^q$; $[v]_i$ denotes the $i-th$ comp

onent of the vector v; $\mathcal{N}(A)$ is the null-space of the matrix A.

2. THE ALGORITHM

We assume that $F : \mathbb{R}^n \to \mathbb{R}^m$, $F \in C^2(\mathbb{R}^n)$ and $\Omega \subset \mathbb{R}^n$ is a nonempty, closed and convex set. We denote $P : \mathbb{R}^n \to \Omega$ the orthogonal projection operator on Ω. We divide the components of F into N blocks,

$$F(x) = (F_1(x), \ldots, F_N(x)),$$

where $F_i : \mathbb{R}^n \to \mathbb{R}^{n_i}, i = 1, \ldots, N$. For all $x \in \mathbb{R}^n$ we denote

$$J_i(x) = F_i'(x) \in \mathbb{R}^{n_i \times n}$$

and we consider the linear manifolds

$$V_i(x) = \{w \in \mathbb{R}^n \,|w = argmin_y \|F_i(x) + J_i(x)(y - x)\|\},$$

$i = 1, \ldots N$.

So, for all $x \in \mathbb{R}^n$, the orthogonal projection of x on $V_i(x), i = 1, \ldots, N$, is

$$\Phi_i(x) = x - J_i(x)^\dagger F_i(x). \tag{2.1}$$

We define

$$\Phi(x) = \Phi_N \circ \ldots \circ \Phi_1(x).$$

Properties of the Moore-Penrose pseudoinverse $J_i(x)^\dagger$ guarantee that $\Phi_i(x)$ is the closest point to x among all points y that minimize $\|F_i(x) + J_i(x)(y - x)\|$. Clearly, if $F_i(x) = 0$, we have that $\Phi_i(x) = x$. So, if $F(x) = 0$, x is a fixed point of Φ. Moreover, if x satisfies (1.2), it is a fixed point of $P \circ \Phi$.

Given $x^0 \in \mathbb{R}^n$, an arbitrary initial point, the successive orthogonal projection method is defined by the recurrence

$$x^{k+1} = P(\Phi(x^k)). \tag{2.2}$$

The following Lemma states the global convergence of the method (2.2) at a linear rate when Φ is an affine function, independently of the compatibility of the system. This result is, probably, well known by people working on projection methods, but we include its proof here for the sake of completeness.

Lemma 2.1. *Assume that F is an affine function. That is*

$$F_i(x) = A_i x + b \text{ for all } x \in \mathbb{R}^n,$$

and that rank $(A_1^T, \ldots, A_N^T) = n$. Then, the sequence generated by (2.2) is globally convergent at a linear rate.

Proof. Clearly,

$$\Phi'(x) = (I - A_N^\dagger A_N) \ldots (I - A_1^\dagger A_1)$$

for all $x \in \mathbb{R}^n$, $i = 1, \ldots N$. For all $w \in \mathbb{R}^n$, we have, since $I - A_i^\dagger A_i$ is the orthogonal projection on $\mathcal{N}(A_i)$, that

$$\|(I - A_i^\dagger A_i)w\| \leq \|w\|, \tag{2.3}$$

and the equality in (2.3) holds only if $w \in \mathcal{N}(A_i)$. Now, take an arbitrary $v \in \mathbb{R}^n$, $\|v\| = 1$. By (2.3), we have that

$$\|(I - A_N^\dagger A_N) \ldots (I - A_1^\dagger A_1)v\| \leq 1$$

and the equality holds only if $v \in \mathcal{N}(A_i)$ for all $i = 1, \ldots, N$. But, by the hypotheses, the intersection of all these null spaces is $\{0\}$, so, since $\|v\| = 1$,

$$\|(I - A_N^\dagger A_N) \ldots (I - A_1^\dagger A_1)v\| < 1.$$

Taking the maximum of $\|(I - A_N^\dagger A_N) \ldots (I - A_1^\dagger A_1)v\|$ for all v in the unitary sphere, we obtain, by compactness, that

$$\|\Phi'(x)\| \equiv \|(I - A_N^\dagger A_N) \ldots (I - A_1^\dagger A_1)\| = r < 1 \tag{2.4}$$

for all $x \in \mathbb{R}^n$. Therefore, Φ is a contraction, that is

$$\|\Phi(x) - \Phi(y)\| \leq r\|x - y\|$$

for all $x, y \in \mathbb{R}^n$. Since P is an orthogonal projection, we also have that

$$\|P(\Phi(x)) - P(\Phi(y))\| \leq \|\Phi(x) - \Phi(y)\| \leq r\|x - y\|$$

for all $x, y \in \mathbb{R}^n$. Therefore, by the Contraction Theorem, the sequence $\{x^k\}$ converges to the unique fixed point of $P \circ \Phi$. $\square$

From now on, we assume that a fixed point x^* of $P \circ \Phi$ exists. So

$$x^* = P(\Phi(x^*)).$$

We define

$$x_0^* = x^*, x_1^* = \Phi_1(x_0^*), x_2^* = \Phi_2(x_1^*), \ldots, x_N^* = \Phi_N(x_{N-1}^*),$$

Clearly, $x^* = P(x_N^*)$.

We want to state sufficient conditions under which the sequence $\{x^k\}$ converges to x^*. Let us define, for all $x \in \mathbb{R}^n$,

$$H_i(x) = [J_i(x)^\dagger e_1]'[F_i(x)]_1 + \ldots + [J_i(x)^\dagger e_{n_i}]'[F_i(x)]_{n_i} \in \mathbb{R}^{n\times n} \tag{2.5}$$

We assume that

$$\text{rank } J_{i+1}(x_i^*) = \text{ rank } J_{i+1}(x^*)$$

for all $i = 1, \ldots, N-1$ and that rank $(J_i(x))$ is constant for all x belonging to a neighborhood V_{i-1} of x_{i-1}^*, $i = 1, \ldots, N$. This implies that $J_i(x)^\dagger$ is continuously differentiable on V_{i-1}. Finally, we assume that

$$\text{rank } (F'(x^*)) = n$$

Theorem 2.2. *With the assumptions stated above, there exist $r \in (0,1)$, $\delta_1, \delta_2, \varepsilon > 0$ such that, if $\|x^0 - x^*\| \le \varepsilon$, $\|H_i(x_{i-1}^*)\| \le \delta_1$ for all $i = 1, \ldots, N$ and $\|J_{i+1}(x_i^*) - J_{i+1}(x^*)\| \le \delta_2$ for all $i = 1, \ldots, N-1$, the sequence $\{x^k\}$ generated by (2.2) converges to x^* and satisfies*

$$\|x^{k+1} - x^*\| \le r\|x^k - x^*\| \tag{2.6}$$

for all $k = 0, 1, 2, \ldots$

Proof. For all $i = 1, \ldots, N$, we have

$$\Phi_i'(x_{i-1}^*) = I - J_i(x_{i-1}^*)^\dagger J_i(x_{i-1}^*) - H_i(x_{i-1}^*).$$

Therefore, using the Chain Rule,

$$\Phi'(x^*) = (I - J_N(x_{N-1}^*)^\dagger J_N(x_{N-1}^*)) \ldots (I - J_1(x_0^*)^\dagger J_1(x_0^*)) + \sum_{i=1}^N D_i H_i(x_{i-1}^*) E_i,$$

where the D_i and E_i are suitable $n \times n$ matrices that depend on the derivatives of F_i at x_{i-1}^*. We define

$$c_1 = \sum_{i=1}^N \|D_i\|\|E_i\|.$$

So, by the constant rank hypothesis and the assumptions on H_i and $J_i(x_{i-1}^*) - J_i(x_0^*)$, we have that there exist $c_2 > 0$ such that

$$\|\Phi'(x^*)\| \le \|(I - J_N(x^*)^\dagger J_N(x^*)) \ldots (I - J_1(x^*)^\dagger J_1(x^*))\| + c_1\delta_1 + c_2\delta_2$$

whenever $\|H_i(x_{i-1}^*)\| \le \delta_1$ for all $i = 1, \ldots, N$ and $\|J_{i+1}(x_i^*) - J_{i+1}(x^*)\| \le \delta_2$ for all $i = 1, \ldots, N-1$. But, using the same arguments that lead to (2.4), we see that exists $r_1 \in (0,1)$ such that

$$\|(I - J_N(x^*)^\dagger J_N(x^*)) \ldots (I - J_1(x^*)^\dagger J_1(x^*))\| \le r_1.$$

So, for small enough δ_1 and δ_2, we have that

$$\|\Phi'(x^*)\| < 1.$$

Since the constant-rank hypothesis implies the continuity of Φ', we have that there exists $r \in (0,1)$, $\varepsilon > 0$ such that, when $\|x - x^*\| \leq \varepsilon$,

$$\|\Phi'(x)\| < r.$$

So, by the Mean Value Theorem, if $\|x - x^*\| \leq \varepsilon$,

$$\|\Phi(x) - \Phi(x^*)\| \leq r\|x - x^*\|.$$

Since P is a projection and x^* is a fixed point of $P(\Phi(x))$, this implies that, whenever $\|x - x^*\| \leq \varepsilon$,

$$\|P(\Phi(x)) - x^*\| = \|P(\Phi(x)) - P(\Phi(x^*)\| \leq \|\Phi(x) - \Phi(x^*)\| \leq r\|x - x^*\|.$$

Therefore, if $\|x^0 - x^*\| \leq \varepsilon$, the sequence $\{x^k\}$ converges to x^*, and (2.6) holds. □

Remarks. The Gauss-Newton method for solving nonlinear least-squares problems is a particular case of the algorithm given by (2.2), where $N = 1$ and $\Omega = \mathbb{R}^n$. Applying Theorem 2.2 to this particular case we obtain the well known result that states local linear convergence of the Gauss-Newton algorithm if the residual at the solution is small enough or if F is not "excessively nonlinear".

In many physical applications it is necessary to fit parameters for which it is known that they begin to some simple convex set, like, say, $\mathbb{R}^n_+$ (the positive orthant of $\mathbb{R}^n$). In these cases, combining a nonlinear least squares solver, like Gauss-Newton or Levenberg-Marquardt (see [22]) with projections on the correct region can improve dramatically the performance of the original method. A real-life problem illustrating this fact can be found in [23].

Comparing Theorem 2.2 with the corresponding convergence result of Cimmino's method (see [7], [10]), we see that the latter does not need "small variation assumptions" on the Jacobians of the blocks, since we have only one Jacobian involved with a fixed point, while in (2.2) we have different Jacobians for the different points $x_0^*, x_1^*, \ldots, x_N^*$ of the limit cycle. So, it can be conjectured that Cimmino's method is more robust than the method of successive projections. However, this type of robustness does not necessarily imply efficiency in practical situations. In fact, the method based on (2.2) tends to progress faster at the first iterations and, many times, obtains very quickly good "approximate solutions". Perhaps the most impressive evidence of this fast initial progress appears in the applications of variations of the successive orthogonal projection method to image reconstruction. See [11], [24], [25], [12], etc.

3. THE STOPPING CRITERION

Kaczmarz's algorithm and its generalizations are greedy methods in the sense that they try to incorporate, as soon as possible, the characteristics of the sets where the solution is desired. As a result, the "first" iterations usually incorporate a lot of information concerning these sets. The increment norm $s^{k+1} \equiv \|x^{k+1} - x^k\|$ is large for k small and relatively small for large k. It seems that most useful information is incorporated to x^k at the "first" iterations and that, ultimately, the algorithm converges at a very slow rate to a "solution" that, in ill-conditioned problems, has no physical meaning.

Based on the observations above, we defined a stopping criterion based on the equality

$$k = \text{ Argmin } \{s_j, j = 1, \ldots, k\}. \tag{3.1}$$

If (3.1) fails to hold during 10 consecutive iterations we judge that no more good progress can be expected from the process, and we stop the execution, defining $x^{\hat{k}}$ as approximate solution, where

$$\hat{k} = \text{Argmin } \{s_j, j = 1, \ldots, k\}. \tag{3.2}$$

As in [10] and [2] we considered the nonlinear Fredholm first kind integral equation

$$\int_0^1 \log \frac{(t-\tau)^2 + 0.04}{(t-\tau)^2 + (0.2 - x(\tau))^2} d\tau = y(t), \tag{3.3}$$

which occurs in inverse gravimetry. For solving (3.3) we choose a basis of piecewise linear functions $\{\phi_j\}_{j=1}^n$ and we take approximations

$$\tilde{x}(\tau) = \sum_{j=1}^{n} x_j \phi_j(\tau).$$

For each observed pair $(t_i, y(t_i))$, $i = 1, \ldots, m$, we have now the equation

$$[F(x)]_i \equiv \int_0^1 \log \frac{(t_i-\tau)^2 + 0.04}{(t_i-\tau)^2 + (0.2 - \tilde{x}(\tau))^2} d\tau - y(t_i) = 0. \tag{3.4}$$

We defined the following "true solution":

$$x_{true}(\tau) = -0.1\exp[-40(\tau - 0.4)^2] - 0.075\exp[-60(\tau - 0.67)^2] + c_1\tau + c_2$$

where c_1 and c_2 are chosen in such a way that $x_{true}(0) = x_{true}(1) = 0$. Finally, using x_{true} we generated $\bar{y}(t_i), i = 1, \ldots, m$, by numerical integration using a Gauss-Legendre six-point formula. In our experiments we used $n = 25$ and $m = 30$. We performed two types of experiments, according to the choice of $y(t_i)$ in (3.4):

(a) $y(t_i) = \bar{y}(t_i)$;

(b) $y(t_i) = \bar{y}(t_i) + \varepsilon_i$,

where the ε_i simulate measurement errors and are pseudo-random and normally distributed with mean 0 and standard deviation $\sigma = 0.002\|\bar{y}(t)\|$. Observe that, although the choice (a) seems to correspond to unperturbed data, a nonnull perturbation is implicit due to the numerical integration rule used to generate y. We took the origin as initial approximation. Since, for this problem, it is known that the solution is negative, we defined

$$\Omega = \{x \in \mathbb{R}^n \mid x \leq 0\}.$$

The true shape of the function $-x_{true}$ is given in Figure 1.

In [10], the results of the application of Cimmino's method, without a specific stopping criterion, are displayed. We compared the experiments reported in [10] with the application of the algorithm described in Section 2 of this paper for the same problems. The results are given in Table 1. Under the fourth column of this table, we report the iteration number of Cimmino's method that produced similar results as $\hat{k}$ iterations of the nonlinear Kaczmarz method method described in this paper. Figures 2 and 3 correspond to the final approximations obtained with this method.We observe that the computational effort used by Cimmino is far larger than the one used by Kaczmarz for obtaining analogous results, if a sequential computer is used for calculations. On the other hand, as it is well known, the advantage of Cimmino's method and its generalizations is the possibility of taking advantage of parallelization. Our codes were written in FORTRAN (single precision) and we used a SUN-Sparc station.

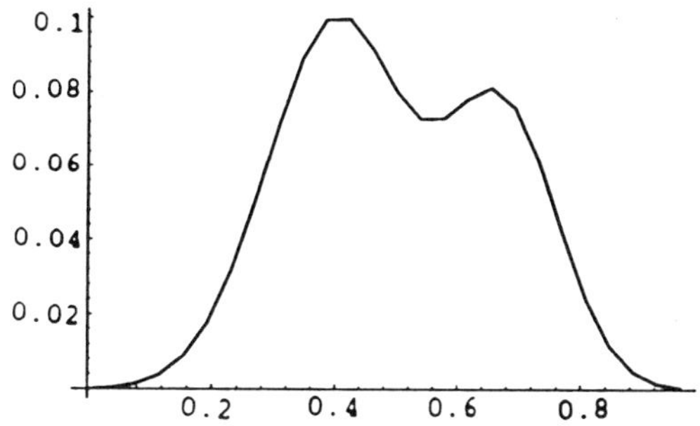

Figure .1: The shape of the "true solution"

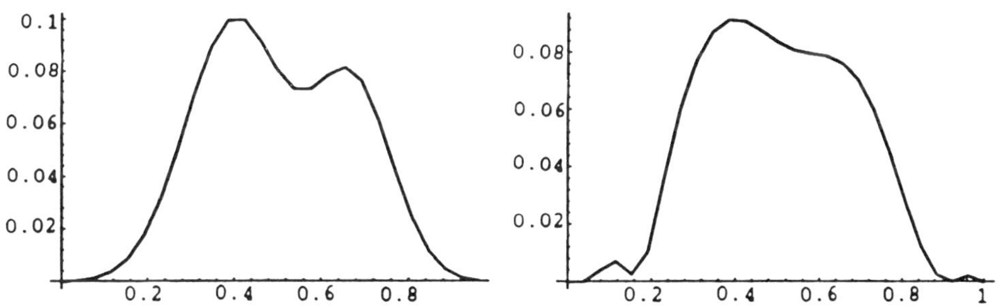

N=10 , after 175 iter.

N=30, after 1974 iter.

Figure .2: Without perturbation

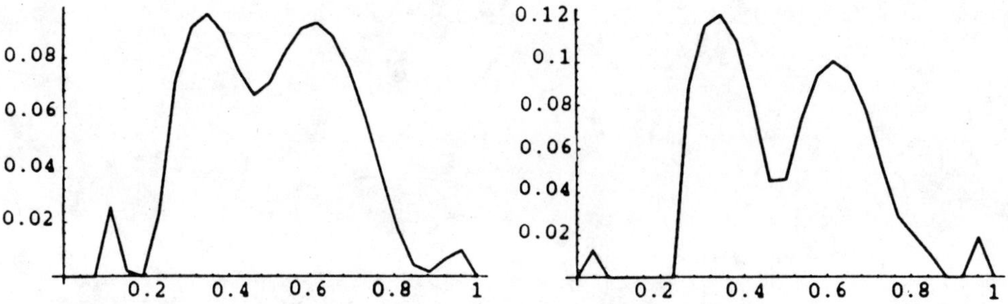

Seed = 53, after 3707 iter. Seed = 1009 , after 2162 iter

Figure .3: With perturbation , N = 30

Table 1. Numerical experiments.

Problem	N	$\hat{k}$	Cimmino
Unperturbed	10	175	1000
Unperturbed	30	1974	10000
Perturbed (seed 53)	30	3707	10000
Perturbed (seed 1009)	30	2162	10000

4. FINAL REMARKS

The success of Kaczmarz-related algorithms for image reconstruction motivated us to study convergence properties of a method of this type, for overdetermined nonfeasible problems. We showed that local convergence to a fixed-point cycle hold under suitable assumptions, that include the classical Gauss-Newton method as a particular case. A practical stopping rule was given, that seems to work well when the problem is very ill-conditioned .

The definition of theoretical-based stopping criteria, based on generalizations of the results of Santos [11] to the nonlinear case is, perhaps, the most interesting open problem related with this research.

ACKNOWLEDGEMENTS

We are indebted to an anonymous referee for useful remarks and to Antonio Moretti for technical assistance. This work was supported by FAPESP (Grant 90-3724-6), FINEP and FAEP-UNICAMP.

REFERENCES

[1] A.N. Tikhonov and V.Y. Arsenin, *"Solution of Ill-posed Problems"*. John Wiley, New York 1977.

[2] C. R. Vogel, "A constrained least squares regularization for nonlinear ill-posed problems". *SIAM Journal on Control and Optimization* 28:34, 1990.

[3] G. H. Golub, M. T. Heath and G. Wahba, "Generalized cross-validation as a method for choosing a good ridge parameter". *Technometrics* 21:215, 1979.

[4] M. Heinkenschloss, "Mesh independence for nonlinear least squares problems with norm constraints". Technical Report 90-18, Department of Mathematical Sciences, Rice University.

[5] J. M. Martínez and S. A. Santos, "A trust-region strategy for minimization on arbitrary domains". *Mathematical Programming* 68:267, 1995.

[6] F. O'Sullivan and G. Wahba, "A cross-validated Bayesian retrieval algorithm for nonlinear remote sensing experiments". *Journal of Computational Physics* 59:441, 1985.

[7] M. A. Diniz-Ehrhardt and J. M. Martínez, "A parallel projection method for overdetermined nonlinear systems of equations". *Numerical Algorithms* 4:241, 1993.

[8] G. Cimmino, "Calcolo approssimato per le soluzione dei sistemi di equazioni lineari". *La Ricerca Scientifica Ser II*, Anno IV 1:326, 1938.

[9] A. R. De Pierro and A. N. Iusem, "A parallel projection method for linear inequalities". *Linear Algebra and its Applications* 64:243, 1985.

[10] M. A. Diniz-Ehrhardt, J. M. Martínez and S. A. Santos, "Parallel projection methods and the resolution of ill-posed problems". *Computers and Mathematics with Applications* 27:11, 1994.

[11] R. J. Santos, *"Iterative Linear Methods and Regularization"*. Ph D Thesis, Department of Applied Mathematics, UNICAMP,Campinas, 1995.

[12] G. T. Herman, *"Image Reconstruction from Projections: The Fundamentals of Computerized Tomography."* Academic Press,New York 1980.

[13] S. Kaczmarz, "Angenaerte Auflösung von Systemen linearer Gleichungen". *Bull. Acad. Polon. Sci. Lett.* A35:355, 1937.

[14] L. G. Gubin, B. T. Polyak and E. V. Raik, "The method of projections for finding the common point of convex sets" *U.S.S.R. Comp. Math. Math. Phys.* 7:1, 1967.

[15] S. F. McCormick, "An iterative procedure for the solution of constrained nonlinear equations with application to optimization problems". *Numerische Mathematik* 23:371, 1975.

[16] S. F. McCormick, "The methods of Kaczmarz and row orthogonalization for solving linear equations an least squares problems in Hilbert spaces". *Indiana University Mathematical Journal* 26:1137, 1977.

[17] K. H. Meyn, "Solution of underdetermined nonlinear equations by stationary iteration methods". *Numerische Mathematik* 42:161, 1983.

[18] J. M. Martínez, "The method of successive orthogonal projections for solving nonlinear simultaneous equations". *Calcolo* 23:93, 1986.

[19] J. M. Martínez, "Solution of nonlinear systems of equations by an optimal projection method". *Computing* 37:59, 1986.

[20] J. M. Martínez, "Solving systems of nonlinear equations by means of an accelerated successive orthogonal projections method". *International Journal of Computational and Applied Mathematics* 16:169, 1986.

[21] P. C. Hansen, "Analysis of discrete ill-posed problems by means of the L-curve". *SIAM Review* 34:561, 1992.

[22] J. E. Dennis and R. B. Schnabel, *"Numerical Methods for Unconstrained Optimization and Nonlinear Equations"*. Prentice Hall, Englewood Cliffs, New Jersey 1983.

[23] M. Mulato and I. Chambouleyron, "Small angle X-ray and neutron scattering of polydisperse systems: determination of the scattering particle size distribution". To appear in *Journal of Applied Crystallograpy.*

[24] Y. Censor, "Row-action methods for huge and sparse systems and their applications". *SIAM Review* 23:444, 1981.

[25] Y. Censor, D. E. Gustafson, A. Lent and H. Tuy, "A new approach to the emission computerized tomography problem: simultaneous calculation of attenuation and activity coefficients". *IEEE Transactions on Nuclear Science* NS-26:2775, 1979.

ON EXACT AUGMENTED LAGRANGIAN FUNCTIONS IN NONLINEAR PROGRAMMING

Di Pillo Gianni and Lucidi Stefano

Dipartimento di Informatica e Sistemistica,
Università di Roma "la Sapienza",
Via Buonarroti 12, 00185 Roma, Italy
e-mail: dipillo@itcaspur.caspur.it

1. INTRODUCTION

The problem considered here is the nonlinear programming problem:

$$\begin{array}{ll} \text{minimize} & f(x) \\ \text{s.t.} & g(x) \leq 0, \end{array} \tag{NLP}$$

where $f : \mathbb{R}^n \to \mathbb{R}$ and $g : \mathbb{R}^n \to \mathbb{R}^m$ are twice continuously differentiable functions. We denote by

$$\mathcal{F} = \{x \in \mathbb{R}^n : \ g(x) \leq 0\}$$

the feasible set of Problem (NLP).

The *Lagrangian function* associated with Problem (NLP) is the function

$$L(x, \lambda) = f(x) + \lambda' g(x),$$

where $\lambda \in \mathbb{R}^m$ is a *multiplier vector.*

A *Karush-Kuhn-Tucker (KKT) pair* is a pair $(\overline{x}, \overline{\lambda}) \in \mathbb{R}^n \times \mathbb{R}^m$ that satisfy the necessary optimality conditions for Problem (NLP) :

$$\nabla_x L(\overline{x}, \overline{\lambda}) = 0, \quad \overline{\lambda}' g(\overline{x}) = 0, \quad \overline{\lambda} \geq 0, \quad g(\overline{x}) \leq 0.$$

An *augmented Lagrangian function* associated with Problem (NLP) is a function $L_a(x, \lambda; \varepsilon)$, where ε is a positive parameter, obtained by adding to $L(x, \lambda)$ a suitable term. An augmented Lagrangian function possesses *exactness properties* if, for some value of ε its unconstrained minimizers correspond to the solutions of the constrained problem (NLP) and the associated multipliers.

A challenging research argument in nonlinear programming is the transformation of the constrained minimization problem (NLP) into the unconstrained minimization of a merit function.

A constrained problem is determined by the interaction of two distinct subproblems: the feasibility subproblem and the subproblem of minimizing the objective function.

Nonlinear Optimization and Applications
Edited by G. Di Pillo and F. Giannessi, Plenum Press, New York, 1996

Therefore a constrained problem can be solved by the unconstrained minimization of a merit function only if this function is able to represent well the combination of the two preceding subproblems. The initial idea in defining merit fuctions was to add to the original objective function penalty terms which weigh the violation of the constraints . This approach has led to the definition of merit functions which are nondifferenziable (see, for example, [1],[9],[5],[6]), or to merit functions which are continuously differentiable but not "exact" in the sense that their minimum points are not minimum points of the original constrained problem ([8]).

The subsequent step in this field has been the introduction of continuously differentiable merit functions which characterize "better" the connections between the feasibility subproblem and the minimizing subproblem: namely continuously differentiable exact merit functions which use penalty terms which consider not only the ammissibility but also other characteristics of the constrained minimum points. A practicable choice has been to define merit functions which present terms which take into account the KKT conditions. Following this line two classes of continuously differentiable exact merit functions have been proposed: the augmented Lagrangian functions and the *penalty functions.* The exact augmented Lagrangian functions are defined on the product space of the problem variables and of the KKT multipliers. This approach of using a space of variables extended to the multipliers is derived, of course, from the idea of using the KKT conditions. On the contrary the exact penalty functions are defined on the same space of the variables of the original constrained problem. However, in order to exploit the KKT conditions, they need the use of a multiplier function, namely a function $\lambda(x)$ which yields an estimate of the KKT multipliers as a function of the variable x. In general a multiplier function is quite expensive from the computational point of view when the number of constraints is large and this may limit somewhat the applicability of the continuously differentiable penalty functions.

We refer to [13], [2] for a description of the more recent results on continuously differentiable exact penalty functions and to [6], [3] for detailed treatments of these functions. Here we restrict our attention to the exact augmented Lagrangian functions with the aim of putting in right perspective a new Lagrangian function recently proposed.

The paper is organized as follows. In Section 2 we state some useful preliminary results, that will enable us to put in a unified framework the matter. In Section 3 we review, in this framework, the developments that lead to augmented Lagrangian functions with better exactness properties. Finally, in Section 4 we describe a new augmented Lagrangian function whose exactness properties can be established under assumptions quite weaker than the ones used before, and whose level sets, under these assumptions, are compact; so that, when dealing with the transformation of a constrained minimization problem into an unconstrained one, it combines very favourable features both from the analytical and the computational point of view.

2. PRELIMINARY RESULTS

In the following we make use of the function $\phi : \mathbb{R} \times \mathbb{R} \to \mathbb{R}$

$$\phi(y,z) = z \max\{y, -\frac{z}{2}\} + \max\{y, -\frac{z}{2}\}^2, \tag{2.1}$$

The function ϕ appears, even if hiddenly, in the expression of all continuously differentiable penalty functions and augmented Lagrangian functions proposed in the literature. Since the function ϕ plays a central role also in the definition of the new Lagrangian function proposed in this paper, we list its properties that are of interest in the following developments. These properties are proved in the Appendix.

Proposition 2.1 *Let ϕ be the function defined by (2.1) then:*

(a) $\phi(y,z)$ *is continuously differentiable for all* $(y,z) \in \mathbb{R} \times \mathbb{R}$, *with derivatives:*

$$\frac{\partial \phi(y,z)}{\partial y} = z + 2\max\{y, -\frac{z}{2}\} \tag{2.2}$$
$$\frac{\partial \phi(y,z)}{\partial z} = \max\{y, -\frac{z}{2}\}; \tag{2.3}$$

(b) *if* $z = 0$ *then we have* $\phi(y,0) = \max\{y,0\}^2$ *for all* $y \in \mathbb{R}$;

(c) *if* $y \le 0$, $z \ge 0$ *and* $yz = 0$ *then* $\phi(y,z) = 0$;

(d) *if* $y \le 0$ *then* $\phi(y,z) \le 0$ *for all* $z \in \mathbb{R}$;

(e) *if* $y \le 0$ *and* $\phi(y,z) = 0$ *then* $y \le 0$, $z \ge 0$ *and* $yz = 0$;

(f) ϕ *can be rewritten in the following ways:*

$$\phi(y,z) = \max\{0, y + \frac{z}{2}\}^2 - \frac{z^2}{4}, \tag{2.4}$$
$$\phi(y,z) = yz + y^2 - \min\{0, y + \frac{z}{2}\}^2. \tag{2.5}$$

In particular, by (5), we recognize that the function

$$L_0(x,\lambda) = f(x) + \sum_{i=1}^{m} \phi(g_i(x), \lambda_i)$$

is an augmented Lagrangian function.

More in general, most of the augmented Lagrangian functions proposed up to now can be represented in the following form:

$$L_G(x,\lambda;\varepsilon) = f(x) + \sum_{i=1}^{m} \frac{1}{s_i(x,\lambda;\varepsilon)} \phi(g_i(x), r_i(x,\lambda;\varepsilon)\lambda_i), \quad +\eta(x,\lambda) \tag{2.6}$$

where the functions $s_i(x,\lambda;\varepsilon)$, $r_i(x,\lambda;\varepsilon)$, for $i = 1,\ldots,m$, and $\eta(x,\lambda)$ are continuously differentiable *w.r.t.* x and λ, and satisfy the following conditions:

(c1) for all $(x,\lambda) \in \mathcal{F} \times \mathbb{R}^m$ and all $\varepsilon > 0$ it results $s_i(x,\lambda;\varepsilon) > 0$;

(c2) for all KKT pairs $(\bar{x},\bar{\lambda})$ of Problem (NLP) and all $\varepsilon > 0$, it results:

$$\frac{r_i(\bar{x},\bar{\lambda};\varepsilon)}{s_i(\bar{x},\bar{\lambda};\varepsilon)} = 1;$$

(c3) for all KKT pairs $(\bar{x},\bar{\lambda})$ of Problem (NLP) it results:

$$\eta(\bar{x},\bar{\lambda}) = 0, \quad \nabla_x \eta(\bar{x},\bar{\lambda}) = 0, \quad \nabla_\lambda \eta(\bar{x},\bar{\lambda}) = 0;$$

(c4) for every KKT pair $(\bar{x}, \bar{\lambda})$ of Problem (NLP) there exist a neighborhood Ω_x of $\bar{x}$, a neighborhood Ω_λ of $\bar{\lambda}$ and a continuous function $\lambda(\cdot)$ such that:

$$\lambda(\cdot) : \Omega_x \to \Omega_\lambda, \quad \lambda(\bar{x}) = \bar{\lambda}, \quad \eta(x, \lambda(x)) = 0 \ \ \forall x \in \Omega_x.$$

The following proposition describes some properties of the augmented Lagragian function $L_G(x, \lambda; \varepsilon)$ which are induced by the properties of the function ϕ and by the conditions (c1)-(c4).

Proposition 2.2 *Let $L_G(x, \lambda; \varepsilon)$ be the function defined by (2.6), and assume that conditions (c1)-(c4) hold; then:*

(a) *$L_G(x, \lambda; \varepsilon)$ is continuously differentiable for all (x, λ) such that $s_i(x, \lambda; \varepsilon) > 0$, $i = 1, \dots, m$, with gradient:*

$$\begin{aligned}
\nabla_x L_G(x, \lambda; \varepsilon) &= \nabla f(x) + \sum_{i=1}^{m} \frac{r_i(x, \lambda; \varepsilon)}{s_i(x, \lambda; \varepsilon)} \nabla g_i(x) \lambda_i \\
&+ \sum_{i=1}^{m} \frac{1}{s_i(x, \lambda; \varepsilon)} [2\nabla g_i(x) + \nabla_x r_i(x, \lambda; \varepsilon)\lambda_i] \max\{g_i(x), -\frac{r_i(x, \lambda; \varepsilon)\lambda_i}{2}\} \\
&- \sum_{i=1}^{m} \frac{1}{s_i(x, \lambda; \varepsilon)^2} \nabla_x s_i(x, \lambda; \varepsilon) \phi(g_i(x), r_i(x, \lambda; \varepsilon)\lambda_i) + \nabla_x \eta(x, \lambda), \\
\nabla_\lambda L_G(x, \lambda; \varepsilon) &= \sum_{i=1}^{m} \frac{1}{s_i(x, \lambda; \varepsilon)} [e_i r_i(x, \lambda; \varepsilon) \\
&\qquad + \nabla_\lambda r_i(x, \lambda; \varepsilon)\lambda_i] \max\{g_i(x), -\frac{r_i(x, \lambda; \varepsilon)\lambda_i}{2}\} \\
&- \sum_{i=1}^{m} \frac{1}{s_i(x, \lambda; \varepsilon)^2} \nabla_\lambda s_i(x, \lambda; \varepsilon) \phi(g_i(x), r_i(x, \lambda; \varepsilon)\lambda_i) + \nabla_\lambda \eta(x, \lambda)
\end{aligned}$$

where e_i is the i-th column of the identity matrix;

(b) *for all $\varepsilon > 0$, every KKT pair $(\bar{x}, \bar{\lambda})$ of Problem (NLP) is a stationary point of $L_G(x, \lambda; \varepsilon)$; moreover, it results $f(\bar{x}) = L_G(\bar{x}, \bar{\lambda}; \varepsilon)$;*

(c) *assume that, for some value of ε, every stationary point of $L_G(x, \lambda; \varepsilon)$ is also a KKT pair for Probelm (NLP); then if $(\bar{x}, \bar{\lambda})$ is a local minimum point of $L_G(x, \lambda; \varepsilon)$, $\bar{x}$ is a local minimum point for Probelm (NLP) and $\bar{\lambda}$ is a corresponding KKT multiplier;*

(d) *assume that, for some value of ε, every stationary point of $L_G(x, \lambda; \varepsilon)$ is also a KKT pair for Problem (NLP); then, if $(\bar{x}, \bar{\lambda})$ is a global minimum point of $L_G(x, \lambda; \varepsilon)$, $\bar{x}$ is a global minimum point for Probelm (NLP) and $\bar{\lambda}$ is a corresponding KKT multiplier, and conversely.*

Proof. Point (a) follows directly from (2.6) and (a) of Proposition 2.1.

Point (b) derives by the expression of the gradient of $L_G(x, \lambda; \varepsilon)$, conditions (c2) and (c3) and by recalling that, if the point $(\bar{x}, \bar{\lambda})$ is a KKT pair of Problem (NLP), then the definition of the max function, (c) of Proposition 2.1 and condition (c1) imply that for $i = 1, \dots, m$:

$$\max\{g_i(\bar{x}), -\frac{r_i(\bar{x}, \bar{\lambda}; \varepsilon)\bar{\lambda}_i}{2}\} = 0, \qquad \phi(g_i(\bar{x}), r_i(\bar{x}, \bar{\lambda}; \varepsilon)\bar{\lambda}_i) = 0.$$

Point (c). Let $(\bar{x}, \bar{\lambda})$ be a local minimum point of $L_G(x, \lambda; \varepsilon)$, therefore there exists a neighborhood $B(\bar{x}, \bar{\lambda})$ of the pair $(\bar{x}, \bar{\lambda})$ such that for all $(x, \lambda) \in B(\bar{x}, \bar{\lambda})$ it results

$$L_G(\bar{x}, \bar{\lambda}; \varepsilon) \leq L_G(\bar{x}, \bar{\lambda}; \varepsilon).$$

By assumption we know that $(\bar{x}, \bar{\lambda})$ is a KKT pair for Problem (NLP) and, hence, by recalling point (b) we have also:

$$L_G(\bar{x}, \bar{\lambda}; \varepsilon) = f(\bar{x}).$$

By (d) of Proposition 2.1, we have that for all $(x, \lambda) \in \mathcal{F} \times R^m \cap B(\bar{x}, \bar{\lambda})$:

$$f(\bar{x}) = L_G(\bar{x}, \bar{\lambda}; \varepsilon) \leq L_G(x, \lambda; \varepsilon) \leq f(x) + \eta(x, \lambda).$$

Now, we can suppose that the neighborhood $B(\bar{x}, \bar{\lambda})$ is sufficiently small, so that $B(\bar{x}, \bar{\lambda}) \subseteq \Omega_x \times \Omega_\lambda$ and condition (c4) holds. Therefore we have:

$$f(\bar{x}) = L_G(\bar{x}, \bar{\lambda}; \varepsilon) \leq L_G(x, \lambda(x); \varepsilon) \leq f(x) + \eta(x, \lambda(x)) = f(x), \forall x \in \Omega_x$$

so that $\bar{x}$ is a local minimum point for Problem (NLP).

Point (d). If $\bar{x}$ is a global minimum point for Probelm (NLP) and $\bar{\lambda}$ is a corresponding KKT multiplier then $(\bar{x}, \bar{\lambda})$ is a KKT pair, so that by point (b) we have $f(\bar{x}) = L_G(\bar{x}, \bar{\lambda}; \varepsilon)$. On the other hand, if $(\bar{x}, \bar{\lambda})$ is a global minimum point of $L_G(x, \lambda; \varepsilon)$, it is also a stationary point of $L_G(x, \lambda; \varepsilon)$; therefore, by assumption, we have that $(\bar{x}, \bar{\lambda})$ is a KKT pair for Problem (NLP), and, by using again point (b), we have that $L_G(\bar{x}, \bar{\lambda}; \varepsilon) = f(\bar{x})$. We can conclude that, in this case, the functions f and L_G take the same value in correspondence at every point that is either a global minimum point for Problem (NLP) or a global minimum point of L_G and this proves point (d) of the proposition. □

The preceding proposition shows that the augmented Lagrangian functions in the class described by (2.6) inherit interesting properties from their expression. Moreover it points out clearly that the crucial point in order to provide an augmented Lagrangian function with exactness properties is to ensure that every stationary point of the considered Lagrangian function is a KKT pair of the Problem (NLP) at least for some value of the penalty parameter.

3. AN OVERALL VIEW ON EXACT AUGMENTED LAGRANGIAN FUNCTIONS

The simplest augmented Lagrangian function belonging to the class described by (2.6) is the one which derives by choosing $s_i(x, \lambda; \varepsilon) = 1$, $r_i(x, \lambda; \varepsilon) = 1$, $i = 1, \ldots, m$ and $\eta(x, \lambda) = 0$. Namely the augmented Lagrangian function $L_0(x, \lambda)$ which is rewritten here:

$$L_0(x, \lambda) = f(x) + \sum_{i=1}^{m} \phi(g_i(x), \lambda_i). \tag{3.1}$$

The function L_0 can be seen as the starting point for describing the augmented Lagrangian approach. In fact it has some interesting properties which are described in the next proposition.

Proposition 3.1 *Let $L_0(x,\lambda)$ be the function defined by (3.1), then:*

(a) *every stationary point of $L_0(x,\lambda)$ is a KKT pair for Problem (NLP) and conversely;*

(b) *every local minimum point of $L_0(x,\lambda)$ is also a local minimum point for Problem (NLP).*

Proof. Point (a) follows from the expression of the gradient of $L_0(x,\lambda)$, namely (recalling point (a) of Proposition 2.1):

$$\begin{aligned}\nabla_x L_0(x,\lambda) &= \nabla f(x) + \sum_{i=1}^{m} \nabla g_i(x)[\lambda_i + 2\max\{g_i(x), -\frac{\lambda_i}{2}\}] \\ \nabla_{\lambda_i} L_0(x,\lambda) &= \max\{g_i(x), -\frac{\lambda_i}{2}\}, \qquad i = 1,\ldots,m.\end{aligned}$$

Point (b) follows from point (a) of the proposition and point (c) of Proposition 2.2.

□

The properties stated in the preceding proposition are not sufficient to guarantee that the $L_0(x,\lambda)$ is an exact merit function. In fact:

- $L_0(x,\lambda)$ has not a global minimum point; it easy to see that it is not bounded from below with respect to the variable λ;

- even if we assume to know a KKT multiplier $\bar{\lambda}$ associated to a global minimum point $\bar{x}$ for Problem (NLP), the function $L_0(x,\bar{\lambda})$ (now depending only on the variable x) may not have a global minimum point.

One of the reasons of the lack of a complete correspondence between its minimum points and those of the constrained problem is that the function $L_0(x,\lambda)$ does not penalize enough the violation of the constraints. In order to obtain a better penalization of the constraints, it is convenient to modify L_0 by introducing a penalty parameter $\varepsilon > 0$, so that we are lead to consider the augmented Lagrangian function $L_1(x,\lambda;\varepsilon)$ given by:

$$L_1(x,\lambda;\varepsilon) = f(x) + \frac{1}{\varepsilon}\sum_{i=1}^{m} \phi(g_i(x), \varepsilon\lambda_i). \tag{3.2}$$

Recalling point (b) of Proposition 2.1 we can easily note that the more the parameter ε is small the more the function $L_1(x,\lambda;\varepsilon)$ tends to weigh the violation of the constraints. In particular it is possible to state the following proposition.

Proposition 3.2 *Let $L_1(x,\lambda;\varepsilon)$ be the function defined by (3.2), then:*

(a) *every stationary point of $L_1(x,\lambda;\varepsilon)$ is a KKT pair for Problem (NLP) and conversely;*

(b) *every local minimum point of $L_1(x,\lambda;\varepsilon)$ is also a local minimum point for Problem (NLP);*

(c) *assume that there exists a compact subset C of $\mathbb{R}^n$ such that all the global minimum points of Problem (NLP) belong to its interior, let $\bar{x}$ be a global minimum point of Problem (NLP) and let $\bar{\lambda}$ be the associated KKT multiplier; then there exists an $\bar{\varepsilon} > 0$ such that, for all $\varepsilon \in (0,\bar{\varepsilon}]$, $\bar{x}$ is an unconstrained global minimum point of $L_1(x,\bar{\lambda};\varepsilon)$ on C.*

Hestenes-Powell-Rockafellar Lagrangian function is an exact merit function. From the algorithmic point of view, the best results concerning this augmented Lagrangian function indicate that, under quite strong assumptions, Problem (NLP) can be tackled by a sequence of unconstrained minimization of the functions $L_1(x, \lambda_k; \varepsilon_k)$ where $\{\lambda_k\}$ is a suitable sequence which converges to a KKT multiplier $\bar{\lambda}$ and ε_k is a decreasing sequence which does not converge necessarily to zero.

The first augmented Lagrangian function truly possessing exactness properties has been introduced in [4]. This merit function can be written in the following way:

$$\begin{aligned} L_2(x, \lambda; \varepsilon) &= f(x) + \frac{1}{\varepsilon} \sum_{i=1}^{m} w_i(\lambda; \varepsilon)\ \phi(g_i(x), \varepsilon q_i(x, \lambda; \varepsilon)\lambda_i) \\ &\quad + \varphi(x, \lambda), \end{aligned} \tag{3.9}$$

where

$$\begin{aligned} w_i(\lambda; \varepsilon) &= 1 + 16\varepsilon\lambda_i^2, \qquad i = 1, \dots, m, \\ q_i(x, \lambda; \varepsilon) &= \frac{1 + 8(\nabla g_i(x)'\nabla_x L(x, \lambda) - 4g_i(x)\lambda_i)}{w_i(\lambda; \varepsilon)}, \qquad i = 1, \dots, m, \\ \varphi(x, \lambda) &= \sum_{i=1}^{m} (\nabla g_i(x)'\nabla_x L(x, \lambda) - 4g_i(x)\lambda_i)^2. \end{aligned}$$

The most characterizing element in the expression of the function $L_2(x, \lambda; \varepsilon)$ is the term $\varphi(x, \lambda)$. This term penalizes further on the fact that the vector λ is different from a KKT multiplier $\bar{\lambda}$ and it performs a "convexification" of the function with respect to the variable λ. In fact it easy to prove the following proposition.

Proposition 3.3 (a) *at every KKT point $\bar{x}$ where the gradients of the active constraints are linearly independent $\varphi(\bar{x}, \lambda) = 0$ if and only if $\lambda = \bar{\lambda}$, where $\bar{\lambda}$ is the KKT multiplier associated to $\bar{x}$;*

(b) *at every feasible point $\tilde{x}$ where the gradients of the active constraints are linearly independent the function $\tilde{\varphi}(\lambda) = \varphi(\tilde{x}, \lambda)$ is strictly convex.*

The term $\varphi(x, \lambda)$ along with the weigths $w_i(\lambda; \varepsilon)$ and $q_i(x, \lambda; \varepsilon)$ allows to guarantee good exactness properties to the Lagrangian function $L_2(x, \lambda; \varepsilon)$ without requiring the knowledge of a KKT multiplier $\bar{\lambda}$. These properties can be established under the following assumptions:

(a1) at every feasible point the gradients of the active constraints are linearly independent;

(a2) there exists a compact subset $\mathcal{A} \times \mathcal{L}$ of $R^n \times R^m$ and a value $\bar{\varepsilon} > 0$ such that for all $\varepsilon \in (0, \bar{\varepsilon}]$ the function $L_2(x, \lambda; \varepsilon)$ admits a stationary point in $\mathcal{A} \times \mathcal{L}$.

In particular, the following proposition summarizes the exactness properties of the Lagrangian function $L_2(x, \lambda; \varepsilon)$ (see [4]).

Proposition 3.4 *Let $L_2(x, \lambda; \varepsilon)$ the function defined by (3.9), then:*

(a) *for all $\varepsilon > 0$, every KKT pair $(\bar{x}, \bar{\lambda})$ of Problem (NLP) is a stationary point of $L_2(x, \lambda; \varepsilon)$;*

(b) *if assumptions* (a1) *and* (a2) *hold, there exists an $\bar{\varepsilon} > 0$ such that for all $\varepsilon \in (0, \bar{\varepsilon}]$:*

Proof. Point (a) and point (b) follow by the same arguments used in the proof of Proposition 3.1.

Now we prove point (c) by contradiction. If the assertion of point (c) is false then for every integer k there must exist an $\varepsilon_k \leq 1/k$ and a point x_k which is a global minimum point of $L_1(x, \bar{\lambda}; \varepsilon_k)$ on $\mathcal{C}$ and such that it results:

$$L_1(x_k, \bar{\lambda}; \varepsilon_k) < L_1(\bar{x}, \bar{\lambda}; \varepsilon_k). \tag{3.3}$$

By using point (c) of Proposition 2.1 we have that $L_1(\bar{x}, \bar{\lambda}; \varepsilon_k) = f(\bar{x})$ and, hence, from (3.3) we obtain:

$$L_1(x_k, \bar{\lambda}; \varepsilon_k) = f(x_k) + \frac{1}{\varepsilon_k} \sum_{i=1}^{m} \phi(g_i(x_k), \varepsilon_k \bar{\lambda}) < f(\bar{x}). \tag{3.4}$$

Since by assumption the set $\mathcal{C}$ is compact, there exists a convergent subsequence which we can relabel $\{x_k\}$, such that $\lim_{k\to\infty} x_k = \hat{x}$. Taking limits in (3.4) we have:

$$\lim_{k\to\infty} L_1(x_k, \bar{\lambda}; \varepsilon_k) \leq f(\bar{x}), \tag{3.5}$$

which implies

$$\phi(g_i(\hat{x}), 0) = 0, \qquad i = 1, \ldots, m, \tag{3.6}$$

$$f(\hat{x}) \leq f(\bar{x}). \tag{3.7}$$

Recalling the expression of function ϕ, given by (2.1), (3.6) and (3.7) yield that $\hat{x}$ is a global minimum point of Problem (NLP) and, hence, the point $\hat{x}$ belongs to the interior of the set $\mathcal{C}$. This implies that, for sufficiently large values of k, x_k is again in the interior of $\mathcal{C}$ and, hence, it is also a stationary point of $L_1(x, \bar{\lambda}; \varepsilon_k)$. Now point (a) of the proposition ensures that $(x_k, \bar{\lambda})$ is a KKT pair for Problem (NLP). Then point (c) of Proposition 2.1 yields $f(x_k) = L_1(x_k, \bar{\lambda}; \varepsilon_k)$. Therefore, recalling (3.4) and for sufficiently large k, we obtain that x_k is a feasible point such that

$$f(x_k) = L_1(x_k, \bar{\lambda}; \varepsilon_k) < f(\bar{x}), \tag{3.8}$$

which contradicts the fact that $\bar{x}$ is a global minimum point for problem (NLP). □

Roughly speaking, point (c) of the preceding proposition ensures that, whenever the original problem admits a solution, the function $L_1(x, \bar{\lambda}; \varepsilon)$, for sufficiently small values of the penalty parameter ε, has an unconstrained global minimum point on the open set $\mathring{\mathcal{C}}$ which is also a global minimum point of Problem (NLP).

The function $L_1(x, \lambda; \varepsilon)$ is the well known Hestenes-Powell-Rockafellar augmented Lagrangian function [11]. In fact, by using point (f) of Proposition 2.1, L_1 can be written in the more familiar forms:

$$L_1(x, \lambda; \varepsilon) = f(x) + \frac{\varepsilon}{4} \sum_{i=1}^{m} \Big[\max\{0, \frac{2}{\varepsilon} g_i(x) + \lambda_i\}^2 - \lambda_i^2 \Big],$$

$$L_1(x, \lambda; \varepsilon) = f(x) + \sum_{i=1}^{m} \Big[g_i(x)\lambda_i + \frac{1}{\varepsilon} g_i(x)^2 - \frac{1}{\varepsilon} \min\{0, g_i(x) + \frac{\varepsilon}{2}\lambda_i\}^2 \Big].$$

However, usually a KKT multiplier $\bar{\lambda}$ is not available; therefore the properties described in Proposition 3.2 do not allow us to transform the original constrained optimization problem into an unconstrained minimization of $L_1(x, \lambda; \varepsilon)$ and, hence, to say that the

- *if $(\bar{x}, \bar{\lambda}) \in \mathcal{A} \times \mathcal{L}$ is a stationary point of $L_2(x, \lambda; \varepsilon)$, the pair $(\bar{x}, \bar{\lambda})$ is a KKT pair for (NLP);*
- *if $(\overline{x}, \overline{\lambda}) \in \mathcal{A} \times \mathcal{L}$ is a local unconstrained minimum point of $L_2(x, \lambda; \varepsilon)$, $\overline{x}$ is a local minimum point of Problem (NLP) and $\overline{\lambda}$ is the corresponding KKT multiplier;*
- *if $\overline{x}$ is a global minimum point of Problem (NLP) belonging to $\mathcal{A}$ and $\overline{\lambda} \in \mathcal{L}$ is the corresponding KKT multiplier, the pair $(\overline{x}, \overline{\lambda})$ is a global minimum point of $L_2(x, \lambda; \varepsilon)$ on $\mathcal{A} \times \mathcal{L}$, and conversely.*

Even if $L_2(x, \lambda; \varepsilon)$ can be considered the first step towards the transformation of a constrained optimization problem in the unconstrained minimization of a continuously differentiable augmented Lagrangian function, this merit function is not completely satisfactory both from the theoretical point of view and the computational one. From the theoretical point of view, there is the difficulty of connecting assumption (a2) with some regularity assumption on the original constrained problem. From the computational point of view, there is the fact that the exactness properties of the function $L_2(x, \lambda; \varepsilon)$ can be established with reference to the compact set $\mathcal{A} \times \mathcal{L}$. Since the function $L_2(x, \lambda; \varepsilon)$ does not present any particular behavior on the boundary of $\mathcal{A} \times \mathcal{L}$, then an unconstrained optimization method employed for the minimization of this augmented Lagrangian function could produce a sequence converging towards stationary points of $L_2(x, \lambda; \varepsilon)$ that are out of $\mathcal{A} \times \mathcal{L}$ and hence, that have no relations with the original constrained problem, or an unbounded sequence.

An augmented Lagrangian function which does not present the drawbacks of the function $L_2(x, \lambda; \varepsilon)$ has been introduced in [12], following the same approach proposed in [4] for exact penalty functions. The expression of this new function is:

$$L_3(x, \lambda; \varepsilon) = f(x) + \frac{1}{\varepsilon} \sum_{i=1}^{m} \frac{1}{b_i(x)} \phi(g_i(x), \varepsilon b_i(x) \lambda_i) + \psi(x, \lambda) \tag{3.10}$$

where

$$\begin{aligned} b_i(x) &= \alpha - g_i(x), \qquad i = 1, \ldots, m, \\ \psi(x, \lambda) &= \sum_{i=1}^{m} (\nabla g_i(x)' \nabla_x L(x, \lambda) + g_i(x)^2 \lambda_i)^2, \end{aligned}$$

with α arbitrary positive constant.

The term $\psi(x, \lambda)$ plays a similar role of the term $\varphi(x, \lambda)$.

Proposition 3.5 (a) *at every KKT point $\bar{x}$ where the gradients of the active constraints are linearly independent $\psi(\bar{x}, \lambda) = 0$ if and only if $\lambda = \bar{\lambda}$, where $\bar{\lambda}$ is the KKT multiplier associated to $\bar{x}$;*

(b) *at every point $\tilde{x}$ where the gradients of the active constraints are linearly independent the function $\tilde{\psi}(\lambda) = \psi(\tilde{x}, \lambda)$ is strictly convex.*

By the preceding proposition we can remark that the only difference between the properties of $\psi(x, \lambda)$ and the ones of $\varphi(x, \lambda)$ is the fact that for the function $\psi(x, \lambda)$ it is not required in (b) that the point x is feasible.

The main novelty in the expression of the augmented Lagrangian function $L_3(x, \lambda; \varepsilon)$ is the presence of the barrier terms $1/b_i(x)$ which go to infinity at the boundary of the following open perturbation of the feasible set:

$$\mathcal{G} = \{x \in R^n : g_i(x) < \alpha, \quad i = 1, \ldots, m\}.$$

We will denote by $\bar{\mathcal{G}}$ the closure of the set $\mathcal{G}$.

As regards the theoretical properties of $L_3(x, \lambda; \varepsilon)$, it is possible to show that, for sufficiently small values of the penalty parameter ε, every minimum point (KKT pair) of the original problem is a minimum point (stationary point) of $L_3(x, \lambda; \varepsilon)$ over $\mathcal{G} \times R^m$ and conversely. These exactness properties can be established under the following assumptions:

(a3) the set $\bar{\mathcal{G}}$ is compact;

(a4) at every $x \in \mathcal{G}$ the gradients of the active constraints are linearly independent;

(a5) at every $x \in \mathcal{G}$ the gradients of the active constraints and of the violated constraints are positive linearly independent.

As mentioned before the function $L_3(x, \lambda; \varepsilon)$ offers some advantages over $L_2(x, \lambda; \varepsilon)$. First of all the required assumptions are formulated as "a priori" constraint qualification assumptions on the original problem. The exactness properties can be established with reference only to the compact set $\bar{\mathcal{G}}$ of R^n without any compactness requirement on the multipliers (see [12]).

Proposition 3.6 *Let $L_3(x, \lambda; \varepsilon)$ the function defined by (3.10), then:*

(a) *for all $\varepsilon > 0$, every KKT pair $(\bar{x}, \bar{\lambda})$ of Problem (NLP) is a stationary point of $L_3(x, \lambda; \varepsilon)$;*

(b) *if assumptions* (a3), (a4) *and* (a5) *hold, there exists an $\bar{\varepsilon} > 0$ such that for all $\varepsilon \in (0, \bar{\varepsilon}]$:*

- *if $(\bar{x}, \bar{\lambda}) \in \mathcal{G} \times R^m$ is a stationary point of $L_3(x, \lambda; \varepsilon)$, the pair $(\bar{x}, \bar{\lambda})$ is a KKT pair for (NLP);*

- *if $(\bar{x}, \bar{\lambda}) \in \mathcal{G} \times R^m$ is a local unconstrained minimum point of $L_3(x, \lambda; \varepsilon)$, $\bar{x}$ is a local minimum point of Problem (NLP) and $\bar{\lambda}$ is the corresponding KKT multiplier;*

- *if $\bar{x}$ is a global minimum point of Problem (NLP) belonging to $\mathcal{G}$ and $\bar{\lambda}$ is the corresponding KKT multiplier, the pair $(\bar{x}, \bar{\lambda})$ is a global minimum point of $L_3(x, \lambda; \varepsilon)$ on $\mathcal{G} \times R^m$, and conversely.*

From the computational point of view, the presence of the barrier terms $1/b_i(x)$ implies that the Lagrangian function $L_3(x, \lambda; \varepsilon)$ goes to infinity at the boundary of the set $\mathcal{G}$. By using this last property and the particular structure of $L_3(x, \lambda; \varepsilon)$ it is possible to show that, for sufficiently small values of the penalty parameter ε, the level set $\{(x, \lambda) \in \mathcal{G} \times R^m : L_3(x, \lambda; \varepsilon) \leq L_3(x_0, \lambda_0; \varepsilon)\}$ is compact for all $(x_0, \lambda_0) \in \mathcal{G} \times R^m$. Therefore many uncostrained minimization algorithms can be easily adapted to produce a sequence $\{(x_k, \lambda_k)\}$ globally convergent to a stationary point of $L_3(x, \lambda; \varepsilon)$ belonging to $\mathcal{G} \times R^m$. Furthermore, these unconstrained minimization algorithms can be combined with an automatic adjustment rule for the penalty parameter ε in order to guarantee that the sequence $\{(x_k, \lambda_k)\}$ globally converges to a KKT point of the original constrained problem.

The successive task of the reasearch in the field of exact augmented Lagrangian functions is to try to enlarge as much as possible the class of constrained optimization problems which can be solved by an unconstrained minimization of a Lagrangian function. In fact Assumptions (a3)-(a5) required to establish the exactness properties of $L_3(x, \lambda; \varepsilon)$ may limit too much its applicabily. For example Assumption (a3) implies that the feasible set of the original constrained problem must be bounded.

In the following section we introduce a new exact augmented Lagrangian function whose exactness properties require assumptions which are quite weaker than the ones used by the Lagrangian function $L_3(x, \lambda; \varepsilon)$. More specifically, Assumption (a3) is no longer necessary and constrained optimization problems with unbounded feasible sets can be tackled by using this new merit function. Assumption (a4) is replaced by a similar one where the linear independence of the gradients of the active constraints is required only in the feasible set. This new assumption seems to be a quite weak requirement needed to ensure the existence and uniqueness of the KKT multipliers. Assumption (a5) is connected to the fact that the original problem is well defined (see [13],[10]). In fact assumption (a5) is a sufficient condition for the nonemptiness of the feasible set. Again in this work we substitute this assumption with a weaker assumption which is still a sufficient condition for the feasible set to be not empty, whereas it is also necessary in the case that the feasible set is given by convex inequalities. Therefore, at least for this class of constraints, the condition used is the weakest possible assumption which guarantees that the original constrained problem is well defined.

Finally, we remark that the new Lagrangian function introduced here has the advantage over the Lagrangian function $L_3(x, \lambda; \varepsilon)$ of having level sets that are compact for *every value* of the penalty parameter ε. This feature is, of course, quite relevant from the computational point of view.

4. AN AUGMENTED LAGRANGIAN FUNCTION WITH IMPROVED EXACTNESS PROPERTIES

In this section we introduce the new augmented Lagrangian function, which possesses improved exactness properties.

Similarly to the function $L_3(x, \lambda; \varepsilon)$, this Lagrangian function is defined, with respect to the variable x, on an open perturbation $\mathcal{S}$ of the feasible set $\mathcal{F}$, that we introduce in the following way. Let $\sigma(x)$ be the function given by:

$$\sigma(x) = \sum_{i=1}^{m} g_i^+(x)^s, \tag{4.1}$$

where $s \geq 2$; $\sigma(x)$ is zero on $\mathcal{F}$ and takes positive values outside $\mathcal{F}$. Then $\mathcal{S}$ is the open set given by:

$$\mathcal{S} = \{x \in \mathbb{R}^n : \sigma(x) < \alpha\}$$

where $\alpha > 0$; the closure of $\mathcal{S}$ is denoted by $\bar{\mathcal{S}}$ and its boundary by $\partial\mathcal{S}$. In connection to the set $\mathcal{S}$ we introduce the function:

$$a(x) = \alpha - \sigma(x), \tag{4.2}$$

which takes positive values on $\mathcal{S}$ and is zero on its boundary.

We can now list all the hypotheses that we use to establish the exactness properties of the new Lagrangian function:

Assumption A1: *Assume that one of the two following conditions is satisfied:*

(a) $\bar{\mathcal{S}}$ *is a bounded set;*

(b) *a point* $\tilde{x} \in \mathcal{F}$ *is known and* $f(x)$ *is radially unbounded on* $\bar{\mathcal{S}}$ *(that is for any* $\{x_k\} \subseteq \bar{\mathcal{S}}$ *such that* $\|x_k\| \to \infty$ *we have* $f(x_k) \to \infty$*).*

Assumption A2: *For every* $x \in \mathcal{F}$ *the gradients* $\nabla g_i(x)$, $i \in \{i' : g_{i'}(x) = 0\}$ *are linearly independent.*

Assumption A3: *Assume that one of the two following conditions is satisfied:*

(a) *at every point* $x \in \mathcal{S}$*, if:*

$$\sum_{i=1}^{m} w_i g_i^+(x) \nabla g_i(x) = 0$$

with $w_i > 0$*, then* $g_i^+(x) = 0$, $i = 1, \ldots, m$*;*

(b) *a point* $\tilde{x} \in \mathcal{F}$ *is known.*

In the sequel, we assume that Assumptions A1 and Assumption A2 hold. Assumption A3 will be invoked explicitly when needed.

As already said in the preceding section, the assumptions listed above are, up to now, the weakest assumptions under which it possible to transform a constrained optimization problem into the unconstrained minimization of a continuously differentiable Lagrangian function. We refer to [13] and [10] for a detailed discussion of the Assumption A1-A3.

In particular, by Assumption A1 it not required that the feasible set is bounded and, hence, the class of nonlinear programming problems that can be tackled by our approach is considerably enlarged. Assumption A1 says that, in the case that a feasible point is known, the boundedness of the feasible set can be replaced by an assumption (quite similar the one usually used in the case of unconstrained optimization) which ensures that all the level sets of the objective function in the open set $\mathcal{S}$ are compact.

Assumptiom A2 guarantees the existence and the uniqueness of the KKT multipliers. This assumption is weaker than Assumption (a4) used in [12] since it requires the linear independence of the gradients of the active constraints only in the feasible set.

Assumption A3 says that either we know a feasible point or we need a regularity assumption on the constraints. Roughly speaking, this assumption imposes a "good behaviour" on the constraints at nonfeasible points in order to ensure the possibility to find a feasible point. Any attempt to solve a constrained minimization problem needs a similar assumption. On the basis of the discussion reported in [13] (see also [10]), Assumption A3 appears very mild in comparision with other ones used in the literature. In particular it is the weakest assumption which ensures that a compact feasible set given by convex inequalities is not empty.

Now we are ready to introduce the new exact augmented Lagrangian function:

$$\begin{aligned} L_a(x, \lambda; \varepsilon) &= f(x) + \frac{1}{\varepsilon} \sum_{i=1}^{m} \frac{1}{p(x, \lambda)} \phi(g_i(x), \varepsilon p(x, \lambda) \lambda_i) \\ &\quad + \psi(x, \lambda), \end{aligned} \tag{4.3}$$

where

$$p(x, \lambda) = \frac{a(x)}{1 + \|\lambda\|^2},$$

the function $\psi(x, \lambda)$ is given by (3.11) and the function $a(x)$ is given by (4.2). Recalling point (a) of Proposition 2.1 it turns out that $L_a(x, \lambda; \varepsilon)$ is continuously differentiable for $(x, \lambda) \in \mathcal{S} \times \mathbb{R}^m$.

This augmented Lagrangian function has a structure similar to the previous ones; in particular the term ψ is the same used in function $L_3(x, \lambda; \varepsilon)$ so that it has the same properties described in Proposition (3.5) and, hence, it plays the same penalization and convexification roles with respect to the variable λ.

The distinguishing element in the expression of the L_a is the term $p(x, \lambda)$. This term presents the following properties:

$$\lim_{x \to \partial \mathcal{S}} p(x, \lambda) = 0 \tag{4.4}$$

$$\lim_{\|\lambda\| \to \infty} p(x, \lambda) = 0. \tag{4.5}$$

Due to (4.4) and (4.5) the term $1/p(x, \lambda)$ plays the role of a barrier term that penalizes both the fact that the variable x is too close to the boundary of $\mathcal{S}$ and the fact that the norm of the vector λ is too large. We recall that the augmented Lagrangian function $L_3(x, \lambda; \varepsilon)$ was based on the barrier terms $1/b_i(x)$ which did not perform this control on the magnitude of the vector λ. The presence and the particular expresssion of the barrier term $1/p(x, \lambda)$ are most effective in order to weaken significantly the assumptions required to establish the exactness properties of the augmented Lagrangian function $L_a(x, \lambda; \varepsilon)$.

As we said before, in the transformation of a constrained minimization problem into the uncostrained minimization of an augmented Lagrangian function it is very important both from the theoretical point of view and from the algorithmic point of view to ensure that the level sets of the augmented Lagrangian function are compact. Roughly speaking, the barrier term $1/p(x, \lambda)$, unlike the previous barrier terms used in the functions $L_2(x, \lambda; \varepsilon)$ and $L_3(x, \lambda; \varepsilon)$, allows the level sets of the augmented Lagrangian function $L_a(x, \lambda; \varepsilon)$ to inherit not only the possible well-behaviour of the constraint functions, but also the property of the objective function. In order to see this fact more formally, we start by redefining the following level set:

$$\Omega(x_0, \lambda_0; \varepsilon) = \{(x, \lambda) \in \mathcal{S} \times \mathbb{R}^m : L_a(x, \lambda; \varepsilon) \le L_a(x_0, \lambda_0; \varepsilon)\}.$$

where $x_0 \in \mathcal{S}$, and $x_0 \in \mathcal{F}$ if a feasible point x_0 is known.

Then, under Assumption A1, it is possible to prove (see [7]) the following proposition.

Proposition 4.1 *For each $\varepsilon > 0$, the level set $\Omega(x_0, \lambda_0; \varepsilon)$ is compact, so that the function $L_a(x, \lambda; \varepsilon)$ admits a global minimum point on $\mathcal{S} \times \mathbb{R}^m$.*

The previous proposition states that the compactness of the level set $\Omega(x_0, \lambda_0; \varepsilon)$ can be ensured (see Assumption A1) both if the feasible set is compact and if the objective function is radially unbounded on $\bar{\mathcal{S}}$ and a feasible point is known. Furthermore, this compactness property holds for every value of the penalty parameter ε.

By using Proposition 4.1, it is possible to establish the exactness properties of the augmented Lagrangian function $L_a(x, \lambda; \varepsilon)$ with reference to the compact level set

$\Omega(x_0, \lambda_0; \varepsilon)$. This fact is very important from the algorithmic point of view, since any unconstrained descent method will produce a sequence $\{(x_k, \lambda_k)\}$ that will remain in the level set $\Omega(x_0, \lambda_0; \varepsilon)$ without being unbounded or being attracted by stationary points not belonging to the compact set $\Omega(x_0, \lambda_0; \varepsilon)$ where the exactness properties of $L_a(x, \lambda_k; \varepsilon)$ hold.

As regards the exactness properties of $L_a(x, \lambda; \varepsilon)$, Proposition 2.2 points out that the crucial property to prove is that, for some value of the penalty parameter ε, every stationary point of the augmented Lagrangian function is a KKT pair of Problem (NLP). In the case of the function $L_a(x, \lambda; \varepsilon)$, this property follows from that fact that, under Assumption A1-A3, it is possible to show that:

- as more the value of the penalty parameter ε is small, as more the stationary points of $L_a(x, \lambda; \varepsilon)$ belonging to the level set $\Omega(x_0, \lambda_0; \varepsilon)$ are close to the feasible set;
- for sufficiently small values of the penalty parameter ε, every stationary point of $L_a(x, \lambda; \varepsilon)$ belonging $\Omega(x_0, \lambda_0; \varepsilon)$ and sufficiently close to a feasible point is a KKT pair of Problem (NLP).

More formally the following two proposition are proved in [7]:

Proposition 4.2 *Let $\{\varepsilon_k\}$ be a sequence of positive numbers converging to zero and let $\{(x_k, \lambda_k)\}$ be a sequence such that $(x_k, \lambda_k) \in \Omega(x_0, \lambda_0; \varepsilon_k)$ and $\nabla L_a(x_k, \lambda_k; \varepsilon) = 0$. Then every limit point $\tilde{x}$ of the sequence $\{x_k\}$ is such that $\tilde{x} \in \mathcal{F}$.*

Proposition 4.3 *Let $\hat{x} \in \mathcal{F}$. Then there exist a number $\varepsilon(\hat{x}) > 0$ and a neighborhood $\mathcal{B}(\hat{x})$ of $\hat{x}$ such that, for all $\varepsilon \in (0, \varepsilon(\hat{x})]$, every stationary point $(\bar{x}, \bar{\lambda})$ satisfying $(\bar{x}, \bar{\lambda}) \in \Omega(x_0, \lambda_0; \varepsilon)$ and $\bar{x} \in \mathcal{B}(\hat{x})$, is a KKT pair of Problem (NPL).*

Now by using Proposition 2.2, Proposition 4.2 and Proposition 4.3 it is straigthforward to prove the following proposition, which states the main exactness properties of $L_a(x, \lambda; \varepsilon)$:

Proposition 4.4 *Let $L_a(x, \lambda; \varepsilon)$ the function defined by (4.3), then:*

(a) *for all $\varepsilon > 0$, every KKT pair $(\bar{x}, \bar{\lambda})$ of Problem (NLP) is a stationary point of $L_a(x, \lambda; \varepsilon)$;*

(b) *there exists an $\bar{\varepsilon} > 0$ such that for all $\varepsilon \in (0, \bar{\varepsilon}]$:*

- *if $(\bar{x}, \bar{\lambda}) \in \Omega(x_0, \lambda_0; \varepsilon)$ is a stationary point of $L_a(x, \lambda; \varepsilon)$, the pair $(\bar{x}, \bar{\lambda})$ is a KKT pair for (NLP);*
- *if $(\bar{x}, \bar{\lambda}) \in \Omega(x_0, \lambda_0; \varepsilon)$ is a local unconstrained minimum point of $L_a(x, \lambda; \varepsilon)$, $\bar{x}$ is a local minimum point of Problem (NLP) and $\bar{\lambda}$ is the corresponding KKT multiplier;*
- *if $\bar{x}$ is a global minimum point of Problem (NLP) and $\bar{\lambda}$ is the corresponding KKT multiplier, the pair $(\bar{x}, \bar{\lambda})$ is a global minimum point of $L_a(x, \lambda; \varepsilon)$ on $\mathcal{S} \times \mathbb{R}^m$, and conversely.*

The preceding proposition shows that augmented Lagrangian $L_a(x, \lambda; \varepsilon)$ presents all the properties to be considered an exact augmented Lagrangian function.

In addition to the properties described in Proposition 4.4, it is possible to state some converse results regarding the local minimum points. In particular, in [7], it is proved that, for sufficiently small values of the penalty parameter ε, every pair $(\bar{x}, \bar{\lambda})$, where $\bar{x}$ belongs to an isolated compact set of local minimum point of Problem (NLP) and $\bar{\lambda}$ is its KKT multiplier, is a local minimum point of the augmented Lagrangian function $L_a(x, \lambda; \varepsilon)$.

Finally we remark that also for the augmented Lagrangian $L_a(x, \lambda; \varepsilon)$ it is possible to define automatic adjustment procedures for the penalty parameter ε, which can be used in many uncostrained minimization algorithms to produce a sequence $\{(x_k, \lambda_k)\}$ globally convergent to a KKT point of the original constrained problem.

REFERENCES

[1] D.P. Bertsekas, *"Constrained Optimization and Lagrange Multipliers Methods"*. Academic Press, New York, 1982.

[2] G. Contaldi, G. Di Pillo, and S. Lucidi, "A continuously differentiable exact penalty function for nonlinear programming problems with unbounded feasible set". *Operations Research Letters*, 14:153–161, 1993.

[3] G. Di Pillo, "Exact penalty methods. In E. Spedicato, (ed.)". *Algorithms for Continuous Optimization: the State of the Art*, pages 1–45. Kluwer Ac. Press, Boston, 1994.

[4] G. Di Pillo and L. Grippo, "An augmented Lagrangian for inequality constraints in nonlinear programming problems". *Journal of optimization theory and applications*, 36:495–519, 1982.

[5] G. Di Pillo and L. Grippo, "On the exactness of a class of nondifferentiable penalty functions". *Journal of optimization theory and applications*, 57:399–410, 1988.

[6] G. Di Pillo and L. Grippo, "Exact penalty functions in constrained optimization". *SIAM Journal on Control and Optimization*, 27:1333–1360, 1989.

[7] G. Di Pillo and S. Lucidi, "An augmented Lagrangian function with improved exactness properties". Technical Report 18-95, DIS-Università di Roma *La Sapienza*, Roma, Italy, 1995.

[8] A.V. Fiacco and G.P. McCormick, *"Nonlinear programming: sequential unconstrained minimization techniques"*. John Wiley & Sons, New York, 1968.

[9] R. Fletcher, *"Practical Methods of Optimization"*. John Wiley & Sons, New York, 1987.

[10] L. Grippo and L. Palagi, "Constructive existence conditions for systems of nonlinear inequalities". Technical Report 13-94, DIS-Università di Roma *La Sapienza*, Roma, Italy, 1994.

[11] M.R. Hestenes, *"Optimization theory. The finite dimensional case"*. Wiley, New York, 1975.

[12] S. Lucidi. New results on a class of exact augmented Lagrangians. *Journal of Optimization Theory and Applicationss*, 58:259–282, 1988.

[13] S. Lucidi, "New results on a continuously differentiable exact penalty function". *SIAM Journal on Optimization*, 2:558–574, 1992.

Appendix. Proof of Proposition 2.1

Points (a), (b) and (c) follow directly from the definition of the function ϕ.

Point (d). If $\max\{y, -z/2\} = y$ the assumption that $y \leq 0$ yields that $-z \leq -z/2 \leq y$ and, hence, $y + z \geq 0$. Therefore we obtain:

$$z \max\{y, -\frac{z}{2}\} + \max\{y, -\frac{z}{2}\}^2 = y(z+y) \leq 0.$$

While if $\max\{y, -z/2\} = -z/2$ then

$$z \max\{y, -\frac{z}{2}\} + \max\{y, -\frac{z}{2}\}^2 = -\frac{z^2}{4} \leq 0.$$

Point (e). By assumption we have that:

$$\phi(y, z) = \max\{y, -\frac{z}{2}\}\left(z + \max\{y, -\frac{z}{2}\}\right) = 0, \tag{A1}$$

which implies that one of the following equalities holds:

$$\max\{y, -\frac{z}{2}\} = 0, \tag{A2}$$

$$\max\{y, -\frac{z}{2}\} = -z. \tag{A3}$$

If (A2) holds, it must be $y \leq 0$, $z \geq 0$ and $yz = 0$. Assume that (A3) holds. In this case first of all we have

$$-\frac{z}{2} \leq \max\{y, -\frac{z}{2}\} = -z$$

which implies that $z \leq 0$. Now if $z = 0$, recalling (A3), we return to the case where (A2) holds. If instead $z < 0$, then (A3) yields that $\max\{y, -z/2\} = y = -z > 0$ but this contradicts the assumption that $y \leq 0$. Therefore, in conclusion, we have that $y \leq 0$, $z \geq 0$ and $yz = 0$.

Point (f). By using the definition (2.1) of function $\phi(y, z)$ we have:

$$\begin{aligned} \phi(y, z) &= z(-\frac{z}{2} + \max\{0, y + \frac{z}{2}\}) + (-\frac{z}{2} + \max\{0, y + \frac{z}{2}\})^2 \\ &= \max\{0, y + \frac{z}{2}\}^2 - \frac{z^2}{4}, \end{aligned}$$

and, hence, the expression (2.4).

As regards (2.5), we observe that

$$(y + \frac{z}{2}) \min\{0, y + \frac{z}{2}\} = \min\{0, y + \frac{z}{2}\}^2;$$

then, by the preceding equality and the definition of $\phi(y, z)$, we obtain:

$$\begin{aligned} \phi(y, z) &= z(y + \max\{0, -y - \frac{z}{2}\}) + (y + \max\{0, -y - \frac{z}{2}\})^2 \\ &= z(y - \min\{0, y + \frac{z}{2}\}) + (y - \min\{0, y + \frac{z}{2}\})^2 \\ &= y^2 + yz - 2(y + \frac{z}{2}) \min\{0, y + \frac{z}{2}\} + \min\{0, y + \frac{z}{2}\}^2 \\ &= y^2 + yz - \min\{0, y + \frac{z}{2}\}^2. \end{aligned}$$

SPACE-TRANSFORMATION TECHNIQUE: THE STATE OF THE ART

Evtushenko Yuri G. and Vitali G. Zhadan

Computing Centre
Russian Academy of Sciences
40 Vavilov Str., 117967 Moscow GSP-1, Russia
e-mail: evtushenko@ccas.ru

Abstract. In this paper we give an overview of some current approaches to LP and NLP based on space transformation technique. A surjective space transformation is used to reduce the original problem with equality and inequality constraints to a problem involving only equality constraints. Continuous and discrete versions of the stable gradient projection method and the Newton method are used for treating the reduced problem. Upon the inverse transformation is applied to the original space, a class of numerical methods for solving optimization problems with equality and inequality constraints is obtained. The following algorithms are presented: primal barrier-projection methods, dual barrier-projection methods, primal barrier-Newton methods, dual barrier-Newton methods and primal-dual barrier-Newton methods. Using special space transformation, we obtained asymptotically stable interior-infeasible point algorithms. The main results about convergence rate analysis are given.

Key words. Linear programming, nonlinear programming, space transformation, surjective mapping, stable gradient projection method, Newton's method, interior point technique.

1. INTRODUCTION

In the past twenty five years quite general and effective space transformation technique has been developed for solving linear programming problems (LP) and nonlinear programming problems (NLP). The idea of this approach commonly occurs in the optimization literature, it came from nonlinear programming and projective geometry. Using a space transformation, the original problem with equality and inequality constraints is reduced to a problem with equality constraints only. Continuous and discrete versions of the stable gradient projection method and the Newton method are applied to the reduced problem. After an inverse transformation to the original space, a class of numerical methods for solving optimization problems with equality and inequality constraints was obtained. The proposed algorithms are based on the numerical integration of systems of ordinary differential equations. Vector fields described by these equations define flows leading to the optimal solution. As a result of the space transformation, the vector fields are changed and additional terms are introduced which serve as a barrier preventing the trajectories from leaving the feasible set. Therefore we call these methods "barrier-projection" and "barrier-Newton" methods. In our algorithms we use the multiplicative barrier functions which are continuous and equal to zero on a boundary. We do not introduce conventional singular barriers and this feature provides a high rate of convergence. In this paper we give a survey of principal results which were published in the last two decades [6]-[18].

In Sect.2 we describe a unified methodology for finding necessary and sufficient optimality conditions in extremal problems with functional equality constraints and nonfunctional inequality constraints. We show how numerous families of algorithms can be developed using various space transformations.

In Sect.3, choosing an exponential space transformation, we obtain the Dikin algorithm [5] from the family of primal barrier-projection methods. This algorithm, however, does not posses the local convergence properties and, as a result, it converges only if starting points belong to relative interior of the feasible set. Furthermore, the convergence rate of a discrete version of the algorithms proves to be weaker than a linear one.

In 1984 N.Karmarkar [25] proposed a special sophisticated step-length rule in the method similar to discrete version of the Dikin affine scaling algorithm. Based on this rule the polynomial complexity was theoretically attained. After this publication an impressive number of papers have been published devoted to further modifications and improvements of the Dikin and Karmarkar algorithms. Many authors were trying to modify and explain these algorithms as classical methods. Various methods were obtained along this direction and the first of our algorithms published in seventies [8], [6], [9] were reinvented. Later on in eighties-nineties, we developed more efficient versions of these methods which are discussed here. These asymptotically stable methods are such that a feasible set is an attractor of the vector fields. They preserve feasibility, but a starting point can be infeasible. They belong to a class of interior-infeasible point algorithms. In Sect.3 we show that among the barrier-projection algorithms there is a method which converges locally and exponentially fast to the optimal solution (in discrete case it converges locally with a linear rate).

In subsequent sections we apply our approach to primal and dual linear programming problems. For the sake of simplicity, we assume that these problems have unique nondegenerate solutions. In Sect.4 we use a nonconventional representation of the dual linear programming problem and we propose a set of algorithms. Upon simplifying the problem and choosing a particular exponential space transformation we arrive at the dual affine scaling method proposed by I.Adler, N.Karmarkar, M.Resende and G.Veiga [1].

In Sect.5 we describe the primal the and dual barrier-Newton methods. The primal-dual interior-infeasible Newton method is set forth in the final Sect.6. For the steepest descent approach an upper bound for the number of iterations is indicated.

2. BASIC APPROACH AND OUTLINE OF A SPACE TRANSFORMATION TECHNIQUE

Define the following NLP problem:

$$\text{minimize } f(x) \text{ subject to } x \in X = \{x \in \mathbb{R}^n : \ g(x) = 0_m, \ x \in P\}, \qquad (2.1)$$

where the functions f and g are twice continuously differentiable, $f(x)$ maps $\mathbb{R}^n$ onto $\mathbb{R}$ and $g(x)$ maps $\mathbb{R}^n$ onto $\mathbb{R}^m$, P is a convex set with nonempty interior, 0_s is the

s-dimensional null vector, 0_{sk} is the $s \times k$ rectangular null matrix.

Important particular cases of (2.1) are a linear programming problem given in standard form

$$\text{minimize } c^T x \text{ subject to } x \in X := \{x \in \mathbb{R}^n : \ b - Ax = 0_m, \ x \geq 0_n\} \tag{2.2}$$

and its dual problem

$$\text{maximize } b^T u \text{ subject to } u \in U := \{u \in \mathbb{R}^m : \ v = c - A^T u \geq 0_n\}, \tag{2.3}$$

where v is the $n-$vector of slack variables; $A \in \mathbb{R}^{mn}(m < n); c, x \in \mathbb{R}^n; b, u \in \mathbb{R}^m$ and $\text{rank}(A) = m$.

We define the relative interior set of X and the interior set of U as:

$$X_0 := \{x \in \mathbb{R}^n : Ax = b, \ x > 0_n\}, \quad U_0 := \{u \in \mathbb{R}^m : \ v = c - A^T u > 0_n\},$$

and assume that these sets are nonempty. We also introduce the following sets:

$$\mathbb{R}^n_+ := \{x \in \mathbb{R}^n : x \geq 0_n\}, \quad \mathbb{R}^n_{++} := \text{int}\, \mathbb{R}^n_+ = \{x \in \mathbb{R}^n : x > 0_n\},$$

$$V := \{v \in \mathbb{R}^n : \text{there exists } u \in \mathbb{R}^m \text{ such that } v = c - A^T u\},$$

$$V_U := \{v \in \mathbb{R}^n : \text{there exists } u \in U \text{ such that } v = c - A^T u\},$$

where int denotes the interior. The set V_U is the image of U under the mapping $v(u) = c - A^T u$. Therefore $V_U = V \bigcap \mathbb{R}^n_+$.

For convenience, assume that the primal feasible set is bounded and both primal and dual problems are nondegenerate, which together imply that optimal solutions x_*, u_* exist and are unique. We split the vectors x_* and $v_* = v(u_*)$ in basic and nonbasic components. Without any loss of generality we assume that

$$x_* = \begin{bmatrix} x_*^B \\ x_*^N \end{bmatrix}, \quad v_* = \begin{bmatrix} v_*^B \\ v_*^N \end{bmatrix}, \quad x_*^B > 0_m, \quad x_*^N = 0_d, \quad v_*^B = 0_m, \quad v_*^N > 0_d,$$

where $d = n - m$.

We denote the components of a vector by using superscripts and the iterate numbers by using subscripts; I_n denotes the identity matrix of the order n; $D(z)$ denotes the diagonal matrix whose entries are the components of the vector z. The dimensionality of $D(z)$ is determined by the dimensionality of z.

In order to construct a family of computational methods for solving the Problems (2.1)-(2.3) we use an approach based on space transformation. We introduce a new n-dimensional space with the coordinates $[y^1, \ldots, y^n]$ and make a differentiable transformation from this space to the original one: $x = \xi(y)$. This surjective transformation maps $\mathbb{R}^n$ onto P or int$P$, i.e. $P = \overline{\xi(\mathbb{R}^n)}$, where $\bar{B}$ is the topological closure of B. Let $\tilde{J}(y) = dx/dy$ be the Jacobian matrix of the transformation $x = \xi(y)$ with respect to y.

Consider the transformed minimization problem

$$\text{minimize } \tilde{f}(y) \ \text{ subject to } y \in Y, \tag{2.4}$$

where $\tilde{f}(y) = f(\xi(y)), \ Y = \{y \in \mathbb{R}^n : \tilde{g}(y) = g(\xi(y)) = 0_m\}$.

Define the Lagrangian functions $L(x, u)$, $\tilde{L}(y, u)$ associated with the Problems (2.1) and (2.4), respectively

$$L(x, u) = f(x) + u^T g(x), \quad \tilde{L}(y, u) = \tilde{f}(y) + u^T \tilde{g}(y).$$

Then the first-order necessary conditions for a local minimum for the Problem (2.4) in the transformed space are

$$\tilde{L}_y(y,u) = \tilde{f}_y(y) + \tilde{g}_y^T(y)u = 0_n, \quad \tilde{g}(y) = 0_m, \tag{2.5}$$

where $\tilde{f}_y = \tilde{J}^T f_x, \tilde{g}_y = g_x \tilde{J}$.

If $\tilde{J}$ is a nonsingular, then there exists an inverse transformation $y = \delta(x)$, so it is possible to return from the y-space to the x-space and we obtain in this way a matrix $J(x) = \tilde{J}(\delta(x))$ which is now a function of x. Using this substitution, we rewrite expressions (2.5) in terms of the variable x. They take the form

$$J^T(x)L_x(x,u) = 0_n, \quad g(x) = 0_m, \quad x \in P. \tag{2.6}$$

Some properties of the nonlinear systems, which are obtained after space transformations, were investigated in [15].

Let $K(x \mid P)$ and $K^*(x \mid P)$, respectively, denote the cone of feasible directions at the point x relative to the set P and its dual:

$$K(x \mid P) = \{z \in \mathbb{R}^n : \quad \exists \lambda(z) > 0 \text{ such that } x + \lambda z \in P, \quad 0 < \lambda \le \lambda(z)\},$$

$$K^*(x \mid P) = \{z \in \mathbb{R}^n : z^T y \ge 0 \quad \forall y \in K(x \mid P)\}.$$

Let $S(x \mid P)$ be a linear hull of the cone $K^*(x \mid P)$. The set of all vectors orthogonal to $S(x \mid P)$ is called orthogonal compliment of $S(x \mid P)$ and is denoted by $S^\perp(x \mid P)$.

We will impose the following condition on the space transformation $\xi(y)$.

Condition A_1. *At each point $x \in P$ the matrix $J(x)$ is defined and the nullspace of $J^T(x)$ coincides with the set $S(x \mid P)$.*

In particular, it follows from this condition that at all interior points $x \in \text{int}P$ the matrix $J(x)$ is non-degenerate, becoming singular only on the boundary of the set P.

Definition 2.1. *Any pair $[x,u]$ is a weak KKT pair for the Problem (2.1) if it satisfies the conditions (2.6).*

It follows from this definition and condition A_1 that $L_x(x_*,u_*) \in S(x_* \mid P)$ at any weak KKT pair. Define the Gram matrix $G(x) = J(x)J^T(x)$. Since the nullspaces of $G(x)$ and $J^T(x)$ coincide, conditions (2.6) for the pair $[x_*,u_*]$ can be rewritten in the form

$$G(x)L_x(x,u) = 0_n, \quad g(x) = 0_m, \quad x \in P. \tag{2.7}$$

Definition 2.2. *A weak KKT pair $[x,u]$ is a KKT pair for the Problem (2.1) if $L_x(x,u) \in K^*(x \mid P)$.*

Let $\text{ri}B$ denote a relative interior of the set B.

Definition 2.3. *A KKT pair $[x,u]$, is a strong KKT pair if $L_x(x_*,u_*) \in \text{ri}K^*(x_* \mid P)$.*

Definition 2.4. *The constraint qualification (CQ) for the Problem (2.1) holds at a point $x \in P$ if all vectors $g_x^i(x)$, $1 \le i \le m$, and any nonzero vector $p \in S(x \mid P)$ are linearly independent. We say that x is a regular point for the Problem (2.1) if the CQ holds at x.*

Theorem 2.1. *Let a regular point x_* be a solution of the Problem (2.1). Then there exists a vector $u_* \in \mathbb{R}^m$ such that $[x_*,u_*]$ forms a weak KKT pair for the Problem (2.1).*

The space transformation described above can be used to derive the second-order sufficient conditions for a point x_* to be an isolated minimum in the Problem (2.1). Introduce a nullspace $N(x) = \{z \in \mathbb{R}_n : g_x(x)J(x)z = 0_m\}$.

Theorem 2.2. *Assume that f and g are twice-differentiable functions and the space transformation $\xi(y)$ satisfies A_1. Sufficient conditions for a point $x_* \in P$ to be an isolated local minimum of the Problem (2.1) are that there exists a strong KKT pair $[x_*, u_*]$ such that $z^\top J^\top(x_*)L_{xx}(x_*, u_*)J(x_*)z > 0$ for every $z \in N(x_*)$ such that $\|J(x_*)z\| \neq 0$.*

If $P = \mathbb{R}^n$ (in other words, the condition $x \in P$ is missing), we can take the trivial space transformation $x = y$. In this case we have

$$J(x) = I_n, \quad N(x) = \{z \in \mathbb{R}_n : g_x(x)z = 0_m\}.$$

$$K(x \mid \mathbb{R}^n) = \mathbb{R}^n, \quad S(x \mid \mathbb{R}^n) = K^*(x \mid \mathbb{R}^n) = \mathrm{ri}K^*(x \mid \mathbb{R}^n) = 0_n.$$

The Theorem 2.2. reduces to the well-known second-order sufficient conditions for an isolated local minimum (see, for example, [20]).

Suppose that the function $\xi(y)$ is such that the matrix $G(x)$ is continuously differentiable. Let $p \in \mathbb{R}^n$ and $G_x(x;p)$ denote a square matrix of order n whose (i,j)-element equals to

$$G_x^{ij}(x;p) = \sum_{k=1}^{n} \frac{\partial G^{ik}(x)}{\partial x^j} p^k.$$

We impose two additional conditions on the space transformation $\xi(y)$:

Condition A_2. *At each point $x \in P$ for any vector $p \in riK^*(x \mid P)$ the matrix $G_x(x;p)$ is symmetric and its nullspace coincides with $S^\perp(x \mid P)$.*

Condition A_3. *If $x \in P$ then $z^T G_x(x;p)z > 0$ for any non-zero vector $z \in S(x \mid P)$ and any vector $p \in riK^*(x \mid P)$.*

Let us consider an important particular case of the Problem (2.1) where $P = \mathbb{R}^n_+$:

$$\text{minimize } f(x) \text{ subject to } x \in X = \{x \in \mathbb{R}^n : \ g(x) = 0_m, x \geq 0_n\}, \tag{2.8}$$

It is convenient for this set to use a component-wise space transformation:

$$x^i = \xi^i(y^i), \quad 1 \leq i \leq n. \tag{2.9}$$

For such a transformation the inverse transformation $y - \delta(x)$ is also component-wise type $y^i = \delta^i(x^i), 1 \leq i \leq n$, and the corresponding matrices $J(x)$ and $G(x)$ are diagonal:

$$J(x) = D(\gamma(x)), \quad \gamma^T(x) = [\gamma^1(x^1), \dots, \gamma^n(x^n)], \quad \gamma^i(x^i) = \dot{\xi}(\delta^i(x^i)),$$

$$G(x) = D(\theta(x)), \quad \theta^T(x) = [\theta^1(x^1), \dots, \theta^n(x^n)], \quad \theta^i(x^i) = [\gamma^i(x^i)]^2, \quad 1 \leq i \leq n.$$

Let $\sigma(x) = \{i : x^i = 0\}$ be a set of active indices at the point $x \in \mathbb{R}^n_+$. In this case

$$K^*(x \mid \mathbb{R}^n_+) = \{z \in \mathbb{R}^n_+ : \text{if } i \notin \sigma(x), \text{ then } z^i = 0, 1 \leq i \leq n\},$$

$$S(x \mid \mathbb{R}^n_+) = \{z \in \mathbb{R}^n : \text{if } i \notin \sigma(x), \text{ then } z^i = 0, 1 \leq i \leq n\},$$

condition A_1 reduces to the following

Condition C_1. *The vector function $\gamma(x)$ is defined at each point $x \in \mathbb{R}^n_+$ and $\gamma^i(x^i) = 0$ if and only if $i \in \sigma(x)$.*

In order to insure conditions A_2 and A_3 we impose the

Condition C$_2$. *The vector function $\theta(x)$ is differentiable in some neighbourhood of $\mathbb{R}^n_+$ and $\dot{\theta}^i(0) > 0$, $1 \le i \le n$.*

As a rule, we perform the following quadratic and exponential transformations

$$x^i = \xi^i(y^i) = \frac{1}{4}(y^i)^2, \quad J(x) = D^{1/2}(x), \quad G(x) = D(x), \tag{2.10}$$

$$x^i = \xi^i(y^i) = e^{y^i}, \quad J(x) = D(x), \quad G(x) = D^2(x). \tag{2.11}$$

In these two cases the Jacobian matrix is singular on the boundary of the set $P = \mathbb{R}^n_+$. These transformations satisfy C_1. The condition C_2 holds only for the quadratic transformation (2.10).

Let e^i denote the n-th order unit vector whose i-th component is equal to one. The CQ for the Problem (2.8) holds at a point x, if all the vectors $g^i(x), 1 \le i \le m$, and all e^j, such that $j \in \sigma(x)$, are linearly independent. The cone $N(x)$ takes the form $N(x) = \{z \in \mathbb{R}^n : g_x(x)D(\gamma(x))z = 0_m\}$.

The strict complementary condition (SCC) holds at a pair $[x, u]$, if $L_{x^i}(x, u) > 0$ for all $i \in \sigma(x)$. From the Theorem 2.2. the following second-order sufficient optimality conditions is obtained.

Theorem 2.3. *Sufficient conditions for a point $x_* \in X$ to be an isolated local minimum of the Problem (2.8) are that there exists a Lagrange multiplier vector u_* such that*

$$D(\gamma(x_*))L_x(x_*, u_*) = 0_n,$$

that the SCC holds at $[x_*, u_*]$ and that

$$z^T D(\gamma(x_*))L_{xx}(x_*, u_*)D(\gamma(x_*))z > 0$$

for all $z \in N(x_*)$, satisfying $\|D(\gamma(x_*))z\| \neq 0$.

Now we will construct numerical methods for solving the Problem (2.1). We use the stable version of the gradient projection method for solving the Problem (2.4). The numerical method is stated as an initial-value problem involving the following system of ordinary differential equations

$$\frac{dy}{dt} = -\tilde{L}_y(y, \tilde{u}(y)), \quad \tilde{L}_y(y, \tilde{u}) = \tilde{f}_y(y) + \tilde{g}_y^\top(y)\tilde{u}, \quad y(0, y_0) = y_0 \in \mathbb{R}^n. \tag{2.12}$$

The function $\tilde{u}(y)$ is chosen to satisfy the following condition

$$\frac{d\tilde{g}}{dt} = \tilde{g}_y\frac{dy}{dt} = -\tau\tilde{g}(y), \quad \tau > 0. \tag{2.13}$$

From this condition we obtain the system of linear algebraic equations

$$\tilde{g}_y(y)\tilde{g}_y^\top(y)\tilde{u}(y) + \tilde{g}_y(y)\tilde{f}_y(y) = \tau\tilde{g}(y),$$

where $\tilde{f}_y = \tilde{J}^\top f_x$, $\tilde{g}_y = g_x\tilde{J}$. By differentiating $\xi(y)$ with respect to y and taking into account (2.12) and (2.13), we have

$$\frac{dx}{dt} = \frac{d\xi}{dy}\frac{dy}{dt} = J(x)\frac{dy}{dt} = -G(x)L_x(x, u(x)), \quad x(0, x_0) = x_0 \in P, \tag{2.14}$$

$$\Gamma(x)u(x) + g_x(x)G(x)f_x(x) = \tau g(x), \tag{2.15}$$

where $\Gamma(x) = g_x(x)G(x)g_x^\top(x)$.

Lemma 2.4. *Let the space transformation $\xi(y)$ satisfy A_1, and let the CQ for the Problem (2.1) hold at a point $x \in P$. Then $\Gamma(x)$ is invertible and positive definite matrix.*

Lemma 2.5. *Let the space transformation $\xi(y)$ satisfy A_1. Then the regular point x_* is an equilibrium state of system (2.14) if, and only if, the pair $[x_*, u_*]$, where $u_* = u(x_*)$, is a weak KKT pair for the Problem (2.1).*

Hence, corresponding to any regular point x_* we can define a corresponding Lagrange multiplier $u(x_*)$ by solving linear algebraic equations (2.15). If a local solution of the original problem (2.1) occurs at a regular point $x_* \in P$, then $[x_*, u(x_*)]$ forms a weak KKT pair for the Problem (2.1) and x_* is an equilibrium state of (2.14).

Let W be a $m \times n$ rectangular matrix whose rank is m. We introduce the pseudoinverse matrix $W^+ = W^\top(WW^\top)^{-1}$ and the orthogonal projector $\pi(W) = I_n - W^+W$. If at a regular point x we define $u(x)$ and substitute it into the right-hand side of (2.14), then (2.14) can be rewritten in the following projective form

$$\frac{dx}{dt} = -J(x)\left\{\pi[g_x(x)J(x)]J^\top(x)f_x(x) + \tau[g_x(x)J(x)]^+g(x)\right\}. \tag{2.16}$$

Let $x(t, x_0)$ denote the solution of the Cauchy Problem (2.14) with an initial condition $x_0 = x(0, x_0)$. In what follows, we assume that the initial-value problem under consideration is always uniquely solvable. A trajectory $x(t, x_0)$ can be continued as long as points on it are regular. By differentiating $f(x(t, x_0))$ with respect to t we arrive at

$$\frac{df}{dt} = - \parallel J^\top(x)L_x(x, u(x)) \parallel^2 + \tau u^\top(x)g(x). \tag{2.17}$$

Hence the objective function $f(x(t, x_0))$ monotonically decreases either on the feasible set X or when the trajectory is close to X, i.e. $\parallel g(x(t, x_0)) \parallel$ is sufficiently small.

The system of ordinary differential equations (2.14), where $u(x)$ is given by (2.15), has the first integral

$$g(x(t, x_0)) = g(x_0)e^{-\tau t}.$$

This means that if $\tau > 0$, (2.14) has a remarkable property: all its trajectories approach the manifold $g(x) = 0_m$, as t tends to infinity, and this manifold is an asymptotically stable attractor (see [7], [10], [29]). Therefore we can call the method (2.16) "the stabe version of the barrier-projection method". If $x_0 \in X$, then the trajectory $x(t, x_0)$ of (2.14) remains on this manifold because $g(x(t, x_0)) \equiv 0_m$ for all $t \geq 0$ and the trajectories of (2.14) coincide with the trajectories of the following system

$$\frac{dx}{dt} = -J(x)\pi[g_x(x)J(x)]J^T(x)f_x(x), \tag{2.18}$$

which can be obtained from (2.16) if we put $\tau = 0$. But in contrast to (2.16) this system is neutrally stable with

respect to equality constraints. It means that if $g(x_0) = c$ then $g(x(t, x_0)) \equiv c$ for all $t \geq 0$ and we have to introduce a correction procedure to remove the violation of constraints. If the condition $x \in P$ is missing, then we can put $\xi(y) = y$, hence $J(x) = I_n$ and (2.18) coincides with well-known gradient projection method [26], which is also neutrally stable.

Note that, according to the Condition A_1, the subspace $S^\perp(x \mid P)$ coincides with the space of the columns of the matrix $J(x)$, and, since the vector $G(x)L_x(x, u(x))$ belongs to this space, the velocity vector $\dot{x}$ always lies in $S^\perp(x \mid P)$. Thus, if x is a boundary point of P, then the vector $\dot{x}$ belongs to the eigen subspace of the space $\mathbb{R}^n$, which coincides with the space $M(x) - x$, where $M(x)$ is the intersection of the support planes of the set P at the point x. If the cone $K^*(x \mid P)$ has a non-empty interior, then this subspace degenerates to a single point (the origin of coordinates).

Let us show that trajectories $x(t, x_0)$ of (2.14) cannot cross the boundary of P. To the contrary, suppose this is not true and a trajectory $x(t, x_0)$ starting inside P leaves P at $t_1 > 0$. Then there exists a vector $p \in K^*(x(t_1, x_0) \mid P)$ such that $p^T \dot{x}(t_1, x_0) < 0$. But the vector $\dot{x}(t, x_0)$ belongs to the orthogonal complement of the subspace $S(x \mid P)$ and the vector p belongs to $S(x \mid P)$. Hence $p^T \dot{x}(t_1, x_0) = 0$ and, consequently, $x(t, x_0) \in P$ for all $t \geq 0$. Thus the matrix $G(x)$ plays the role of a "barrier" preventing $x(t, x_0)$ from intersecting the boundary of P. The trajectory $x(t, x_0)$ can approach the boundary points only in the limit as $t \to +\infty$. If the initial point x_0 lies on the boundary, then the entire trajectory of system (2.14) belongs to the boundary of P.

By applying the Euler method for solving system (2.16), we obtain

$$x_{k+1} = x_k - \alpha_k J(x_k) \left\{ \pi[g_x(x_k)J(x_k)] J^\top(x_k) f_x(x_k) + \tau [g_x(x_k)J(x_k)]^+ g(x_k) \right\}, \quad (2.19)$$

where a stepsize $\alpha_k > 0$.

Each equilibrium point x_* of the system (2.16) is a fixed point of iterations (2.19), i.e. $x_k = x_*$ implies $x_{k+1} = x_*$, and if iterates (2.19) converge to a regular point x_*, then the pair $[x_*, u(x_*)]$ satisfies conditions (2.7).

Theorem 2.6. *Let $[x_*, u_*]$ be a weak KKT pair of the Problem (2.1), where the CQ and the second-order sufficiency conditions of the Theorem 2.2. hold. Let the space transformation $\xi(y)$ satisfy the Conditions A_1-A_3 and $\tau > 0$. Then x_* is an asymptotically stable equilibrium state of the system (2.16); there exists a positive number α_* such that for any fixed $0 < \alpha_k < \alpha_*$ the sequence $\{x_k\}$, generated by (2.19), converges locally with a linear rate to x_* while the corresponding sequence $\{u_k\}$, where $u_k = u(x_k)$, converges to u_*.*

The proof of this theorem is given in [15]. It is based on the Lyapunov linearization principle. The conditions for asymptotic stability to be valid are expressed in terms of eigenvalues determined by a matrix arisen from the right-hand side of (2.16) linearized about the equilibrium point x_*. The asymptotic stability of the point x_* implies the local, exponentially fast convergence of a trajectory $x(t, x_0)$ to the optimal solution x_*. The corresponding statement about the linear convergence of discrete versions follows from Theorem 2.3.7 (see [7]).

The Theorem 2.6., being applied to the Problem (2.8), gives the the following statement:

Theorem 2.7. *Let $[x_*, u_*]$ be a weak KKT pair of the Problem (2.8), where the CQ and the second-order sufficiency conditions of the Theorem 2.3. hold. Let the componentwise space transformation (2.9) satisfy the Conditions C_1, C_2 and $\tau > 0$. Then x_* is an asymptotically stable equilibrium state of system (2.16); there exists a positive number α_* such that for any fixed $0 < \alpha_k < \alpha_*$ the sequence $\{x_k\}$, generated by (2.19), converges locally with a linear rate to x_* while the corresponding sequence $\{u_k\}$ converges to u_*.*

This theorem was proved in [10], [12].

The preceding results and algorithms admit straightforward extensions for problems involving general functional inequality constraints by using space dilation. Consider a problem

$$\text{minimize } f(x), \quad \text{subject to } x \in X = \{x \in \mathbb{R}^n : g(x) = 0_m, h(x) \leq 0_c\}, \tag{2.20}$$

where $h(x)$ maps $\mathbb{R}^n$ into $\mathbb{R}^c$.

Space transformation approach can be used in this case by extension of the space and by converting the inequality constraints to equalities. We introduce an additional variable $p \in \mathbb{R}^c$, define $q = m + c$, combine primal, dual variables and all constraints:

$$z = \begin{bmatrix} x \\ p \end{bmatrix} \in \mathbb{R}^{n+c}, \quad w = \begin{bmatrix} u \\ v \end{bmatrix} \in \mathbb{R}^q, \quad \Phi(z) = \begin{bmatrix} g(x) \\ h(x) + p \end{bmatrix}.$$

Then the original Problem (2.20) is transformed into the equivalent problem

$$\text{minimize } f(x), \quad \text{subject to } z \in Z = \{z \in \mathbb{R}^{n+c} : \Phi(z) = 0_q, p \in \mathbb{R}^c_+\}. \tag{2.21}$$

Here P is a positive orthant in $\mathbb{R}^c$. This problem is similar to (2.8). In order to take into account the constraint $p \geq 0_c$ we introduce a surjective componentwise differentiable mapping $\xi : \mathbb{R}^c \to \mathbb{R}^c_+$ and make the space transformation $p = \xi(y)$, where $y \in \mathbb{R}^c$, $\overline{\xi(\mathbb{R}^c)} = \mathbb{R}^c_+$. Let $\xi_y(y)$ be the square $c \times c$ Jacobian matrix of the mapping $\xi(y)$ with respect to y. We assume that it is possible to define the inverse transformation $y = \psi(p)$ and hence we obtain the $c \times c$ Jacobian and Gram diagonal matrices:

$$J(p) = \xi_y(y) \mid_{y=\psi(p)} = D(\gamma(p)), \quad G(p) = J(p)J^\top(p) = D(\theta(p)).$$

Combining variables and constraints for the reduced problem, let us define

$$\hat{z} = \begin{bmatrix} x \\ y \end{bmatrix} \in \mathbb{R}^{n+c}, \quad \hat{\Phi}(\hat{z}) = \begin{bmatrix} g(x) \\ h(x) + \xi(y) \end{bmatrix}, \quad \hat{\Phi}_{\hat{z}} = \begin{bmatrix} g_x & 0_{mc} \\ h_x & \xi_y \end{bmatrix}.$$

The Problem (2.21) can be formulated as follows:

$$\text{minimize } f(x), \quad \text{subject to } \hat{z} \in \hat{Z} = \{\hat{z} \in \mathbb{R}^{n+c} : \hat{\Phi}(\hat{z}) = 0_q\}. \tag{2.22}$$

In the last Problem we have only equality constraints, therefore we can use the numerical method described above. After an inverse transformation to the space of x and p we obtain from (2.14), (2.15)

$$\frac{dz}{dt} = -\tilde{G}(p)L_z(z, w(z)). \tag{2.23}$$

Here

$$L(z,w) = f(x) + w^\top \Phi(z), \quad L_z(z,w) = f_z(z) + \Phi_z^\top(z)w,$$
$$\Phi_z(z)\tilde{G}(p)L_z(z, w(z)) = \tau\Phi(z), \tag{2.24}$$
$$\tilde{G}(p) = \begin{bmatrix} I_n & 0_{nc} \\ 0_{cn} & G(p) \end{bmatrix}, \quad \Phi_z = \left[\Phi_x \middle| \begin{matrix} 0_{mc} \\ I_c \end{matrix}\right], \quad \Phi_x = \begin{bmatrix} g_x \\ h_x \end{bmatrix}, \quad f_z = \begin{bmatrix} f_x \\ 0_c \end{bmatrix}.$$

System (2.23) can be rewritten in the more detailed form

$$\frac{dx}{dt} = -L_x(x, w(z)), \quad \frac{dp}{dt} = -G(p)v(z), \tag{2.25}$$

where the function $w^\top(z) = [u^\top(z), v^\top(z)]$ is found from the following linear system of q equations

$$\Gamma(z)w(z) + \Phi_x(x)f_x(x) = \tau\Phi(z), \quad \Gamma(z) = \Phi_x(x)\Phi_x^\top(x) + \begin{bmatrix} 0_{mm} & 0_{mc} \\ 0_{cm} & G(p) \end{bmatrix}.$$

Condition (2.24) can be written as

$$\frac{dg(x)}{dt} = -\tau g(x), \quad \frac{d(h(x)+p)}{dt} = -\tau(h(x)+p).$$

Therefore system (2.25) has two first integrals:

$$g(x(t,z_0)) = g(x_0)e^{-\tau t},$$

$$h(x(t,z_0)) + p(t,z_0) = (h(x_0)+p_0)e^{-\tau t}, \quad z_0^\top = [x_0^\top, p_0^\top]. \tag{2.26}$$

Similarly to (2.17) we obtain

$$\frac{df}{dt} = -\parallel L_x \parallel^2 - \parallel J^\top(p)v \parallel^2 + \tau[u^\top g + v^\top(h+p)].$$

Consider the simplified version of method (2.25). Suppose that along the trajectories of system (2.25) the following condition holds

$$h(x(t,z_0)) + p(t,z_0) \equiv 0_c.$$

¿From this equality we can define p as a function of h. We exclude from system (2.25) the additional vector p and integrate the system which does not employ this vector:

$$\frac{dx}{dt} = -L_x(x, w(x)), \tag{2.27}$$

where

$$\Gamma(x)w(x) + \Phi_x(x)f_x(x) = \tau \begin{bmatrix} g(x) \\ 0_c \end{bmatrix}, \tag{2.28}$$

$$\Gamma(x) = \Phi_x(x)\Phi_x^\top(x) + \begin{bmatrix} 0_{mm} & 0_{mc} \\ 0_{cm} & G(-h(x)) \end{bmatrix}.$$

Along the trajectories of (2.27) we have

$$\frac{dg}{dt} = -\tau g(x), \quad \frac{dh}{dt} = -G(-h(x))v(x), \tag{2.29}$$

$$\frac{df}{dt} = -\parallel L_x(x,w(x)) \parallel^2 - \parallel J^\top(-h(x))v(x) \parallel^2 + \tau u^\top(x)g(x).$$

Let us show that the solution $x(t,x_0)$ does not leave the set X for any $t > 0$, if $x_0 \in X$. Suppose to the contrary that this is not true and $h^i(x(t,x_0)) > 0$ for some $t > 0$. Then there is an earlier instant $0 < t_1 < t$ such that $h^i(x(t_1,x_0)) = 0$ and $\dot{h}^i(x(t_1,x_0)) > 0$. This contradicts (2.29) since $\theta^i(0) = 0$. Hence $x(t,x_0) \in X$ for all $t \geq 0$. Thus the matrix $G(-h(x))$ plays the role of a "barrier" preventing $x(t,x_0)$ from intersecting the hypersurface $h^i(x) = 0$.

The method (2.27) is closely related to (2.14). Let us consider the Problem (2.8). We have the two alternatives ensuing from (2.14) or (2.27). The main body of computational work required when using any numerical integration method is to evaluate the right-hand sides of equations for various values of x. This could be done by solving the linear system (2.15) of m equations or the system (2.28) of $m+n$ equations, respectively. One might expect that introducing a slack variable p increases the computational work considerably. However, with allowance made for the the simple structure of (2.28), we can reduce the computational time by using the Frobenius formula for an inverse matrix. Upon some transformations we find that (2.27), (2.28) can be written in the form of (2.14), (2.15) respectively, if we put

$$G(x) = D(\mu(x)), \quad \mu^i(x^i) = \theta^i(x^i)/[1+\theta^i(x^i)], \quad 1 \leq i \leq n.$$

Therefore the performances of both methods which seem at a glance to be unrelated are in fact very similar.

3. PRIMAL BARRIER-PROJECTION METHODS FOR SOLVING LP PROBLEMS

Applying the stable barrier-projection method (2.14) for solving the Problem (2.2), we obtain the following continuous and discrete versions

$$\frac{dx}{dt} = -G(x)[c - A^T u(x)], \quad x(0, x_0) = x_0, \tag{3.1}$$

$$x_{k+1} = x_k - \alpha_k G(x_k)[c - A^T u_k], \quad u_k = u(x_k), \tag{3.2}$$

where the function $u(x)$ is found from the linear equation (2.15) which can be rewritten as

$$AG(x)A^T u(x) - AG(x)c = \tau(b - Ax). \tag{3.3}$$

By differentiating the objective function with respect to t we obtain

$$c^T \frac{dx}{dt} = - \| J(x)(c - A^T u(x)) \|^2 + \tau u^T(x)(b - Ax).$$

The system of ordinary differential equations (3.1) has the first integral

$$Ax(t, x_0) = b + (Ax_0 - b)e^{-\tau t}. \tag{3.4}$$

Under the nondegeneracy assumption all feasible points are regular, and each weak KKT pair $[x, u(x)]$ is such that x is an equilibrium state of system (3.1). The pair $[x, u(x)]$ is a strong KKT pair if and only if $x = x_*$.

Denote

$$\alpha_* = 2/\mu^*, \quad \mu^* = \max\left[\tau, \max_{m+1 \le i \le n} \dot{\theta}^i(x_*^i)v_*^i\right], \quad \mu_* = \min\left[\tau, \min_{m+1 \le i \le n} \dot{\theta}^i(x_*^i)v_*^i\right].$$

Here μ^* and μ_* are, respectively, the largest and smallest eigenvalues of the first approximation matrix of the right-hand side of (3.1) at the optimal solution x_*.

Theorem 3.1. *Let x_*, u_* be unique nondegenerate solutions of the Problems (2.2) and (2.3) respectively. Assume that the componentwise space transformation $\xi(y)$ satisfy the Conditions C_1, C_2 and $\tau > 0$. Then the following statements are true:*

1. *The point x_* is an asymptotically stable equilibrium state of system (3.1).*
2. *The solutions $x(t, x_0)$ of system (3.1) converge locally exponentially fast to the optimal point x_*.*
3. *For any fixed $0 < \alpha_k < \alpha_*$ the sequence $\{x_k\}$, generated by (3.2), converges locally with a linear rate to x_* while the corresponding sequence $\{u_k\}$ converges to u_*.*
4. *All extreme feasble points of X are unstable equilibrium points of (3.1) and (3.2), except the optimal solution x_*.*

The exponential rate of convergence of the solution $x(t, x_0)$ of (3.1) to the equilibrium point x_* implies that there exist a neighbourhood $\Delta(x_*)$ of x_* and a constant $C > 0$ such that

$$\| x(t, x_0) - x_* \| \le C \| x_0 - x_* \| e^{-\mu_* t}$$

for all $x_0 \in \Delta(x_*)$, $t > 0$.

In (3.1) we can take a starting vector x_0 such that $\| Ax_0 - b \| \neq 0$. Moreover, if the Condition C_2 holds then the components of x_0 corresponding to nonbasic components of the vector x_* may be negative. In this case the entire trajectory $x(t, x_0)$ is infeasible. Nevertheless, owing to the local convergence property, the trajectory $x(t, x_0)$ converges to x_*, if $\| x_0 - x_* \|$ is sufficiently small. It follows from (3.4) that, if initially $Ax_0 = b$, then $Ax(t, x_0) \equiv b$ for all $t \geq 0$. Still further if $x_0 \geq 0_n$, then the trajectory $x(t, x_0)$ of (3.1) remains in the feasible set X and the objective function monotonically decreases along $x(t, x_0)$. (3.1) preserves feasibility, hence it is an interior point method. If C_2 holds, then this method allows to start the computation from an infeasible point. Therefore we can call it an interior-infeasible method.

If we use the quadratic and exponential space transformations, then (3.1) and (3.2) can be cast in the form

$$\frac{dx}{dt} = D^\beta(x)(A^T u(x) - c), \quad AD^\beta(x)A^T u(x) = AD^\beta(x)c + \tau(b - Ax), \tag{3.5}$$

$$x_{k+1} = x_k + \alpha_k D^\beta(x_k)[A^T u_k - c], \tag{3.6}$$

where $\beta = 1$ for (2.10), and $\beta = 2$ for (2.11).

The continuous version of (3.5) had been studied by Smirnov [27]. He proved that the following estimate

$$\| x^N(t, x_0) \| = \begin{cases} O\left(e^{-\mu_* t}\right), & \text{if } \beta = 1, \\ O\left(t^{1/(1-\beta)}\right), & \text{if } \beta > 1. \end{cases}$$

holds for the vector of nonbasic components. Hence the trajectories of (3.5) derived with the help of the quadratic space transformation locally converge faster than the trajectories of (3.5), derived with the help of the exponential transformation. Therefore in our codes we have mainly used the quadratic space transformation.

It was shown that if β is odd, then x_* is an asymptotically stable point of (3.5) and the local convergence takes place. If β is even, then the optimal point x_* is unstable and trajectories converge to x_* provided that $x_0 \in \Delta(x_*) \bigcap X_0$.

If we set $\beta = 2, \tau = 0$, then (3.5) leads to the Dikin affine scaling method

$$\frac{dx}{dt} = D^2(x)(A^T u(x) - c), \quad AD^2(x)A^T u(x) = AD^2(x)c. \tag{3.7}$$

Discrete and continuous versions of (3.7) were investigated in numerous papers (see, for example, [3], [4], [5], [23], [24], [30], [31]). It is worthy of note that the optimal solution x_* proves to be unstable equilibrium point and therefore (3.7) does not possess the local convergence property. In the latter case a starting point x_0 should meet a constraint $x_0 \in \Delta(x_*) \bigcap X_0$ for the convergence to take place.

If we set $\beta = 1, \tau = 0$ then (3.5) yields

$$\frac{dx}{dt} = D(x)(A^T u(x) - c), \quad AD(x)A^T u(x) = AD(x)c. \tag{3.8}$$

Since the quadratic space transformation is used here, the Condition C_2 holds and we have the exponential rate of convergence (in discrete case a linear rate). However owing to the simplification $\tau = 0$ a starting point x_0 must be such that $x_0 \in \Delta(x_*), Ax_0 = b$. The convergence property of the method (3.8) is more attractive, than (3.7). Therefore the method (3.8) is considered in the West as essential improvement of Dikin-Karmarkar method. The book [22] claims that (3.8) "is identical with the gradient flow for linear programming proposed and studied extensively by Faybusovich (1991). Independently this flow was also studied by Herzel, Recchioni and Zirilli (1991)... Starting from the seminal work of Khachian and Karmarkar, there has been a lot of progress in developing interior point algorithms for linear programming and nonlinear programming, due to Bayer, Lagarias and Faybusovich".

In point of fact, (3.8) has been proposed and analyzed as far back as 1977 in [9], and later on a comprechensive exposition of this method was described in numerous our papers and in the book [7] appeared in English in 1985 and three years before in Russian.

Still further, the nonlocal convergence analysis of (3.8) was carried out in [14]. We assumed that a condition $\sum_{i=1}^{n} x^i = 1$ was introduced among the other equality constraints. In this case the Lyapunov function

$$V(x) = \sum_{i=1}^{m} x_*^i [\ln x_*^i - \ln x^i] \tag{3.9}$$

decreases along the trajectories of the system (3.8) insofar as

$$\frac{dV}{dt} = c^T(x_* - x) \leq 0 \tag{3.10}$$

everywhere on the feasible set X. Using (3.10) we evaluated a number of iterations required for finding ϵ-solution of (2.2), using stepsizes α_k determined by the steepest descent approach.

Numerous papers have been published in the West on the interior point techniques [3], [4], [19], [21], [23], [24], [30]. Some remarks deserve to be made concerning the main differences between the methodology adopted in these papers and our original approach:

1. Along with the LP problems we considered the NLP problems as well.

2. We developed asymptotically stable interior-infeasible-point algorithms. Our analysis was not confined to the interior point technique. For this reason the current points are in general allowed to be infeasible, however if the initial or current points are feasible, then the corresponding trajectory remains in the feasible set.

3. In all proposed methods multiplicative barriers were used and we did not resort to singular penalties.

4. The steepest descent approach was employed in computations. The trajectory could move along the boundary of the feasible set.

All these items can be considered as advantages of presented approach.

4. DUAL BARRIER-PROJECTION METHODS FOR SOLVING LP PROBLEMS

By extension of the space and by converting the inequality constraints to equalities, we transform the original dual problem (2.3) into the following equivalent problem

$$\max_{u,y}(b^T u), \text{ subject to } \xi(y) + A^T u - c = 0_n, \tag{4.1}$$

where $v = \xi(y) \in \mathbb{R}_+^n, \overline{\xi(\mathbb{R}^n)} = \mathbb{R}_+^n$.

The Problem (4.1) is similar to (2.22). System (2.25) being applied to solving (4.1) can be rewritten in terms of u and v as follows

$$\frac{du}{dt} = b - Ax(u,v), \quad \frac{dv}{dt} = -G(v)x(u,v), \tag{4.2}$$

where $\Phi(v)x(u,v) = A^T b + \tau(v + A^T u - c)$ and $\Phi(v) = G(v) + A^T A$.

If $u_0 \in U$, then we can get rid of the equation for v and this way simplify systems (4.2). In this case, (4.2) can be expressed as

$$\frac{du}{dt} = b - Ax(u), \quad \left(G(v(u)) + A^T A\right) x(u) = A^T b, \tag{4.3}$$

where $u(0, u_0) = u_0 \in U$.

For this system we obtain the following inequality

$$b^T \frac{du}{dt} = \|b - Ax(u)\|^2 + x^T(u)G(v(u))x(u) \geq 0.$$

Hence the objective function of the dual problem monotonically increases on a feasible set.

By applying the Euler numerical integration method we obtain the following iterative algorithm

$$u_{k+1} = u_k + \alpha_k(b - Ax_k), \quad v_{k+1} = v_k - \alpha_k G(v_k)x_k, \tag{4.4}$$

$$\left(G(v_k) + A^T A\right) x_k = A^T b + \tau \left(v_k + A^T u_k - c\right).$$

Similarly for the system (4.3) we have

$$u_{k+1} = u_k + \alpha_k (b - Ax_k), \quad \left(G(v_k) + A^T A\right) x_k = A^T b, \tag{4.5}$$

where $v_k = v(u_k)$. Both variants solve the primal and dual problems simultaneously.

Denote

$$\alpha_* = 2/\lambda^*, \quad \lambda^* = \max\left[\tau, \max_{1 \leq i \leq m} \dot{\theta}(0)x_*^i\right], \quad \lambda_* = \min\left[\tau, \min_{1 \leq i \leq m} \dot{\theta}(0)x_*^i\right],$$

where λ^* and λ_* are respectively maximum and minimum eigenvalues of the matrix of the equation of the first approximation about the optimal solution u_*.

Theorem 4.1. *Let x_* and u_* be unique nondegenerate solutions of the Problems (2.2) and (2.3) respectively and let $v_* = c - A^T u_*$. Assume that the componentwise space transformation $\xi(y)$ satisfies conditions C_1, C_2 and $\tau > 0$. Then the following statements are true:*

1. *The pair $[u_*, v_*]$ is an asymptotically stable equilibrium state of system (4.2).*
2. *The solutions $u(t, z_0), v(t, z_0)$ of system (4.2) converge locally exponentially fast to the pair $[u_*, v_*]$. The corresponding function $x(u(t, z_0), v(t, z_0))$ converges to the optimal solution x_* of the primal Problem (2.2).*
3. *The point u_* is an asymptotically stable equilibrium state of system (4.3).*
4. *The solutions $u(t, u_0)$ of system (4.3) converge locally exponentially fast to the optimal solution u_* of the dual Problem (2.3). The corresponding function $x(u(t, u_0))$ converges to the optimal solution x_* of the primal Problem (2.2).*

5. *For any fixed* $0 < \alpha_k < \alpha_*$ *the sequence* $\{u_k, v_k\}$ *generated by (4.4) converges locally with a linear rate to* $[u_*, v_*]$ *while the corresponding sequence* $\{x_k\}$ *converges to* x_*.

6. *There exists an* $\alpha_* > 0$ *such that for any fixed* $0 < \alpha_k < \alpha_*$ *the sequence* $\{u_k\}$ *generated by (4.5) converges locally with a linear rate to* u_* *while the corresponding sequence* $\{x_k\}$ *converges to* x_*.

The method (4.3) was proposed and studied in 1977 (see [9]). Method (4.2) was given in [11]. In [16] we describe nonlocal convergence analysis of dual method. We supposed the Problem (2.2) is such that $Ae = 0_m$, where e is a vector of ones in $\mathbb{R}^n$. We used the Lyapunov function similar to (3.9):

$$V(u) = \sum_{i=m+1}^{n} v_*^i \left[\ln v_*^i - \ln v^i(u)\right].$$

Let P be a full rank $d \times n$ matrix such that $AP^T = 0_{md}$. Therefore the columns of P^T are linearly independent and form a basis for the nullspace of A. We partition A as $A = [B, N]$, where the square matrix B is nonsingular. We can now write the matrix P as

$$P = \left[-N(B^T)^{-1} \mid I_d\right].$$

The definitions of the sets V and V_U can be rewritten as follows

$$V = \{v \in \mathbb{R}^n : P(v-c) = 0_d\}, \quad V_U = \left\{v \in \mathbb{R}^n_+ : P(v-c) = 0_d\right\}.$$

Let $\bar{x} \in \mathbb{R}^n$ be an arbitrary vector which satisfies the constraint $A\bar{x} = b$. Then

$$\max_{u \in U} b^T u = \max_{u \in U} \bar{x}^T A^T u = \max_{v \in V_U} \bar{x}^T (c - v) = \bar{x}^T c - \min_{v \in V_U} \bar{x}^T v.$$

Hence the solution of the dual Problem (2.3) can be substituted by the following equivalent minimization problem

$$\min_{v \in V_U} \bar{x}^T v.$$

Applying the stable barrier-projection method (3.1) to this problem, we obtain

$$\frac{dv}{dt} = G(v)\left(P^T x(v) - \bar{x}\right), \tag{4.6}$$

$$PG(v)P^T x(v) = PG(v)\bar{x} + \tau P(c - v). \tag{4.7}$$

If a point v is such that the matrix $PG(v)P^T$ is invertible, then we can solve the linear equation (4.7) and obtain

$$x(v) = \left(PG(v)P^T\right)^{-1}\left(PG(v)\bar{x} + \tau P(c - v)\right).$$

Let $H(v) = G^{1/2}(v)$ and introduce the pseudoinverse and projection matrices

$$(PH)^+ = (PH)^T (PGP^T)^{-1} \quad (PH)^\sharp = (PH)^+ PH.$$

The system (4.6), (4.7) can be rewritten in the following projective form

$$\frac{dv}{dt} = H\left[\tau (PH)^+ P(c - v) - \left(I_n - (PH)^\sharp\right) H\bar{x}\right]. \tag{4.8}$$

The first vector in the square brackets belongs to the nullspace of AH^{-1} and the second vector belongs to the row space of this matrix. Furthermore

$$P\frac{dv}{dt} = \tau P(c-v), \quad P(c-v(t,v_0)) = P(c-v_0)e^{-\tau t}.$$

Hence, the trajectories $v(t,v_0)$ approach the set V as $t \to \infty$.

If $v_0 \in V_U$ and $v_0 > 0$, then the entire trajectory does not leave the feasible set V_U, the objective function $\bar{x}^T v(t,v_0)$ is a monotonically decreasing function of t and (4.8) can be rewritten as follows

$$\frac{dv}{dt} = -G(v)\left(I_n - P^T\left(PG(v)P^T\right)^{-1} PG(v)\right)\bar{x}, \quad v_o \in riV_U. \tag{4.9}$$

Theorem 4.2. *Suppose that the conditions of the Theorem 4.1. hold. Then:*

1. *the point v_* is an asymptotically stable equilibrium point of system (4.6);*
2. *the solutions $v(t,v_0)$ of (4.8) converge locally to v_* with an exponential rate of convergence;*
3. *there exists an $\alpha_* > 0$ such that for any fixed $0 < \alpha_k < \alpha_*$ the discrete version*

$$v_{k+1} = v_k - \alpha_k G(v_k)\left(\bar{x} - P^T x_k\right), \quad x_k = x(v_k) \tag{4.10}$$

converges locally with a linear rate to v_* while the corresponding sequence $\{x_k\}$ converges to x_*.

Since for system (4.9) $P\dot{v} = 0_d$, it follows that the vector $\dot{v}$ belongs to nullspace of P which coincides with the row space of A. Therefore there exists a vector $\lambda \in \mathbb{R}^m$ such that

$$\dot{v} = A^T\lambda. \tag{4.11}$$

If $v > 0_n$, then after left multiplying both sides of (4.11) with $AG^{-1}(v)$ and in view of (4.9) we obtain

$$\lambda = -\left(AG^{-1}(v)A^T\right)^{-1} A\bar{x} = -\left(AG^{-1}(v)A^T\right)^{-1} b.$$

Hence, on the set riV_U the method (4.9) takes the form

$$\frac{dv}{dt} = -A^T\left(AG^{-1}(v)A^T\right)^{-1} b, \quad v_0 \in riV_U.$$

In $u-$space this method can be written as

$$\frac{du}{dt} = \left(AG^{-1}(v(u))A^T\right)^{-1} b, \quad u_0 \in U_0.$$

If we use the quadratic and exponential space transformations (2.10), (2.11) we obtain

$$\frac{du}{dt} = \left(AD^{-1}(v(u))A^T\right)^{-1} b, \quad u_0 \in U_0, \tag{4.12}$$

and

$$\frac{du}{dt} = \left(AD^{-2}(v(u))A^T\right)^{-1} b, \quad u_0 \in U_0, \tag{4.13}$$

respectively. The system (4.13) coincides with the continuous version of the dual affine scaling method proposed by I.Adler, N.Karmarkar, M.Resende and G.Veiga in 1989 (see [1]).

According to the Theorem 4.2., the solution of (4.6) converges locally with an exponential rate to the equilibrium point $v_* = v(u_*)$. Therefore the solutions of (4.12) also converge to the point u_* on the set U_0.

The discrete version of (4.12) consists of the iteration

$$u_{k+1} = u_k + \alpha_k \left(AD^{-1}(v_k)A^T\right)^{-1} b, \quad u_0 \in U_0, \tag{4.14}$$

where $v_k = v(u_k)$. Taking into account the Theorem 2.3.7 from [7] we conclude that the exponential rate of convergence of (4.12) insures local linear convergence of the discrete variant (4.14) if the step length α_k is sufficient small.

5. PRIMAL AND DUAL BARRIER-NEWTON METHODS

Let us consider the Problem (2.1) supposing that all functions $f(x), g^i(x), 1 \leq i \leq m$, are at least twice continuously differentiable. Equation (2.15) can be written as

$$g_x G(x) L_x(x, u(x)) = \tau g(x). \tag{5.1}$$

If the regular point x is such that

$$G(x) L_x(x, u(x)) = 0, \tag{5.2}$$

then $[x, u(x)]$ is a weak KKT pair for the Problem (2.1). In section 2 we actually used a method of successive approximation for finding a solution of this nonlinear equation. Now we will apply the Newton method for this purpose. The continuous version of the Newton method leads to the initial value problem for the following system of ordinary differential equations

$$\Lambda(x) \frac{dx}{dt} = -G(x) L_x(x, u(x)), \quad x(0, x_0) = x_0, \tag{5.3}$$

where $\Lambda(x)$ is a Jacobian matrix of the mapping $G(x)L_x(x, u(x))$ with respect to x. By following the trajectories satisfying (5.3), we can theoretically obtain a solution of the system of nonlinear equations (5.2). In practice, we build the iterative procedures using a discretization of dynamical systems.

Along the trajectories of (5.3) we have

$$G(x(t, x_0)) L_x(x(t, x_0), u(x(t, x_0))) = e^{-t} G(x_0) L_x(x_0, u(x_0)),$$

$$g(x(t, x_0)) = \frac{e^{-t}}{\tau} g_x(x(t, x_0)) G(x_0) L_x(x_0, u(x_0)).$$

If the trajectories of (5.3) can be continued as $t \to +\infty$ and the norm of the vector

$$g_x(x(t, x_0)) G(x_0) L_x(x_0, u(x_0))$$

is bounded, then the pair $[x(t, x_0), u(x(t, x_0))]$ converges to a weak KKT pair.

The brief analysis of method (5.3) for the case where $P = \mathbb{R}^n_+$ is given in [12]. Here we apply the primal barrier-Newton method (5.3) for solving the linear programming Problem (2.2). In this case we have

$$\Lambda(x) = [I_n - H(x)] D(\dot{\theta}(x)) D(c - A^T u(x)) + \tau H(x),$$

$$H(x) = D(\theta(x)) A^T (A D(\theta(x)) A^T)^{-1} A.$$

Using relation (5.1) we obtain that the system (5.3) has first integrals

$$D(\theta(x(t,x_0)))(c - A^T u(x(t,x_0))) = D(\theta(x_0))(c - A^T u(x_0))e^{-t},$$

$$b - Ax(t,x_0) = (b - Ax_0)e^{-t}.$$

Introduce a Lebesgue level set in $\mathbb{R}^n$

$$\Omega = \{x \in \mathbb{R}^n_+ : \|Ax - b\| \le \|Ax_0 - b\|, \ 0_n \le D(\theta(x))(c - A^T u(x)) \le D(\theta(x_0))v_0\},$$

where x_0 is an initial point in (5.3), $v_0 = c - A^T u_0$, $u_0 = u(x_0)$.

Theorem 5.1. *Suppose that the set Ω is compact and contains a unique stationary point x_*. Assume that the space transformation (2.9) satisfies conditions C_1, C_2 and is such that the matrix $\Lambda(x)$ is nonsingular everywhere on Ω. If the starting point x_0 is such that $x_0 > 0_n$, $v_0 > 0_n$, then*

$$\lim_{t\to\infty} x(t,x_0) = x_*, \quad \lim_{t\to\infty} u(x(t,x_0)) = u_*, \tag{5.4}$$

where x_*, u_* are the solutions of the Problems (2.2) and (2.3), respectively.

Integrating (5.3) using the Euler method, we obtain the following iterative process:

$$x_{k+1} = x_k - \alpha\Lambda^{-1}(x_k)D(\theta(x_k))(c - A^T u(x_k)), \tag{5.5}$$

where $\alpha > 0$ is a stepsize.

If the conditions of the Theorem 3.1. hold and the space transformation function satisfies conditions C_1, C_2, then the matrix $\Lambda(x_*)$ is nonsingular. Therefore if the stepsize α is fixed and $0 < \alpha < 2$, then the discrete version (5.5) locally converges to the point x_* with at least linear rate. If matrix $\Lambda(x)$ satisfies the Lipschitz condition in a neighborhood of x_* and $\alpha = 1$, then the sequence $\{x_k\}$ converges quadratically to x_*.

Let a function $x(u)$ be defined from (4.3). Substituting this function in feasibility condition we obtain the following nonlinear system

$$Ax(u) - b = 0_m.$$

Applying the Newton method for solving this system, we obtain the following continuous and discrete versions

$$Q(u)\frac{du}{dt} = Ax(u) - b, \tag{5.6}$$

$$u_{k+1} = u_k + \alpha Q^{-1}(u_k)(Ax_k - b), \tag{5.7}$$

where

$$Q(u) = -A\left(D(\theta(v)) + A^T A\right)^{-1} D(\dot\theta(v))D(x(u))A^T.$$

Theorem 5.2. *Let the mapping $\xi(y)$, defined in section 4, be such that conditions C_1 and C_2 hold. Then the matrix $Q(u_*)$ is nonsingular, the solution u_* of the Problem (2.3) is an asymptotically stable equilibrium state of the system (5.6), the discrete version (5.7) converges locally to the point u_* with at least linear rate if $0 < \alpha < 2$.*

The more detailed information about this method can be found in [17].

6. PRIMAL-DUAL BARRIER-NEWTON METHODS

Here we construct a primal-dual barrier-Newton method for solving the Problem (2.8). Introduce an additional mapping

$$\phi(z) = \left[\phi^1(z^1), \dots, \phi^n(z^n)\right].$$

and assume that $\phi(z)$ satisfies the Conditions C_1 and C_2.

Then the necessary optimality conditions (2.7) for the Problem (2.8) can be rewritten in the form

$$D(\theta(x))\phi(L_x(x,u)) = 0_n, \quad g(x) = 0_m, \quad x \in \mathbb{R}^n_+. \tag{6.1}$$

For solving this system we use the continuous version of Newton's method. The computation process is described by the system of ordinary differential equations

$$W(x,u)\begin{pmatrix} \dot{x} \\ \dot{u} \end{pmatrix} = -\begin{pmatrix} \alpha D(\theta(x))\phi(L_x(x,u)) \\ \tau g(x) \end{pmatrix}, \tag{6.2}$$

where $\alpha > 0, \tau > 0, W$ is a square $(n+m)^2$ matrix,

$$W(x,u) = \begin{pmatrix} M & D(\theta(x))D(\dot{\phi})g_x^T \\ g_x & 0_{mm} \end{pmatrix}, \quad M = D(\dot{\theta})D(\phi) + D(\theta)D(\dot{\phi})L_{xx}. \tag{6.3}$$

Lemma 6.1. *Let $[x_*, u_*]$ be a weak KKT pair, where the conditions of the Theorem 2.3. are satisfied. Assume that x_* is a regular point for the Problem (2.8) and the functions $\theta(x), \phi(z)$ satisfy C_1, C_2 . Then the matrix $W(x_*, u_*)$ is nonsingular.*

Let $x(t,z_0), u(t,z_0)$ denote the solutions of the Cauchy problem (6.2) with initial conditions $x_0 = x(0,z_0), u_0 = u(0,z_0), z_0^T = [x_0^T, u_0^T]$. Using this notation, we rewrite the system of equations (6.2) as

$$W(z)\frac{dz}{dt} = -D(\gamma)\,\mathbb{R}(z), \quad z(0,z_0) = z_0, \tag{6.4}$$

where γ has the first n components equal to α and all other components equal to τ. We denote $\gamma_* = \min[\alpha, \tau]$.

Theorem 6.2. *Suppose that the conditions of the Lemma 6.1. hold. Then for any $\alpha > 0, \tau > 0$ the pair $z_*^T = [x_*^T, u_*^T]$ is an asymptotically stable equilibrium point of system (6.4). If stepsize h_k is fixed and $0 < h_k < 2/\gamma_*$ then the discrete version*

$$z_{k+1} = z_k - h_k W^{-1}(z_k)D(\gamma)\,\mathbb{R}(z_k) \tag{6.5}$$

locally converges to the point z_* with at least linear rate. If $W(z)$ satisfies a Lipschitz condition in a neighbourhood of z_* and $h_k = \alpha = \tau = 1$, then the sequence $\{z_k\}$ converges quadratically to z_*.

Let us use a homogeneous function $\theta(x)$ of order λ, i.e. $\theta^i(x) = (x^i)^\lambda, 1 \le i \le n$. In this case we modify (6.2) and use the following system

$$\tilde{W}(x,u)\begin{pmatrix} \dot{x} \\ \dot{u} \end{pmatrix} = -\begin{pmatrix} \alpha D(x)\phi(L_x(x,u)) \\ \tau g(x) \end{pmatrix}, \tag{6.6}$$

where $\alpha > 0, \tau > 0, \tilde{W}$ is a square matrix

$$\tilde{W}(x,u) = \begin{pmatrix} M & D(x)D(\dot{\phi})g_x^T \\ g_x & 0_{mm} \end{pmatrix}, \quad M = \lambda D(\phi) + D(x)D(\dot{\phi})L_{xx}.$$

Lemma 6.3. *Let x be a regular point, and let the pair $[x,u]$ be such that $x^i \neq 0, L_{x^i}(x,u) \neq 0$ for all $1 \leq i \leq n$, and $M(x,u)$ is nonsingular. Then $W(x,u)$ is nonsingular.*

Define the nonnegative Lyapunov function

$$F(x,u) = \|D(\theta(x))\phi(L_x(x,u))\| + \|g(x)\|$$

and introduce two sets:

$$\Omega_0 = \left\{[x,u] : F(x,u) \leq F(x_0,u_0), \quad x \geq 0_n, \ L_x(x,u) \geq 0_n\right\},$$

$$\tilde{\Omega}_0 = \left\{[x,u] \in \Omega_0 : \quad x > 0_n, \quad L_x(x,u) > 0_n\right\}.$$

Theorem 6.4. *Suppose that the set Ω_0 is bounded and contains the unique KKT pair $[x_*, u_*]$. Suppose also that for any pair $[x,u] \in \tilde{\Omega}_0$ the conditions of the Lemma 6.3. are satisfied. Then all trajectories of (6.4) starting from a pair $[x_0, u_0] \in \tilde{\Omega}_0$, converge to $[x_*, u_*]$.*

The system of ordinary differential equations (6.4) has the first integrals

$$D(\theta(x(t,z_0)))\phi(L_x(x(t,z_0),u(t,z_0))) = D(\theta(x_0))\phi(L_x(x_0,u_0))e^{-\alpha t}, \tag{6.7}$$

$$g(x(t,z_0)) = g(x_0)e^{-\tau t}. \tag{6.8}$$

The solutions of (6.4) belong to Ω_0 and are therefore bounded. The right-hand sides of (6.7), (6.8) are strictly positive and tend to zero only as $t \to \infty$. By moving along the trajectories of (6.4) we do not violate nonnegativity of x and L_x. Therefore the trajectories do not cross the boundary of the set Ω_0. The transformation functions $\theta(x)$ and $\phi(v)$ thus play the role of the multiplicative barriers preserving nonnegativity. All trajectories that emanate from $\tilde{\Omega}_0$ remain in the interior of Ω_0. According to La Salle's Invariance Principle [2] the solutions $x(t,z_0), u(t,z_0)$ can be prolonged as $t \to \infty$, the positive limit set of the solution is a compact connected set contained in Ω_0 and coincides with the equilibrium pair $[x_*, u_*]$, which is unique on Ω_0.

Now we apply the Newton method (6.6) to LP problem. For the sake of simplicity we consider the case, where $\varphi(v) = v$ and $\lambda = \alpha/\tau$. Introduce new vectors $\mu \in \mathbb{R}^m, \eta \in \mathbb{R}^n$:

$$\mu = \left(AD(x)D^{-1}(v)A^T\right)^{-1} b, \quad \eta = D^{-1}(v)A^T\mu.$$

Let e be the vector of ones in $\mathbb{R}^n$. Now the system (6.6) can be written as

$$\frac{dx}{dt} = \tau D(x)[\eta - e], \quad \frac{du}{dt} = \alpha\mu. \tag{6.9}$$

Theorem 6.5. *Let $[x_*, u_*]$ be a nondegenerate optimal pair for the Problems (2.2) and (2.3). Assume that Ω is bounded. Let the starting point $x_0 \in \mathbb{R}^n_{++}$ and $u_0 \in U_0$ be interior, then the trajectories of (6.9) are such that:*

1. *the matrix $AD(x(t,z_0))D^{-1}(v(t,z_0))A^T$ is nondegenerate $\forall t \geq 0$;*

2. *$z(t, z_0) \in \Omega$ and $v(t, z_0) \in V$ for all $t > 0$;*

3. *the objective function $b^T u(t, z_0)$ of the dual problem increases monotonically;*

4. *the pair $[x(t, z_0), u(t, x_0)]$ is bounded and converges to $[x_*, u_*]$ as $t \to \infty$;*

5. *all components of vectors $D^\lambda(x(t, z_0))v(t, z_0)$, $Ax(t, z_0)$ change monotonically and*

$$D^\lambda(x(t, z_0))v(t, z_0) = e^{-\alpha t} D^\lambda(x_0)v_0,$$

$$Ax(t, z_0) - b = e^{-\tau t}(Ax_0 - b).$$

By applying the Euler numerical integration method to system (6.9) we obtain the simplest discrete version of the method:

$$x_{k+1} = D(x_k)(e + \tau_k(\eta_k - e)), \quad u_{k+1} = u_k + \alpha_k \mu_k. \tag{6.10}$$

Note, that if we set in (6.9), (6.10) $\alpha = 1, \tau = 0$ and $x = e$, then we obtain (4.12) and (4.14), respectively.

We specify three classes of procedures for determining the step lengths:

1. step lengths are fixed and small enough, hence the discrete process (6.10) is close to a continuous one (6.9);

2. stepsizes are close to one and therefore the discrete process has properties of Newton's method;

3. stepsizes are chosen from steepest descent conditions or from another auxiliary optimization problem.

The investigation of all these cases can be found in [18]. Here we concider the third approach to stepsize choice, which proved to be computationally the most effecient.

The iterates produced by algorithm (6.10) are well-defined if vectors x_k, v_k are strictly positive for all k. In order to ensure the positiveness of x_{k+1} and v_{k+1} we have to choose the step lengths α_k and τ_k such that

$$e \geq \alpha_k \eta_k, \quad e \geq \tau_k(e - \eta_k).$$

It is now straightforward to verify that non-negativity conditions hold if α_k and τ_k satisfy

$$0 < \alpha_k \leq \alpha_k^* = 1/[\eta_k^*]_+, \quad 0 < \tau_k \leq \tau_k^* = 1/[1 - \eta_*^k]_+,$$

where $[\alpha]_+ = \max[0, \alpha], \eta_k^*$ and η_*^k are maximal and minimal components of the vector η_k respectively. Define the steps $\alpha_k = \omega\alpha_k^*, \tau_k = \omega\tau_k^*$, where $0 < \omega < 1$ is a safety factor.

Introduce two functions

$$\Phi(x, u) = x^T v(u) + \| Ax - b \|, \quad \phi(\alpha_k, \tau_k) = \Phi(x_{k+1}, u_{k+1}).$$

The steepest descent stepsizes $\bar{\alpha}_k, \bar{\tau}_k$ are found from the solution of the following auxiliary problem

$$\phi(\bar{\alpha}_k, \bar{\tau}_k) = \min_{0 \leq \alpha_k \leq \omega\alpha_k^*} \min_{0 \leq \tau_k \leq \omega\tau_k^*} \phi(\alpha_k, \tau_k).$$

Here ϕ is a bilinear function of α and τ. In [18] this problem was solved analytically. Denote $\| \eta \|_\infty = \max_{1 \leq i \leq n} |\eta^i|$. Introduce the function

$$\mathcal{K}(x, v) = \left\lceil \frac{1 + C}{\omega} \ln \frac{\Phi(x, v)}{\epsilon} \right\rceil,$$

where $\lceil a \rceil$ is the least integer larger than or equal to a and $\epsilon > 0$.

Theorem 6.6. *Let $x_0 \in \mathbb{R}^n_{++}, u_0 \in U_0$ and suppose that the sequence $\{x_k, u_k\}$ generated by algorithm (6.10) with steepest descent rule is such that $\|\eta_k\|_\infty \leq C$ for all k. Then the sequence $\{x_k, u_k\}$ converges to $[x_*, u_*]$ at finite number of steps or at least superlinearly and the function $\Phi(x_k, v_k)$ becomes less then ϵ in at most $\mathcal{K}(x_0, v_0)$ iterations.*

Method (6.10) posseses local convergence property, therefore it is posible take starting points outside the positive orthant, but in this case we must take into account that the matrix $AD(x)D^{-1}(v)A^T$ is singular on some manifold and we must complete the stepsize rule. A phase portrait analysis of (6.9), proof of the Theorem 6.6. and illustrative computational example can be found in [13], [18].

Algorithm (6.10) has one important disadvantege connected with the necessity to know a starting point $u_0 \in U_0$. It is possible to get rid of this restriction if we use the barrier-Newton method in the extended space of variables x, u, v. The simplest version of the method is described by following system of $2n + m$ differential equations

$$\frac{dx}{dt} = \tau G(A^T\zeta - c), \quad \frac{dv}{dt} = \alpha(c - v - A^T\zeta), \quad \frac{du}{dt} = \alpha(\zeta - u), \tag{6.11}$$

where $\zeta = (AGA^T)^{-1}[b - Ax + AGc], G = D(x)D^{-1}(v)$. The essential difference in the requirements on the initial conditions between systems (6.9) and (6.11) is that in the latter case we impose only the simplest restrictions: $x_0 > 0_n, v_0 > 0_n$.

Let $Ax_0 = b$. If we set in (6.11) $\alpha = 0, \tau = 1$ and $v = e$ then we obtain (3.8). The generalization of systems (6.9) and (6.11) for the LP problem with box constraints is given in [28].

ACKNOWLEDGEMENT

The Research has been supported by the Russian Scientific Fund (grant N. 94-01-01379), which is here gratefully acknowledged.

REFERENCES

[1] I.Adler, N.Karmarkar, M.G.C.Resende and G.Veiga, "An Implementation of Karmarkar's Algorithm for Linear Programming". *Math. Programming* 44, 297-335, 1989.

[2] A. Bacciotti, *"Local stability of nonlinear control systems"*. Series on Advances in Mathematics for Applied Sciences 8, World Scientific Publishing Co. Ptc. Ltd., Singapore, 1992.

[3] E. Barnes, "A variation on Karmarkar's algorithm for solving linear programming problems". *Math. Programming* 36, 174-182, 1986.

[4] D. Bayer and J. Lagarias, "The nonlinear geometry of linear programming. Affine and projective scaling trajectories". *Trans. Amer. Math. Soc.* 314, pp. 499-526, 1989.

[5] I.I.Dikin, "Iterative solution of problems of linear and quadratic programming". *Sov. Math. Dokl.* 8, pp. 674-675, 1967.

[6] Yu.G.Evtushenko, "Two numerical methods of solving nonlinear programming problems". *Sov. Math. Dokl.* 215 (2), pp. 420-423, 1974.

[7] Yu.G.Evtushenko, *"Numerical Optimization Techniques. Optimization Software"*. Inc. Publications Division, New York., 1985.

[8] Yu.G.Evtushenko and V.G.Zhadan, "Numerical methods for solving some operations research problems", *U.S.S.R. Comput. Maths. Math. Phys.* 13 (3), pp. 56-77, 1973.

[9] Yu.G.Evtushenko and V.G.Zhadan, "A relaxation method for solving problems of non-linear programming". *U.S.S.R. Comput. Maths. Math. Phys.* 17 (4), pp. 73-87, 1977.

[10] Yu.G.Evtushenko and V.G.Zhadan, *"Barrier-projective and barrier-Newton numerical methods in optimization (the nonlinear programming case)"*. Computing Centre of the USSR Academy of Sciences, Reports in Comput. Math., (in Russian), 1991.

[11] Yu.G.Evtushenko and V.G.Zhadan, *"Barrier-projective and barrier-Newton numerical methods in optimization (the linear programming case)"*, Computing Centre of the Russian Academy of Sciences, Reports in Comput. Math., (in Russian), 1992.

[12] Yu.G.Evtushenko and V.G.Zhadan, "Stable Barrier-Projection and Barrier-Newton Methods in Nonlinear Programming". *Optimization Methods and Software* 3 (1-3), pp. 237-256, 1994.

[13] Yu.G.Evtushenko and V.G.Zhadan, "Stable Barrier-Projection and Barrier-Newton Methods for Linear and Nonlinear Programming. In *"Algorithms for Continuous Optimization"*, (Edited by E.Spedicato), NATO ASI Series, 255-285, Kluwer Academic Publishers, 1994.

[14] Yu.G.Evtushenko and V.G.Zhadan, "Stable Barrier-Projection and Barrier-Newton Methods in Linear Programming". *Computational Optimization and Applications* 3 (4), pp. 289-303, 1994.

[15] Yu.G.Evtushenko and V.G.Zhadan, "Barrier-Projective Methods for Non-Linear Programming". *Comput. Math. and Math. Physics* 34 (5), pp. 579-590, 1994.

[16] Yu.G.Evtushenko and V.G.Zhadan, "Dual Barrier-Projective Methods in Linear Programming". *An Inter. Journal Computers and Mathematics with Applications*, 1995 (to be published).

[17] Yu.G.Evtushenko and V.G.Zhadan, "Dual Barrier-Projective and Barrier-Newton Methods for Linear Programming Problems". *Comput. Math. and Math. Physics*, (to be published).

[18] Yu.G.Evtushenko, V.G.Zhadan and A.P.Cherenkov, "Application of Newton's Method to Solving Linear Programming Problems". *Comput. Math. and Math. Physics* 35 (6), pp. 850-866, 1995, (English version is to be published).

[19] L.E.Faybusovich, "Hamiltonian structure of dynamical systems which solve linear programming problems". *Physica D* 53, pp. 217-232, 1991.

[20] A. Fiacco and G. McCormic, *"Nonlinear programming: Sequential unconstrained minimization techniques"*. John Wiley & Sons, New York, 1968.

[21] C. Gonzaga, "Path following methods for linear programming", *SIAM Review* 34, pp. 167-224, 1992.

[22] U. Helmke and J. B. Moore, *"Optimization and Dynamical Systems"*, Springer-Verlag. 1994.

[23] S.Herzel, M.C.Recchioni and F.Zirilli, "A quadratically convergent method for linear programming". *Linear Algebra and its Applications* 152, pp. 255-289, 1991.

[24] M. Kallio, "On gradient projection for linear programming". Working paper 94, Yale School of Organization and Management, 1986.

[25] N. Karmarkar, "A new polynomial-time algorithm for linear programming". *Combinatorica*, No. 4, pp. 373-395, 1984.

[26] J. Rosen, "The gradient projection method for nonlinear programming, part 1, linear constraints". *SIAM J. Applied Math.* 8, pp. 181-217, 1960.

[27] G. Smirnov, "Convergence of barrier-projection methods of optimization via vector Lyapunov functions". *Optimization Methods and Software* 3 (1-3), pp. 153-162, 1994.

[28] A.I.-A.Stanenevichus and L.V.Sherbak, "New Variants of Barrier-Newton Method for Solving Linear Programming Problem". *Comput. Math. and Math. Physics* 35 (12),pp. 1796-1807, 1995.

[29] K. Tanabe, "A geometric method in nonlinear programming". *Journal of Optimization Theory and Applications* 30, pp. 181- 210, 1980.

[30] R. Vanderbei, M. Meketon and B. Freedman, "A modification of Karmarkar's linear programming algorithm". *Algorithmica* 1, pp. 395-407, 1986.

[31] Wei Zi-luan, "An interior point method for linear programming". *Journal of Computing Mathematics*, Oct., pp. 342-350, 1987.

INEXACT NEWTON METHODS FOR SEMISMOOTH EQUATIONS WITH APPLICATIONS TO VARIATIONAL INEQUALITY PROBLEMS

Facchinei Francisco[1], Fischer Andreas[2] and Kanzow Christian[3]

[1] Dipartimento di Informatica e Sistemistica
Università di Roma "La Sapienza"
Via Buonarroti 12, I-00185 Roma, Italy
e-mail: soler@peano.dis.uniroma1.it

[2] Institute of Numerical Mathematics
Technical University of Dresden
D-01062 Dresden, Germany
e-mail: fischer@math.tu-dresden.de

[3] Institute of Applied Mathematics
University of Hamburg
Bundesstrasse 55, D-20146 Hamburg, Germany
e-mail: kanzow@math.uni-hamburg.de

Abstract. We consider the local behaviour of inexact Newton methods for the solution of a semismooth system of equations. In particular, we give a complete characterization of the Q-superlinear and Q-quadratic convergence of inexact Newton methods. We then apply these results to a particular semismooth system of equations arising from variational inequality problems, and present a globally and locally fast convergent algorithm for its solution.

Key words. Semismoothness, inexact Newton methods, variational inequality problems, global convergence, superlinear convergence, quadratic convergence.

1. INTRODUCTION

Consider the nonlinear system of equations

$$G(x) = 0,$$

with $G : \mathbb{R}^n \to \mathbb{R}^n$. Solving this system of equations is a well-understood problem if the operator G is continuously differentiable, see, e.g., [6]. However, many mathematical problems and several applications lead to a system of equations with a nonsmooth operator G; see, e.g., [23]. Of course, these problems are much more difficult to solve, but there is a growing interest in finding efficient methods which are able to handle them.

Nonlinear Optimization and Applications
Edited by G. Di Pillo and F. Giannessi, Plenum Press, New York, 1996

In this paper we focus on the case in which the system $G(x) = 0$ is just semismooth (see next section). We consider the use of extensions of the classical inexact Newton method which, in the smooth case, is known to be among the most efficient algorithms for the solution of large-scale systems. The convergence theory presented here completes the recent results by Martínez and Qi [20]. In particular, we give a complete characterization of the Q-superlinear and Q-quadratic convergence of any sequence generated by the nonsmooth inexact Newton method. These characterizations generalize the classical results by Dembo, Eisenstat and Steihaug [4] from the smooth to the semismooth case.

One of the source problems which leads to nonsmooth equations is the variational inequality problem, VI(X, F) for short. This is to find a vector x^* in a feasible set $X \subseteq \mathbb{R}^n$ such that

$$F(x^*)^T(x - x^*) \geq 0 \qquad \forall x \in X,$$

where $F : \mathbb{R}^n \to \mathbb{R}^n$. Using a function introduced by Fischer [10], we will reformulate the optimality conditions of VI(X, F) as a semismooth system of equations and apply an inexact Newton method to this particular system. This algorithm will be shown to enjoy good global and local convergence properties. In particular, we are able to establish global convergence for a wide class of problems as well as to prove Q-superlinear and Q-quadratic convergence results without assuming strict complementarity at the solution.

The organization of the paper is as follows. In the next section we will shortly review some basic properties of semismooth functions. In Sect. 3 we present the local convergence results for an inexact Newton method applied to a general semismooth equation $G(x) = 0$. This theory will be used in Sect. 4 in order to show some strong convergence properties of a new algorithm for solving variational inequality problems.

Notation. We say that a function $G : \mathbb{R}^n \to \mathbb{R}^n$ is a C^k-function if G is k times continuously differentiable. A C^k-function G is called an LC^k-function if its kth derivative is locally Lipschitz-continuous. The Jacobian of a C^1-function $G : \mathbb{R}^n \to \mathbb{R}^n$ at a point $x \in \mathbb{R}^n$ is denoted by $G'(x)$. Finally, $\|\cdot\|$ indicates the Euclidean norm or the subordinate matrix norm.

2. PROPERTIES OF SEMISMOOTH FUNCTIONS

Let $G : \mathbb{R}^n \to \mathbb{R}^n$ be a locally Lipschitzian function. Then G is almost everywhere differentiable by Rademacher's theorem. Let us denote by D_G the set of points at which G is differentiable and by $G'(x; d)$ the directional derivative of G at x in the direction d. Then the *B-subdifferential* of G at a point $x \in \mathbb{R}^n$ is defined by

$$\partial_B G(x) := \{H \in \mathbb{R}^{n \times n} : \exists \{x^k\} \to x \text{ with } x^k \in D_G, \text{ and } G'(x^k) \to H\}.$$

The *generalized Jacobian* by Clarke [2] is the convex hull of the B-subdifferential:

$$\partial G(x) := \operatorname{co} \partial_B G(x).$$

The following definition of a semismooth operator is due to Qi and Sun [25] and generalizes a concept by Mifflin [21] from functionals to vector-valued functions. The definition is also closely related to a similar concept suggested by Kummer [19].

Definition 2.1. *Let $G : \mathbb{R}^n \to \mathbb{R}^n$ be locally Lipschitzian at $x \in \mathbb{R}^n$. G is said to be* semismooth *at x if the limit*

$$\lim_{H \in \partial G(x+tv'), v' \to v, t \downarrow 0} Hv'$$

exists for every $v \in \mathbb{R}^n$.

We note that semismooth functions are known to be directionally differentiable, and that the directional derivative of G at x in the direction v is equal to the limit in Definition 2.1.. Furthermore it can be shown that G is semismooth at x if and only if

$$Hd - G'(x;d) = o(\|d\|).$$

This latest property motivates the following definition; see [25].

Definition 2.2. *Suppose that $G : \mathbb{R}^n \to \mathbb{R}^n$ is semismooth at $x \in \mathbb{R}^n$. Then G is said to be* strongly semismooth *at x if, for any $d \to 0$ and for any $H \in \partial G(x+d)$, we have*

$$Hd - G'(x;d) = O(\|d\|^2).$$

Note that every C^1-function is semismooth and that every LC^1-function is strongly semismooth. For more details on semismooth functions, we refer the reader to [21,24,25,12].

The following definition of a BD-regular vector plays a crucial role in establishing fast local convergence results of several iterative methods.

Definition 2.3. *We say that a Lipschitzian function $G : \mathbb{R}^n \to \mathbb{R}^n$ is BD-regular at a point $x \in \mathbb{R}^n$ if all elements in the B-subdifferential $\partial_B G(x)$ are nonsingular.*

The following result was proved by Qi [24].

Proposition 2.1. *Assume that $G : \mathbb{R}^n \to \mathbb{R}^n$ is semismooth and that $x^* \in \mathbb{R}^n$ is a BD-regular solution of the system $G(x) = 0$. Then there is a neighbourhood Ω of x^* and a constant $c > 0$ such that, for all $x \in \Omega$ and all $H \in \partial_B G(x), H$ is nonsingular and*

$$\|H^{-1}\| \leq c.$$

The next result is due to Pang and Qi [23] and plays an important role in establishing the superlinear rate of convergence of certain Newton-type methods.

Proposition 2.2. *Assume that $G : \mathbb{R}^n \to \mathbb{R}^n$ is semismooth at $x \in \mathbb{R}^n$. Then*

$$\lim_{h \to 0, H \in \partial G(x+h)} \frac{\|G(x+h) - G(x) - Hh\|}{\|h\|} = 0.$$

A corresponding result, for strongly semismooth functions, was established by Facchinei and Kanzow [8], see also Fischer [12].

Proposition 2.3. *Assume that $G : \mathbb{R}^n \to \mathbb{R}^n$ is strongly semismooth at $x \in \mathbb{R}^n$ and directionally differentiable in a neighbourhood of x. Then*

$$\limsup_{h \to 0, H \in \partial G(x+h)} \frac{\|G(x+h) - G(x) - Hh\|}{\|h\|^2} < \infty.$$

The following is a characterization theorem for superlinear convergence due to Pang and Qi [23]. It generalizes the famous characterization theorem by Dennis and Moré [5] from the smooth to the nonsmooth case.

Theorem 2.4. *Let $G : \mathbb{R}^n \to \mathbb{R}^n$ be a locally Lipschitz-continuous function in the open convex set $D \subseteq \mathbb{R}^n$. Let $\{x^k\} \subset D$ be any sequence converging to $x^* \in D$ with $x^k \neq x^*$ for all k. If G is semismooth and BD-regular at x^*, then $\{x^k\}$ converges Q-superlinearly to x^* and $G(x^*) = 0$ if and only if*

$$\lim_{k\to\infty} \frac{\|G(x^k) + H_k d^k\|}{\|d^k\|} = 0,$$

where $H_k \in \partial_B G(x^k)$ and $d^k = x^{k+1} - x^k$.

There is a similar result, due to Facchinei and Kanzow [8], which characterizes the quadratic rate of convergence.

Theorem 2.5. *Let $G : \mathbb{R}^n \to \mathbb{R}^n$ be a locally Lipschitz-continuous function in the open convex set $D \subseteq \mathbb{R}^n$. Let $\{x^k\} \subset D$ be any sequence converging to $x^* \in D$ with $x^k \neq x^*$ for all k. If G is strongly semismooth and BD-regular at x^*, and directionally differentiable in a neighbourhood of x^*, then $\{x^k\}$ converges Q-quadratically to x^* and $G(x^*) = 0$ if and only if*

$$\limsup_{k\to\infty} \frac{\|G(x^k) + H_k d^k\|}{\|d^k\|^2} < \infty,$$

where $H_k \in \partial_B G(x^k)$ and $d^k = x^{k+1} - x^k$.

Finally, we state a result which will be useful in Sect. 4.

Proposition 2.6. *Let $G : \mathbb{R}^n \to \mathbb{R}^n$ be semismooth and $x^* \in \mathbb{R}^n$ be a BD-regular solution of the system $G(x) = 0$. Suppose that there are two sequences $\{x^k\}$ and $\{d^k\}$ such that*

$$x^k \to x^* \text{ and } \lim_{k\to\infty} \frac{\|x^k + d^k - x^*\|}{\|x^k - x^*\|} = 0.$$

Then

$$\lim_{k\to\infty} \frac{\|G(x^k + d^k)\|}{\|G(x^k)\|} = 0.$$

Proof. The result is actually due to Facchinei and Soares [9, Lemma 5.5], where, however, it has been stated only under the stronger assumption that all elements in the generalized Jacobian $\partial G(x^*)$ are nonsingular. It is not difficult to see that their proof goes through also under the weaker BD-regularity assumption, so we omit the details here. □

3. NONSMOOTH INEXACT NEWTON METHODS

First assume that $G : \mathbb{R}^n \to \mathbb{R}^n$ is a smooth function and consider the following inexact Newton method for solving the system of nonlinear equations $G(x) = 0$.

Algorithm 3.1. *(Smooth inexact Newton method)*

(S.0) *Let* $x^0 \in \mathbb{R}^n, \eta_0 \geq 0$ *and set* $k = 0$.

(S.1) *If* $G(x^k) = 0$, *stop.*

(S.2) *Find a step* $d^k \in \mathbb{R}^n$ *such that*

$$G'(x^k)d^k = -G(x^k) + r^k,$$

where the residual vector $r^k \in \mathbb{R}^n$ *satisfies the condition*

$$\|r^k\| \leq \eta_k \|G(x^k)\|.$$

(S.3) *Choose* $\eta_{k+1} \geq 0$, *set* $x^{k+1} := x^k + d^k, k := k+1$ *and go to (S.1).*

We summarize the main convergence results in the following theorem. For its proof, the reader is referred to the classical paper by Dembo, Eisenstat and Steihaug [4].

Theorem 3.1. *Let* $x^* \in \mathbb{R}^n$ *be a solution of the system* $G(x) = 0$. *Assume that* G *is continuously differentiable in a neighbourhood of* x^* *and that the Jacobian* $G'(x^*)$ *is nonsingular. Then the following statements hold:*

(a) *Let* $\bar{\eta} \in (0,1)$ *be arbitrary. Then there is an* $\varepsilon > 0$ *such that, if* $\|x^0 - x^*\| \leq \varepsilon$ *and* $\eta_k \leq \bar{\eta}$ *for all* k, *the sequence* $\{x^k\}$ *generated by Algorithm 3.1. is well-defined and converges Q-linearly to the solution* x^*.

(b) *If the sequence* $\{x^k\}$ *generated by Algorithm 3.1. converges to the solution* x^*, *then the rate of convergence is Q-superlinear if and only if* $\|r^k\| = o(\|G(x^k)\|)$.

(c) *If the sequence* $\{x^k\}$ *generated by Algorithm 3.1. converges to the solution* x^* *and if* G' *is Lipschitz-continuous in a neighbourhood of* x^*, *then the rate of convergence is Q-quadratic if and only if* $\|r^k\| = O(\|G(x^k)\|^2)$.

We note that the assumption $\|r^k\| = o(\|G(x^k)\|)$ in Theorem 3.1. is satisfied if the forcing sequence $\{\eta_k\}$ goes to 0. The assumption $\|r^k\| = O(\|G(x^k)\|^2)$ is satisfied if $\eta_k = O(\|G(x^k)\|)$.

Next we assume that the operator $G : \mathbb{R}^n \to \mathbb{R}^n$ is locally Lipschitz-continuous. The following algorithm is a generalization of the smooth inexact Newton method 3.1. to the nonsmooth case based on the generalized Jacobian by Clarke [2].

Algorithm 3.2. *(Nonsmooth inexact Newton method)*

(S.0) *Let* $x^0 \in \mathbb{R}^n, \eta_0 \geq 0$ *and set* $k = 0$.

(S.1) *If* $G(x^k) = 0$, *stop.*

(S.2) *Select an element* $H_k \in \partial_B G(x^k)$. *Find a step* $d^k \in \mathbb{R}^n$ *such that*

$$H_k d^k = -G(x^k) + r^k,$$

where the residual vector $r^k \in \mathbb{R}^n$ *satisfies the condition*

$$\|r^k\| \leq \eta_k \|G(x^k)\|.$$

(S.3) *Choose* $\eta_{k+1} \geq 0$, *set* $x^{k+1} := x^k + d^k, k := k+1$ *and go to (S.1).*

The following theorem contains the corresponding convergence properties.

Theorem 3.2. *Assume that G is semismooth in a neighbourhood of x^* and that x^* is a BD-regular solution of the system $G(x) = 0$. Then the following statements hold:*

(a) There are numbers $\bar{\eta} > 0$ and $\varepsilon > 0$ such that, if $\|x^0 - x^*\| \leq \varepsilon$ and $\eta_k \leq \bar{\eta}$ for all k, the sequence $\{x^k\}$ generated by Algorithm 3.2. is well-defined and converges Q-linearly to the solution x^*.

(b) If the sequence $\{x^k\}$ generated by Algorithm 3.2. converges to the solution x^*, then the rate of convergence is Q-superlinear if and only if $\|r^k\| = o(\|G(x^k)\|)$.

(c) If the sequence $\{x^k\}$ generated by Algorithm 3.2. converges to the solution x^* and if G is strongly semismooth in a neighbourhood of x^*, then the rate of convergence is Q-quadratic if and only if $\|r^k\| = O(\|G(x^k)\|^2)$.

Proof. Part (a) has been shown by Martínez and Qi [20]. So we come to part (b). First assume that $\|r^k\| = o(\|G(x^k)\|)$. Actually, under this assumption, it has also been shown in [20] that $\{x^k\}$ converges Q-superlinearly to x^*. Here, however, we give a different proof of this sufficient part by exploiting the characterization in Theorem 2.4..

We first note that, in view of the boundedness of the sequence $\{H_k\}$, there is a constant $c_1 > 0$ such that

$$\|G(x^k) - r^k\| = \|H_k d^k\| \leq c_1 \|d^k\|,$$

where the equality comes from the inexact Newton equation. Since $\|r^k\| = o(\|G(x^k)\|)$ by assumption, we therefore have

$$\|G(x^k)\| \leq c_2 \|d^k\|$$

for some constant $c_2 > 0$ and all k sufficiently large. Hence we get

$$\frac{\|r^k\|}{\|d^k\|} \leq c_2 \frac{\|r^k\|}{\|G(x^k)\|} \to 0.$$

Since $r^k = G(x^k) + H_k d^k$, the Q-superlinear convergence of the sequence $\{x^k\}$ to x^* now follows from Theorem 2.4..

We now prove the converse direction. So assume that $\{x^k\}$ converges to x^* Q-superlinearly. Let us denote by $e^k := x^k - x^*$ the error vector at the kth iterate. From the inexact Newton equation $G(x^k) + H_k d^k = r^k$, we obtain the identity

$$r^k = [G(x^k) - G(x^*) - H_k e^k] + [H_k e^{k+1}].$$

Dividing both sides by $\|e^k\|$, we obtain from Proposition 2.2., the boundedness of the sequence $\{H_k\}$ and the assumed Q-superlinear convergence of $\{x^k\}$ to x^* that

$$\frac{r^k}{\|e^k\|} \to 0.$$

In view of Proposition 3 in [23], however, there is a constant $c_3 > 0$ such that

$$\|G(x^k)\| \geq c_3 \|e^k\|$$

for all k sufficiently large. Hence we have $\|r^k\| = o(\|G(x^k)\|)$.

The proof of part (c) is similar to the one of part (b). Instead of Proposition 2.2. and Theorem 2.4. one has to use Proposition 2.3. and Theorem 2.5.. We omit the details here. □

We stress that, concerning the linear convergence, there is one major difference in the Theorems 3.1. and 3.2.: In the smooth case we can take $\bar{\eta} \in (0,1)$ arbitrarily in order to prove local Q-linear convergence, whereas in the nonsmooth case Theorem 3.2. just states that there exists an $\bar{\eta} > 0$ such that the sequence $\{x^k\}$ converges Q-linearly if $\eta_k \leq \bar{\eta}$ holds for all k. Martínez and Qi [20] showed by a counterexample that it is in general not sufficient to take $\bar{\eta} \in (0,1)$ arbitrarily in the nonsmooth case in order to prove Q-linear convergence of the sequence $\{x^k\}$. Note, however, that the superlinear and quadratic convergence results are the same for the smooth and nonsmooth inexact Newton methods.

4. APPLICATION TO VARIATIONAL INEQUALITIES

In the previous section we studied the local behaviour of truncated Newton schemes for the solution of semismooth systems of equations. We saw that the main results on the convergence rate of this class of methods go through from the smooth to the semismooth case. However, combining global and superlinear convergence in the semismooth case turns out to be a much more difficult task than in the smooth case, see [20] and the remarks at the end of the previous section.

In this section we consider a particular semismooth system of equations derived from the optimality conditions of variational inequalities. Using the structure of this system and the local theory developed in Sect. 3, we show how it is possible to overcome the difficulties just mentioned and to define, under mild assumptions, a globally convergent inexact Newton algorithm which compares favourably with existing algorithms for the same class of problems.

We consider a variational inequality VI(X,F) as introduced in Sect. 1. Here we assume that the set X is given by

$$X := \{x \in \mathbb{R}^n \,|\, g(x) \geq 0, h(x) = 0\},$$

where $g : \mathbb{R}^n \to \mathbb{R}^m$ and $h : \mathbb{R}^n \to \mathbb{R}^p$. Instead of reviewing the large number of existing solution methods for VI(X,F), we refer the interested reader to the survey papers [14,13] and the references therein.

Consider the following Karush-Kuhn-Tucker (KKT) optimality conditions of VI(X,F):

$$\begin{aligned} F(x) - g'(x)^T y + h'(x)^T z &= 0, \\ h(x) &= 0, \\ g(x) \geq 0, y \geq 0, y^T g(x) &= 0. \end{aligned} \tag{4.1}$$

If x^* solves VI(X,F) and if a certain constraint qualification holds (e.g., the linear independence of the gradients of the active constraints), then multiplier vectors $y^* \in \mathbb{R}^m$ and $z^* \in \mathbb{R}^p$ exist such that the vector $w^* := (x^*, y^*, z^*) \in \mathbb{R}^n \times \mathbb{R}^m \times \mathbb{R}^p$ is a KKT-point of VI(X,F), i.e., satisfies the KKT-conditions (4.1). Conversely, if all component functions g_i are concave and all component functions h_j are affine, so that X is a convex set, then the x-part of every KKT-point $w^* := (x^*, y^*, z^*)$ is a solution of VI(X,F), see [14]. Moreover, if X is a polyhedral set, then the KKT-conditions are both necessary and sufficient for a point to be a solution of VI(X,F) without assuming any constraint qualification.

We now want to rewrite the KKT-conditions as a nonlinear system of equations. To this end, we make use of the function $\varphi : \mathbb{R}^2 \to \mathbb{R}$ defined by

$$\varphi(a,b) := \sqrt{a^2+b^2} - a - b.$$

This function was introduced by Fischer [10] in 1992, and, since then, has become quite popular in the fields of linear and nonlinear complementarity, constrained optimization and variational inequality problems, see, e.g., [3,8,9,10,11,12,15,16,17,18,26,27]. The main property of this function is the following characterization of its zeros:

$$\varphi(a,b) = 0 \Longleftrightarrow a \geq 0, b \geq 0, ab = 0.$$

For this reason, the KKT-conditions (4.1) can equivalently be written as the nonlinear system of equations

$$\Phi(w) := \Phi(x,y,z) = 0, \tag{4.2}$$

where $\Phi : \mathbb{R}^n \times \mathbb{R}^m \times \mathbb{R}^p \to \mathbb{R}^n \times \mathbb{R}^m \times \mathbb{R}^p$ is defined by

$$\Phi(w) := \Phi(x,y,z) := \begin{pmatrix} L(x,y,z) \\ h(x) \\ \varphi(g(x),y) \end{pmatrix}$$

with

$$\begin{aligned} L(x,y,z) &:= F(x) - g'(x)^T y + h'(x)^T z, \\ \varphi(g(x),y) &:= (\varphi(g_1(x),y_1),\ldots,\varphi(g_m(x),y_m))^T \in \mathbb{R}^m. \end{aligned}$$

Note that the function φ is not differentiable in the origin, so that the system (4.2) is a nonsmooth reformulation of the KKT-conditions (4.1). However, it can easily be seen that Φ is a locally Lipschitz-continuous operator under suitable assumptions on F, g and h. More precisely, the following stronger properties can be shown, see [7].

Proposition 4.1. *The following statements hold:*

(a) Assume that F is a C^1-function and g, h are C^2-functions. Then Φ is semismooth.

(b) Assume that F is an LC^1-function and g, h are LC^2-functions. Then Φ is strongly semismooth.

Exploiting results concerning the composition of semismooth functions [21] and strongly semismooth functions [12], respectively, Proposition 4.1. can be shown to hold under weaker differentiability assumptions on F, g and h. However, we omit these details for the sake of simplicity. Now define the merit function $\Psi : \mathbb{R}^n \times \mathbb{R}^m \times \mathbb{R}^p \to \mathbb{R}$ by

$$\Psi(w) := \frac{1}{2}\Phi(w)^T\Phi(w) = \frac{1}{2}\|\Phi(w)\|^2.$$

Its smoothness properties are summarized in the following result, see [7].

Proposition 4.2. *Assume that F is a C^1-function and g, h are C^2-functions. Then Ψ is continuously differentiable, and $\nabla\Psi(w) = H^T\Phi(w)$ with $H \in \partial\Phi(w)$ being arbitrary.*

We note that the differentiability of the merit function Ψ is quite surprising and plays a crucial role in the globalization strategy of the following inexact Newton method applied to the system of equations $\Phi(w) = 0$.

Algorithm 4.1. *(Nonsmooth inexact Newton method for VI(X, F))*

(S.0) (Data)
Choose $w^0 = (x^0, y^0, z^0) \in \mathbb{R}^n \times \mathbb{R}^m \times \mathbb{R}^p, \rho > 0, \kappa > 2, \sigma \in (0, 1/2), \eta_0 \geq 0, \varepsilon \geq 0$ *and set* $k = 0$.

(S.1) (Termination criterion)
If $\|\nabla\Psi(w^k)\| \leq \varepsilon$, *stop.*

(S.2) (Search direction calculation)
Select an element H_k *in* $\partial_B\Phi(w^k)$. *Find a solution* d^k *of the system*

$$H_k d = -\Phi(w^k) + r^k \tag{4.3}$$

such that $\|r^k\| \leq \eta_k\|\Phi(w^k)\|$. *If this is not possible or if the condition*

$$\nabla\Psi(w^k)^T d^k \leq -\rho\|d^k\|^\kappa \tag{4.4}$$

is not satisfied, set $d^k = -\nabla\Psi(w^k)$.

(S.3) (Line search)
Find the smallest $i^k \in \{0, 1, 2, \ldots\}$ *such that*

$$\Psi(w^k + 2^{-i^k} d^k) \leq \Psi(w^k) + \sigma 2^{-i^k} \nabla\Psi(w^k)^T d^k. \tag{4.5}$$

(S.4) (Update)
Set $w^{k+1} = w^k + 2^{-i^k} d^k$, $k \leftarrow k + 1$, *choose* $\eta_k \geq 0$ *and go to (S.1).*

The stopping criterion at Step (S.1) can be substituted by any other suitable criterion without changing the properties of the algorithm. Note that the above algorithm makes use of the existence of $\nabla\Psi$ in several places, namely in the termination criterion, in test (4.4) in order to find out whether or not the inexact Newton direction d^k is a "good" descent direction, in taking $d^k = -\nabla\Psi(w^k)$ if this test is not satisfied, and, finally, in the Armijo line search test.

In what follows, as usual in analyzing the behaviour of algorithms, we shall assume that $\varepsilon = 0$ and that the algorithm produces an infinite sequence of points. The following result summarizes the main convergence properties of the algorithm.

Theorem 4.3. *Let* F *be a* C^1*-function and* g *and* h *be* C^2*-functions. Assume that* $\|r^k\| \leq \eta_k\|\Phi(w^k)\|$, *where* η_k *is a sequence of positive numbers such that* $\eta_k \leq \bar{\eta}$ *for every* k *with an arbitrary* $\bar{\eta} \in (0, 1)$. *Then the following assertions are valid:*

(a) Each accumulation point of the sequence $\{w^k\}$ *generated by the algorithm is a stationary point of* Ψ.

(b) If the sequence $\{w^k\}$ *has an accumulation point, let us say* w^*, *which is an isolated KKT-point of VI(X, F), then* $\{w^k\} \to w^*$.

(c) If the sequence $\{w^k\}$ *has an accumulation point, let us say* w^*, *which is a BD-regular solution of the system* $\Phi(w) = 0$, *then* $\{w^k\} \to w^*$. *Moreover, if* $\{\eta_k\} \to 0$, *then* $\{w^k\}$ *converges to* w^* *Q-superlinearly. Furthermore, if* $\eta_k = O(\|\Phi(w^k)\|)$, *if* F *is an* LC^1*-function and if* g *and* h *are* LC^2*-functions, then the convergence rate is Q-quadratic.*

Proof. First, part (a) will be proved by contradiction. Without loss of generality, it can be assumed that the direction is always given by (4.3). In fact, if, for an infinite set of indices K, we have $d^k = -\nabla\Psi(w^k)$ for all $k \in K$, then any limit point is a stationary point of Ψ by well-known results (see, e.g., Proposition 1.16 in [1]).

Suppose now, renumbering if necessary, that $\{w^k\} \to w^*$ and that $\nabla\Psi(w^*) \neq 0$. Since the direction d^k always satisfies (4.3), we get

$$\|\Phi(w^k) - r^k\| = \|H_k d^k\| \leq \|H_k\| \, \|d^k\| \tag{4.6}$$

which yields

$$\|d^k\| \geq \frac{\|\Phi(w^k) - r^k\|}{\|H_k\|}. \tag{4.7}$$

Note that $\|H_k\|$ cannot be 0. Otherwise, (4.6) would imply $\Phi(w^k) - r^k = 0$ which, in turn, since $\|r^k\| \leq \eta_k\|\Phi(w^k)\|$ with $\eta_k \leq \bar{\eta} < 1$, is only possible if $\|\Phi(w^k)\| = 0$, so that w^k would be a stationary point and the algorithm would have stopped.

We now show that

$$0 < \gamma_1 \leq \|d^k\| \leq \gamma_2 \tag{4.8}$$

for some positive γ_1 and γ_2. In fact, if, for some subsequence K, $\{\|d^k\|\}_K \to 0$ we have from (4.7) that $\{\|\Phi(w^k) - r^k\|\}_K \to 0$ because H_k is bounded on the bounded sequence $\{w^k\}$ by known properties of the generalized Jacobian. But then, by continuity and the fact that

$$\|\Phi(w^k) - r^k\| \geq \|\Phi(w^k)\| - \|r^k\| \geq \|\Phi(w^k)\| - \eta_k\|\Phi(w^k)\| \geq (1 - \bar{\eta})\|\Phi(w^k)\|,$$

we get $\Phi(w^*) = 0$. Thus, w^* is a solution of the variational inequality problem. This contradicts the assumption $\nabla\Psi(w^*) \neq 0$. On the other hand, $\|d^k\|$ cannot be unbounded because, taking into account that $\nabla\Psi(w^k)$ is bounded and $\kappa > 2$, this would contradict (4.4).

Then, since (4.5) holds at each iteration and Ψ is bounded from below, we have that $\{\Psi(w^{k+1}) - \Psi(w^k)\} \to 0$ which implies, by the line search test,

$$\{2^{-i^k}\nabla\Psi(w^k)^T d^k\} \to 0. \tag{4.9}$$

We want to show that 2^{-i^k} is bounded away from 0. Suppose the contrary. Then, subsequencing if necessary, we have that $\{2^{-i^k}\} \to 0$ so that at each iteration the stepsize is reduced at least once and (4.5) gives

$$\frac{\Psi(w^k + 2^{-(i^k-1)}d^k) - \Psi(w^k)}{2^{-(i^k-1)}} > \sigma\nabla\Psi(w^k)^T d^k. \tag{4.10}$$

With regard to (4.8), we can assume, subsequencing if necessary, that $\{d^k\} \to \bar{d} \neq 0$, so that, passing to the limit in (4.10), we get

$$\nabla\Psi(w^*)^T\bar{d} \geq \sigma\nabla\Psi(w^*)^T\bar{d}. \tag{4.11}$$

On the other hand, we also have, by (4.4), that $\nabla\Psi(w^*)^T\bar{d} \leq -\rho\|\bar{d}\|^\kappa < 0$, which contradicts (4.11). Hence 2^{-i^k} is bounded away from 0. But then (4.9) and (4.4) imply $\{d^k\} \to 0$ which contradicts (4.8). Therefore, $\nabla\Psi(w^*) = 0$ must be valid.

The proof of point (b) is identical to the proof of Theorem 3.1 (b) in [3] and can therefore be omitted. We note that part (b) can also be proved by exploiting Lemma 4.10 in Moré and Sorensen [22]. We now pass to the proof of point (c).

The fact that the whole sequence $\{w^k\}$ converges to w^* follows by part (b) noting that the BD-regularity assumption implies, by [23, Proposition 3], that w^* is an isolated solution of the system $\Phi(w) = 0$ and hence also an isolated KKT-point of VI(X, F). Then, we first prove that locally the direction is always the solution of system (4.3) and then that eventually the stepsize of one satisfies the line search test (4.5), so that the algorithm eventually reduces to the undamped inexact Newton method and the assertions on the convergence rate readily follow from Theorem 3.2. and Proposition 4.1..

Since $\{w^k\}$ converges to a BD-regular solution of the system $\Phi(w) = 0$, we have, by Proposition 2.1., that there exists a positive number C such that $\|H_k^{-1}\| \leq C$ for every k. Then, taking into account that we can write, for every vector v, $\|v\| \leq \|H_k^{-1}\|\|H_k v\|$, we get

$$m\|v\| \leq \|H_k v\|, \qquad \forall k,\ \forall v \in \mathbb{R}^n, \tag{4.12}$$

where $m = 1/C$. Furthermore, we also have, from the boundedness of the generalized Jacobian on bounded sets, that there is a positive constant M such that

$$\|H_k v\| \leq M\|v\|. \qquad \forall k,\ \forall v \in \mathbb{R}^n, \tag{4.13}$$

We now note that, since H_k is nonsingular for k sufficiently large, system (4.3) always admits a solution in the sense that $\|r^k\| \leq \eta_k\|\Phi(w^k)\|$. We want to show that this solution satisfies, for some positive ρ_1, the condition

$$\nabla\Psi(w^k)^T d^k \leq -\rho_1\|d^k\|^2. \tag{4.14}$$

Using (4.3), (4.12), and (4.13), we can write

$$m\|d^k\| \leq \|H_k d^k\| = \|\Phi(w^k) - r^k\| \leq M\|d^k\|$$

which, taking into account that η_k tends to 0 so that $\|r^k\| = o(\|\Phi(w^k)\|)$, gives

$$\frac{m}{2}\|d^k\| \leq \|\Phi(w^k)\| \leq 2M\|d^k\| \tag{4.15}$$

for all k sufficiently large. But then, since $\nabla\Psi(w^k)$ can be written as $H_k^T\Phi(w^k)$ by Proposition 4.2., we get from (4.3) and (4.15)

$$\begin{aligned} \nabla\Psi(w^k)^T d^k &= -\|\Phi(w^k)\|^2 + \Phi(w^k)^T r^k \\ &\leq -\frac{m^2}{4}\|d^k\|^2 + o(\|\Phi(w^k)\|^2) \\ &= -\frac{m^2}{4}\|d^k\|^2 + o(\|d^k\|^2) \leq -\frac{m^2}{8}\|d^k\|^2 \end{aligned} \tag{4.16}$$

for all k sufficiently large.

Then (4.14) follows from (4.16) by taking $\rho_1 \leq m^2/8$. But now, noting that $\{\|d^k\|\} \to 0$, it is easy to see that (4.14) eventually implies (4.4) for any $\kappa > 2$ and any positive ρ.

To complete the proof of the theorem it only remains to show that eventually the stepsize determined by the Armijo test (4.5) is 1, that is, that eventually $i^k = 0$. The statements on the local rate of convergence then follow immediately from Theorem 3.2.. The main step in order to show that $i^k = 0$ is eventually accepted by the Armijo test (4.5) is to show that there is a constant $\gamma > 0$ such that eventually

$$\Psi(w^k) + \sigma\nabla\Psi(w^k)^T d^k \geq \gamma\Psi(w^k). \tag{4.17}$$

Taking into account the definition of Ψ, the fact that we can write $\nabla\Psi(w^k) = H_k^T\Phi(w^k)$ with H_k being the matrix from (4.3) and the fact that d^k satisfies (4.3) with $\|r^k\| \leq \eta_k\|\Phi(w^k)\|$, we obtain from the Cauchy-Schwarz-inequality:

$$\begin{aligned}
\Psi(w^k) + \sigma\nabla\Psi(w^k)^T d^k &= \Psi(w^k) - \sigma\|\Phi(w^k)\|^2 + \sigma\Phi(w^k)^T r^k \\
&\geq \Psi(w^k) - \sigma\|\Phi(w^k)\|^2 - \sigma\|\Phi(w^k)\|\,\|r^k\| \\
&\geq \Psi(w^k) - \sigma(1+\eta_k)\|\Phi(w^k)\|^2 \\
&= (1 - 2\sigma - 2\sigma\eta_k)\Psi(w^k).
\end{aligned}$$

Hence there is a constant $\gamma > 0$ such that (4.17) holds for all k sufficiently large because $\sigma \in (0, 1/2)$ and $\eta_k \to 0$ by assumption. Using part (b) of Theorem 3.2. with $G = \Phi$, we get from Proposition 2.6. that the condition

$$\Psi(w^k + d^k) \leq \gamma\Psi(w^k)$$

is satisfied for all k sufficiently large. In particular, it then follows from (4.17) that the Armijo-test (4.5) is eventually satisfied with $i^k = 0$. This completes the proof. □

We stress that, for the global convergence part of the above theorem, it was sufficient to choose $\bar{\eta} \in (0,1)$ arbitrarily.

The above convergence theorem raises two questions:

- Under what assumptions is a KKT-point w^* of VI(X,F) a BD-regular solution (and hence an isolated solution) of the system $\Phi(w) = 0$?
- Under what assumptions is a stationary point of Ψ a KKT-point of VI(X,F) (and hence a solution of VI(X,F) itself under certain assumptions)?

The remaining part of this section is devoted to these questions.

Let $w = (x,y,z) \in \mathbb{R}^n \times \mathbb{R}^m \times \mathbb{R}^p$ be an arbitrary vector, and let us introduce the following two index sets:

$$\begin{aligned}
I_0(w) &:= \{i \in I \,|\, g_i(x) = 0 \text{ and } y_i \geq 0\}, \\
I_+(w) &:= \{i \in I \,|\, g_i(x) = 0 \text{ and } y_i > 0\}.
\end{aligned}$$

Note that, if $w = w^*$ is a KKT-point of VI(X,F), then $I_0(w^*)$ is the index set of all active constraints, whereas $I_+(w^*)$ contains the indices of the strongly active contraints. Obviously, we have $I_+(w) \subseteq I_0(w)$ at an arbitrary point w, and $I_+(w^*) = I_0(w^*)$ if and only if strict complementarity holds at a KKT-point w^*.

The following result gives a sufficient condition for a KKT-point w^* of VI(X,F) to ba a BD-regular solution of the equation $\Phi(w) = 0$ and therefore guarantees that parts (b) and (c) of Theorem 4.3. are applicable under the assumptions of this result.

Theorem 4.4. *Let F be a C^1-function and g and h be C^2-functions. Let $w^* := (x^*, y^*, z^*) \in \mathbb{R}^n \times \mathbb{R}^m \times \mathbb{R}^p$ be a KKT-point of VI(X,F). Suppose that*

(a) $v^T\nabla_x L(w^*)v > 0$ for all $v \in \mathbb{R}^n, v \neq 0$, such that $\nabla h_j(x^*)^T v = 0\,(j \in J)$ and $\nabla g_i(x^*)^T v = 0\,(i \in I_+(w^*))$,

(b) the gradients $\nabla h_j(x^*)\,(j \in J)$ and $\nabla g_i(x^*)\,(i \in I_0(w^*))$ are linearly independent.

Then all elements $H \in \partial\Phi(w^*)$ are nonsingular, in particular, w^* is a BD-regular solution of the system $\Phi(w) = 0$.

Proof. See Jiang [16] or [7]. □

We note that the above theorem holds without assuming strict complementarity. Hence, Theorem 4.3. guarantees local Q-superlinear and Q-quadratic convergence of the sequence $\{w^k\}$ even if this assumption is not satisfied.

We now turn to the question under what assumptions a stationary point of Ψ is a KKT-point of VI(X, F). We first consider the general problem.

Theorem 4.5. *Let F be a C^1-function and g and h be C^2-functions. Let $w^* := (x^*, y^*, z^*) \in \mathbb{R}^n \times \mathbb{R}^m \times \mathbb{R}^p$ be a stationary point of Ψ. Then w^* is a KKT-point of VI(X, F) if one of the following two assumptions holds:*

(1) (a) $v^T \nabla_x L(w^*) v > 0$ for all $v \in \mathbb{R}^n, v \neq 0$, such that $\nabla h_j(x^*)^T v = 0\,(j \in J)$ and $\nabla g_i(x^*)^T v = 0\,(i \in I_0(w^*))$, and

(b) the gradients $\nabla h_j(x^*)\,(j \in J)$ and $\nabla g_i(x^*)\,(i \in I_0(w^*))$ are linearly independent, or

(2) (a) $v^T \nabla_x L(w^*) v > 0$ for all $v \in \mathbb{R}^n, v \neq 0$, such that $\nabla h_j(x^*)^T v = 0\,(j \in J)$ and $\nabla g_i(x^*)^T v = 0\,(i \in I_+(w^*))$, and

(b) the gradients $\nabla h_j(x^*)\,(j \in J)$ and $\nabla g_i(x^*)\,(i \in I_+(w^*))$ are linearly independent.

Proof. See [7]. □

Note that the difference in assumptions (4.1) and (4.2) of Theorem 4.5. is that assumption (4.1) (a) is weaker than assumption (4.2) (a), but that assumption (4.1) (b) is stronger than assumption (4.2) (b).

If F is monotone, i.e., if

$$(x - y)^T [F(x) - F(y)] \geq 0 \qquad \forall x, y \in \mathbb{R}^n,$$

and if the feasible set X is polyhedral, one can prove the following result.

Theorem 4.6. *Assume that F is a monotone C^1-function and that X is polyhedral. Let w^* be a stationary point of Ψ. Then w^* is a KKT-point of VI(X, F) if one of the following conditions is satisfied:*

(1) The set X is bounded.

(2) The set X is contained in the nonnegative orthant $\mathbb{R}^n_+$.

Proof. See [7]. □

In particular, if the polyhedral set X is given in the standard form

$$X = \{x \in \mathbb{R}^n \,|\, Ax = b, x \geq 0\}$$

for some matrix $A \in \mathbb{R}^{m \times n}$ and a vector $b \in \mathbb{R}^m$, then assumption (4.2) of Theorem 4.6. is satisfied. Theorem 4.6. therefore says that, in most cases, our algorithm is able to solve monotone variational inequalities over polyhedral sets (recall that in this case the KKT-conditions and problem VI(X, F) itself are completely equivalent).

5. CONCLUDING REMARKS

In this paper, we first investigated the local rate of convergence of inexact Newton methods for a semismooth system of equations, and gave a complete characterization of its Q-superlinear and Q-quadratic convergence. We then applied these results to variational inequalities and presented a globally and locally fast convergent algorithm for the solution of VI(X, F).

We think that this method has several interesting properties, e.g.

- it just requires the inexact solution of one linear system at each iteration,
- it converges (at least) Q-superlinearly even if strict complementarity is not satisfied at a solution,
- it has stronger global convergence properties than many other methods, see [14,13].

We therefore believe that this algorithm could be a very powerful and efficient tool for solving variational inequalities, especially large-scale ones. Preliminary numerical results obtained for a set of small test problems are very promising.

REFERENCES

[1] D.P. Bertsekas, *"Constrained Optimization and Lagrange Multiplier Methods"*. Academic Press, New York, 1982.

[2] F.H. Clarke, *"Optimization and Nonsmooth Analysis"*. Wiley, New York, 1983 (reprinted by SIAM, Philadelphia, 1990).

[3] T. De Luca, F. Facchinei and C. Kanzow: *"A semismooth equation approach to the solution of nonlinear complementarity problems"*. DIS Technical Report 01.95, Università di Roma "La Sapienza", Roma, Italy, January 1995 (revised July 1995).

[4] R.S. Dembo, S.C. Eisenstat and T. Steihaug, *"Inexact Newton methods"*. SIAM Journal on Numerical Analysis 19, pp. 400–408, 1982.

[5] J.E. Dennis, Jr., and J.J. Moré, *"A characterization of the superlinear convergence and its application to quasi-Newton methods"*. Mathematics of Computation 28, pp. 549–560, 1974.

[6] J.E. Dennis, Jr., and R.B. Schnabel, *"Numerical Methods for Unconstrained Optimization and Nonlinear Equations"*. Prentice Hall, Englewood Cliffs, NJ, 1983.

[7] F. Facchinei, A. Fischer and C. Kanzow: *A semismooth Newton method for variational inequalities: Theoretical results and preliminary numerical experience.* Preprint MATH-NM-22-1995, Institute of Numerical Mathematics, Technical University of Dresden, Dresden, Germany, 1995.

[8] F. Facchinei and C. Kanzow, *"A nonsmooth inexact Newton method for the solution of large-scale nonlinear complementarity problems"*. Preprint 95, Institute of Applied Mathematics, University of Hamburg, Hamburg, Germany, May 1995.

[9] F. Facchinei and J. Soares: *A new merit function for nonlinear complementarity problems and a related algorithm.* DIS Technical Report 15.94, Università di Roma "La Sapienza", Roma, Italy, 1994. To appear in SIAM J. on Optimization.

[10] A. Fischer, *"A special Newton-type optimization method"*. Optimization 24, pp. 269–284, 1992.

[11] A. Fischer, *"A special Newton-type method for positive semidefinite linear complementarity problems"*. Preprint MATH-NM-04-1992, Technical University of Dresden, Dresden, Germany, 1992 (revised 1993). Journal of Optimization Theory and Applications, to appear.

[12] A. Fischer, *"Solution of monotone complementarity problems with locally Lipschitzian functions"*. Preprint MATH-NM-9-1995, Institute of Numerical Mathematics, Technical University of Dresden, Dresden, Germany, May 1995.

[13] M. Fukushima, *"Merit functions for variational inequality and complementarity problems"*. Technical Report, Nara Institute of Science and Technology, Nara, Japan, June 1995.

[14] P.T. Harker and J.-S. Pang, *"Finite-dimensional variational inequality and nonlinear complementarity problems: a survey of theory, algorithms and applications"*. Mathematical Programming 48, pp. 161–220, 1990.

[15] C. Geiger and C. Kanzow, *"On the resolution of monotone complementarity problems"*. Preprint 82, Institute of Applied Mathematics, University of Hamburg, Hamburg, Germany, April 1994 (revised February 1995). Computational Optimization and Applications, to appear.

[16] H. Jiang, *"Local properties of solutions of nonsmooth variational inequalities"*. Optimization 33, pp. 119–132, 1995.

[17] H. Jiang and L. Qi, *"A nonsmooth equations approach to nonlinear complementarities"*. Applied Mathematics Report AMR 94/31, School of Mathematics, University of New South Wales, Sydney, Australia, October 1994.

[18] C. Kanzow, *"Global convergence properties of some iterative methods for linear complementarity problems"*. Preprint 72, Institute of Applied Mathematics, University of Hamburg, Hamburg, Germany, June 1993 (revised 1994). SIAM Journal on Optimization, to appear.

[19] B. Kummer, *"Newton's method for non-differentiable functions"*. In "Mathematical Research, Advances in Mathematical Optimization", J. Guddat et al. (eds.) Akademie, Verlag, Berlin, Germany, pp. 114–125, 1988.

[20] J.M. Martínez and L. Qi, *"Inexact Newton methods for solving nonsmooth equations"*. Applied Mathematics Report 93/9, School of Mathematics, University of New South Wales, Sydney, Australia, 1993 (revised April 1994).

[21] R. Mifflin, *"Semismooth and semiconvex functions in constrained optimization"*. SIAM Journal on Control and Optimization 15, pp. 957–972, 1977.

[22] J.J. Moré and D.C. Sorensen, *"Computing a trust region step"*. SIAM Journal on Scientific and Statistical Computing 4, pp. 553–572, 1983.

[23] J.-S. Pang and L. Qi, *"Nonsmooth equations: motivation and algorithms"*. SIAM Journal on Optimization 3, pp. 443–465, 1993.

[24] L. Qi, *"A convergence analysis of some algorithms for solving nonsmooth equations"*. Mathematics of Operations Research 18, pp. 227–244, 1993.

[25] L. Qi and J. Sun, *"A nonsmooth version of Newton's method"*. Mathematical Programming 58, pp. 353–368, 1993.

[26] P. Tseng, *"Growth behaviour of a class of merit functions for the nonlinear complementarity problem"*. Technical Report, Department of Mathematics, University of Washington, Seattle, May 1994 (revised March 1995). Journal of Optimization Theory and Applications, to appear.

[27] N. Yamashita and M. Fukushima, *"Modified Newton methods for solving semismooth reformulations of monotone complementarity problems"*. Technical Report TR-IS-95021, Nara Institute of Science and Technology, Nara, Japan, May 1995.

ACCESSING REALISTIC MIXED COMPLEMENTARITY PROBLEMS WITHIN MATLAB

Ferris Michael C.[1] and Rutherford Thomas F.[2]

[1] Computer Sciences Department
University of Wisconsin
Madison, Wisconsin 53706
The work of this author was based on research supported by
the Department of Energy grant DE-FG03-94ER61915,
the Air Force Office of Scientific Research grant F49620-94-1-0036
and the National Science Foundation grant CCR-9157632.
e-mail: ferris@cs.wisc.edu

[2] Department of Economics
The University of Colorado
Boulder, Colorado 80309
The work of this author was based on research supported by
the Department of Energy grant DE-FG03-94ER61915.

Abstract. This paper describes a suite of programs that allow realistic complementarity problems to be accessed from within the MATLAB programming environment. The suite of programs uses data generated from the GAMS/MCP library and makes function and Jacobian evaluations available directly to the MATLAB user. All the required data is stored in a binary encoding of the problems, completely independent of GAMS. Use of the programs is described here, along with a list of all the problems that are currently available in binary form. Details on how the binary files are generated from GAMS are also given.

Key words. Mixed complementarity problems, algorithms, modeling.

1. INTRODUCTION

In recent years there has been an enormous amount of interest in developing algorithms for solving nonlinear and mixed complementarity problems. MATLAB is a package of routines for numeric computation and visualization that provides an excellent environment for prototyping such algorithms. Many basic tools, such as factorizations, matrix multiplication and addition and eigenstructure analysis are provided both in a dense and sparse matrix setting. Furthermore, other tools for matrix manipulation and visualization enable many complex algorithms to be succinctly and quickly implemented, and underlying structure to be identified.

In order to exploit these features while prototyping algorithms for complementarity problems, there needs to be a library of test problems readily available within MATLAB for testing purposes.

A suite of problems that arise from realistic applications, with reasonable size and real data, is already available within the GAMS/MCP setting as described in [3] and [7]. Other problems [11] are currently under development within this setting. This has led to broad testing of some of algorithms via the GAMS/MCP interface described in [10]. However, most of the algorithms remain largely untested in this computational setting.

This paper describes a suite of programs that are designed to make the GAMS/MCP complementarity problems readily accessible within the MATLAB [12] programming environment, without the overhead of trying to write specialized code to evaluate the underlying nonlinear functions. All of the programs are publicly available and can be obtained by anonymous ftp from `ftp.cs.wisc.edu:math-prog/matlab/` or from the authors. All the problems contained in the GAMS library and MCPLIB are freely available in the form of machine specific binary files; the data contained within these files are accessed by the aforementioned programs. The third part of the interface is a program that allows a binary file to be generated from any GAMS/MCP model. It is anticipated that this will allow the suite of problems to grow and become even more comprehensive.

The actual complementarity problems that we will generate are in the MCP format, that is, to find $z \in [l, u]$ with

$$\text{MCP}: \quad \begin{aligned} &f(z) = w - v \\ &w \geq 0, \; w^T(z - l) = 0 \\ &v \geq 0, \; v^T(u - z) = 0. \end{aligned}$$

To completely specify this problem, the function f and the values of l and u (which may possibly be infinite) must be given. For algorithm development, it is usually necessary to provide starting values for z and routines that give an evaluation of $f(z)$ and $J(z)$, the function and its Jacobian at any given point.

The interface we provide consists of a large number of problem files and several mex functions, that give the MATLAB user all of the above. Thus, the routines and the data that are provided completely specify the underlying complementarity problem. The programs essentially consist of a function and Jacobian evaluator for all of the complementarity problems. While the data necessary for these evaluations was originally obtained directly from the GAMS/MCP interface, the function evaluators are completely independent of GAMS; each problem is encoded within a single machine dependent binary file. This format was chosen essentially for speed: we believe it crucial that the function evaluations can be carried out as efficiently and quickly as possible. The binary files are freely available from the same sources as the programs.

The remainder of the paper is organized as follows. In the next section, we give a brief description of the programs, or more correctly, the `mex` files that we provide to access the problem data, and give some simple examples of how we expect these functions will be used. In the following section, a list of all the currently available problems, with some size and density information is given. Section 4 describes how the binary files were originally generated and gives details on some further advanced features of the interface. In particular, the technique a user might invoke for generating binary files from a new GAMS file is outlined. We conclude with some suggestions of possible future extensions of this work.

2. USING THE INTERFACE

The first step is to obtain the mex files and the binary problem files either from the authors or by anonymous ftp from `ftp.cs.wisc.edu:math-prog/matlab`. The interface can be used without knowing anything about the models or GAMS/MCP by just obtaining copies of the relevant binary files for your machine architecture running MATLAB. The problem files are contained in two machine dependent compressed files labeled "gamsmat" and "mcplib". The first file contains the binary files for the problems from the GAMS library, and the second contains those for MCPLIB. The original gams source files, typically having extensions `gms`, provide a valuable source of documentation for the problems. The `mex` functions can be found in the compressed file "mexfiles". The form of compression and the file extensions are different for each machine architecture. The `mex` files should be placed on the current search path of MATLAB.

Simple use of the interface consists of three `mex` files that enable MATLAB to extract from the binary file the specified starting point, the bounds on the variables, and a function and Jacobian evaluation at any given point. These mex files are `cpsetup`, `cpfun`, and `cpjac`.

These three functions are now described in more detail for a specific problem. We assume that a file `josephy.bin` is in the current MATLAB directory.

In a directory containing the relevant binary files (which should always have the extension `bin`), MATLAB should be invoked. To set up the complementarity problem in MATLAB, the following call is specified:

```
matlab> global wsvec cpfname
matlab> [z,l,u] = cpsetup('josephy');
```

This call performs the following operations. Firstly, two vectors are put into the global workspace. The vector `cpfname` contains a string which holds the name of the binary file currently in use. The vector `wsvec` is a workspace vector for use by the interface routines that is dynamically allocated within the `cpsetup` routine and is used by the other mex functions for workspace in evaluating the functions and Jacobian. Both of these vectors are generated and stored in the MATLAB workspace by `cpsetup`. After initializing these vectors, `cpsetup` determines the initial starting point `z`, and the lower and upper bounds `l,u` that specify the mixed complementarity problem (MCP). Note that the first starting point that is encountered in the GAMS problem file is the one that is returned by `cpsetup`. If any other starting point is required, the user should set it explicitly after this call, using the function `cpstart`.

After setting up the problem, to evaluate the function f of the MCP, the following call is needed:

```
matlab> f = cpfun(z);
```

The vector `f` contains the values of the function evaluated at the point `z`.

To evaluate the Jacobian `J` corresponding to the MCP, the following call is needed, which also must come after the initialization of the problem using `cpsetup`:

```
matlab> [f,J] = cpjac(z);
```

The matrix `J` is a MATLAB sparse matrix containing the evaluation of $\nabla f(z)$. Note that the (dense) vector `f` is output at the same time containing an evaluation of $f(z)$. This vector should be identical to that given by `cpfun`.

At this point, any of the MATLAB utilities can be used to formulate an algorithm, inspect the sparsity pattern of the Jacobian or modify the current iterate. We now give three examples of the use of the interface.

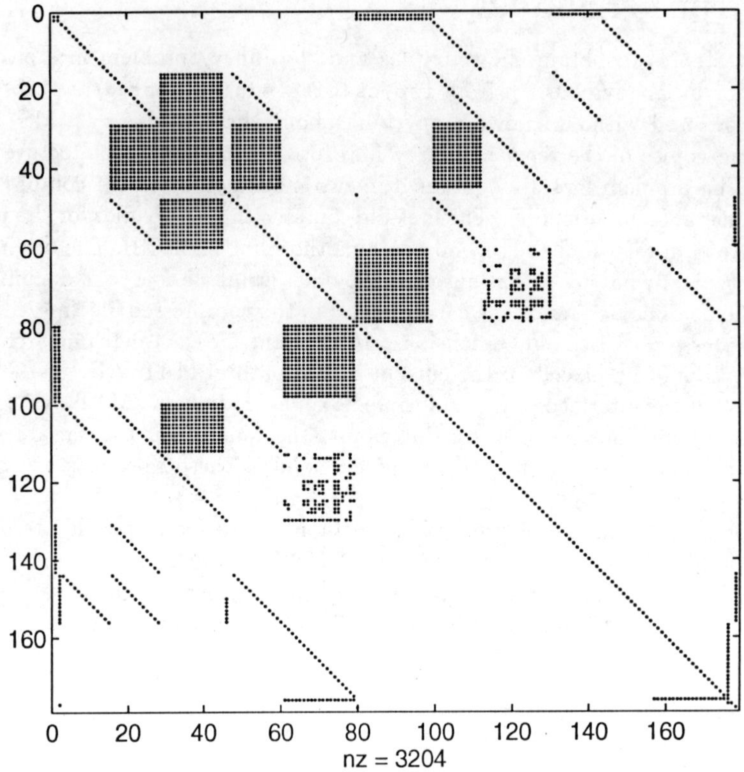

Figure 1: Jacobian of model gemmge

In the first example, we show how the interface may be effective for visualization of problem structure. This in turn may lead to improved modeling, or to algorithms that can exploit the underlying structure of many complementarity problems.

```
matlab> global wsvec cpfname
matlab> [z,l,u] = cpsetup('gemmge');
matlab> [f,J] = cpjac(z);
matlab> spy(J);
```

The resulting plot of the Jacobian of this model is given in Figure 1. The postscript file used here was generated after the "spy" command, by invoking "print -dps gemmge". At this stage, we can demonstrate some of the potentially useful features MATLAB for algorithmic design. For large scale problems, many algorithms are limited by their ability to solve large sparse systems of equations. Frequently, reordering of the matrices improves sparsity of the factors without creating numerical difficulties. MATLAB has several popular orderings included in its toolbox, for example the minimum degree ordering and the reverse Cuthill-McKee ordering. The resulting plots from applying these orderings to the "gemmge" Jacobian are given in Figures 2 and 3, along with a plot showing the resulting fill in the *LU* factors. Minimum degree clearly outperforms reverse Cuthill-McKee on this problem.

The second and third examples are matlab "m" files that use the interface routines to implement a simple projected gradient and projected Newton solver for complementarity problems. The first step in generating these algorithms is to set up a merit function.

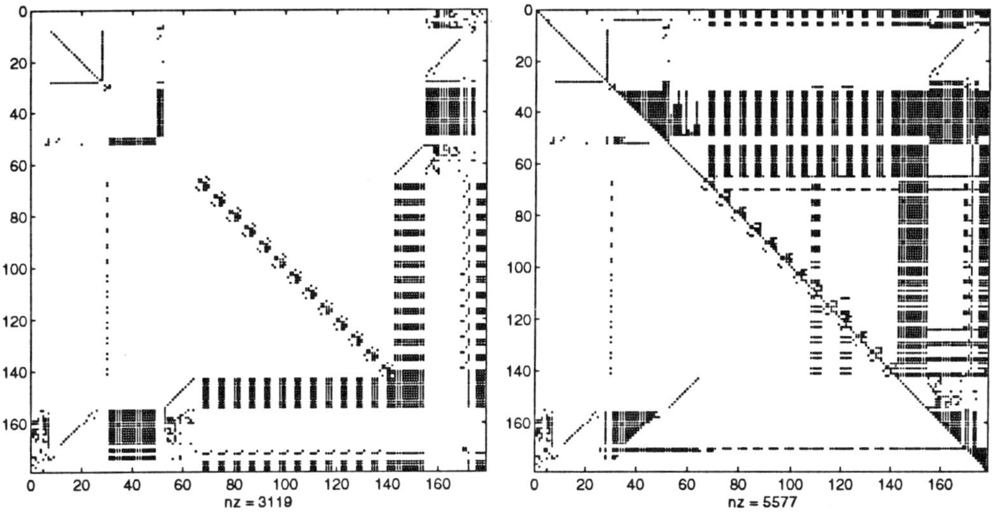

Figure 2: Jacobian reordered by Minimum Degree and resulting sparsity of L+U

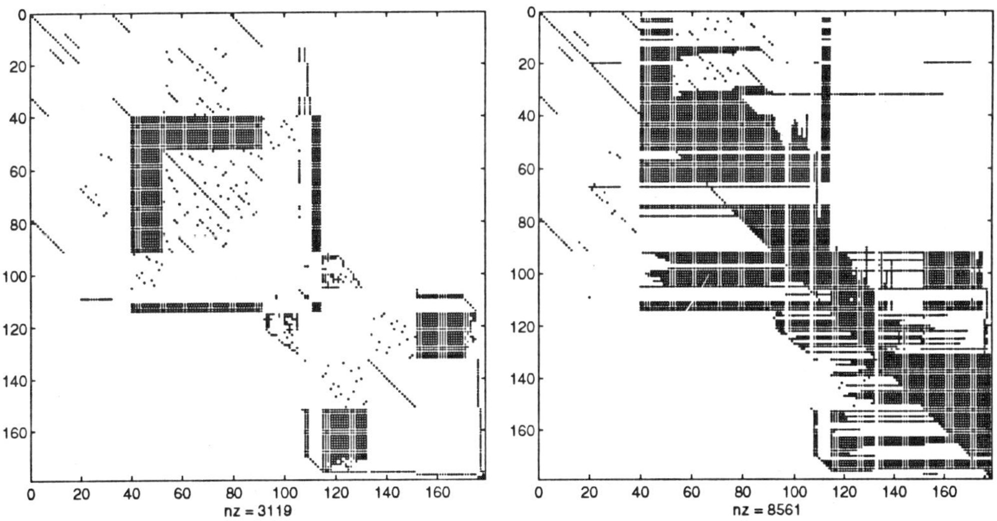

Figure 3: Jacobian reordered by Reverse Cuthill-McKee and resulting sparsity of L+U

The file displayed in Figure 4, "merit.m" evaluates

$$\sqrt{\sum_{i:z_i<=l_i} (f_i(z))_-^2 + \sum_{i:z_i>=u_i} (f_i(z))_+^2 + \sum_{i:l_i<z_i<u_i} f_i(z)^2}$$

assuming it is given the current values of z, l, u and $f(z)$. The subscripts + and − correspond to the positive and negative parts of f respectively. Note that the global variables wsvec and cpfname are not needed within this routine. Once this function has

```
function norm = merit (z, f, l, u)

norm = 0;
for i=1:length(z)
    if (z(i) <= l(i))
      dt = min(0,f(i)); norm = norm + dt*dt;
    elseif (z(i) >= u(i))
      dt = max(0,f(i)); norm = norm + dt*dt;
    else
      norm = norm + f(i)*f(i);
    end
end
norm = sqrt(norm);
```

Figure 4: MATLAB code to evaluate merit function

been implemented, the code in Figure 5 can be used to set up a simple projected gradient algorithm for solving the MCP. These matlab codes can be invoked at the MATLAB command line as follows.

```
matlab> global wsvec cpfname
matlab> [z,l,u] = cpsetup('gemmge');
matlab> znew = pgrad(z,l,u);
```

A more sophisticated code is based on the projected Newton preprocessor implemented as part of PATH [8, 9]. A full description is given in [6], but the file in Figure 6, "pnewt.m" contains all of the main points. This algorithm can be invoked in the same manner as the pgrad algorithm, except that pgrad is replaced by pnewt. Note that it is easy to update the "m" file to test out the effectiveness of least squares approaches for solving the system of equations defining the search direction, or to implement proximal perturbations in the problem of finding the Newton direction.

These examples demonstrate the ease with which algorithms can be prototyped and tested using the MATLAB interface. Indeed, Tseng [16] has generated a suite of algorithms for complementarity problems which were easily modified to use this interface. However, the results using these codes are perhaps not illuminating, since all of these codes fail on a large subset of the models. It is currently unclear whether this is due to the simple implementation of the algorithms, or due to properties of the algorithms themselves. While it is easy to prototype algorithms on many problems using the MATLAB interface, it is important to realize that practical codes for complementarity problems, such as PATH [8], MILES [14], SMOOTH [4], NE/SQP [13], SEMISMOOTH [5] and QPCOMP [2] have many features added that make them much more effective than a simple MATLAB encoding would demonstrate. A detailed comparison of the above

```
function z = pgrad(z,l,u)
global wsvec cpfname
step_tol = 1e-2;

[f,J] = cpjac(z); residual = merit(z,f,l,u),

for iter=1:10
  alpha = 1.0;
  while (alpha > step_tol)
    znew = min(u,max(l,z - alpha*f));
    fnew = cpfun(znew); resnew = merit(znew,fnew,l,u);
    if (resnew < residual)
      break; end
    alpha = alpha * 0.5;
  end

  if (alpha <= step_tol)
    disp('linesearch failure'); return; end

  z = znew; f = fnew; residual = resnew,
  if (residual < 1e-6)
    return; end
end;
```

Figure 5: MATLAB code for projected gradient algorithm

```
function z = pnewt(z,l,u)
global wsvec cpfname
step_tol = 1e-2;

[f,J] = cpjac(z); residual = merit(z,f,l,u),

for iter=1:10
  basiscount = 0; needed = [];
  for i=1:length(z)
    if ~((z(i) <= l(i) & f(i) >= 0.) | (z(i) >= u(i) & f(i) <= 0.))
      basiscount = basiscount+1; needed(basiscount) = i;
    end
  end

  r = f(needed); B = J(needed,needed); d = B\r;
  dd = sparse(size(z)); dd(needed) = d;

  alpha = 1.0;
  while (alpha > step_tol)
    znew = min(u,max(l,z - alpha*dd));
    fnew = cpfun(znew); resnew = merit(znew,fnew,l,u);
    if (resnew < residual)
      break; end
    alpha = alpha * 0.5;
  end

  if (alpha <= step_tol)
    disp('linesearch failure'); return; end

  z = znew; residual = resnew,
  if (residual < 1e-6)
    return; end
  [f,J] = cpjac(z);
end;
```

Figure 6: MATLAB code for projected Newton algorithm

codes has been given in [1]. Discerning how to use the interface in the best manner possible remains the topic of further research.

3. PROBLEM CHARACTERISTICS

The current set of problems arise from two sources, the library of GAMS test problems and MCPLIB. The following two tables correspond to these two sources. Further documentation on the problems can be found in [15] and [7] respectively.

Note that the following abbreviations are used when referring to the type of the problem:

MCP	General mixed complementarity problem
MPSGE	General economic equilibrium from MPSGE
LCP	Linear complementarity problem
NCP	Nonlinear complementarity problem
LMCP	Linear mixed complementarity problem
NE	Nonlinear equations
NLP	Optimality conditions of a nonlinear program

4. FILE GENERATION & ADVANCED OPTIONS

The interface between GAMS/MCP and MATLAB is created using CPLIB [10] and a single binary file. The binary file is machine specific, being written using the I/O library facilities of GAMS. Many platforms are supported by this library. If the user has access to GAMS/MCP, it is possible to generate new binary files incorporating any GAMS/MCP model using the following procedure.

To generate a binary file, a GAMS/MCP model is created as described in [10]. A line is added to the `gms` file after the model specification that reads

```
option mcp = matlab;
solve josephy using mcp;
```

where `josephy` is the name of the model. A GAMS solver "matlab" is invoked at the solve instruction that dumps all the relevant information into a machine specific binary file, "matlab.bin" in the current directory. This file can then be renamed and used independently of GAMS.

We conclude this section with a complete syntactical description of the currently available `mex` and `m` files.

```
[z,l,u] = cpsetup('josephy');
```

The input argument is a string, which corresponds to a file "josephy.bin" that is located in the current directory. The three output arguments are required and contain starting values for z, along with the problem data l and u. Values of $\pm 10^{20}$ should be treated as infinite. This routine must be called before any other, and will write the vectors `wsvec` and `cpfname` into the matlab workspace. Typically, these vectors should be made global. The returned starting point is the first one encountered in the associated `gms` file.

```
z = cpstart('josephy',k);
```

This function returns the `k`th starting point for the problem named "josephy" from the `gms` file. An error occurs if the requested starting point is not found in this m-file.

Model	Type	n	nnz	density
cafemge	MPSGE	101	900	8.82%
cammcp	NCP	242	1621	2.77%
cammge	MPSGE	128	1227	7.49%
cirimge	MPSGE	9	33	40.74%
co2mge	MPSGE	208	1463	3.38%
dmcmge	MPSGE	170	1594	5.52%
ers82mcp	MCP	232	1552	2.88%
etamge	MPSGE	114	848	6.53%
finmge	MPSGE	153	1915	8.18%
gemmcp	MCP	262	2793	4.07%
gemmge	MPSGE	178	3441	10.86%
hansmcp	NCP	43	398	21.53%
hansmge	MPSGE	43	503	27.20%
harkmcp	NCP	32	131	12.79%
harmge	MPSGE	11	60	49.59%
kehomge	MPSGE	9	75	92.59%
kormcp	MCP	78	423	6.95%
mr5mcp	NCP	350	1687	1.38%
nsmge	MPSGE	212	1408	3.13%
oligomcp	NCP	6	21	58.33%
sammge	MPSGE	23	117	22.12%
scarfmcp	NCP	18	150	46.30%
scarfmge	MPSGE	18	181	55.86%
shovmge	MPSGE	51	375	14.42%
threemge	MPSGE	9	77	95.06%
transmcp	LCP	11	34	28.10%
two3mcp	NCP	6	29	80.56%
unstmge	MPSGE	5	25	100.00%
vonthmcp	NCP	125	760	4.86%
vonthmge	MPSGE	80	594	9.28%
wallmcp	NE	6	25	69.44%

Table 1: GAMSLIB Models

Model	Type	n	nnz	density
bertsekas	NCP	15	74	32.89%
billups	NCP	1	1	100.00%
bert_oc	LMCP	5000	21991	0.09%
bratu	NLP	5625	33749	0.11%
choi	NCP	13	169	100.00%
colvdual	NLP	20	168	42.00%
colvnlp	NLP	15	113	50.22%
cycle	LCP	1	1	100.00%
ehl_k60	MCP	61	3721	100.00%
ehl_k80	MCP	81	6561	100.00%
ehl_kost	MCP	101	10201	100.00%
explcp	LCP	16	152	59.38%
freebert	MCP	15	74	32.89%
gafni	MCP	5	25	100.00%
hanskoop	NCP	14	129	65.82%
hydroc06	NE	29	222	26.40%
hydroc20	NE	99	838	8.55%
josephy	NCP	4	16	100.00%
kojshin	NCP	4	16	100.00%
mathinum	NCP	3	9	100.00%
mathisum	NCP	4	14	87.50%
methan08	NE	31	225	23.41%
nash	MCP	10	100	100.00%
obstacle	LMCP	2500	14999	0.24%
opt_cont	LMCP	288	4928	5.94%
pgvon105	NCP	105	796	7.22%
pgvon106	NCP	106	898	7.99%
pies	MCP	42	183	10.37%
powell	NLP	16	203	79.30%
powell_mcp	NCP	8	54	84.38%
scarfanum	NCP	13	98	57.99%
scarfasum	NCP	14	109	55.61%
scarfbnum	NCP	39	361	23.73%
scarfbsum	NCP	40	614	38.38%
sppe	NCP	27	110	15.09%
tobin	NCP	42	243	13.78%

Table 2: MCPLIB Models

```
[f,domerr] = cpfun(z);
```

The input argument z is a vector containing the point at which the function is to be evaluated. The function has a variable number of output arguments. The first output argument is required and is a vector of the same size as z containing an evaluation of $f(z)$. The second argument domerr is optional and contains the number of domain violations that occurred during the function evaluation. If the second output argument is omitted and a domain violation occurs, then a MATLAB error is generated.

```
[f,J,domerr] = cpjac(z);
```

The input argument z and first output argument f are identical to those of the function cpfun. The domerr output argument is optional and acts exactly as described above for cpfun. The required output J is an evaluation of $\nabla f(z)$ and is returned in MATLAB sparse matrix form.

```
\begin{verbatim}
[f,J,domerr] = cpfunjac(z,flag);
```

The first input argument z and first output argument f are identical to those of the function `cpfun`. The domerr output argument is optional and acts exactly as described above for `cpfun`. The second input argument $flag$ is 1 if J should contain an evaluation of $\nabla f(z)$ (which will be returned in MATLAB sparse matrix form), and is 0 if no such evaluation should be performed. Currently, this is implemented as an m-file that simply calls `cpfun` or `cpjac`.

```
t = cptype(v);
```

In conjunction with the gms files that were used to generate the binary files, it may sometimes be helpful to extract "name" information corresponding to variables in the model. The following function gives limited access to these names. Furthermore, for models that were originally generated as an MPSGE model [15], the output vector t holds variable type indicators, with the following meanings:

t(i)	=	1	production sector
		2	commodity price
		3	consumer income
		0	auxiliary variable (no structural information)

The input vector v is a mask indicating what names are required (v(i)=0 if that name is to be skipped). One use of this function is to mask out the auxiliary variables and corresponding equations from the Jacobian. The resulting matrix typically consists of a symmetric part and a skew symmetric part, corresponding to the production and consumer portions of the model.

5. CONCLUSIONS

The interface consists of a freely available suite of programs and binary files that make a substantial library of problems available within MATLAB. The data, function and Jacobian evaluations can be used independently of the GAMS/MCP program that was used to generate the binary problem files. However, the corresponding gms files provide a nice documentation of the underlying models.

The binary format means that the function and Jacobian evaluations can be executed quickly. Furthermore, many commercial enterprises may be willing to provide such binary files for algorithmic testing since it enables their actual data to remain anonymous. This could be crucial in enlarging the suite of available problems.

Clearly, the MATLAB interface is just one technique for accessing the data from the binary files. Future developments of the programs could make a similar Fortran or C interface available for these problems. It may also become necessary to allow the MATLAB user to access more of the data in the binary file from within MATLAB, when designing algorithms to better exploit structure. This needs to be determined after experimentation with the current interface.

REFERENCES

[1] S. C. Billups, S. P. Dirkse, and M. C. Ferris, "A comparison of algorithms for large scale mixed complementarity problems". Mathematical Programming Technical Report 95–16, Computer Sciences Department, University of Wisconsin, Madison, Wisconsin, 1995. Available from ftp://ftp.cs.wisc.edu/math-prog/tech-reports/.

[2] S. C. Billups and M. C. Ferris, "QPCOMP: A quadratic program based solver for mixed complementarity problems". Mathematical Programming Technical Report 95–09, Computer Sciences Department, University of Wisconsin, Madison, Wisconsin, 1995. Available from ftp://ftp.cs.wisc.edu/math-prog/tech-reports/.

[3] A. Brooke, D. Kendrick, and A. Meeraus, *"GAMS: A User's Guide"*. The Scientific Press, South San Francisco, CA, 1988.

[4] C. Chen and O. L. Mangasarian, "A class of smoothing functions for nonlinear and mixed complementarity problems". *Computational Optimization and Applications, forthcoming*, 1995.

[5] T. De Luca, F. Facchinei, and C. Kanzow, "A semismooth equation approach to the solution of nonlinear complementarity problems". Submitted to Mathematical Programming, 1995.

[6] S. P. Dirkse, *"Robust Solution of Mixed Complementarity Problems"*. PhD thesis, Computer Sciences Department, University of Wisconsin, Madison, Wisconsin, 1994. Available from ftp://ftp.cs.wisc.edu/math-prog/tech-reports/.

[7] S. P. Dirkse and M. C. Ferris, "MCPLIB: A collection of nonlinear mixed complementarity problems". *Optimization Methods and Software*, 5:319–345, 1995.

[8] S. P. Dirkse and M. C. Ferris, "The PATH solver: A non-monotone stabilization scheme for mixed complementarity problems". *Optimization Methods and Software*, 5:123–156, 1995.

[9] S. P. Dirkse and M. C. Ferris, "A pathsearch damped Newton method for computing general equilibria". *Annals of Operations Research, forthcoming*, 1995.

[10] S. P. Dirkse, M. C. Ferris, P. V. Preckel, and T. Rutherford, "The GAMS callable program library for variational and complementarity solvers". Mathematical Programming Technical Report 94-07, Computer Sciences Department, University of Wisconsin, Madison, Wisconsin, 1994. Available from ftp://ftp.cs.wisc.edu/math-prog/tech-reports/.

[11] M. C. Ferris and J. S. Pang, "Engineering and economic applications of complementarity problems". Discussion Papers in Economics 95–4, Department of Economics, University of Colorado, Boulder, Colorado, 1995. Available from ftp://ftp.cs.wisc.edu/math-prog/tech-reports/.

[12] MATLAB. *User's Guide*. The MathWorks, Inc., 1992.

[13] J. S. Pang and S. A. Gabriel, "NE/SQP: A robust algorithm for the nonlinear complementarity problem". *Mathematical Programming*, 60:295–338, 1993.

[14] T. F. Rutherford, "MILES: A mixed inequality and nonlinear equation solver". Working Paper, Department of Economics, University of Colorado, Boulder, 1993.

[15] T. F. Rutherford, "Extensions of GAMS for complementarity problems arising in applied economic analysis". *Journal of Economic Dynamics and Control, forthcoming*, 1995.

[16] P. Tseng, "Matlab implementations of complementarity codes". Private Communication, 1995.

MERIT FUNCTIONS FOR VARIATIONAL INEQUALITY AND COMPLEMENTARITY PROBLEMS

Fukushima Masao

Graduate School of Information Science
Nara Institute of Science and Technology
Ikoma, Nara 630-01, Japan
e-mail: fuku@is.aist-nara.ac.jp

Abstract. This paper presents an overview of the recent effort in reformulating the variational inequality problem and the complementarity problem as an equivalent optimization problem with certain desirable properties. Merit functions used to formulate such equivalent optimization problems turn out to be useful in designing a globally convergent algorithm and provide error bounds for those problems under appropriate conditions.

Key Words. Variational inequality problem, complementarity problem, optimization problem, merit function, gap function.

1. INTRODUCTION

In recent years, much effort has been made to reformulate a variational inequality (VI) as an equivalent optimization problem with certain desirable properties, thereby designing new descent algorithms for solving the VI and deriving error bounds which are useful in estimating how far any given point is to the solution set of the VI. In this paper, we overview such attempts to the VI and the complementarity problem (CP) which is an important subclass of the VI.

The *variational inequality (VI)* here considered consists in finding a point $x \in S$ such that

$$\langle F(x), y - x \rangle \geq 0, \quad \forall\, y \in S, \tag{1.1}$$

where S is a nonempty closed convex subset of $\mathbb{R}^n$, F is a continuous mapping from $\mathbb{R}^n$ into $\mathbb{R}^n$, and $\langle \cdot, \cdot \rangle$ denotes the inner product in $\mathbb{R}^n$. In the special case where $S = \{x \in \mathbb{R}^n \mid x_i \geq 0, i = 1, \cdots, n\}$, the VI (1.1) is equivalent to the following *complementarity problem (CP)*:

$$F(x) \geq 0, \;\; x \geq 0, \;\; \langle F(x), x \rangle = 0. \tag{1.2}$$

When F is an affine mapping such that $F(x) = Mx + q$ with an $n \times n$ matrix M and an n-vector q, the CP becomes the *linear complementarity problem (LCP)*:

$$Mx + q \geq 0, \;\; x \geq 0, \;\; \langle Mx + q, x \rangle = 0. \tag{1.3}$$

A comprehensive treatment of the VI and CP may be found in [3,17,32,37].

Nonlinear Optimization and Applications
Edited by G. Di Pillo and F. Giannessi, Plenum Press, New York, 1996

We say that the mapping $F : \mathbb{R}^n \to \mathbb{R}^n$ is *monotone* if

$$\langle F(x) - F(x'), x - x' \rangle \geq 0, \quad \forall x, x' \in \mathbb{R}^n,$$

strictly monotone if

$$\langle F(x) - F(x'), x - x' \rangle > 0, \quad \forall x, x' \in \mathbb{R}^n \ (x \neq x'),$$

and *strongly* (or *uniformly*) *monotone* with modulus $\mu (> 0)$ if

$$\langle F(x) - F(x'), x - x' \rangle > \mu \|x - x'\|^2, \quad \forall x, x' \in \mathbb{R}^n.$$

Clearly any strongly monotone mapping is strictly monotone and any strictly monotone mapping is monotone. The following relations are well known [34, Theorem 5.4.3]: Let F be continuously differentiable. Then F is monotone if and only if the Jacobian matrix $\nabla F(x)$ is positive semidefinite for all x; F is strictly monotone if $\nabla F(x)$ is positive definite for all x; and F is strongly monotone if and only if $\nabla F(x)$ is uniformly positive definite, which is equivalent to saying that the minimum eigenvalues of the symmetric matrices $\nabla F(x) + \nabla F(x)^{\top}$ are bounded away from zero on $\mathbb{R}^n$, where $\top$ denotes transposition. Throughout the paper, we suppose that F is continuously differentiable everywhere.

2. MERIT FUNCTIONS FOR VI

A real-valued function f on $\mathbb{R}^n$ is called a *merit function* for the VI (1.1) if the set of global minimizers of f on a given set $X \subset \mathbb{R}^n$ coincides with the set of solutions to the VI, where X is usually the set S involved in the VI itself or the entire space $\mathbb{R}^n$. If f is a merit function for the VI, then the VI can be reformulated as the following optimization problem:

$$\text{minimize} \ \ f(x) \ \ \text{subject to} \ \ x \in X. \tag{2.1}$$

Desirable properties the merit function f is expected to have may be listed as follows:

- f is differentiable.
- Any stationary point or local minimum of f on X is also a global minimum of f on X.
- f provides a global (respectively, local) error bound for the VI, i.e., for any given point x, the distance from x to the solution set of the VI can be bounded by the value of $f(x)$ multiplied by some positive constant on the whole space $\mathbb{R}^n$ (respectively, on some neighborhood of the solution set).

2.1. Classical gap functions

The first merit function for the VI is the *gap function* introduced by Auslender [2] and Hearn [18]. This function is defined by

$$g(x) = \sup\{\langle F(x), x - y \rangle \ : \ y \in S\}. \tag{2.2}$$

If the set S is bounded, then the function g is finite everywhere. Otherwise, there may exist a point x such that $g(x) = +\infty$.

It is not difficult to see that the gap function g has the following properties:

- $g(x) \geq 0$ for all $x \in S$.
- $g(x) = 0$ and $x \in S$ if and only if x is a solution to the VIP.

These two properties imply that the VIP (1.1) is equivalent to the optimization problem

$$\text{minimize } \; g(x) \quad \text{subject to} \; x \in S. \tag{2.3}$$

Hence g is a merit function for the VI. Note that this equivalence does not require any monotonicity assumption on the mapping F.

Now let

$$\Omega(x) = \arg\max \{\langle F(x), x - y\rangle \; : \; y \in S\}.$$

Then the function g is differentiable at x if and only if

$$\nabla F(x)y = \nabla F(x)y' \quad \forall y, y' \in \Omega(x). \tag{2.4}$$

If (2.4) holds, the gradient of g at x is actually given by

$$\nabla g(x) = F(x) - \nabla F(x)(y - x)$$

for any $y \in \Omega(x)$. Clearly, the condition (2.4) is satisfied whenever the set $\Omega(x)$ is a singleton. As pointed out by Auslender [2, p. 139], the latter condition is met when the set S is "strongly" convex in the sense that, for all $x, x' \in S$ such that $x \neq x'$ and for all $\lambda \in (0,1)$, there exists a constant $r > 0$ such that

$$\| \lambda x + (1-\lambda)x' - z \| \leq r \quad \Longrightarrow \quad z \in S.$$

Unfortunately, the strong convexity of S is too restrictive a condition to be met in many practical applications. For instance, it is never satisfied by a polyhedral convex set. Thus we cannot in general expect that the gap function g is differentiable everywhere.

Since the gap function g yields an optimization problem equivalent to the VI, it is quite natural to try to solve the VI by iteratively decreasing the function g. Under the monotonicity assumption on F, Hearn [18] and Marcotte [28] claim that, if $\nabla g(x)$ exists and $g(x) > 0$, then the vector $d = y - x$ with $y \in \Omega(x)$ is a descent direction of g at x. However, since g is hardly differentiable everywhere as mentioned above, the Frank-Wolfe method that uses d and g as the search direction and the merit function, respectively, does not seem to be applicable to a general VI. Taking into account the nondifferentiability of g, Marcotte [28] proposed a bundle-type descent algorithm for the gap function g and proved its convergence under monotonicity assumption on F. On the other hand, Marcotte and Dussault [29] proposed using the gap function g as a merit function to globalize the Newton method that solves at each iteration the following linearized VI: Given x^k, find a point $z \in S$ such that

$$\langle \bar{F}(z, x^k), y - z\rangle \geq 0, \quad \forall y \in S, \tag{2.5}$$

where $\bar{F}(\cdot, x^k) : \mathbb{R}^n \to \mathbb{R}^n$ is the affine mapping defined by

$$\bar{F}(z, x^k) = F(x^k) + \nabla F(x^k)^\top (z - x^k). \tag{2.6}$$

Assuming the monotonicity of F and the compactness of S, Marcotte and Dussault [29] showed that the vector $d^k = \bar{z}(x^k) - x^k$ with $\bar{z}(x^k)$ being a solution to the linearized VI (2.5) is a feasible descent direction of g at x^k and that the Newton method with exact line search on g is globally convergent to a solution of the VI (1.1). Moreover, assuming the strong monotonicity of F, the polyhedral convexity of S and the strict complementarity condition at the solution, they showed that the method is quadratically convergent. Another iterative method that uses the gap function g as a merit function has also been proposed by the same authors [30] for the VI (1.1) in which the constraint set S is a compact polyhedral convex set specified by a system of linear inequalities.

The following "dual" gap function has also been used to reformulate the VI (1.1):

$$h(x) = \sup\{\langle F(y), x - y\rangle \, : \, y \in S\}. \tag{2.7}$$

When F is monotone, it follows from the definition of functions g and h that

$$g(x) \geq h(x), \quad \forall x \in \mathbb{R}^n, \tag{2.8}$$

because the monotonicity of F implies

$$\langle F(x), x - y\rangle \geq \langle F(y), x - y\rangle, \quad \forall x, y \in \mathbb{R}^n.$$

Since $g(x) = 0$ and $x \in S$ if and only if x solves the VI (1.1) as mentioned above, and since $h(x) \geq 0$ clearly holds for each $x \in S$, the inequality (2.8) indicates that $h(x) = 0$ and $x \in S$ hold for any solution x of the VI (1.1), provided that F is monotone. Namely, when F is monotone, any solution x to (1.1) is also a solution of the problem

$$\text{minimize} \quad h(x) \quad \text{subject to} \quad x \in S. \tag{2.9}$$

Moreover, when F is strictly monotone, the unique solution of (1.1), if exists, uniquely minimizes the function h over S. Notice that the function h is a convex function, which may take value $+\infty$ unless some additional assumption like compactness is imposed on S. The evaluation of h, however, generally requires maximization of a nonconcave function, whereas g can be evaluated by maximizing a linear function over S. Though convex, the function h may not be differentiable in general. Nguyen and Dupuis [33] proposed a special cutting plane method for minimizing the function h. (Some properties of the dual gap function h are summarized in [25].)

2.2. Regularized gap functions

Whether or not the general VI can be reformulated as a differentiable optimization problem had long been an open problem until Fukushima [13] and Auchmuty [1] independently and almost simultaneously gave an affirmative answer to it. Fukushima [13] considered the *regularized gap function* defined by

$$f_\alpha(x) = \max_{y \in S} \left\{ \langle F(x), x - y\rangle - \frac{\alpha}{2}\|y - x\|^2 \right\}, \tag{2.10}$$

where α is a positive parameter. (The Euclidean norm used in the definition (2.10) may be replaced by any norm induced by a positive definite symmetric matrix [13]. We adopt the above definition here only for simplicity of notation.) Since the definition (2.10) of f_α consists of maximizing a strongly concave function over a closed convex set, the supremum is always attained and hence the function f_α is finite-valued everywhere. This is in contrast with the gap function g, which may take value $+\infty$ somewhere. Moreover, since the maximizer on the right-hand side of (2.10) is unique, the function f_α is guaranteed to be differentiable at any point. Specifically, if we denote the unique maximizer by $y_\alpha(x)$, then the gradient of f_α is given by

$$\nabla f_\alpha(x) = F(x) - [\nabla F(x) - \alpha I](y_\alpha(x) - x). \tag{2.11}$$

Note that

$$y_\alpha(x) = \Pi_S[x - \alpha^{-1}F(x)], \tag{2.12}$$

where Π_S denotes the projection onto S. It is well known that x solves the VI (1.1) if and only if $x = \Pi_S[x - \alpha^{-1}F(x)]$. Since the function f_α can be expressed as

$$f_\alpha(x) = \frac{\alpha}{2}\left\{\|\alpha^{-1}F(x)\|^2 - \|y_\alpha(x) - (x - \alpha^{-1}F(x))\|^2\right\}, \tag{2.13}$$

it is not difficult to deduce from (2.12) and (2.13) that, similar to the gap function g, the regularized gap function f_α possesses the following properties:

- $f_\alpha(x) \geq 0, \ \forall x \in S$.
- $f_\alpha(x) = 0$ and $x \in S$ if and only if x solves the VI (1.1).

Therefore the VI (1.1) is equivalent to

$$\text{minimize} \ \ f_\alpha(x) \ \ \text{subject to} \ \ x \in S. \tag{2.14}$$

This fact implies that f_α serves as a merit function for the VI. A remarkable feature of the regularized gap function f_α is of course that it is continuously differentiable. In general, the function f_α need not be convex even though the mapping F has some desirable properties like the strong monotonicity. (This is also the case for the gap function g.) So the function f_α may have local minima or stationary points which do not minimize $f_\alpha(x)$ over S globally, i.e., do not solve the VI (1.1). Since most of the iterative minimization methods are only guaranteed to converge to local minima or stationary points, it is important to know when a local minimum or a stationary point of f_α over S is actually a global minimum. In this regard, Fukushima [13] proved the following result:

- If the Jacobian $\nabla F(x)$ is positive definite on S, then any stationary point of (2.14) solves the VI (1.1).

Moreover, Taji, Fukushima and Ibaraki [46] showed that, if F is strongly monotone with modulus $\mu > 0$ and if $\alpha < 2\mu$, then the following error bound is obtained for the VI (1.1):

$$\|x - x^*\| \leq \sqrt{\frac{f_\alpha(x)}{\mu - \alpha/2}}, \qquad \forall \ x \in S, \tag{2.15}$$

where x^* is the unique solution of the VI (1.1).

The minimization formulation (2.14) naturally suggests that a globally convergent algorithm for solving the VI (1.1) be developed on the basis of the merit function f_α. Fukushima [13] showed that, if $\nabla F(x)$ is positive definite on S, then the vector

$$d^k = y_\alpha(x^k) - x^k$$

is a feasible descent direction of f_α at each x^k. A descent algorithm that utilizes the vector d^k as a search direction and performs line search for the merit function f_α was developed in [13]. Note that, in this 'first-order' method, we need not evaluate the gradient $\nabla f_\alpha(x)$. This could be desirable because the evaluation of $\nabla f_\alpha(x)$ requires the evaluation of $\nabla F(x)$, which may be expensive in some practical problems. If one is willing to use $\nabla F(x)$, then it is possible to develop a 'second-order' method of Newton type. Specifically, Taji, Fukushima and Ibaraki [46] showed that the vector

$$d^k = \bar{z}(x^k) - x^k,$$

where $\bar{z}(x^k)$ is a solution of the linearized VI (2.5), is a feasible descent direction of f_α at each x^k, provided that F is strongly monotone with modulus μ and α is chosen small enough to satisfy $\alpha < 2\mu$. Note that, under the strong monotonicity assumption, the solution $\bar{z}^k$ of the linearized VI (2.5) exists uniquely. It was shown in [46] that this algorithm is globally convergent under the above-mentioned assumptions and that the rate of convergence is quadratic under additional assumptions such as the polyhedral convexity of S and the strict complementarity at the solution. This algorithm shares much in common with the Newton method of Marcotte and Dussault [29], which uses the gap function g as a merit function. It is worth mentioning, however, that using the regularized gap function f_α enables us to adopt inexact line search with Armijo rule [46].

Similar but more general classes of regularized gap functions have been considered by several authors [1,25,38,49,54,55]. Those functions are typically derived by replacing the quadratic term in the definition (2.10) of f_α by a more general convex term. In particular, Larsson and Patriksson [25] recently presented a unified framework of regularized gap functions derived from the class of Auchmuty's merit functions [1]. Error bound results that generalize the condition (2.15) have been established in [25,55]. Descent methods, which are extensions of the one given in Fukushima [13], have been proposed in [25,38,49,54,55]. In particular, Zhu and Marcotte [54] proposed a modified descent algorithm which uses varying parameter α in the regularized gap function f_α. A remarkable feature of the latter algorithm is that its convergence requires only monotonicity of F (see also [38]).

2.3. Nonlinearly constrained VI

By definition, the evaluation of the regularized gap function f_α requires computing the projection of a point x onto the closed convex set S. In general, however, this is not an easy task unless S has a certain tractable structure such as polyhedral convexity. For the VI (1.1) with the set S given by

$$S = \{x \,:\, c_i(x) \leq 0, \ \ i = 1, \cdots, m\}, \tag{2.16}$$

where $c_i : \mathbb{R}^n \to \mathbb{R}$ are continuously differentiable convex functions, Taji and Fukushima [44,45] recently proposed the following modification of f_α:

$$\tilde{f}_\alpha(x) = \max_{y \in T(x)} \left\{ \langle F(x), x - y \rangle - \frac{\alpha}{2} \|y - x\|^2 \right\}, \tag{2.17}$$

where $T(x)$ is a polyhedral approximation of S at x, defined by

$$T(x) = \{\, y \,:\, c_i(x) + \langle \nabla c_i(x), y - x \rangle \leq 0, \ \ i = 1, \cdots, m\}. \tag{2.18}$$

It is easy to see that, by the convexity of c_i, the polyhedral convex set $T(x)$ always contains S, and hence

$$\tilde{f}_\alpha(x) \geq f_\alpha(x), \quad \forall x \in \mathbb{R}^n. \tag{2.19}$$

Moreover, the value of $\tilde{f}_\alpha(x)$ is finite for all $x \in \mathbb{R}^n$ and is obtained by computing the projection of a point onto the polyhedral convex set $T(x)$. Assuming the Slater's constraint qualification, Taji and Fukushima [44] proved the following results:

- $\tilde{f}_\alpha(x) \geq 0, \ \forall x \in S$.
- $\tilde{f}_\alpha(x) = 0$ and $x \in S$ if and only if x solves the VI (1.1).

Moreover, it follows from (2.15) and (2.19) that, if F is strongly monotone with modulus $\mu > 0$ and if $\alpha < 2\mu$, then

$$\|x - x^*\| \le \sqrt{\frac{\tilde{f}_\alpha(x)}{\mu - \alpha/2}}, \qquad \forall\, x \in S.$$

The above mentioned properties of $\tilde{f}_\alpha$ reveal that the VI (1.1) is equivalent to

$$\text{minimize} \;\; \tilde{f}_\alpha(x) \;\; \text{subject to} \;\; x \in S. \tag{2.20}$$

The function $\tilde{f}_\alpha$ has a very similar feature to the regularized gap function f_α. An important difference between f_α and $\tilde{f}_\alpha$, however, lies in the differentiability. Although the maximizer, say $\tilde{y}_\alpha(x)$, on the right-hand side of (2.17) is unique, the function $\tilde{f}_\alpha$ need not be differentiable, unlike f_α, even though F is differentiable. Instead, as shown in [44], under the once and twice continuous differentiability of F and c_i, respectively, the function $\tilde{f}_\alpha$ is directionally differentiable in any direction d and its directional derivative $\tilde{f}'_\alpha(x;d)$ is given by

$$\tilde{f}'_\alpha(x;d) = \min_{\lambda \in \Lambda(x)} \langle F(x) - [L(x,\lambda) - \alpha I](\tilde{y}_\alpha(x) - x), d \rangle, \tag{2.21}$$

where the set $\Lambda(x)$ is defined by

$$\begin{aligned}\Lambda(x) \;\; = \;\; \{\lambda \in \mathbb{R}^m \, : \, & F(x) + \alpha(\tilde{y}_\alpha(x) - x) + \sum_{i=1}^m \lambda_i \nabla c_i(x) = 0, \;\; \lambda_i \ge 0, \\ & \lambda_i[c_i(x) + \langle \nabla c_i(x), \tilde{y}_\alpha(x) - x \rangle] = 0, \;\; i = 1, \cdots, m\}\end{aligned}$$

and the matrix $L(x,\lambda)$ is defined by

$$L(x,\lambda) = \nabla F(x) + \sum_{i=1}^m \lambda_i \nabla^2 c_i(x). \tag{2.22}$$

Note that $\Lambda(x)$ is the set of optimal Lagrange multipliers in the maximization problem that appears in (2.17). It follows from (2.21) that the function $\tilde{f}_\alpha$ is actually differentiable at x, provided that the set $\Lambda(x)$ is a singleton, which is true under the linear independence of active constraint gradients and the strict complementarity condition. Since $\tilde{f}_\alpha$ is only directionally differentiable in general, we say that a point $x \in S$ is a stationary point of (2.20) if it satisfies

$$\tilde{f}'_\alpha(x; y - x) \ge 0, \quad \forall y \in S.$$

The following result was proved in [44].

- If the Jacobian $\nabla F(x)$ is positive definite on S, and c_i are twice continuously differentiable for all i, then any stationary point of problem (2.20) solves the VI (1.1).

Like the regularized gap function f_α, the function $\tilde{f}_\alpha$ can be used to develop a globally convergent algorithm for solving the VI (1.1) with the set S given by (2.16). Taji and Fukushima [44] proposed an iterative method that uses a search direction

$$d^k = \tilde{y}_\alpha(x^k) - x^k,$$

where $\tilde{y}_\alpha(x^k)$ is the unique solution of the quadratic programming problem that appears in the definition (2.17) of $\tilde{f}_\alpha$. In this case, the generated sequence need not be contained in the set S, so that it is not appropriate to use the function $\tilde{f}_\alpha$ itself as a merit function in the line search. In [44], the following exact penalty function for problem (2.20) is used to determine a step size along the direction d^k:

$$\theta_{\alpha,r}(x) = \tilde{f}_\alpha(x) + r\sum_{i=1}^{m} \max(0, c_i(x)), \tag{2.23}$$

where r is a sufficiently large positive parameter. Under the positive definiteness of the Jacobian $\nabla F(x)$, this algorithm is shown to be globally convergent to the solution of the VI (1.1).

As an extension of the methods of [29] and [46], Taji and Fukushima [45] proposed a Newton-type method that solves at each iteration the following linearized VI: Given the current point x^k and an arbitrary $\lambda^k \in \Lambda(x^k)$, find a point $z \in T(x^k)$ such that

$$\langle \tilde{F}(z, x^k, \lambda^k), y - z\rangle \geq 0, \quad \forall y \in T(x^k), \tag{2.24}$$

where $\tilde{F}(\cdot, x^k, \lambda^k) : \mathbb{R}^n \to \mathbb{R}^n$ is the affine mapping defined by

$$\tilde{F}(z, x^k, \lambda^k) = F(x^k) + L(x^k, \lambda^k)^{\mathsf{T}}(z - x^k), \tag{2.25}$$

where $L(x^k, \lambda^k)$ is the matrix given by (2.22). Let $\tilde{z}(x^k, \lambda^k)$ solve the linearized VI (2.24). Then, under the strong monotonicity of F, the search direction

$$d^k = \tilde{z}(x^k, \lambda^k) - x^k$$

is shown to be a descent direction of the exact penalty function $\theta_{\alpha,r}$ defined by (2.23), provided that the parameter α involved in $\tilde{f}_\alpha$ is chosen small enough. Some convergence theorems for this algorithm are given in [45].

2.4. Unconstrained optimization formulations

The equivalent optimization formulations discussed so far are all constrained problems, each of which is to minimize a certain merit function over the feasible set S of the original VI. Recently Yamashita and Fukushima [51] considered further applying the Moreau-Yosida regularization to some gap functions. The resulting functions, which may not be easy to evaluate in general, certainly possess nice theoretical properties. Aside from the differentiability, those functions are shown to provide global error bounds for the VI under the strong monotonicity only.

More recently, Peng [39] considered the function

$$M_\alpha(x) = f_\alpha(x) - f_{1/\alpha}(x), \tag{2.26}$$

where f_α is the regularized gap function defined by (2.10) and α is a parameter such that $0 < \alpha < 1$, and showed the following results:

- $M_\alpha(x) \geq 0$ for all $x \in \mathbb{R}^n$.
- $M_\alpha(x) = 0$ if and only if x solves the VI (1.1).

Thus the VI (1.1) is equivalent to the unconstrained minimization problem

$$\text{minimize} \;\; M_\alpha(x) \quad \text{subject to} \;\; x \in \mathbb{R}^n. \tag{2.27}$$

By definition, it is clear that the differentiability of M_α is inherited from f_α. In fact, the function M_α shares much in common with f_α. Among others, the following result is very important.

- If the Jacobian $\nabla F(x)$ is positive definite for each x, then any stationary point of (2.27) solves the VI (1.1).

Moreover, Peng [39] proved that M_α satisfies the following inequalities:

$$(1-\alpha)\|r_1(x)\|^2 \leq M_\alpha(x) \leq \frac{1-\alpha}{\alpha}\|r_1(x)\|^2, \quad \forall x \in \mathbb{R}^n, \tag{2.28}$$

where r_1 is the "natural" residual for the VI defined by $r_1(x) = x - y_1(x)$. The inequalities (2.28) in particular imply that the function $\sqrt{M_\alpha}$ provides an error bound for the VI if $\|r_1\|$ does. For the VI, it is known that the latter is true if F is strongly monotone and Lipschitz continuous on $\mathbb{R}^n$ [35]. In [40], it is shown that, in the case of the CP, the function M_α is reduced to a merit function called the implicit Lagrangian [27] (see §3.2). Thus M_α may be regarded as a generalization of the implicit Lagrangian for the CP to the VI.

The results obtained in [39,40] have further been extended by Yamashita, Taji and Fukushima [53], who considered the function

$$g_{\alpha\beta}(x) = f_\alpha(x) - f_\beta(x), \tag{2.29}$$

where f_α and f_β are regularized gap functions with parameters α and β such that $0 < \alpha < \beta$. (In [53], the functions f_α and f_β are defined by (2.10) with the quadratic term replaced by a more general convex term.) The function $g_{\alpha\beta}$ is called the *D-gap function*, where "D" stands for the word "difference". Similar to the function M_α, the following properties hold for the D-gap function $g_{\alpha\beta}$:

- $g_{\alpha\beta}(x) \geq 0, \quad \forall x \in \mathbb{R}^n$.
- $g_{\alpha\beta}(x) = 0$ if and only if x solves the VI (1.1).

Therefore the VI (1.1) is equivalent to the unconstrained minimization problem

$$\text{minimize } g_{\alpha\beta}(x) \quad \text{subject to } x \in \mathbb{R}^n. \tag{2.30}$$

The function $g_{\alpha\beta}$ is differentiable whenever F is differentiable. Also we have the following result:

- If the Jacobian $\nabla F(x)$ is positive definite for each x, then any stationary point of (2.30) solves the VI (1.1).

Moreover, it is shown in [53] that $g_{\alpha\beta}$ satisfies the following inequalities:

$$\frac{\beta-\alpha}{2}\|x - y_\beta(x)\|^2 \leq g_{\alpha\beta}(x) \leq \frac{\beta-\alpha}{2}\|x - y_\alpha(x)\|^2, \quad \forall x \in \mathbb{R}^n,$$

where $y_\alpha(x)$ is the unique maximizer in the definition (2.10) of f_α and $y_\beta(x)$ is defined similarly for f_β. Also it is shown in [53] that, if F is strongly monotone and if either F is Lipschitz continuous or S is compact, then the function $\sqrt{g_{\alpha\beta}}$ provides an error bound for the VI.

3. MERIT FUNCTIONS FOR CP

Since the CP is a special case of the VI such that the feasible set S is the nonnegative orthant in $\mathbb{R}^n$, any merit function for the VI can in principle be used as a merit function for the CP. For example, the regularized gap function yields an equivalent optimization problem with simple nonnegativity constraints on the variables. Moreover, remarkable progress has recently been made in the study of merit functions that lead to unconstrained optimization reformulations of the CP. In this section, we first review optimization formulations with simple bound constraints and then discuss two important classes of merit functions that lead to unconstrained optimization reformulations of the CP.

3.1. Optimization formulations with simple bound constraints

The regularized gap function f_α for the VI can be specialized to the CP as follows:

$$\begin{aligned} f_\alpha(x) &= \langle F(x), x - [x - \frac{1}{\alpha}F(x)]_+\rangle - \frac{\alpha}{2}\|[x - \frac{1}{\alpha}F(x)]_+ - x\|^2 \\ &= \langle F(x), x\rangle + \frac{\alpha}{2}\left(\|[x - \frac{1}{\alpha}F(x)]_+\|^2 - \|x\|^2\right), \end{aligned} \tag{3.1}$$

where $[z]_+$ denotes the vector with components $\max(z_i, 0), i = 1, \cdots, n$. The CP is equivalent to the optimization problem

$$\text{minimize} \;\; f_\alpha(x) \quad \text{subject to} \;\; x \geq 0. \tag{3.2}$$

Because the function f_α can easily be evaluated and the optimization problem (3.2) has the simple bound constraints on the variables only, globally convergent algorithms that utilize the regularized gap function, originally developed for the VI, may be specialized to the CP in an effective manner [43].

For the LCP (1.3), Friedlander, Martínez and Santos [11] considered the optimization problem

$$\text{minimize} \;\; [\rho\|Mx + q - z\|^2 + (\langle x, z\rangle)^p] \quad \text{subject to} \;\; x \geq 0, \;\; z \geq 0, \tag{3.3}$$

where $\rho > 0$ and $p > 1$ are arbitrary constants. Clearly this problem is equivalent to the LCP (1.3) in the sense that a global optimal solution of (3.3) with zero objective value is a solution of the LCP and vice versa. It was shown in [11] that, under some condition, which is implied by the positive semidefiniteness of M, any stationary point of problem (3.3) is a solution of the LCP (1.3). Using a similar idea, the same authors [12] proposed a bound constrained optimization reformulation of a linearly constrained VI.

For the CP (1.2), Moré [31] considered the following optimization problem, which is similar to the problem (3.3) for the LCP:

$$\text{minimize} \;\; \|F(x) - z\|^2 + \sum_{i=1}^{n}(x_i z_i)^2 \quad \text{subject to} \;\; x \geq 0, \;\; z \geq 0. \tag{3.4}$$

In particular, it was shown in [31] that any stationary point of problem (3.4) comprises a solution of the CP (1.2) under a regularity condition, which is satisfied, for example, if $\nabla F(x)$ is a positive definite matrix or an M-matrix. Moré [31] also proposed a trust region method for solving (3.4) and discussed its global and superlinear convergence properties.

Yet another bound constrained optimization formulation of the CP (1.2) is the following problem:

$$\text{minimize} \;\; \sum_{i=1}^{n}[\min(x_i, F_i(x))]^2 \quad \text{subject to} \;\; x \geq 0. \tag{3.5}$$

It is clear that this problem is equivalent to the CP (1.2). It was shown in [36] that, under a regularity condition, a stationary point of (3.5) is a solution to the CP (1.2). Unlike the optimization reformulations mentioned in this subsection, the objective function in problem (3.5) is not differentiable. Nevertheless, the algorithms proposed by Pang and Gabriel [36,14] are shown to have global and superlinear convergence properties under some regularity assumptions.

3.2. Implicit Lagrangian

Mangasarian and Solodov [27] introduced the following merit function, which is now commonly called the *implicit Lagrangian*, for the CP:

$$M_\alpha(x) = \langle F(x), x\rangle + \frac{\alpha}{2}\Big(\|[x - \frac{1}{\alpha}F(x)]_+\|^2 - \|x\|^2 + \|[F(x) - \frac{1}{\alpha}x]_+\|^2 - \|F(x)\|^2\Big), \qquad (3.6)$$

where α is a parameter such that $0 < \alpha < 1$. They showed that

- $M_\alpha(x) \geq 0$ for all $x \in R^n$.
- $M_\alpha(x) = 0 \iff x$ solves the CP (1.2).

Therefore the CP (1.2) is equivalent to the unconstrained optimization problem

$$\text{minimize } M_\alpha(x) \quad \text{subject to } x \in \mathbb{R}^n. \qquad (3.7)$$

Comparing the definitions (3.1) and (3.6), one may notice some similarity between the regularized gap function and the implicit Lagrangian. In fact, these two functions can be treated in a unified manner [48]. As observed by Peng and Yuan [40], in particular, the implicit Lagrangian M_α can be represented as the difference of two regularized gap functions (see (2.26)). Like f_α, the implicit Lagrangian M_α is differentiable and the gradient of M_α can be expressed as

$$\begin{aligned}\frac{1}{\alpha}\nabla M_\alpha(x) &= \Big(I - \frac{1}{\alpha}\nabla F(x)\Big)\Big([x - \frac{1}{\alpha}F(x)]_+ - x\Big) \\ &\quad + \Big(\nabla F(x) - \frac{1}{\alpha}I\Big)\Big([F(x) - \frac{1}{\alpha}x]_+ - F(x)\Big).\end{aligned}$$

Yamashita and Fukushima [50] proved the following result:

- If $\nabla F(x)$ is positive definite for every $x \in \mathbb{R}^n$, then any stationary point of M_α is a solution to the CP (1.2).

This result was recently improved by Jiang [19], who showed that the positive definiteness of $\nabla F(x)$ can be replaced by the weaker condition that $\nabla F(x)$ is a P-matrix.

Concerning error bounds for the CP, Luo et al. [26] showed that there exist $\kappa > 0$ and $\delta > 0$ such that

$$M_\alpha(x) \geq \kappa\,\mathrm{dist}(x, X^*)^2 \quad \forall\, x \in \{x : M_\alpha(x) \leq \delta\},$$

where X^* is the solution set of the CP. Moreover, Yamashita and Fukushima [50] showed that, if F is strongly monotone and Lipschitz continuous, then the following inequality holds for some $\kappa > 0$:

$$M_\alpha(x) \geq \kappa\|x - x^*\|^2 \quad \forall\, x \in \mathbb{R}^n,$$

where x^* denotes the unique solution to the CP. Note that the last inequality in particular implies that, if F is strongly monotone and Lipschitz continuous, then the level sets of the implicit Lagrangian M_α are bounded. More on the error bound results for the CP, the reader may refer to [48].

To solve a differentiable CP, we may directly apply any gradient-based algorithm to the equivalent minimization problem (3.7) (see [27]). As an alternative to the gradient-based algorithms, a simple descent method that does not require evaluating $\nabla F(x)$ has been proposed in [50].

3.3. NCP functions

Recently the following function $\Psi : \mathbb{R}^n \to \mathbb{R}$ has attracted the attention of many researchers in the field of complementarity problems:

$$\Psi(x) = \frac{1}{2}\sum_{i=1}^{n} \varphi(x_i, F_i(x))^2, \tag{3.8}$$

where $\varphi : \mathbb{R}^2 \to \mathbb{R}$ is defined by

$$\varphi(a, b) = \sqrt{a^2 + b^2} - a - b. \tag{3.9}$$

This function was first introduced by Fischer [7] but attributed to Burmeister. As can be seen easily, the function φ satisfies

$$\varphi(a, b) = 0 \quad \Longleftrightarrow \quad a \geq 0,\ b \geq 0,\ ab = 0, \tag{3.10}$$

so that the CP is equivalent to the system of equations

$$\Phi(x) = 0, \tag{3.11}$$

where $\Phi(x) = (\varphi(x_1, F_1(x)), \varphi(x_2, F_2(x)), \cdots, \varphi(x_n, F_n(x)))^\top$. Hence the CP can be cast as the following unconstrained optimization problem:

$$\text{minimize } \Psi(x) \quad \text{subject to } x \in \mathbb{R}^n. \tag{3.12}$$

Alternatively, one may obtain another reformulation of the CP by considering the augmented system of equations

$$\hat{\Phi}(x, z) = 0, \tag{3.13}$$

where $\hat{\Phi}(x, z) = (F(x) - z, \varphi(x_1, z_1), \varphi(x_2, z_2), \cdots, \varphi(x_n, z_n))^\top$. This is clearly equivalent to (3.11) and yields the following unconstrained optimization problem:

$$\text{minimize } \hat{\Psi}(x, z) \quad \text{subject to } x \in \mathbb{R}^n,\ z \in \mathbb{R}^n, \tag{3.14}$$

where

$$\hat{\Psi}(x, z) = \frac{1}{2}\left(\|F(x) - z\|^2 + \sum_{i=1}^{n} \varphi(x_i, z_i)^2\right).$$

Any function satisfying the condition (3.10) may be called an *NCP-function* [10]. Besides the above-mentioned function φ, various NCP functions are known to date (see, e.g., [21,22,24]). In this section, however, we will confine ourselves to the NCP function φ defined by (3.9), because it enjoys many favorable properties in the complementarity problem. First it is noted that Ψ is differentiable. (Note that $\varphi(x_i, F_i(x))$ is not differentiable at a point x such that $x_i = F_i(x) = 0$.) The gradient of Ψ is given by

$$\nabla \Psi(x) = [D_a(x) + \nabla F(x) D_b(x)]\ \Phi(x), \tag{3.15}$$

where $D_a(x)$ and $D_b(x)$ are diagonal matrices with nonpositive diagonal elements $a_i(x)$ and $b_i(x)$, $i = 1, \cdots, n$, respectively, which are given by

$$(a_i(x), b_i(x)) = \begin{cases} \left(\frac{x_i}{\sqrt{x_i^2+F_i(x)^2}} - 1, \frac{F_i(x)}{\sqrt{x_i^2+F_i(x)^2}} - 1\right), & \text{if } (x_i, F_i(x)) \neq (0,0) \\ (\xi_i - 1, \eta_i - 1), & \text{if } (x_i, F_i(x)) = (0,0) \end{cases}$$

for some ξ_i and η_i such that $\xi_i^2 + \eta_i^2 \leq 1$.

Geiger and Kanzow [15] proved the following results:

- If $\nabla F(x)$ is positive semidefinite for every $x \in \mathbb{R}^n$, then any stationary point of Ψ is a solution to the CP (1.2).
- If F is strongly monotone, then the level sets of Ψ are bounded.

Comparing these with the corresponding results for the implicit Lagrangian, we see that the NCP function Ψ enjoys these properties under weaker assumptions than the implicit Lagrangian. (In [50], a counter-example is provided to show that the positive definiteness of $\nabla F(x)$ cannot be replaced even by the strict monotonicity of F in order to guarantee that any stationary point of M_α solves the CP .) Facchinei and Soares [5] pointed out that the monotonicity and the strong monotonicity assumptions in the above results may further be replaced by the weaker conditions that F is a P_0-function and that F is a uniformly P-function (see also [4]). Tseng [47] studied growth properties of various NCP functions including Ψ and $\hat{\Psi}$.

As mentioned above, the functions Φ and $\hat{\Phi}$ are not differentiable. In spite of this fact, a number of Newton-type methods have been developed to solve the nonsmooth equation reformulations (3.11) and (3.13) for the LCP [8,9] and the CP [4,5,6,20,22,52]. A key property for such algorithms to converge superlinearly is the semismoothness of the functions Φ and $\hat{\Phi}$. The semismoothness of vector functions was introduced in [41] and used in the analysis of generalized Newton methods for nonsmooth equations [41,42]. In the present case, the function Φ is actually shown to be semismooth [5,20]. A direction-finding problem in a Newton-type method for (3.11) may therefore be given as

$$V_k d = -\Phi(x^k), \tag{3.16}$$

where V_k is an $n \times n$ matrix chosen from the generalized Jacobian $\partial\Phi(x^k)$. If a solution d^k of (3.16) is obtained, then the next iterate x^{k+1} may be determined by line search for the merit function Ψ. A possible trouble with this approach is that the matrix V_k is not necessarily nonsingular, and hence the linear equation (3.16) may not have a solution, even though the function F is monotone. In order to guarantee the nonsingularity of V_k, we need to assume that $\nabla F(x)$ is positive definite or F is a uniform P-function [20]. A remedy for this trouble is to use the steepest descent direction for the merit function Ψ when the system (3.16) is not solvable or d^k does not afford a direction of sufficient decrease for Ψ [4,5,6,20]. Another cure is to slightly perturb V_k to obtain a positive definite matrix, so that the modified direction-finding subproblem is always consistent [8,9,52]. Under appropriate conditions, those methods are globally convergent to a solution of the LCP or the CP. Moreover, it can be shown that, under some regularity assumptions, they have a superlinear or quadratic convergence property.

4. CONCLUDING REMARKS

In this article, we have reviewed the recent developments of merit functions for the VI and CP. In closing, we point out that some of those results can be extended to more general classes of problems. For example, for the generalized complementarity problem (GCP)

$$F(x) \geq 0, \quad G(x) \geq 0, \quad \langle F(x), G(x) \rangle = 0 \tag{4.1}$$

with $F : \mathbb{R}^n \to \mathbb{R}^n$ and $G : \mathbb{R}^n \to \mathbb{R}^n$, Tseng, Yamashita and Fukushima [48] showed that the regularized gap function f_α and the implicit Lagrangian M_α can be derived in a unified manner, and extended a number of basic results about those merit functions for the CP. (In [48], the GCP (4.1) is considered under a more general setting where the vector inequality is defined in terms of a closed convex cone and its polar cone.) For the GCP (4.1), Kanzow and Fukushima [23] also extended various results known for the merit function Ψ with the NCP function φ defined by (3.9) for the CP. On the other hand, Giannessi [16] considered the quasi-variational inequality (QVI), which consists in finding a point $x \in S(x)$ such that

$$\langle F(x), y - x \rangle \geq 0, \quad \forall y \in S(x), \tag{4.2}$$

where $S(x)$ is given by

$$S(x) = \{y \in Y(x) \, : \, c(x,y) \in C\},$$

with a point-to-set mapping $Y(x) : \mathbb{R}^n \to \mathbb{R}^n$, a function $c : \mathbb{R}^n \times \mathbb{R}^n \to \mathbb{R}^m$ and a closed convex cone $C \subseteq \mathbb{R}^m$. When $Y(x)$ and $c(x,y)$ are independent of x, the QVI (4.2) reduces to an ordinary VI. In [16], attempts were made to extend some gap functions for the ordinary VI to the QVI (4.2).

ACKNOWLEDGMENTS

I would like to thank Christian Kanzow, Kouichi Taji and Nobuo Yamashita for their comments on an earlier version of this paper.

REFERENCES

[1] G. Auchmuty, "Variational principles for variational inequalities". *Numerical Functional Analysis and Optimization*, 10, pp. 863–874, 1989.

[2] A. Auslender, "Optimisation: Méthodes Numériques". Masson, Paris, 1976.

[3] R.W. Cottle, J.-S. Pang and R.E. Stone, "The Linear Complementarity Problem". Academic Press, San Diego, 1992.

[4] T. De Luca, F. Facchinei and C. Kanzow, "A semismooth equation approach to the solution of nonlinear complementarity problems". Technical Report, Dipartimento di Informatica e Sistemistica, Università di Roma "La Sapienza", Rome, Italy, January 1995.

[5] F. Facchinei and J. Soares, "A new merit function for nonlinear complementarity problems and a related algorithm", Technical Report, Dipartimento di Informatica e Sistemistica, Università di Roma "La Sapienza", Rome, Italy, December 1994.

[6] F. Facchinei and J. Soares, "Testing a new class of algorithms for nonlinear complementarity problems". In *"Variational Inequality and Network Equilibrium Problems"*, F. Giannessi and A. Maugeri (eds.), Plenum Press, New York, pp. 69–83, 1995.

[7] A. Fischer, "A special Newton-type optimization method". *Optimization*, 24, pp. 269–284, 1992.

[8] A. Fischer, "On the local superlinear convergence of a Newton-type method for LCP under weak conditions". Technical Report, Institute of Numerical Mathematics, Technical University of Dresden, Dresden, Germany, March 1994.

[9] A. Fischer, "A Newton-type method for linear complementarity problems". To appear in *Journal of Optimization Theory and Applications.*

[10] A. Fischer, "An NCP-function and its use for the solution of complementarity problems." In *"Recent Advances in Nonsmooth Optimization"*, D.-Z. Du, L. Qi and R. S. Womersley (eds.), World Scientific Publishers, Singapore, pp. 88-105, 1995.

[11] A. Friedlander, J.M. Martínez and S.A. Santos, "Resolution of linear complementarity problems using minimization with simple bounds". Technical Report 66/93, Department of Applied Mathematics, State University of Campinas, Campinas SP, Brazil, November 1993.

[12] A. Friedlander, J.M. Martínez and S.A. Santos, "A new strategy for solving variational inequalities in bounded polytopes". Technical Report 02/94, Department of Applied Mathematics, State University of Campinas, Campinas SP, Brazil, January 1994.

[13] M. Fukushima, "Equivalent differentiable optimization problems and descent methods for asymmetric variational inequality problems". *"Mathematical Programming"*, 53, pp. 99–110, 1992.

[14] S.A. Gabriel and J.-S. Pang, "An inexact NE/SQP method for solving the nonlinear complementarity problem". *Computational Optimization and Applications*, 1, 1992, pp. 67–91.

[15] C. Geiger and C. Kanzow, "On the resolution of monotone complementarity problems". To appear in *Computational Optimization and Applications.*

[16] F. Giannessi, "Separation of sets and gap functions for quasi-variational inequalities". In *"Variational Inequalities and Network Equilibrium Problems"*, F. Giannessi and A. Maugeri (eds.), Plenum Press, New York, pp. 101–121, 1995.

[17] P.T. Harker and J.-S. Pang, "Finite-dimensional variational inequality and nonlinear complementarity problems: A survey of theory, algorithms and applications". *Mathematical Programming*, 48, 1990, pp. 161–220.

[18] D.W. Hearn, "The gap function of a convex program". *Operations Research Letters*, 1, pp. 67–71, 1982.

[19] H. Jiang, "Unconstrained minimization approaches to nonlinear complementarities". AMR 94/33, School of Mathematics, The University of New South Wales, Sydney, Australia, October 1994.

[20] H. Jiang and L. Qi, "A new nonsmooth equations approach to nonlinear complementarities". AMR94/31, School of Mathematics, The University of New South Wales, Sydney, Australia, October 1994.

[21] C. Kanzow, "Nonlinear complementarity as unconstrained optimization". To appear in *Journal of Optimization Theory and Applications.*

[22] C. Kanzow, "Some equation-based methods for the nonlinear complementarity problem". *Optimization Methods and Software*, 3, pp. 327–340, 1994.

[23] C. Kanzow and M. Fukushima, "Equivalence of the generalized complementarity problem to differentiable unconstrained minimization". TR-IS-95001, Nara Institute of Science and Technology, Nara, Japan, January 1995.

[24] C. Kanzow and H. Kleinmichel, "A class of Newton-type methods for equality and inequality constrained optimization". *Optimization Methods and Software*, 5, 1995, pp. 173-198.

[25] T. Larsson and M. Patriksson, "A class of gap functions for variational inequalities". *Mathematical Programming*, 64, pp. 53–79, 1994.

[26] Z.-Q. Luo, O.L. Mangasarian, J. Ren and M. V. Solodov, "New error bounds for the linear complementarity problem". To appear in *Mathematics of Operations Research.*

[27] O.L. Mangasarian and M.V. Solodov, "Nonlinear complementarity as unconstrained and constrained minimization". *Mathematical Programming*, 62, pp. 277–297, 1993.

[28] P. Marcotte, "A new algorithm for solving variational inequalities with application to the traffic assignment problem". *Mathematical Programming*, 33, pp. 339–351, 1985.

[29] P. Marcotte and J.P. Dussault, "A note on a globally convergent Newton method for solving monotone variational inequalities". *Operations Research Letters*, 6, pp. 35–42, 1987.

[30] P. Marcotte and J.P. Dussault, "A sequential linear programming algorithm for solving monotone variational inequalities." *SIAM Journal on Control and Optimization*, 27, pp. 1260–1278, 1989.

[31] J.J. Moré, "Global methods for nonlinear complementarity problems". Preprint MCS-P429-0494, Argonne National Laboratory, Argonne, Illinois, April 1994.

[32] A. Nagurney, *"Network Economics: A Variational Inequality Approach"*, Kluwer Academic Publishers, Boston, 1993.

[33] S. Nguyen and C. Dupuis, "An efficient method for computing traffic equilibria in networks with asymmetric transportation costs". *Transportation Science*, 18, pp. 185–202, 1984.

[34] J.M. Ortega and W.C. Rheinboldt, "Iterative Solution of Nonlinear Equations in Several Variables", Academic Press, New York, 1970.

[35] J.-S. Pang, "A posteriori error bounds for the linearly-constrained variational inequality problem". *Mathematics of Operations Research*, 12, pp. 474–484, 1987.

[36] J.-S. Pang and S.A. Gabriel, "NE/SQP: A robust algorithm for the nonlinear complementarity problem". *Mathematical Programming*, 60, pp. 295–337, 1993.

[37] J.-S. Pang, "Complementarity problems". In *"Handbook of Global Optimization"*, R. Horst and P. Pardalos (eds.), Kluwer Academic Publishers, Boston, pp. 271–338, 1995.

[38] M. Patriksson, "On the convergence of descent methods for monotone variational inequalities". *Operations Research Letters*, 16, pp. 265–269, 1994.

[39] J.M. Peng, "Equivalence of variational inequality problems to unconstrained optimization". Technical Report, State Key Laboratory of Scientific and Engineering Computing, Academia Sinica, Beijing, China, April 1995.

[40] J.M. Peng and Y.X. Yuan, "Unconstrained methods for generalized complementarity problems". Technical Report, State Key Laboratory of Scientific and Engineering Computing, Academia Sinica, Beijing, China, December 1994.

[41] L. Qi and J. Sun, "A nonsmooth version of Newton's method". *Mathematical Programming*, 58, pp. 353–368, 1993.

[42] L. Qi, "Convergence analysis of some algorithms for solving nonsmooth equations". *Mathematics of Operations Research*, 18, 1993, pp. 227–244.

[43] K. Taji and M. Fukushima, "Optimization based globally convergent methods for the nonlinear complementarity problem". *Journal of the Operations Research Society of Japan*, 37, pp. 310–331, 1994.

[44] K. Taji and M. Fukushima, "A new merit function and a successive quadratic programming algorithm for variational inequality problems". To appear in *SIAM Journal on Optimization.*

[45] K. Taji and M. Fukushima, "A globally convergent Newton method for solving variational inequality problems with inequality constraints". In *Recent Advances in Nonsmooth Optimization*, D.-Z. Du, L. Qi and R. S. Womersley (eds.), World Scientific Publishers, Singapore, pp. 405-417, 1995.

[46] K. Taji, M. Fukushima and T. Ibaraki, "A globally convergent Newton method for solving strongly monotone variational inequalities". *Mathematical Programming*, 58, pp. 369–383, 1993.

[47] P. Tseng, "Growth behaviour of a class of merit functions for the nonlinear complementarity problem". Technical Report, Department of Mathematics, University of Washington, Seattle, May 1994.

[48] P. Tseng, N. Yamashita and M. Fukushima, "Equivalence of complementarity problems to differentiable minimization: A unified approach". To appear in *SIAM Journal on Optimization.*

[49] J.H. Wu, M. Florian and P. Marcotte, "A general descent framework for the monotone variational inequality problem". *Mathematical Programming*, 61, pp. 281–300, 1993.

[50] N. Yamashita and M. Fukushima, "On stationary points of the implicit Lagrangian for nonlinear complementarity problems". *Journal of Optimization Theory and Applications*, 84, pp. 653–663, 1995.

[51] N. Yamashita and M. Fukushima, "Equivalent unconstrained minimization and global error bounds for variational inequality problems". TR-IS-94034, Nara Institute of Science and Technology, Nara, Japan, November 1994.

[52] N. Yamashita and M. Fukushima, "Modified Newton methods for solving semismooth reformulations of monotone complementarity problems". TR-IS-95021, Nara Institute of Science and Technology, Nara, Japan, May 1995.

[53] N. Yamashita, K. Taji and M. Fukushima, "Unconstrained optimization reformulations of variational inequality problems". TR-IS-95024, Nara Institute of Science and Technology, Nara, Japan, August 1995.

[54] D.L. Zhu and P. Marcotte, "Modified descent methods for solving the monotone variational inequality problem". *Operations Research Letters*, 14, pp. 111–120, 1993.

[55] D.L. Zhu and P. Marcotte, "An extended descent framework for variational inequalities". *Journal of Optimization Theory and Applications*, 80, pp. 349–366, 1994.

SOME REMARKS ON PENALIZATION FOR VARIATIONAL INEQUALITIES AND GENERALIZATIONS

Giannessi Franco

Department of Mathematics
University of Pisa
Via F. Buonarroti 2, 56127 Pisa, Italy
e–mail: gianness@gauss.dm.unipi.it

Abstract. By using separation arguments some remarks are made on possible ways for extending penalization methods to Quasi–Variational Inequalities and their generalizations.

Key words. Variational Inequality, Quasi–Variational Inequality, Penalization, Separation.

1. INTRODUCTION

The method of penalization, originally conceived for constrained extremum problems, has been extended to Variational Inequalities [1,2,6,7,10]. The separation approach, originally conceived for constrained extremum problems, has been extended to Variational and Quasi–Variational Inequalities [4,5,12]. The present note tries to connect penalization and separation schemes for Variational and Quasi–Variational Inequalities.

Let Ξ be a Hilbert space, $F : \Xi \to \Xi$, $X : \Xi \rightrightarrows \Xi$, $g : X(y) \times X(y) \to \mathbb{R}^m$. Consider a Quasi–Variational Inequality (for short, QVI) in the form: find $y \in K(y)$, such that:

$$\langle F(y), x - y \rangle \geq 0, \quad \forall x \in K(y) := \{x \in X(y) : g(y; x) \in \mathcal{C}\} \tag{1.1}$$

where $\mathcal{C} \subset \mathbb{R}^m$ is a closed and convex cone with apex at the origin, and $K(y)$ is assumed to be nonempty $\forall y \in \Xi$.

In [5,12] a QVI is associated with a separation scheme, extending the approach proposed for a Variational Inequality (for short, VI) in [4]. It starts with the obvious remark that $y \in K(y)$ is a solution of (1.1) iff the system (in the unknown x):

$$u := \langle F(y), y - x \rangle > 0, \quad v := g(y; x) \in \mathcal{C}, \quad x \in X(y) \tag{1.2}$$

is infeasible. The space where (u,v) runs is the *image space* associated to (1.1), and the set

$$\mathcal{K}(y) := \{(u,v) \in \mathbb{R} \times \mathbb{R}^m : u = \langle F(y), y - x \rangle > 0, \quad v = g(y;x), \ x \in X(y)\}$$

is the *image* of (1.1). System (1.2) is associated with the set $\mathcal{H} := \{(u,v) \in \mathbb{R} \times \mathbb{R}^m : u > 0, \ v \in \mathcal{C}\}$. Another obvious remark is that the impossibility of (1.2) holds if and only if $\mathcal{H} \cap \mathcal{K}(y) = \emptyset$. The above mentioned separation scheme aims to show such a disjunction. In [5,12] it has been shown that a wide class of gap functions can be obtained as a by–product of such a separation scheme. The purpose of the present note is to prove that in a analogous way a penalization approach can be derived for (1.1). Concepts and notations of [4,5,12] will be here understood.

2. SEPARATION AND PENALIZATION

A class of separation functions is now considered; they are of the following type $w : \Xi \times \mathbb{R} \times \mathbb{R}^m \times \Omega \to \mathbb{R}$ with

$$w(y;u,v;\omega) := u + \gamma(y;v;\omega) \tag{2.1}$$

where $\gamma : \Xi \times \mathbb{R}^m \to \mathbb{R}$ and where the set of parameters Ω is such that, $\forall y \in X(y)$, the relation

$$\text{lev}_{>0} w \supset \mathcal{H}, \tag{2.2a}$$

holds and all the subsequent operations are possible; the level set is considered with respect to (u,v) only. The above class, which is described by the parameter ω, is a subclass of that considered in [5]. Under this condition, (2.1) is a weak separation function in the sense of [5]. However, for some purposes the class (2.1) may be "poor"; for instance, even if $\mathcal{K}$ is convex, no separation between $\mathcal{H}$ and $\mathcal{K}$ may be guaranteed; further conditions might be necessary, as

$$\forall y \in X(y) \backslash K(y) \ , \quad \bigcap_{\omega \in \Omega} \text{lev}_{>0} w = \text{cl}\, \mathcal{H}. \tag{2.2b}$$

Now assume that $K(y)$ be closed $\forall y \in \Xi$, γ be continuous in $\Xi \times \mathbb{R}^m \times \Omega$ and such that: for each $\omega \in \Omega$

$$\gamma(y;v;\omega) \begin{cases} = 0 \ , & \forall v \in \mathbb{R}^m & , \text{if } y \in \text{int}\, K(y), \\ = 0 \ , & \forall v \in \mathcal{C} & , \text{if } y \in \text{frt}\, K(y), \\ < 0 \ , & \forall v \in \mathbb{R}^m \backslash \mathcal{C} & , \text{if } y \in \text{frt}\, K(y), \\ > 0 \ , & \forall v \in \mathcal{C} \backslash \{0\} & , \text{if } y \in X(y) \backslash K(y), \end{cases} \tag{2.3a}$$

where $\{0\}$ is cut off to embed the linear case; moreover $\exists \omega_0 \in \Omega$ such that:

$$\lim_{\omega \to \omega_0} \gamma(y;v;\omega) = -\infty \ , \quad \forall v \in \mathbb{R}^m \backslash \mathcal{C} \ , \quad \forall y \in \text{frt}\, K(y), \tag{2.3b}$$

$$\lim_{\omega \to \omega_0} \gamma(y;v;\omega) = +\infty \ , \quad \forall v \in \mathcal{C}, \qquad \forall y \in X(y) \backslash K(y). \tag{2.3c}$$

Proposition 1. The function γ, defined by (2.3), satisfies the condition (2.2a) $\forall y \in K(y)$, $\forall \omega \in \Omega$.

Proof. $(u,v) \in \mathcal{H} \Leftrightarrow (u > 0,\ v \in \mathcal{C})$. Hence, because of (2.3a), $\forall y \in K(y)$ and $\forall \omega \in \Omega$ we have:

$$w(y;u,v;\omega) = u + \gamma(y;v;\omega) \geq u > 0\ ,\ \forall (u,v) \in \mathcal{H},$$

so that (2.2a) is verified. □

Now consider the system (in the unknown x):

$$\langle F(y), y - x\rangle + \gamma(y; g(y;x);\omega) > 0\ ,\ x \in X(y), \tag{2.4}$$

and a neighbourhood of ω_0 as

$$\Omega(\epsilon) := \begin{cases} \{\omega \in \Omega : \|\omega - \omega_0\| < \epsilon\}, & \text{if } \|\omega_0\| < \infty, \\ \{\omega \in \Omega : \|\omega\| > \epsilon\}, & \text{if } \|\omega_0\| = \infty, \end{cases}$$

where $\epsilon \in \mathbb{R}_+ \setminus \{0\}$.

Proposition 2. Let γ be the function defined by (2.3). Assume that there exists $\bar{\epsilon} > 0$ such that $\forall \omega \in \Omega(\bar{\epsilon})\ \exists y_\omega \in X(y_\omega)$ such that (2.4) is impossible. Then, $\exists \tilde{\epsilon} \in]0, \bar{\epsilon}]$ such that, $\forall \omega \in \Omega(\tilde{\epsilon}), y_\omega$ belongs to $K(y_\omega)$ and makes (1.2) impossible.

Proof. Let $B(y;r_y)$ be a closed sphere with center at y and radius r_y such that $B(y;r_y) \cap K(y) \neq \emptyset$, $\forall y \in X(y)$. There exists $L(y)$ such that:

$$|\langle F(y), x - y\rangle| \leq L(y)\ ,\ \forall x \in X(y) \cap B(y;r_y). \tag{2.5}$$

Because of (2.3) $\exists \tilde{\epsilon} \in]0, \bar{\epsilon}]$ such that $\forall \omega \in \Omega(\tilde{\epsilon})$ we have:

$$\gamma(y_\omega; g(y_\omega;x);\omega) \begin{cases} > L(y_\omega)\ , \quad \forall x \in K(y_\omega) & , \text{if } y_\omega \notin K(y_\omega), \\ \\ < -L(y_\omega)\ ,\ \forall x \in X(y_\omega) \backslash K(y_\omega) & , \text{if } y_\omega \in \operatorname{frt} K(y_\omega), \end{cases} \tag{2.6}$$

where $\bar{\epsilon}$ and y_ω are those of the assumption. This implies that, $\forall \omega \in \Omega(\bar{\epsilon})$, the system

$$\langle F(y_\omega), y_\omega - x\rangle > -\gamma(y_\omega; g(y_\omega;x);\omega),\ x \in X(y_\omega)$$

is impossible, so that, $\forall \omega \in \Omega(\bar{\epsilon})$, we find

$$\langle F(y_\omega), y_\omega - x\rangle \leq -\gamma(y_\omega; g(y_\omega;x);\omega),\ \forall x \in X(y_\omega). \tag{2.7}$$

Ab absurdo, suppose that the condition "$y_\omega \in K(y_\omega)\ \forall \omega \in \Omega(\tilde{\epsilon})$" be false. Then, for those $\omega \in \Omega(\tilde{\epsilon})$ such that $y_\omega \notin K(y_\omega)$, using (2.5), the 1–st of (2.6) and (2.7), we find a contradiction:

$$-L(y_\omega) \leq \langle F(y_\omega), y_\omega - x\rangle \leq -\gamma(y_\omega; g(y_\omega;x);\omega) < -L(y_\omega)\ ,\ \forall x \in K(y_\omega) \cap B(y_\omega; r_{y_\omega}).$$

Hence $y_\omega \in K(y_\omega)$, $\forall \omega \in \Omega(\tilde{\epsilon})$. To show the 2–nd part of the thesis we distinguish, for each $\omega \in \Omega(\tilde{\epsilon})$, two cases. Suppose that $y_\omega \in \operatorname{int} K(y_\omega)$. Then (2.4) becomes

$$\langle F(y_\omega), y_\omega - x \rangle > 0 \ , \ x \in X(y_\omega),$$

and, being $K(y_\omega) \subseteq X(y_\omega)$, its infeasibility implies that of (1.2). Indeed, in this case – since $\operatorname{int} K(y_\omega) \subseteq \operatorname{int} X(y_\omega)$ – the impossibility of (2.4) implies $F(y_\omega) = 0$. Now suppose that $y_\omega \in \operatorname{frt} K(y_\omega)$. Because of (2.3) the impossibility of (2.4) means the impossibility both of system

$$\langle F(y_\omega), y_\omega - x \rangle > 0 \ , \ x \in K(y_\omega),$$

and of system

$$\langle F(y_\omega), y_\omega - x \rangle + \gamma(y_\omega; g(y_\omega; x); \omega) > 0 \ , \ x \in X(y_\omega) \backslash K(y_\omega).$$

The impossibility of the former implies that of (1.2), and this completes the proof. Note that the impossibility of the latter – which can be deduced directly from (2.5) and (2.6) at $x \in X(y_\omega) \backslash K(y_\omega)$ – can receive the following interpretation: (1.1) may be false in $X(y) \backslash K(y)$; hence a penalization of (1.1) allows us to search for a solution of (1.1) in the relaxed domain $X(y)$. □

As a consequence of Proposition 2, we have that a solution of the inequality: find $y \in X(y)$ such that

$$\langle F(y), x - y \rangle - \gamma(y; g(y; x); \omega) \geq 0 \ , \ \forall x \in X(y), \tag{2.8}$$

with ω close enough to ω_0, is a solution of (1.1) too.

Starting with a weak separation function, exterior exact penalization has been achieved. Same goal may be pursued with a strong separation function [5].

Inequality (2.8) may receive special forms. For instance, assume that a QVI be given, as in Sect.1, in the form:

$$\langle F(y), x - y \rangle \geq 0 \ , \ \forall x \in K(y), \tag{2.9}$$

where again $K(y) \subseteq X(y)$, but now $K(y)$ is not specified as in (1.1). Let us introduce a function $\Psi : \Xi \to \mathbb{R}$ directionally derivable, and assume that its directional derivative at y in the direction $x - y$ be such that:

$$\Psi'(y; x - y) \begin{cases} = 0 \ , \ \forall x \in X(y) & , \text{if } y \in \operatorname{int} K(y), \\ = 0 \ , \ \forall x \in K(y) & , \text{if } y \in \operatorname{frt} K(y), \\ > 0 \ , \ \forall x \in X(y) \backslash K(y) & , \text{if } y \in \operatorname{frt} K(y), \\ < 0 \ , \ \forall x \in K(y) & , \text{if } y \in X(y) \backslash K(y). \end{cases} \tag{2.10}$$

Note that $\Psi'(y; x - y) \geq 0$ is satisfied by every solution of (2.10). If in the inequality (1.1) we set $m = 1$, $\mathcal{C} = \mathbb{R}_+$, $\Omega = \mathbb{R}_+$, $\omega_0 = +\infty$, and

$$g(y; x) = \Psi'(y; x - y) \ , \ \gamma(y; g(y; x); \omega) = -\omega g(y; x) = -\omega \Psi'(y; x - y),$$

so that γ fulfils (2.3), then (2.9) can be embedded in (1.1) and (2.8) becomes:

$$\langle F(y), x - y \rangle + \omega \Psi'(y; x - y) \geq 0 \ , \ \forall x \in X(y), \tag{2.11}$$

and has a more "variational aspect" than (2.8). The above way is equivalent to replace $K(y)$ with the set $\{x \in K(y) : \Psi'(y; x - y) \geq 0\} = K(y)$, and then apply the preceding separation scheme with w linear.

One might desire to handle a Ψ which is differentiable on $X(y)$. This is attainable if we weaken condition (2.3a), in the sense that we renounce to distinguish between interior and frontier of $K(y)$ and to control v outside $\mathcal{C}$. In this order of ideas (2.3a) can be replaced with

$$\gamma(y; v; \omega) \begin{cases} = 0 \ , & \forall v \in \mathcal{C} \quad , \text{ if } y \in K(y), \\ > 0 \ , & \forall v \in \mathcal{C} \quad , \text{ if } y \notin K(y). \end{cases} \tag{2.12}$$

Such a weakening of γ – which can be interpreted as less opportunities of separating in the sense of [5,12] – does not guarantee any longer the thesis of Proposition 2, namely exact penalization. However, we can obtain penalization, even if not necessarily exact. Under condition (2.12), (2.10) becomes:

$$\langle \Psi'(y), x - y \rangle \begin{cases} = 0 \ , & \forall x \in K(y) \quad , \text{ if } y \in K(y), \\ > 0 \ , & \forall x \in K(y) \quad , \text{ if } y \notin K(y), \end{cases} \tag{2.13}$$

where now Ψ' denotes gradient.

The following functions are examples of Ψ. Set $I := \{1, \ldots, m\}$, $g := (g_1, \ldots, g_m)$, $g_i^-(y; x) := \min\{0, g_i(y; x)\}$ $i \in I$. The functions

$$\alpha \|x - \mathrm{proj}_{K(y)} x\| + \beta \|x - \mathrm{proj}_{K(y)} x\|^2 \ , \quad \alpha, \beta \geq 0, \tag{2.14}$$

$$\alpha \|x - \mathrm{proj}_{K(y)} x\| + \beta \sum_{i=1}^{m} [g_i^-(y; x)]^2 \ , \qquad \alpha, \beta \geq 0, \tag{2.15}$$

where proj denotes projection, play the role of Ψ. In the case where $K(y)$ is independent of y – namely QVI becomes VI – the function (2.14) at $\beta = 0$ and (2.15) at $\alpha = 0$ collapse to those classically considered in the literature (see, for instance, [1, page 152] and [2,7]); in this particular case the functions (2.14) and (2.15) fulfil (2.12). However, in (2.13) $\Psi'(y)$ may be replaced with an operator, say, $G(y)$, which is not necessarily the gradient of any functional; in such a case (2.11) becomes:

$$\langle F(y) + \omega G(y), x - y \rangle \geq 0 \ , \ \forall x \in X(y), \tag{2.16}$$

and we are faced with a QVI having an operator which is a pencil of the given operator and the penalization one.

Since in case (2.12) we cannot replace (1.1) with only one (penalized) QVI, it is natural to try to set up a sequence of QVI, such that a corresponding sequence of their solutions converge in some sense to a solution of (1.1). More precisely, we can conceive to construct a sequence $\{y_{\omega_r}\}_{r=1}^{\infty}$ such that y_{ω_r} be a solution of (2.8) (where now γ is supposed to satisfy (2.12)) at $\omega = \omega_r$ and $\lim\limits_{r \to +\infty} \omega_r = \omega_0$; we ask for conditions on (1.1) and on γ such that any (or least one) limit point of the above sequence be a solution of (1.1). This way is an exterior penalization; if we start with a strong separation function – instead of a weak one –, then interior penalization is met.

If $X(y)$ is open, then in (2.8) we have the equality and, in case (2.16), we are led to find y_{ω_r} as solution of the equation

$$F(y) + \omega_r G(y) = 0. \tag{2.17}$$

When (1.1) collapses to a VI, the method based on the resolution of (2.17) is the classic exterior penalization method for VI (see, for instance, [1, page 153]; the assumptions, under which any limit point of $\{y_{\omega_r}\}$ is a solution of VI, cover the VI originated by a strictly convex, but not merely convex, minimization problem; for the connections between this problem and VI see [11]).

Let us now consider an example, by setting $\Xi = \mathbb{R}$, $F(y) = 2y+1$, $X = \mathbb{R}$, $m = 1$, $g(y;x) = x$. Then (1.1) becomes

$$(2y+1)(x-y) \geq 0 \ , \ \forall x \geq 0. \tag{2.18}$$

If we apply the classic exterior penalization, we are led to solve (2.17). By choosing (2.15) with $\alpha = 0$ and identifying β with ω, we find $G(y) = 2y$ if $y < 0$ and $G(y) = 0$ if $y \geq 0$. Then (2.17) is equivalent to the union of systems:

$$\begin{cases} 2y + 1 + 2y\omega = 0 \\ y < 0 \end{cases} \qquad \begin{cases} 2y + 1 = 0 \\ y \geq 0 \end{cases} \qquad (\omega > 0)$$

and we achieve the solution $y_\omega = -1/2(1+\omega)$, $\omega > 0$, which fulfils (2.18) at limit, as $\omega \to +\infty$. If we apply (2.10), we achieve exact penalization. It is easy to find a Ψ which satisfies (2.10): $\Psi(y) = -y$, if $y < 0$ and $\Psi(y) = 0$ if $y \geq 0$. Thus we have:

$$\Psi'(y, x-y) = \begin{cases} 0 & , \text{if } y \geq 0 \text{ and } x \geq 0 \text{ with } x \neq y, \\ -(x-y) & , \text{if } y = 0 \text{ and } x < 0 \text{ or } y < 0 \text{ and } x \geq 0, \end{cases}$$

and, correspondingly, (2.11) becomes:

$$(2y+1)(x-y) + \omega \begin{pmatrix} 0 \\ -(x-y) \end{pmatrix} \geq 0, \ \forall x \in \mathbb{R}.$$

The unique solution of this inequality is easily found to be $y = 0$.

3. FURTHER DEVELOPMENTS

In the preceding sections we have discussed an extension to VI and QVI of penalization for constrained optimization. More appropriately, we should think of penalization approach to generalized systems; indeed the source for both constrained optimization and VI or QVI is a generalized system as (1.2), and separation of sets can be used to investigate them.

Such an approach arises several questions. One has been pointed out in the preceding section about the sequence $\{y_{\omega_r}\}_{r=1}^{\infty}$. In this context it would be useful to connect penalization (2.8) – and its particular cases – with the theory of gap functions for VI and QVI [1,5,12]. In the latter we always operate in the domain $K(y)$, while in the former the domain is replaced with a relaxation; separation of sets can be considered as a background for both theories. Another obvious question concerns with the existence of solutions to (2.8) and uniqueness.

The difference between a penalization of type (2.3) and the classic one – namely, (2.12) or, more generally, (2.10) – can be appreciated in the case where the operator is a gradient (in this case QVI gives a 1–st order optimality condition for a constrained extremum where the feasible region depends on the unknown): the latter requires that y_{ω_r} be a stationary point of the penalized functional in the sense that its gradient be zero (equation (2.17) must hold; now F and G are supposed to be gradients); in the former y_{ω_r} must merely make non–negative the directional derivative of the penalized functional (namely, (2.11) must hold); hence in the latter $X(y)$ is tipically open, while in the former not necessarily.

Several special cases are of great interest. When $\Xi = \mathbb{R}^n$, for instance the cases of F linear, of K independent of y (namely, VI) especially polyhedral, of K convex cone (implicit and ordinary Complementarity Systems) are very important for the applications. When Ξ is not finite dimensional, the penalization approach has already shown to be useful [2,6,8,9,10]; extensions to QVI are of great interest. Penalization might help also in developing the case where the unknown y depends on time through a delay.

When F is the gradient of a functional $f : \Xi \to \mathbb{R}$, the preceding penalization scheme can be applied to find the stationary points of the problem

$$\min f(x) \quad , \quad \text{s.t. } \; x \in K(x). \tag{3.1}$$

If a stationary point is also a minimization point, then the present penalization is connected with that for constrained optimization. The study of (3.1) is interesting by itself; it can be seen as the search for a generalized fixed–point of the point–to–set map K which minimizes f.

Recently a lot of attention has been devoted to several generalizations of VI. One of them is called Generalized Variational Inequality (in short, GVI); see, for instance, [1, page 120]. If in (1.1) we modify the definition of F and let it be a point–to–set map, namely $F : \Xi \rightrightarrows \Xi$, then (1.1) becomes: find $y \in K(y)$ and $z \in F(y)$ such that:

$$\langle z, x - y \rangle \geq 0 \; , \; \forall y \in K(y). \tag{3.2}$$

This inequality, which can be called Generalized Quasi–Variational Inequality (in short, GQVI), embraces a GVI. Now (1.2) is replaced by

$$u := \langle z, y - x\rangle > 0,\ v := g(y; x) \in \mathcal{C}\ ,\ x \in X, \tag{3.3}$$

and $\mathcal{K}(y)$ by

$$\mathcal{K}(y, z) := \{(u, v) \in \mathbb{R} \times \mathbb{R}^m : u = \langle z, y - x\rangle\ ,\ v = g(y; x),\ x \in X(y)\}$$

with $y \in K(y)$ and $z \in F(y)$; $\mathcal{H}$ does not change. The preceding remarks on penalization spread to (3.2). If there is a functional $f : \Xi \to \mathbb{R}$ such that $F(x)$ is a generalized differential (for instance, the classic subdifferential), then (3.2) can be seen as the 1–st order necessary optimality condition of problem (3.1) where now f is nondifferentiable.

The case where, in (3.2), y is replaced by $T(y)$, with $T : \Xi \to \Xi$, is also of interest and can embedded in a separation scheme. When K is independent of y, such an inequality has already been investigated and shown to be equivalent to the so–called Implicit Complementarity System [8]. It is useful to investigate the penalization scheme for VI and its generalizations when K is a cone and to connect it to the penalization (as well as the gap or merit functions) conceived for Complementarity Systems and their generalizations.

Another interesting generalization is that where (1.1) is replaced with

$$\langle F(y), x - y\rangle \geq \varphi(y) - \varphi(x)\ ,\ \forall x \in K(y), \tag{3.4}$$

where $\varphi : \Xi \to \mathbb{R}$ is given. All the preceding remarks spread to (3.4). This inequality may be the 1–st order necessary optimality condition of the problem

$$\min[f(x) + \varphi(x)]\ ,\ \text{s.t. } x \in K(x), \tag{3.5}$$

if $K(y)$ is convex, φ is convex, f is differentiable, and $F(x) = f'(x)$. In facts, if y is a minimum point of (3.5), then $\exists z \in \partial\varphi(y)$, depending on $x - y$ (beside y), such that:

$$\langle f'(y) + z, x - y\rangle \geq 0\ ,\ \forall x \in K(y), \tag{3.6}$$

where $\partial\varphi$ denotes the (classic) subdifferential of φ. $\forall z \in \partial\varphi(y)$ we have:

$$\langle z, y - x\rangle \geq \varphi(y) - \varphi(x)\ ,\ \forall x \in K(y). \tag{3.7}$$

From (3.6) and (3.7) we deduce the inequality

$$\langle f'(y), x - y\rangle \geq \varphi(y) - \varphi(x)\ ,\ \forall x \in K(y),$$

which is a particular case of (3.4).

Several other formats and topics may be taken into consideration. Within the former ones we may quote a vector Variational Inequality [4, page 167], and general models for equilibrium problems [3]; among the latter ones it is interesting to connect penalization with duality for VI and QVI.

REFERENCES

[1] A. Auslender, "Optimization. Méthodes numériques". Masson, Paris, 1976.

[2] A. Bensoussan and P. Kenneth, "Sur l'analogie entre les méthodes de régularization et de pénalisation". Revue d' Informatique et de Recherche Opérationelle, N. 13, 1969.

[3] E. Blum and W. Oettli, "From Optimization and Variational inequalities to equilibrium problems". Proceedings of the 57-th Conference of the Indian Mathem. Soc., Published in "The Mathematics Student", Vol. 63, N. 1–4, pp. 1–23, 1993.

[4] F. Giannessi, "Theorems of the alternative, quadratic programs and complementarity problems". In "Variational Inequalities and complementarity problems", R.W. Cottle et al. (eds.), J. Wiley, New York, pp. 151–186, 1980.

[5] F. Giannessi, "Separation of sets and gap functions for Quasi–Variational Inequalities". In "Variational Inequalities and network equilibrium problems", F. Giannessi and A. Maugeri (Ed.s), Plenum, New York, pp. 101–121, 1995.

[6] R. Glowinski and P. Le Tallec, "Augmented Lagrangian and operator–splitting methods in nonlinear Mechanics". SIAM Studies in Appl. Mathematics, Philadelphia, 1989.

[7] J. Gwinner, "On the penalty method for constrained Variational Inequalities". In "Optimization Theory and Applications", J.–B. Hiriart–Hurruty, W. Oettli and J. Stoer (Ed.s), Lecture Notes in Pure and Applied Mathematics, Vol. 86, Marcel Dekker, New York, pp. 197–211, 1983.

[8] G. Isac, "Complementarity problems". Springer–Verlag, Berlin, 1992.

[9] D. Kinderleherer and G. Stampacchia, "An Introduction to Variational Inequalities and their applications". Academic Press, New York, 1980.

[10] J.-L. Lions, "Quelques méthodes de résolution des problèmes aux limites non linéaires". Dunod, Paris, 1969.

[11] O. G. Mancino and G. Stampacchia, "Convex programming and Variational Inequalities". Journal of Optimization Theory and Applications, Vol. 9, pp. 3–23, 1972.

[12] A. Maugeri, "Variational and Quasi–Variational Inequalities in network flow models. Recent developments in theory and algorithms". In "Variational Inequalities and network equilibrium problems", F. Giannessi and A. Maugeri (Ed.s), Plenum, New York, pp. 195–211, 1995.

GLOBALLY CONVERGENT ONLINE MINIMIZATION ALGORITHMS FOR NEURAL NETWORK TRAINING

Grippo Luigi

Dipartimento di Informatica e Sistemistica
Università di Roma " La Sapienza"
Via Buonarroti 12, 00185 Roma, Italy
e-mail: grippo@newton.dis.uniroma1.it

Abstract. In this paper we consider the application of unconstrained minimization methods to the solution of learning problems in neural networks and we define new globally convergent online algorithms based on a reformulation of the original problem as an equivalent constrained problem in an extended space. For the solution of this problem we propose an alternating direction technique employing as merit function a differentiable exact augmented Lagrangian function of suitable structure.

Key words. Neural networks, machine learning, training algorithms, unconstrained minimization

1. INTRODUCTION

An artificial neural network, (see, e. g., [5], [23], [28], [37]), can be viewed as a system (a physical system or a mathematical model) of a special structure that approximates an unknown real mapping

$$\Phi : X \to \mathbb{R}.$$

Then, the *learning problem* for a neural network of given topology can be formulated as the problem of determining a vector of unknown parameters $x \in \mathbb{R}^n$ in a structured model, on the basis of a given set of data

$$T = \{(\xi_j, v_j), \quad j = 1, \dots, M\},$$

where $v_j \in \mathbb{R}$ is some estimate of $\Phi(\xi_j)$, possibly corrupted by a random noise.

In case of noiseless data we are faced, in general, with a difficult multidimensional nonlinear approximation problem, whereas in the presence of noise we have a typical nonlinear regression problem. We refer, e.g. to [11], [16], [37] for a discussion of the assumptions needed to give theoretical foundations to the learning problem, in connection with various network structures. Here we assume that the problem of minimizing the squared error between the observations and the actual output of a given network makes sense, and we are concerned with numerical algorithms for computing stationary points of the error function.

Nonlinear Optimization and Applications
Edited by G. Di Pillo and F. Giannessi, Plenum Press, New York, 1996

If we denote by $x \in \mathbb{R}^n$ the vector of parameters and by $\Psi(x;\xi)$ the actual output of the network corresponding to an input $\xi \in X$, we can define the learning problem as the (nonlinear) least square problem:

$$\min_{x \in \mathbb{R}^n} \; f(x) := \sum_{j=1}^{M} (v_j - \Psi(x;\xi_j))^2 . \tag{1.1}$$

Algorithms that solve problem (1.1) are known as *training algorithms* and the set T of patterns (ξ_j, v_j) is referred to as the *training set.* If problem (1.1) can be solved successfully, then the network can be used in the *recall phase* as a system that yields instantaneously an output $y = \Psi(x;\xi)$ approximating $\Phi(\xi)$, in correspondence to any input $\xi \in X$, which may be not the element of a training pair. The extent to which this is achieved defines the *generalization* capability of the network.

In concrete applications, we can distinguish two situations, which may have a different impact in the definition of solution algorithms for problem (1.1). The first case is when the training phase is temporally separated from the recall phase, and hence we can assume that the whole training set is available before that learning is started (*offline learning*). The second case is when the training set is formed (at least partially) during real time operations of the network (*online learning*). We note that this distinction is referred only to the times at which the training pairs are assumed to be available and not to the actual use of the network, which is typically intended as an online device that can also"*learn offline to be adaptive online*" [35].

As regards training algorithms, a distinction is usually made between *batch* techniques, which use at each step the knowledge of the global error function f, and *online algorithms* (or *pattern search* methods), which use only the knowledge of a single error term $(v_j - \Psi(x;\xi_j))^2$ at time. In particular, the "classical" technique for solving learning problems is the so called *backpropagation* (BP) algorithm [34]. In the batch version (*batch* BP) it can be viewed as an heuristic implementation of the gradient method for minimizing f; in the online version (*online* BP) it uses the negative gradient of a single error term at each step and some heuristic for choosing the stepsize (called *learning rate*).

It is easily seen that a batch technique can be considered, essentially, as an unconstrained minimization algorithm and hence many efficient methods are available for computing stationary points of f in batch mode, even for very large dimensional problems. In particular, the use of conjugate gradient methods has been suggested since the early papers on neural networks and good results have been reported by employing the Polak-Ribière [3] or the Shanno conjugate gradient method [35]. Viable alternatives could be truncated Newton methods [6], [9], [30] or some new convergent implementation of the Polak-Ribière method [15], [21]. In this connection, we remark that there is no apparent reason for preferring the use of heuristics in the computation of the stepsizes to the adoption of an efficient line search technique. We note also that nonmonotone linesearch techniques [19] [20] could be of value for escaping from steep sided valleys in the contours of the objective function, which represent a typical difficulty in training problems.

For online learning problems, batch methods are not suitable, since the objective function is not known when training is started, and hence online methods have to be adopted. Online methods are also preferred, in practice, for offline learning, at least when the number M of training pairs is very large and the training set is highly redundant. In fact, in the presence of a large and redundant training set, it could be wasteful to spend much time in computing exactly the gradient of f when far from the solution; in addition, it would seem that an online procedure introduces some sort of randomness that may help escaping from local minimizers [5].

Thus there exists a considerable interest in developing online algorithms (both for online and for offline learning), which can conciliate the good convergence properties of batch methods with the needs of rapid real-time learning. However, when we turn our attention to online methods, we no more have at our disposal a well established theory, on which computational techniques can be based. In most of neural network applications the online BP algorithm (and its many variations) has been employed on a purely heuristic base. Only recently, convergence results have been obtained by giving explicit rules for the stepsizes, which, under suitable assumptions, may ensure convergence towards stationary points. In particular, stepsize rules for nonconvex problems have been established in [14], [36], by using stochastic approximation ideas and in [27], [29], by employing deterministic approaches. The main difficulty with these approaches is that convergence can be enforced only by imposing that the stepsizes (learning rate) converge to zero and also that the boundedness of the sequence must be assumed in order to ensure the existence of accumulation points. The use of linesearches employing only function values, in connection with online methods, has been considered in [17], where convergence is established for a *hybrid online-batch* BP technique, which requires only derivative information on a component function at time for the construction of the search directions, but also uses, periodically, the knowledge of the overall objective function for imposing bounds on the stepsizes. This technique may improve the behaviour of online BP (and actually does) in case of offline learning with a moderately large training set, but still has the disadvantage that the whole objective function has to be evaluated, which is expensive when M is very large and is unsuitable in case of "truly" online learning.

In the present paper we propose a different approach, which may allow us, in principle, to construct globally convergent online algorithms that do not enforce the learning rate to converge to zero, but also avoid the need of evaluating the whole objective function at each step.

The starting point is that of reformulating the unconstrained problem as a constrained problem in an extended space, whose elements are vectors formed with multiple "copies" of the original variables. In order to illustrate this possibility, let us assume that the problem can be put into the form:

$$\min_{x \in \mathbf{R}^n} f(x) := \sum_{j=1}^{N} f_j(x),$$

where we can assume (when convenient) that each component function f_j is formed by aggregating several error terms. In particular, in case of online learning, we can identify, for instance, f_1 with the sum of the terms which are known when training starts, whereas the remaining components may represent the contributions of patterns that become available at later subsequent times. In offline learning problems, we can suppose that the error terms are subdivided into different groups; in this case f_1 may represent the error associated to a subset of nonredundant patterns that ideally we would process in a batch mode.

Now we can rewrite our problem as an equivalent constrained problem of the form

$$\begin{aligned} &\text{minimize } \sum_{j=1}^{N} f_j(y_j) \\ &\text{subject to} \\ &y_{j+1} - y_1 = 0, \quad j = 1, \ldots, N-1 \end{aligned}$$

where $y_j \in \mathbb{R}^n$ for $j = 1, \ldots, N$. A similar reformulation has been exploited, in a rather different context, in a number of papers (see, e.g., [2], [4], [25], [7], [10], [12], [13]), in connection with decomposition of convex problems solved through the so called alternating direction augmented Lagrangian method. In our case, however, the convexity assumption is not required and the classical augmented Lagrangian function will be replaced by a differentiable exact augmented Lagrangian [1], [31], [33]. More specifically, under the assumption that the functions f_j are sufficiently smooth and coercive, it will be shown that the originary problem can solved via the unconstrained minimization of a differentiable function of the form

$$\begin{aligned} S(y,\lambda) \; &:= \sum_{j=1}^{N} f_j(y_j) + \sum_{j=1}^{N-1} \lambda_j'(y_{j+1} - y_1) + \sum_{j=1}^{N-1} \left[c + \tau\|\lambda_j\|^2\right] \|y_{j+1} - y_1\|^2 \\ &+ \eta \sum_{j=1}^{N-1} \|\nabla f_{j+1}(y_{j+1}) + \lambda_j\|^2, \end{aligned}$$

with respect to *both* the problem variables *and* the Lagrange multiplier estimates $\lambda_j \in \mathbb{R}^n$, for given suitable values of the parameters η, τ and c. We note that the use of an exact Lagrangian function (of a somewhat different form) in the decomposition of constrained optimization problems has been suggested for the first time in [26]. In comparison with the classical augmented Lagrangian approach, we have now that a single unconstrained problem has to be solved, and, more importantly, that the solution of the constrained problem, together with the associated multiplier, constitutes a minimum point rather than a saddle point of the Lagrangian function as in the classical case. This feature turns out to be important in our context, since we can perform linesearches with respect to both the problem variables and the multipliers with reference to the same merit function.

In this paper we will attempt to define globally convergent online algorithms by applying the block partitioning technique studied in [22] to the unconstrained minimization of the augmented Lagrangian S. In fact, it is proved in [22] that a globally convergent algorithm for minimizing an unconstrained (non convex) function can be constructed by employing suitable linesearches with respect to separate blocks of variables. Then the structure of S allow us to perform linesearches (with respect to both y and λ) in different spaces, which enforce a monotonic decrease of S, without requiring a complete *a priori* knowledge of the augmented Lagrangian function. It is easily seen, as an example, that a linesearch along a descent direction in the y_1−space, for fixed values of the remaining variables will not require the knowledge of f_j, for $j > 1$ and similarly, for $j > 1$ we can perform, for instance, a linesearch in the variables (y_{j+1}, λ_j) without requiring the knowledge of f_i for $i \neq j$. The interactions between these processes are taken into account in the term

$$\sum_{j=1}^{N-1} \lambda_j'(y_{j+1} - y_1) + \left[c + \tau\|\lambda_j\|^2\right] \|y_{j+1} - y_1\|^2$$

which does not depend on the problem functions.

A cycle is completed when all variables have been updated and then a new sequence of linesearches can be performed. In the limit, if all variables are updated infinitely often, we have that the N "copies" y_j will converge to the same point, which is a stationary point of the original objective function.

This procedure can be made the basis for constructing various online algorithms with deterministic convergence properties. However, in order to define a technique to be adopted in online learning we must give rules for deriving an estimate of the solution in the original x-space, for otherwise the process will converge through independent sequences of points in different spaces.

A first important observation is that, after a search in a subspace, we can attempt to replace with the new point obtained, say y_j^k, the current initial estimate y_{j+1}^{k-1} for the next step. The convenience of this replacement can be decided, once that f_{j+1} is available, by comparing the corresponding values of S and can be viewed, in a certain sense, as a "redundancy test". In fact, in case of redundancy, we may expect that the point that yields a reduction in the component function f_j would produce a similar reduction in the value of f_{j+1}. When this is possible for all components, we will obtain the same pattern in every subspace and the algorithm will resemble, essentially, an online BP algorithm, where now the stepsizes are computed through suitable linesearches.

In the general case, however, we may need to derive a real-time estimate of the solution from the different copies y_j^k.

One possibility could be simply that of taking as x-estimate at each step an (arbitrary) affine combination of the different copies; as the vectors y_j converge to the same limit point, this estimate will converge to a stationary point of f. This criterion would resemble a "syncronization step" of the form considered in [8] in the context of parallel optimization.

An alternative approach, which will be considered in the sequel in more detail, could be that of taking one variable, say y_1, as the current estimate of the solution. This can be justified as long as we give a special meaning to the term f_1 that may represent, for instance, the most relevant knowledge that we want to process essentially in a batch mode. However, we cannot wait for the completion of a whole cycle, for, otherwise, the advantages of an online learning process would be lost. A solution could be that of performing one (or more) searches in the y_1-space, each time that a new step in the other spaces is performed. In this way we can take into account the effect of the most recent patterns, while continuing a batch learning process in the y_1-space, with repeated presentations of a subset of patterns.

In conclusion, we can say that the augmented Lagrangian reformulation may favour, in principle, a greater flexibility in devising online algorithm, while preserving the guarantee of an ultimate convergence. However, the real advantages (if any) over the current approaches have to be assessed. The present work is essentially devoted to establish some basic and very preliminary result.

The paper is organized as follows. In Section 2 we state the problem and we prove the exactness of the augmented Lagrangian function; in Section 3 we formulate an algorithm model and we outline a convergence proof on the basis of the results of [22]; finally in Section 4 we discuss implementation problems and possible developments.

2. THE AUGMENTED LAGRANGIAN FUNCTION

We consider the problem

$$\begin{array}{l} \text{minimize} \quad f(x) \\ x \in \mathbb{R}^n \end{array} \tag{2.1}$$

where $f : \mathbb{R}^n \to \mathbb{R}$ is an objective function expressed as the sum of N *component functions* $f_j : \mathbb{R}^n \to \mathbb{R}$ for $j = 1, \ldots, N$, that is

$$f(x) := \sum_{j=1}^{N} f_j(x).$$

We suppose that the following assumption holds.

Assumption 1 *For every $j \in \{1, \ldots, N\}$ the function $f_j : \mathbb{R}^n \to \mathbb{R}$ is twice continuously differentiable and satisfies:*

$$\lim_{\|x\| \to \infty} f_j(x) = \infty. \qquad \square$$

Remark 1

We note that in neural network applications, Assumption 1 may be not satisfied, because of the form of the error functions, which typically exhibit saturations at infinity. In this case, the simplest possibility is that of adding to f a perturbation term of the form $\varepsilon\|x\|^2$, where ε is a small positive parameter. This term can be distributed among the component functions by redefining these function as:

$$\tilde{f}_j(x) := f_j(x) + \frac{\varepsilon}{N}\|x\|^2.$$

We observe, in passing, that in some cases this device may be also useful for escaping from a local minimizer, since we can attempt to start a new minimization with a smaller value of ε. $\square$

Given $x^0 \in \mathbb{R}^n$ we denote by $\mathcal{L}_0$ the level set

$$\mathcal{L}_0 := \{x \in \mathbb{R}^n : f(x) \leq f(x^0)\}.$$

As a consequence of Assumption 1 we have that the set $\mathcal{L}_0$ is compact and that Problem (2.1) admits a global minimizer on $\mathbb{R}^n$, which is a stationary point of f in $\mathcal{L}_0$.
Problem (2.1) can be equivalently rewritten as a constrained problem of the form:

$$\begin{aligned} &\text{minimize } \sum_{j=1}^{N} f_j(y_j) \\ &\text{subject to} \\ &y_{j+1} - y_1 = 0, \quad j = 1, \ldots, N-1 \end{aligned} \tag{2.2}$$

where $y_j \in \mathbb{R}^n$ for $j = 1, \ldots, N$. We denote by $y \in \mathbb{R}^{nN}$ the vector with components $y_j \in \mathbb{R}^n$, that is, we set

$$y' = (y_1', \ldots y_N').$$

For the solution of Problem (2.2) we introduce a *differentiable exact augmented Lagrangian function* of suitable structure.
Define first a vector of *multipliers* $\lambda \in \mathbb{R}^{n(N-1)}$ with vector components $\lambda_j \in \mathbb{R}^n$:

$$\lambda' = (\lambda_1', \ldots \lambda_{N-1}')$$

and let $L : \mathbb{R}^{nN} \times \mathbb{R}^{n(N-1)} \to \mathbb{R}$ be the *Lagrangian function* associated to Problem (2.2), that is:

$$L(y, \lambda) := \sum_{j=1}^{N} f_j(y_j) + \sum_{j=1}^{N-1} \lambda_j'(y_{j+1} - y_1). \tag{2.3}$$

It is easily seen that the Lagrange multipliers corresponding to a local solution of Problem (2.2) can be uniquely determined from the conditions

$$\nabla f_{j+1}(y_{j+1}) + \lambda_j = 0, \quad j = 1, \dots, N-1 \tag{2.4}$$

Then, by introducing suitable penalty terms on the optimality conditions for Problem (2.2) we can construct an augmented Lagrangian function $S : \mathbb{R}^{nN} \times \mathbb{R}^{n(N-1)} \to \mathbb{R}$ of the following form:

$$\begin{aligned} S(y, \lambda) \; &:= \sum_{j=1}^{N} f_j(y_j) + \sum_{j=1}^{N-1} \lambda_j'(y_{j+1} - y_1) + \sum_{j=1}^{N-1} \left[c + \tau \|\lambda_j\|^2\right] \|y_{j+1} - y_1\|^2 \\ &+ \eta \sum_{j=1}^{N-1} \|\nabla f_{j+1}(y_{j+1}) + \lambda_j\|^2, \end{aligned} \tag{2.5}$$

where η, τ and c are given parameters.

The properties of exactness of this function, which has essentially the same structure of the augmented Lagrangian functions studied in [1], [31], [32], can be established by using quite standard arguments in the field of exact penalization. However we will give, for completeness, direct proofs of the correspondence with the original unconstrained problem, which will be also useful for imposing conditions on the parameters.

First of all we indicate under which conditions the unconstrained minimization of S on $\mathbb{R}^{nN} \times \mathbb{R}^{n(N-1)}$ has a solution. Let us define a point $(y^0, \lambda^0) \in \mathbb{R}^{nN} \times \mathbb{R}^{n(N-1)}$ by assuming:

$$y_j^0 = x^0, \quad j = 1, \dots, N, \quad \lambda_j^0 = 0, \quad j = 1, \dots, N-1 \tag{2.6}$$

so that we have

$$S(y^0, \lambda^0) = f(x^0) + \eta \sum_{j=1}^{N-1} \|\nabla f_{j+1}(x^0)\|^2.$$

Now define a function $\phi : \mathbb{R}^{nN} \times \mathbb{R}^{n(N-1)} \to \mathbb{R}$ by letting

$$\phi(y, \lambda) := L(\lambda, y) + \tau \sum_{j=1}^{N-1} \|\lambda_j\|^2 \|y_{j+1} - y_1\|^2 + \eta \sum_{j=1}^{N-1} \|\nabla f_{j+1}(y_{j+1}) + \lambda_j\|^2, \tag{2.7}$$

whence it follows that we can write

$$S(y, \lambda) = \phi(y, \lambda) + c \sum_{j=1}^{N-1} \|y_{j+1} - y_1\|^2, \tag{2.8}$$

and

$$\phi(y^0, \lambda^0) = S(y^0, \lambda^0).$$

Then consider the level sets:

$$\mathcal{R}_0 := \{(y, \lambda) \in \mathbb{R}^{nN} \times \mathbb{R}^{n(N-1)} : \phi(y, \lambda) \le \phi(y^0, \lambda^0)\} \tag{2.9}$$

$$\begin{aligned} \mathcal{W}_0 : \; &= \{(y, \lambda) \in \mathbb{R}^{nN} \times \mathbb{R}^{n(N-1)} : S(y, \lambda) \le S(y^0, \lambda^0)\} \\ &= \{(y, \lambda) \in \mathbb{R}^{nN} \times \mathbb{R}^{n(N-1)} : S(y, \lambda) \le \phi(y^0, \lambda^0)\}. \end{aligned} \tag{2.10}$$

We can state the following proposition.

Proposition 2.1 *Let $\tau > 0$, $\eta > 0$ and and $c \ge 0$ be given numbers. Then:*

(i) *the set $\mathcal{R}_0$ is compact;*

(ii) *the set $\mathcal{W}_0$ is a compact subset of $\mathcal{R}_0$;*

(iii) *the function S admits a global minimizer on* $\mathbb{R}^{nN} \times \mathbb{R}^{n(N-1)}$.

Proof. From the expression of ϕ, noting that

$$\lambda_j'(y_{j+1} - y_1) + \tau\|\lambda_j\|^2\|y_{j+1} - y_1\|^2 \geq -\|\lambda_j\|\|y_{j+1} - y_1\| + \tau\|\lambda_j\|^2\|y_{j+1} - y_1\|^2 \geq -\frac{1}{4\tau}$$

we obtain immediately the inequality

$$\phi(y, \lambda) \geq \sum_{j=1}^{N} f_j(y_j) + \eta \sum_{j=1}^{N-1} \|\nabla f_{j+1}(y_{j+1}) + \lambda_j\|^2 - \frac{(N-1)}{4\tau}, \qquad (2.11)$$

To prove the assertions we must show that $\mathcal{R}_0$ is bounded. Suppose the contrary and let $\{(y^k, \lambda^k)\}$ be an unbounded sequence of points in $\mathcal{R}_0$. Assume first that $\{y^k\}$ is unbounded, so that there exists a $j \in \{1, \ldots, N\}$ and a subsequence, which we relabel $\{(y^k, \lambda^k)\}$, such that $\|y_j^k\| \to \infty$. Then, from Assumption 1 and (2.11) we have that $\{\phi(y^k, \lambda^k)\}$ goes to infinity and this yields a contradiction. Now assume that $\{y^k\}$ is bounded and hence that there exists an index j and a subsequence (call it again $\{(y^k, \lambda^k)\}$) such that $\|\lambda_j^k\| \to \infty$. As $\{y^k\}$ is bounded and ∇f is continuous, this implies that $\|\nabla f_{j+1}(y_{j+1}^k) + \lambda_j^k\|^2 \to \infty$ and thus we have again that $\{\phi(y^k, \lambda^k)\}$ goes to infinity, which is a contradiction. It can be concluded that $\mathcal{R}_0$ is bounded and hence that (i) is verified. On the other hand, by (2.8) we have that $S(y, \lambda) \leq S(y^0, \lambda^0)$ implies $\phi(y, \lambda) \leq \phi(y^0, \lambda^0)$. This proves that the closed set $\mathcal{W}_0$ is a subset of $\mathcal{R}_0$ and hence that (ii) and (iii) are verified. □

We give now the expression of the gradient of S. In order to simplify notation, let us define the matrices

$$B_j(\eta) := I + 2\eta\nabla^2 f_{j+1}(y_{j+1}), \quad j = 1, \ldots, N-1 \qquad (2.12)$$

$$C_j(\tau) := I + 2\tau\lambda_j(y_{j+1} - y_1)', \quad j = 1, \ldots, N-1. \qquad (2.13)$$

Then, after straightforward manipulations we get the following expressions of the partial gradients with respect to the subvectors y_j and λ_j.

$$\nabla_{y_1} S(y, \lambda) = \nabla f_1(y_1) - \sum_{j=1}^{N-1} \lambda_j - 2\sum_{j=1}^{N-1} \left[c + \tau\|\lambda_j\|^2\right](y_{j+1} - y_1) \qquad (2.14)$$

and for $j = 1 \ldots, N-1$

$$\nabla_{y_{j+1}} S(y, \lambda) = B_j(\eta)(\nabla f_{j+1}(y_{j+1}) + \lambda_j) + 2\left(c + \tau\|\lambda_j\|^2\right)(y_{j+1} - y_1) \qquad (2.15)$$

$$\nabla_{\lambda_j} S(y, \lambda) = 2\eta\left(\nabla f_{j+1}(y_{j+1}) + \lambda_j\right) + C_j(\tau)(y_{j+1} - y_1). \qquad (2.16)$$

We are now ready to establish the required properties of exactness. In the next proposition we relate stationary points of S to stationary points of f.

Proposition 2.2 *For every value of the parameters τ, η and c, if $\bar{x}$ is a stationary point of f, then the point $(\bar{y}, \bar{\lambda})$ defined by*

$$\bar{y}_j = \bar{x}, \quad j = 1, \ldots, N \quad \bar{\lambda}_j = -\nabla f_{j+1}(\bar{x}) \quad j = 1, \ldots, N-1$$

is a stationary point of S and we have $S(\bar{y}, \bar{\lambda}) = f(\bar{x})$.
Conversely, for any given $\tau > 0$ and $\eta > 0$ there exists a number $c^* > 0$, such that for every $c \geq c^*$, if $(\bar{y}, \bar{\lambda})$ is a stationary point of S in the level set $\mathcal{W}_0$ we have:

$$\bar{y}_{j+1} - \bar{y}_1 = 0, \quad \bar{\lambda}_j = -\nabla f_{j+1}(\bar{y}_{j+1}) \quad j = 1, \ldots, N-1$$

and

$$\nabla f(\bar{y}_1) = 0, \quad f(\bar{y}_1) = S(\bar{y}, \bar{\lambda})$$

Proof. The first assertion follows immediately from the expressions (2.14) (2.15) (2.16) of the partial gradients of S.
In order to prove the second assertion, let us choose a sufficiently large number $c^* > 0$ such that for all $c \geq c^*$ the matrices

$$D_j(y,\lambda) := 2\eta I - \frac{B_j(\eta)C_j(\tau)}{2\,(c+\tau\|\lambda_j\|^2)}, \quad j = 1,\dots N-1$$

are non singular on the compact set $\mathcal{R}_0$. Then, let $c \geq c^*$ and suppose that $(\bar{y},\bar{\lambda})$ is a stationary point of S in $\mathcal{W}_0$. By Proposition 2.1 we have that $(\bar{y},\bar{\lambda})$ is in the compact set $\mathcal{R}_0$ and therefore the assumption made on c implies that the matrices $D_j(\bar{y},\bar{\lambda})$ are non singular.
Now, recalling (2.15), for $j = 1,\dots,N-1$ from $\nabla_{y_{j+1}} S(\bar{y},\bar{\lambda}) = 0$ we get

$$\bar{y}_{j+1} - \bar{y}_1 = -\frac{B_j(\eta)}{2\left(c+\tau\|\bar{\lambda}_j\|^2\right)}\left(\nabla f_{j+1}(\bar{y}_{j+1}) + \bar{\lambda}_j\right),$$

so that, from $\nabla_{\lambda_j} S(\bar{y},\bar{\lambda}) = 0$ and (2.16) we obtain

$$D_j(\bar{y},\bar{\lambda})\left(\nabla f_{j+1}(\bar{y}_{j+1}) + \bar{\lambda}_j\right) = 0.$$

As the matrices $D_j(\bar{y},\bar{\lambda})$ are non singular this implies

$$\nabla f_{j+1}(\bar{y}_{j+1}) + \bar{\lambda}_j = 0, \tag{2.17}$$

whence it follows:

$$\bar{\lambda}_j = -\nabla f_{j+1}(\bar{y}_{j+1}) \quad j = 1,\dots,N-1.$$

Therefore, again by $\nabla_{y_{j+1}} S(\bar{y},\bar{\lambda}) = 0$ we have

$$\bar{y}_{j+1} - \bar{y}_1 = 0, \quad j = 1,\dots,N-1, \tag{2.18}$$

so that $f(\bar{y}_1) = S(\bar{y},\bar{\lambda})$.
As $\nabla_{y_1} S(\bar{y},\bar{\lambda}) = 0$, equation (2.18) implies, by (2.14) and (2.17), that $\nabla f(\bar{y}_1) = 0$ and this concludes the proof. □

The next proposition establishes the correspondence between global minimizers of f and S and follows immediately from Proposition 2.2.

Proposition 2.3 *For any given $\tau > 0$ and $\eta > 0$, let $c^* > 0$ be the number considered in Proposition 2.2. Then, for every $c \geq c^*$ we have:*

(i) *if $x^* \in \mathbb{R}^n$ is a global minimizer of f, the point (y^*,λ^*) defined by*

$$y^*{}_j = x^*, \quad j = 1,\dots,N \quad \lambda^*{}_j = -\nabla f_{j+1}(x^*) \quad j = 1,\dots,N-1$$

is a global minimizer of S;

(ii) *if $(y^*,\lambda^*) \in \mathbb{R}^{nN} \times \mathbb{R}^{n(N-1)}$ is a global minimizer of S we have:*

$$y^*{}_{j+1} - y^*{}_1 = 0, \quad \lambda^*{}_j = -\nabla f_{j+1}(y^*{}_{j+1}) \quad j = 1,\dots,N-1$$

and y_1^* is a global minimizer of f.

Proof. Let $c \geq c^*$ and define the set:

$$\mathcal{W}_0^* := \{(y, \lambda) \in \mathbb{R}^{nN} \times \mathbb{R}^{n(N-1)} : S(y, \lambda) \leq f(x^0)\}.$$

Suppose that $x^* \in \mathbb{R}^n$ is a stationary point of f in $\mathcal{L}_0$. By Proposition 2.2 the point (y^*, λ^*) defined as in (i) must be a stationary point of S and we have $S(y^*, \lambda^*) = f(x^*) \leq f(x^0)$, so that (y^*, λ^*) belongs to $\mathcal{W}_0^*$. On the other hand, if (y^*, λ^*) is a stationary point of S in $\mathcal{W}_0^*$, again by Proposition 2.2 we have that y_1^* is a stationary point of f such that $f(y_1^*) = S(y^*, \lambda^*) \leq f(x^0)$. It can be concluded that stationary points of f in $\mathcal{L}_0$ are in a one-to-one correspondence with stationary points of S in $\mathcal{W}_0^*$ and that f and S take the same value at each of these corresponding pairs. This obviously implies that the global minimizers of the two functions (on the level sets $\mathcal{L}_0$ and $\mathcal{W}_0^*$, and hence on the entire spaces) are in the same correspondence. □

Using standard arguments, it also possible to show that local minimizers of S yield local minimizers of f. This is proved formally in the next proposition.

Proposition 2.4 *For any given $\tau > 0$ and $\eta > 0$, let $c^* > 0$ be the number considered in Proposition 2.2. Then, for every $c \geq c^*$, if (y^*, λ^*) is a local minimizer of S in $\mathcal{W}_0$, we have:*

$$y^*{}_{j+1} - y^*{}_1 = 0, \quad \lambda^*{}_j = -\nabla f_{j+1}(y^*{}_{j+1}) \quad j = 1, \ldots, N-1$$

and y_1^* is a local minimizer of f.

Proof. Let $c \geq c^*$ and suppose that (y^*, λ^*) is a local minimizer of S in $\mathcal{W}_0$. Thus there exists a neighbourhood Ω of (y^*, λ^*) such that

$$S(y^*, \lambda^*) \leq S(y, \lambda), \quad \text{for all} \quad (y, \lambda) \in \Omega. \tag{2.19}$$

But (y^*, λ^*) is also a stationary point of S and thus from Proposition 2.2 we have $\nabla f(y_1^*) = 0$, $y^*{}_{j+1} = y_1^*$ and $\lambda_j^* = -\nabla f_{j+1}(y^*{}_{j+1})$, for $j = 1, \ldots, N-1$, so that $S(y^*, \lambda^*) = f(y_1^*)$. Now, given $y_1^* \in \mathbb{R}^n$ we can choose, by continuity, a sufficiently small neighbourhood of y_1^* in $\mathbb{R}^n$, say U, such that for every $x \in U$ the point $(y(x), \lambda(x))$ with components $y_j(x) = x$, for $j = 1, \ldots, N$ and $\lambda_j(x) = -\nabla f_{j+1}(x)$ for $j = 1, \ldots, N-1$ remains in Ω. By definition of S we have $S(y(x), \lambda(x)) = f(x)$ and therefore, from (2.19) we get

$$f(y_1^*) \leq f(x) \quad \text{for all} \quad x \in U,$$

which proves our assertion. □

3. A PARTITIONING ALGORITHM MODEL

In this section we define an algorithm model for the minimization of S, which makes use of a partitioning algorithm, derived from the results of [22]. We suppose that the problem variables are partitioned into the subvectors:

$$z_1 := y_1 \quad \text{and} \quad z_{j+1} := (y'_{j+1} \quad \lambda'_j)' \quad \text{for} \quad j = 1, \ldots, N-1,$$

and we set $z = (z'_1, \ldots, z'_j, \ldots, z'_N)'$. In correspondence to the partition of z, the function value is indicated by

$$S(y, \lambda) = S(z) = S(z_1, \ldots, z_j, \ldots, z_N).$$

The gradient of S with respect to z is denoted by $\nabla S(z)$ and $\nabla_j S(z)$ is the gradient of S with respect to z_j, for $j = 1, \ldots, N$, that is:

$$\nabla_1 S(z) := \nabla_{y_1} S(y, \lambda) \qquad \nabla_{j+1} S(z) := \begin{pmatrix} \nabla_{y_{j+1}} S(y, \lambda) \\ \\ \nabla_{\lambda_j} S(y, \lambda) \end{pmatrix} \qquad j = 1 \ldots, N-1.$$

The algorithm we will define consists of successive iterations that produce a sequence of points $\{z^k\}$, starting from a given initial point z^0. Each iteration (indexed by k) consists of a cycle of inner steps (indexed by j) that update, one at time, the N components z_j^k of z^k and produce repeated updates of the component z_1 according to suitable rules. As anticipated in Section 1, we suppose that the successive values of z_1 are taken as "real-time" estimates of the solution x of the originary problem.

During the inner steps we can perform either a linesearch in the z_j–space, along a search direction d_j, for fixed values of the remaining variables, which replaces z_j with a new point $z_j + \alpha_j d_j$, or a "redundancy test" that may have the effect of replacing z_j with a vector obtained in the preceding step.

We suppose that, in correspondence to every point z produced by the algorithm, and for every choice of an index $j \in \{1, \ldots, N\}$, there exist search directions d_j satisfying the following condition.

Assumption 2 *There exist numbers $c_j > 0$, $j = 1, \ldots, N$ such that for every z and j we have $d_j' \nabla_h S(z) \le -c_j \|d_j\| \|\nabla_j S(z)\|$.* □

We note that Assumption 2 is satisfied, in particular, if we set $c_j = 1$ and: $d_j := -\nabla_j S$.

First we define the linesearch procedure, which is based on the techniques studied in [24] [18] and can be viewed as a modified Armijo rule. In order to simplify notation, we indicate by $S_j(z_j)$ the value of S as a function of z_j and we tacitly assume that the remaining variables are held fixed to their current values.

Linesearch procedure (LS)

Data: $j \in \{1, \ldots, N\}$, z, d_j, $\rho_j > 0$, $\sigma_j \in (0, 1)$, $\delta_j > 0$.

If $\nabla_j S(z) \neq 0$ then:

(a) choose $\Delta \ge \rho_j \dfrac{|d_j' \nabla_j S(z)|}{\|d_j\|^2}$

(b) compute $\alpha = \max_h \{\sigma_j^h \Delta : h = 0, 1, \ldots\}$ such that

$$S_j(z_j + \alpha d_j) \le S_j(z_j) - \delta_j \alpha^2 \|d_j\|^2$$ □

Remark 2 We note that Algorithm LS stated above constitutes a simplified model used for establishing convergence results. In real computations, both the acceptance rules and the interpolation criterion can be improved and extended under several aspects, while preserving the essential features of our model. In particular, it is known that the choice of a good initial estimate for the stepsize can be very important for reducing the number of function evaluations during the linesearch. □

It can be proved (see [22]) that Algorithm LS terminates in a finite number of steps; more precisely, we can state the following proposition.

Proposition 3.1 *For every $j \in \{1, \ldots, N\}$ and $z \in \mathcal{W}_0$, if $\nabla_j S(z) \neq 0$ there exist a finite integer m such that Algorithm* LS *terminates with a positive number $\alpha = \sigma_j^m \Delta$ satisfying the acceptability condition at Step* (b). □

Next we introduce what we have called a "redundancy test", which is aimed at improving the current estimate of a component z_r before starting a linesearch from it. We define this as a *replacement procedure.*

Replacement procedure (RP)

Data: $r \in \{1,\ldots,N\}$, $s \in \{1,\ldots,N\}$, z, $\bar{z}_s$, $\theta_r > 0$
If:

$$S(z_1,\ldots,z_{r-1},\bar{z}_s,z_{r+1}\ldots,z_N) \le S(z_1,\ldots,z_{r-1},z_r,z_{r+1}\ldots,z_N) - \theta_r\|\bar{z}_s - z_r\|^2$$

then set $z_r = \bar{z}_s$. □

We are now ready to state the following algorithm model.

Partitioning Algorithm

Step 0. Given $z^0 \in \mathbb{R}^{nN}$; set $k = 0$.

Step 1. Set $z = z^k$ and perform the following inner cycle.

(i) Set $j = 1$ and compute $\bar{z}_1 = z_1 + \alpha d_1$ by means of procedure LS; update z_1 and z and set $z(k,1) = z$;

(ii) for $j = 2,\ldots,N$:

replace, if possible, z_j with $\bar{z}_1$ by means of procedure RP with $r = j$, $s = 1$ and update z;

compute $\bar{z}_j = z_j + \alpha d_j$ by means of procedure LS; update z_j and z;

replace, if possible, z_1 with $\bar{z}_j$ by means of procedure RP with $r = 1$, $s = j$ and update z;

compute $\bar{z}_1 = z_1 + \alpha d_1$ by means of procedure LS; update z_1 and z;

set $z(k,j) = z$.

Step 2. If $z = z^k$ stop; else set $z^{k+1} = z$, $k = k+1$ and go to Step 1. □

In the next proposition we outline a convergence proof, which follows with minor modifications from that given in [22]

Proposition 3.2 *Let $\{z^k\}$ be the sequence of points generated by Algorithm 1; then either the algorithm terminates at Step 2 for some index ν and in this case we have $\nabla S(z_\nu) = 0$, or it produces an infinite sequence of points in $\mathcal{W}_0$, whose limit points are stationary points of S.*

Proof. First we observe that the algorithm terminates at z_ν when every step in a cycle is a null step; because of Assumption 2 and the instructions of Procedure LS, this is possible only when $\nabla_j S(z_\nu) = 0$ for all j. Therefore, we can assume that an infinite sequence is produced.

By the instructions of the Algorithm and the assumptions made, it is easily seen that the sequence of values of S is strictly decreasing and satisfies

$$S(z^{k+1}) = S(z(k,N)) \le \ldots \le S(z(k,j)) \le \ldots \le S(z(k,1)) \le S(z^k).$$

This implies that the points z^k and $z(k,j)$ remain on the level set $\mathcal{W}_0$ and that the sequences $\{S(z^k)\}$ and $\{S(z(k,j))\}$ converge to the same limit. Then, by compactness of $\mathcal{W}_0$ we have also that there exists a limit point $\bar{z}$ of $\{z^k\}$ in $\mathcal{W}_0$. Moreover, the instructions of Procedures LS and RP, together with the convergence of the sequences $\{S(z^k)\}$ and $\{S(z(k,j))\}$ imply that, for all j we have:

$$\lim_{k\to\infty} \|z(k,j+1) - z(k,j)\| = 0, \quad \lim_{k\to\infty} \|z(k,j) - z^k\| = 0.$$

This implies that if $\{z^k\}_K$ is a subsequence converging to $\bar{z}$ we will have also that the subsequences $\{z(k,j)\}_K$ will converge to $\bar{z}$. Similarly, it easily seen that the vectors z produced during the inner cycle, either as a result of a linesearch or as a result of the replacement procedure, will converge to the same limit for $k \in K$. Using this fact, recalling the instructions of the linesearch, and using standard arguments of steplength analysis, it can be shown that at a limit point $\bar{z}$ we have $\nabla_j S(\bar{z}) = 0$, for all j, which establishes our thesis. □

The preceding algorithm, allow us to construct a sequence of points $\{x^k\}$ formed with the subsequent updates of the component z_1. On the basis of Proposition 3.2, under the assumption that the penalty parameter in S is sufficiently large, we can assert that the limit points of $\{x^k\}$ will be stationary points of the originary objective function $f(x)$.

The same conclusion can be reached also in correspondence to different partitions of the problem variables and to different criteria for constructing x^k.

4. IMPLEMENTATION PROBLEMS

In the model algorithm described in the preceding section we have left unspecified many points that can be important from a computational point of view.

A first point is the selection of the penalty coefficient c in the augmented Lagrangian. In principle, recalling the proof of Proposition 2.2 we can give an explicit estimate, in terms of bounds on the second order derivatives; in practice, however, it could be convenient to start from small values for c and to give rules for updating this parameter in a way that exactness is enforced. We refer, for instance, to [32], [1], [33] for examples of these schemes.

A second point is the presence of first order derivatives in the objective function, which implies the need of evaluating the second order derivatives of the problem functions for computing the gradient of S. However, we can note from the expression of ∇S that only a term of the form $\nabla^2 f_{j+1}(y_{j+1})(\nabla f_{j+1}(y_{j+1}) + \lambda_j)$ has to be evaluated and this can be done with enough precision by means of two gradient evaluations [32]. Alternatively, we can can also attempt to use, whenever possible, an approximate gradient direction with respect to (y_{j+1}, λ_j) by omitting at all the term containing the second order derivatives. However, as shown in [6] the computation of second order derivatives along a search direction can be performed at a low computational cost by means of the reverse mode technique of automatic differentiation. As third order derivatives can be ignored for computing a Newton-type direction in the minimization of an exact penalty function (see, e.g., [1], [32]) this implies also that the use of a truncated Newton method for computing the search directions could be of value.

Finally there remains the need of evaluating from a computational point of view the conceptual framework proposed here and, possibly, the definition of alternative criteria for constructing online schemes.

REFERENCES

[1] D.P. Bertsekas, *"Constrained Optimization and Lagrange Multiplier Methods"*. Academic Press, New York, 1982.

[2] D.P. Bertsekas and J.N. Tsitsiklis, *"Parallel and Distributed Computation"*. Prentice-Hall International Editions, Englewood Cliffs, New Jersey, 1989.

[3] C. Charalambous, "Conjugate gradient algorithm for efficient training of artificial neural networks". *IEE Proceedings Part G*, 139:301–310, 1992.

[4] G. Chen and M. Teboulle, "A proximal-based decomposition method for convex minimization problems". *Mathematical Programming*, 64:81–101, 1994.

[5] A. Cichocki and R. Unbehauen, *"Neural Networks for Optimization and Signal Processing"*. John Wiley and Sons, Chichester, 1993.

[6] L.C.W. Dixon, "Neural networks and unconstrained optimization". In E. Spedicato, editor, *"Algorithms for continuous optimization: the state of the art"*, NATO ASI Series, Kluwer Academic Publisher, Dordrecht, 1993.

[7] J. Eckstein and M. Fukushima, "Some reformulations and applications of the alternating direction method of multipliers". In D.W. Hearn W.W. Hager and P.M. Pardalos, editors, *Large Scale Optimization: the State of the Art*, Kluwer Academic Publisher, Dordrecht, 1994.

[8] M.C. Ferris and O.L. Mangasarian, "Parallel variable distribution". *SIAM Journal on Optimization*, 4:815–832, 1994.

[9] R. Fletcher, "An overview of unconstrained optimization". In E. Spedicato, editor, *"Algorithms for continuous optimization: the state of the art"*, NATO ASI Series, Kluwer Academic Publisher, Dordrecht, 1993.

[10] M. Fortin and R. Glowinski, *"Augmented Lagrangian Methods: Applications to the Solution of Boundary-Valued Problems"*. North-Holland, Amsterdam, 1983.

[11] J.H. Friedman, "An overview of predictive learning and function approximation". In V. Cherkassky, J.H. Friedman, and H. Wechsler, editors, *From Statistics to Neural Networks*", NATO ASI Series, Springer-Verlag, Berlin, 1991.

[12] M. Fukushima, "Application of the alternating direction method of multipliers to separable convex programming problems". *Computational Optimization and Applications*, 1:93–111, 1992.

[13] D. Gabay and B. Mercier, "A dual algorithm for the solution of nonlinear variational problems via finite-element approximations". *Computers and Mathematics with Applications*, 2:17–40, 1976.

[14] A. A. Gaivoronski, "Convergence analysis of parallel back-propagation algorithm for neural networks". *Optimization Methods and Software*, 4:117–134, 1994.

[15] J.C. Gilbert and J. Nocedal, "Global convergence of conjugate gradient methods for optimization". *SIAM Journal on Optimization*, 2(1):21–42, 1992.

[16] F. Girosi, "Regularization theory, radial basis functions and networks". In V. Cherkassky, J.H. Friedman, and H. Wechsler, editors, *From Statistics to Neural Networks*, NATO ASI Series, Springer-Verlag, Berlin, 1991.

[17] L. Grippo, "A class of unconstrained minimization methods for neural network training". *Optimization Methods and Software*, 4:135–150, 1994.

[18] L. Grippo, F. Lampariello, and S. Lucidi, "Global convergence and stabilization of unconstrained minimization methods without derivatives". *Journal of Optimization Theory and Applications*, 56(3):385–406, 1988.

[19] L. Grippo, F. Lampariello, and S. Lucidi, "A truncated Newton method with nonmonotone linesearch for unconstrained optimization". *Journal of Optimization Theory and Applications*, 60:401–419, 1989.

[20] L. Grippo, F. Lampariello, and S. Lucidi, "A class of nonmonotone stabilization methods in unconstrained optimization". *Numerische Mathematik*, 59:779–805, 1991.

[21] L. Grippo and S. Lucidi, "A globally convergent version of the Polak-Ribière conjugate gradient method". Technical Report R. 08-95, DIS, Roma, Italy, 1995.

[22] L. Grippo and M. Sciandrone, "Globally convergent partitioning techniques in unconstrained optimization". Technical Report R.11-95, DIS, Roma, Italy, 1995.

[23] J. Hertz, A. Krogh, and R. G. Palmer, *"Introduction to the theory of neural computation"*. Addison-Wesley, Redwood City, California, 1991.

[24] R. De Leone, M. Gaudioso, and L. Grippo, "Stopping criteria for linesearch methods without derivatives". *Mathematical Programming*, 30:285–300, 1984.

[25] R.De Leone, R.R.Meyer, S.Kontogiorgis, A.Zakarian, and G.Zaker, "Coordination in coarse-grained decomposition". *SIAM Journal on Optimization*, 4:777–793, 1994.

[26] A. De Luca and G. Di Pillo, "Exact augmented lagrangian approach to multilevel optimization of large-scale systems". *Int.J. of Systems Science*, 18:157–176, 1987.

[27] Z.-Q. Luo and P. Tseng, "Analysis of an approximate gradient projection method with applications to the backpropagation algorithm". *Optimization Methods and Software*, 4:85–102, 1994.

[28] O. L. Mangasarian, "Mathematical programming in neural networks". *ORSA Journal on Computing*, 5:347–363, 1993.

[29] O. L. Mangasarian and M. V. Solodov, "Serial and parallel backpropagation convergence via nonmonotone perturbed minimization". *Optimization Methods and Software*, 4:103–116, 1994.

[30] S.G. Nash and J. Nocedal, "A numerical study of the limited memory BFGS method and the truncated-Newton method for large scale optimization". *SIAM Journal on Optimization*, 14:358–372, 1991.

[31] G. Di Pillo and L. Grippo, "A new class of augmented Lagrangians in nonlinear programming". *SIAM Journal on Control and Optimization*, 17(5):618–628, 1979.

[32] G. Di Pillo, L. Grippo, and F. Lampariello, "A method for solving equality constrained optimization problems by unconstrained minimization". In K. Iracki, K. Malanowski, and S. Walukiewicz, editors, *Proc. 9th IFIP Conference on Optimization Techniques*, Springer Verlag, Berlin, 1980.

[33] G. Di Pillo and S. Lucidi, "On exact augmented Lagrangian functions for nonlinear programming problems". In the present Volume.

[34] D. E. Rumelhart, G. E. Hinton, and R. J. Williams, "Learning internal representation by error backpropagation". In Rumelhart and McClelland, editors, *"Parallel Distributed Processing-Explorations in the Microstructure of Cognition"*, MIT press, Cambridge, 1986.

[35] P.J. Werbos, "Supervised learning: can it escape its local minimum?" In V. Roychowdhury, K.-Y. Siu, and A. Orlitsky, editors, *"Theoretical advances in neural computation and learning*, Kluwer Academic Publishers", Norwell, Ma., 1994.

[36] H. White, "Some asymptotic results for learning in single hidden-layer feedforward network models". *Journal of the American Statistical Association*, 84:1003–1013, 1989.

[37] H. White, *"Artificial neural networks"*. Blackwell, Cambridge, Massachusetts, 1992.

SEMISMOOTHNESS AND SUPERLINEAR CONVERGENCE IN NONSMOOTH OPTIMIZATION AND NONSMOOTH EQUATIONS

Jiang Houyuan, Qi Liqun, Chen Xiaojun and Sun Defeng

School of Mathematics
University of New South Wales
Sydney 2052, Australia
e-mail: jiang@solution.maths.unsw.edu.au
e-mail: L.Qi@unsw.edu.au

Abstract. This paper discusses recent developments on superlinear convergence results in nonsmooth optimization and nonsmooth equations. The concept of semismoothness has been shown to play a key role to obtain superlinear convergence for algorithms solving nonsmooth problems. Using semismoothness and nonsingularity conditions, generalized Newton methods have been proved locally and superlinearly convergent for solving systems of nonsmooth equations. Moreover, nonsmooth equation based generalized Newton methods have been successfully used for solving nonlinear complementarity and related problems. Various quasi-Newton methods have been proposed for different nonsmooth equations. But, more work on quasi-Newton methods is needed. The idea of generalized Newton methods was further applied to nonsmooth unconstrained and constrained optimization, for instance, LC^1 problems and nonsmooth convex optimization.

Key words. Newton's method, quasi-Newton methods, nonsmooth equations, nonsmooth optimization, semismoothness, superlinear convergence.

1. INTRODUCTION

Nonsmooth optimization and nonsmooth equations are important research topics in mathematical programming fields. Rockafellar [84] reviewed a number of nonsmooth optimization problems with special structures.

The standard version of nonlinear optimization problem is

$$\begin{array}{ll} \min & f(x) \\ \text{s.t.} & g(x) \leq 0 \\ & h(x) = 0, \end{array} \tag{1.1}$$

where $f : \mathbb{R}^n \to \mathbb{R}, g : \mathbb{R}^n \to \mathbb{R}^m$ and $h : \mathbb{R}^n \to \mathbb{R}^l$.

Different Newton-type methods have been developed and shown efficient in practice for (1.1) when all functions in (1.1) are twice continuously differentiable. One of the important features of Newton-type methods is their rapid local convergence. Superlinear (or quadratic) convergence of those methods can be achieved by using a quadratic model to approximate the original problem. Moreover, the quadratic model should have some association with the Hessian matrices directly or indirectly. However, when functions in (1.1) are not twice continuously differentiable, Hessian matrices are not available. Therefore, classical Newton-type methods are not applicable for (1.1) whenever functions in (1.1) do not possess second-order differentiability, and new approaches are needed. Recent development in nonsmooth equations provides us such possibilities.

The system of nonlinear equations is of the following form

$$H(x) = 0, \tag{1.2}$$

where $H : \mathbb{R}^n \to \mathbb{R}^n$. It is well known [13] that Newton-type methods can achieve local superlinear (or quadratic) convergence if H is continuously differentiable. Similar to the above argument for (1.1), classical Newton-type methods are not applicable any more for (1.2) if H is not continuously differentiable.

Due to the recent success of nonsmooth analysis [81,10,52,79,58,45,73,66,63], it has been shown that generalized Newton-type methods are still convergent superlinearly for solving (1.1) and (1.2) in the case that some involved functions in (1.1) are not twice differentiable and H in (1.2) is not differentiable. The present paper is intended to address the recent progress in solving nonsmooth versions of (1.1) and (1.2). In particular, we shall pay special attention to superlinear convergence results based on semismoothness.

In this paper we will consider the following four problems:

(i) Newton's methods for solving nonsmooth equations and their applications to nonlinear complementarity problems, variational inequalities and the KKT system of (1.1);

(ii) Newton's methods for solving (1.1) where f, g and h are continuously differentiable and $\nabla f, \nabla g$ and ∇h are locally Lipschitz;

(iii) Quasi-Newton methods for solving nonsmooth equations and applications to nonlinear complementarity problems, variational inequalities and the KKT system of (1.1);

(iv) Proximal Newton methods for solving nondifferentiable convex optimization problems.

2. NEWTON'S METHODS FOR NONSMOOTH EQUATIONS

The early study of nonsmooth equations can be traced back to [48,49]. Mangasarian [48,49] formulated the KKT system of (1.1) as a system of nonsmooth equations, see Example 2.3 below. The system of nonsmooth equations (1.2) arises from many applications. Pang and Qi [63] reviewed eight problems in the studies on optimization, which can be formulated as systems of nonsmooth equations. In what follows, we present three main problems as the motivation for the study of nonsmooth equations (1.2).

Example 2.1. Nonlinear complementarity problems

Let $p : D \to \mathbb{R}^n$ be a given function assumed to be continuously differentiable on the open set $D \subseteq \mathbb{R}^n$ containing the nonnegative orthant $\mathbb{R}^n_+$. The nonlinear complementarity problem, denoted by NCP(p), is to find a vector x such that

$$x \geq 0, \quad p(x) \geq 0, \qquad x^T p(x) = 0.$$

See [33,60]. There are many ways to formulate this problem as a system of nonsmooth equations. The following two formulations are frequently used in the recent literature.

Pang's formulation is

$$H(x) = \min(x, p(x)), \tag{2.1}$$

where "min" denotes the componentwise minimum operator, see [58].

Fischer's formulation is

$$H(x) = (H_1(x), \cdots, H_n(x))^T, \text{ with } H_i(x) = \sqrt{x_i^2 + (p_i(x))^2} - x_i - p_i(x), \; i = 1, 2, \cdots, n, \tag{2.2}$$

see [19].

Clearly, x solves the NCP(p) if and only x is a zero of H defined by (2.1) or (2.2).

Example 2.2. Variational inequalities

The NCP(p) is a special case of the variational inequality problem defined over a closed convex set. Let C be a closed convex subset of $\mathbb{R}^n$ and p be defined on the open set $D \subseteq \mathbb{R}^n$ containing C. The variational inequality problem, denoted by VI(C, p), is to find a vector $x^* \in C$ such that

$$(y - x^*)^T p(x^*) \geq 0, \quad \text{for all} \;\; y \in C.$$

See [33]. It is clear that the VI(C, p) reduces to the NCP(p) if C is the nonnegative orthant of $\mathbb{R}^n$. When p is the gradient mapping of the real-valued function $f : \mathbb{R}^n \to \mathbb{R}$ and C is represented as

$$C = \{x \mid g(x) \leq 0, \; h(x) = 0, \; x \in \mathbb{R}^n\},$$

where g and h are convex functions, the VI(C, p) becomes the stationary point problem of the optimization problem (1.1) with the constraint set C.

The VI(C, p) can be derived as a system of nonsmooth equations in several ways. For example, let H be defined by

$$H(x) = x - \Pi_C(x - p(x)),$$

where Π_C denotes the projection operator to the set C. It can be verified that x solves the VI(C, p) if and only if x satisfies $H(x) = 0$, see, for example [33].

Example 2.3. The KKT system

The Karush-Kuhn-Tucker (KKT for short) system for (1.1) is

$$\begin{gathered}
\nabla f(x) + \nabla g(x) u + \nabla h(x) v = 0 \\
u \geq 0, \; g(x) \leq 0 \\
u^T g(x) = 0 \\
h(x) = 0.
\end{gathered}$$

By formulating the NCP part of the KKT system into a system of nonsmooth equations as done in Example 2.1, we have nonsmooth KKT equations. See [72] for different nonsmooth KKT equations.

There have been developed some extensions of Newton and quasi-Newton methods; Josephy [40] for strongly regular generalized equations which were extensively discussed by Robinson [77,78], Pang and Chan [61] for variational inequalities and complementarity problems. Some early studies of solving systems of nonsmooth equations by Newton's methods include Kojima and Shindo [43], Pang [58], Robinson [80] and Kummer [44]. Subsequently, a number of generalized Newton methods were proposed for nonsmooth equations and related problems, see, for example [34,59,66,73,63,45,19,62,76]. Some methods solve nonlinear subproblems at each step, see [40,80,58,76,91]. In some special cases, these nonlinear subproblems can be solved. Superlinear convergence results were established based on this. Qi and Sun [73] analysed a generalized Newton method for solving (1.2), in which a system of linear equations is solved at each step. Superlinear convergence of their generalized Newton method is based on a key concept of so-called semismoothness. Independently, Kummer [45] presented a general analysis of superlinear convergence for generalized Newton methods of nonsmooth equations.

Suppose that H of (1.2) is locally Lipschitz. By Rademacher's Theorem, H is almost everywhere differentiable. Let

$$D_H = \{x \mid H \text{ is differentiable at } x\}.$$

Then the generalized Jacobian of H at x can be defined by

$$\partial H(x) = \text{conv}\partial_B H(x),$$

where

$$\partial_B H(x) = \{ \lim_{\substack{x^j \to x \\ x^j \in D_H}} \nabla H(x^j)^T \}.$$

The generalized Newton method for solving (1.2) can be defined as follows: Having the vector x^k, compute x^{k+1} by

$$x^{k+1} = x^k - V_k^{-1} H(x^k), \tag{2.3}$$

where $V_k \in \partial H(x^k)$.

It is clear that (2.3) reduces to the classic Newton method for a solution of equations if H is continuously differentiable. Moreover, The sequence $\{x^k\}$ generated by (2.3) is locally superlinearly convergent to a solution x^* if $\nabla H(x^*)$ is nonsingular. However, in general the iterative method (2.3) is not superlinearly convergent for nonsmooth equations (1.2). The following example shows that (2.3) is not even locally convergent, see [44] for another counterexample.

Example 2.4. Let $H : \mathbb{R} \to \mathbb{R}$ be defined by

$$H(x) = \begin{cases} x^2 \sin\frac{1}{x} + 2x & \text{if } x \neq 0, \\ 0 & \text{if } x = 0. \end{cases}$$

It is easy to verify that H is locally Lipschitz and differentiable everywhere, that H is continuously differentiable at any point $x \neq 0$, and that the equation $H(x) = 0$ has a solution $x = 0$ at which Clarke's generalized subdifferential is the interval [1, 3]. Let N be any positive integer and $x = \pm\frac{1}{2N\pi}$. Then the next iterate generated by (2.3) is $x^+ = x - \frac{H(x)}{\nabla H(x)}$. By some calculations, we have $x^+ = -x$. Consequently, it implies that the generalized Newton method (2.3) can be divergent no matter how close x is to the solution point $x = 0$ (i.e. for sufficiently large N).

In order to establish convergence of the generalized Newton method (2.3), let us introduce the important concept of semismoothness. Let H be directionally differentiable at x. H is said to be semismooth at x if

$$Vd - H'(x;d) = o(\|d\|), \quad d \to 0$$

and H is called strongly semismooth at x if

$$Vd - H'(x;d) = O(\|d\|^2), \quad d \to 0,$$

where $V \in \partial H(x+d)$. Semismoothness was originally introduced by Mifflin [52] for functionals. Semismooth functionals play an important role in the global convergence theory of nonsmooth optimization, see [65]. Qi and Sun [73] extended the concept of semismoothness to vector-valued functions. We refer the reader to [52,73,66] for more properties of semismooth functions.

Using semismoothness, Qi and Sun [73] presented the following convergence theorem for the generalized Newton method (2.3).

Theorem 2.1. *Suppose that $H(x^*) = 0$ and that all $V \in \partial H(x^*)$ are nonsingular. Then the generalized Newton method (2.3) is Q-superlinearly convergent in a neighborhood of x^* if H is semismooth at x^*, and quadratically convergent if H is strongly semismooth at x^*.*

Before the paper [73] (the first version was finished in 1989) was published, Kummer [45] independently discussed sufficient and necessary conditions for the convergence of Newton method based on generalized derivatives. One of the conditions for guaranteeing convergence (see Theorem 2 of [45]) is that for any $V \in \partial H(x+d)$, $d \to 0$,

$$H(x+d) - H(x) - Vd = o(\|d\|). \tag{2.4}$$

Since H is locally Lipschitz continuous, from [85] we know that if $H'(x;d)$ exists, then $H'(x;d)$ coincides with the B-derivative of H at x; i.e.,

$$\lim_{d\to 0} \frac{H(x+d) - H(x) - H'(x;d)}{\|d\|} = 0.$$

So, if $H'(x;d)$ exists, then (2.4) implies that for any $V \in \partial H(x+d)$, $d \to 0$,

$$Vd - H'(x;d) = o(\|d\|).$$

Hence (2.4) implies the semismoothness of H at x if $H'(x;d)$ exists. On the other hand, the semismoothness of H at x implies (2.4) since $H'(x;d)$ exists in this case.

Note that the nonsingularity of $\partial H(x^*)$ in the above theorem is somewhat restrictive in some cases. Qi [66] presented a modified version of (2.3) which may be stated as follows

$$x^{k+1} = x^k - V_k^{-1} H(x^k), \tag{2.5}$$

where $V_k \in \partial_B H(x^k)$. The difference of this version from (2.3) is that V_k is chosen from $\partial_B H(x^k)$ rather than the convex hull of $\partial_B H(x^k)$. Analogous to Theorem 2.1., Qi [66] established the following result.

Theorem 2.2. *Suppose that $H(x^*) = 0$ and that all $V \in \partial_B H(x^*)$ are nonsingular. Then the generalized Newton method (2.5) is Q-superlinearly convergent in a neighborhood of x^* if H is semismooth at x^*, and quadratically convergent at x^* if H is strongly semismooth at x^*.*

For inexact version of (2.3) and (2.5) and their superlinear convergence theorems, see [50].

Pang and Qi [63] generalized the superlinear convergence results of Dennis-Moré [12] for smooth equations.

Theorem 2.3. *Assume that H is semismooth at x^* and that all elements in $\partial_B H(x^*)$ are nonsingular. Let $\{x^k\} \subseteq D$ be any sequence that converges to x^* with $x^k \neq x^*$ for all k. Then $\{x^k\}$ converges Q-superlinearly to x^* and $H(x^*) = 0$ if and only if*

$$\lim_{k\to\infty} \frac{\|H(x^k) + V_k d^k\|}{\|d^k\|} = 0,$$

where $V_k \in \partial_B H(x^k)$ and $d^k = x^{k+1} - x^k$.

Clearly, Theorems 2.1., 2.2. and 2.3. generalize the convergence results of classical Newton method for smooth equations without assuming differentiability of H. On the other hand, those convergence results can be applied to some important mathematical programming problems such as nonlinear complementarity problems, variational inequalities, the KKT equations of (1.1) and other related problems.

As an example, we consider the NCP(p). Assume that p is smooth. Note that H is piecewise smooth, hence semismooth if H is defined by (2.1). Although H is not piecewise smooth if H is defined by (2.2), it is still semismooth if p is continuously differentiable or even strongly semismooth if p is twice continuously differentiable, see [72,39,18]. Therefore, the above convergence results can be applied to the NCP(p).

It should be pointed out that one of interesting properties for Fischer's reformulation of the NCP(p) is that the least square $||H||^2$ is continuously differentiable on $\mathbb{R}^n$ although H itself is not in general, see [28,18,22]. On the other hand, this property does not hold for the reformulation (2.1). Moreover, Fischer's reformulation enjoys many other fine properties, and provides a valuable tool for designing efficient algorithms for solving the NCP(p) and related problems. This aspect of research attracts a number of authors' attention, for example, De Luca, Facchinei, Fischer, Fukushima, Jiang, Kanzow, Qi, Soares, Tseng, Yamashita, etc.. Extensive study of Fischer's reformulation-based methods of the NCP(p) includes papers [11,17,18,20,21,23,39,42,90,94]. Numerical results reported in [11,17,39] showed that those approaches are very promising. For other generalized Newton methods for complementarity and variational inequality problems, see [1,14,27,42,62,88,89,91,92,93] and references therein.

The above convergence results have also been successfully applied to quasi-variational inequalities by Outrata and Zowe [57].

There are also some discussions on global convergence of generalized Newton methods for solving (1.2) [30,58,63,66,68,70].

3. NEWTON'S METHODS FOR LC^1 OPTIMIZATION

We call a function $\phi : \mathbb{R}^n \to \mathbb{R}^m$ an LC^1 function if it is differentiable and its derivative function is locally Lipschitz. We call (1.1) an *LC^1* optimization problem if f, g and h are LC^1 functions.

One method for solving (1.1) is to solve the following linearly constrained quadratic program

$$\min \quad \nabla f(x^k)^T(x - x^k) + \frac{1}{2}(x - x^k)^T G_k (x - x^k)$$

$$\text{s.t. } g(x^k) + \nabla g(x^k)^T(x - x^k) \le 0$$

$$h(x^k) + \nabla h(x^k)^T(x - x^k) = 0 \tag{3.1}$$

successively. Here G_k is an $n \times n$ matrix. This method is called an approximate Newton method or an SQP (sequential quadratic programming) method. If f is twice continuously differentiable and if G_k is exactly the second-order derivative of the Lagrangian with respect to x at x^k, it reduces to the classical Newton method.

In the literature of such approximate Newton or SQP methods for solving nonlinear programming problems, though calculations of most of these methods only need first derivative data, the superlinear convergence proofs require among other things twice continuously differentiable objective and constraint functions, sometimes even Lipschitz second-order derivatives of these functions. However, second-order differentiability may not hold for some problems, for example, nonlinear minimax problems [35], stochastic programs [6,74], augmented Lagrangians [83], semi-infinite programs [35] and some differentiable penalty function methods for constrained optimization problems [35].

Some generalized second-order necessary and sufficient conditions were derived for LC^1 optimization in [35], which turn out to be fundamental for LC^1 optimization.

Qi [67] proposed an approximate Newton or SQP method for solving LC^1 optimization problems and established local superlinear convergence of his method under the assumptions that the derivatives of the objective and constraint functions are semismooth, and (1.1) satisfies the generalized second-order sufficiency conditions, the strict complementarity condition and the linear independence constraint qualification at a solution point. Han and Sun [32] modified Qi's results without assuming the strict complementarity condition.

Let $L(x,u,v) = \nabla f(x) + \nabla g(x)u + \nabla h(x)v$ and $z = (x,u,v)$. A generalization of Robinson's [78] strong second-order sufficient condition is

$$d^T V d > 0, \quad \forall 0 \neq d \in T(x)$$

for any V such that $(V\ \nabla g(x)\ \nabla h(x)) \in \partial L(x,u,v)$,

$$T(x) = \{d \mid \nabla g_i(x)^T d = 0 \text{ for } i \in I_1(x) \text{ and } \nabla h(x)^T d = 0\}$$

and

$$I_1(x) = \{i | g_i(x^*) = 0, u_i^* > 0\}.$$

Theorem 3.1. *Suppose that $z^* = (x^*, u^*, v^*) \in \mathbb{R}^n \times \mathbb{R}^m \times R^l$ is a KKT point of (1.1) and satisfies the generalized strong second-order sufficient conditions and the linear independence constraints qualification. Suppose that ∇f, ∇g and ∇h are semismooth at x^*. Let c be a suitable positive number. If there exists V_k satisfying $(V_k\ \nabla g(x^k)\ \nabla h(x^k)) \in \partial L(x^k, u^k, v^k)$ such that*

$$\|G_k - V_k\| \le \frac{1}{4c} \quad \forall\ k,$$

then the method (3.1) is well defined and the sequence $\{x^k\}$ generated by (3.1) Q-linearly converges to z^* in a neighborhood of z^*. If furthermore,

$$\lim_{k\to\infty} \frac{\|(G_k - V_k)(x^{k+1} - x^k)\|}{\|z^{k+1} - z^k\|} = 0, \tag{3.2}$$

then the convergence is Q-superlinear.

The Maratos effect is known to be an undesirable feature for many SQP algorithms in nonlinear programming. Much effort has been devoted to the study of conditions for avoiding it. Basically, this phenomenon refers to non-attainment of the unit stepsize when some line search rule is applied. The phenomenon results in failure of rapid convergence. However, Facchinei [15] has proved that the Maratos effect does not occur for unconstrained LC^1 optimization if some conditions hold.

Pang and Qi [64] presented a generalized Newton method with line search for solving a linearly constrained LC^1 convex optimization problem. The global and local superlinear convergence analysis only requires semismoothness and nonsingularity conditions at a solution. Jiang and Qi [38] studied a generalized Newton method with a trust region strategy for the problem considered in [64]. Chen [4] studied a BFGS-SQP method for solving LC^1 linearly constrained convex optimization problem and discussed its global and superlinear convergence.

LC^1 problems have found many applications in the NCP(p) [18], the VI(C,p) [24,25], the extended linear-quadratic programming (ELQP for short) problem and differentiable exact penalty functions [16].

In [6], Chen, Qi and Womersley generalized the results in [74] by applying the approximate Newton method with line search to solve quadratic stochastic programs with recourse, which is an LC^1 convex optimization problem. For large scale optimization problems, we cannot solve the quadratic programming subproblem exactly at each step of the SQP method. In [7], Chen and Womersley studied a parallel inexact Newton method with a line search for LC^1 optimization problems and applied this method to solve quadratic stochastic programs with recourse. At each step the quadratic programming subproblem is solved to the required accuracy. Global convergence and local superlinear convergence were established.

4. QUASI-NEWTON METHODS FOR NONSMOOTH EQUATIONS

The main disadvantage of Newton's method for solving (1.2) is that the user must supply formulae from which the generalized Jacobian matrix can be evaluated. This disadvantage can be avoided by using quasi-Newton methods.

Quasi-Newton methods for solving (1.2) are based on

$$x^{k+1} = x^k + d^k, \; d^k = -B_k^{-1}H(x^k), \tag{4.1}$$

where $B_k \in \mathbb{R}^{n\times n}$ satisfies the quasi-Newton equation

$$B_{k+1}d^k = y^k, \; y^k = H(x^{k+1}) - H(x^k). \tag{4.2}$$

B_k can be updated at every iteration by a quasi-Newton formula. The following Broyden formula

$$B_{k+1} = B_k - \frac{(y^k - B_k d^k)(d^k)^T}{(d^k)^T d^k} \tag{4.3}$$

is often used.

Josephy [41] established superlinear convergence of quasi-Newton methods for generalized equations. A nonlinear subproblem has to be solved at each inner iteration. Kojima and Shindo [43] developed a Broyden's quasi-Newton method for solving piecewise smooth equations, and proved superlinear convergence of their method if the nonsingularity of the Jacobians of the active C^1 pieces holds at a solution point. Their method could be computationally less efficient since it requires storing a potentially large number of matrices, see [43,37].

Among other works [3,5,8,9,29,51] on quasi-Newton methods for solving nonsmooth equations, Ip and Kyparisis [37] analyzed convergence of the above quasi-Newton methods. Superlinear convergence of quasi-Newton methods were established by Dennis and Moré [12,13] under the assumption that ∇H is Lipschitz continuous. Ip and Kyparisis [37] relaxed the Lipschitz continuity of ∇H to the Lipschitz continuity of B-derivative of H, which implies the strong differentiability of H (see [56]), at a solution point. In particular, Ip and Kyparisis [37] proved the following theorem for the Broyden formula.

Theorem 4.1. *Let $H : \mathbb{R}^n \to \mathbb{R}^n$ be Lipschitz continuous and B-differentiable on an open, convex set $D \subseteq \mathbb{R}^n$, and suppose that $H(\cdot, d)$ is Lipschitz continuous at $x^* \in D$ (with respect to the Euclidean vector norm), that $H(x^*) = 0$, and $\nabla H(x^*)^{-1}$ exists. Then, there exist $\varepsilon > 0$ and $\delta > 0$ such that, if $||x^0 - x^*|| \le \varepsilon$ and $||B_0 - \nabla H(x^*)|| \le \delta$, then the sequence $\{x^k\}$ generated by (4.1) and (4.3) is well defined and converges superlinearly to x^*.*

The strong differentiability condition in the above theorem is somewhat restrictive. But this assumption is not easy to be weakened. Actually, discussions at the Erice and Vienna workshops in June, 1995 show that many researchers believe that in general superlinear convergence cannot be achieved for quasi-Newton methods for solving nonsmooth equations if the strong differentiability condition is not assumed at a solution point. If H is not strongly differentiable at the solution, when the sequence $\{x^k\}$ approaches the solution x^*, the corresponding $V_k \in \partial H(x^k)$ can be very different. By Theorem 2.3., we need

$$\lim_{k\to\infty} \frac{||(B_k - V_k)d^k||}{||d^k||} = 0$$

to achieve superlinear convergence of quasi-Newton methods, where B_k satisfies the quasi-Newton equation (4.2). This condition is hardly achievable.

Currently, some mixed strategies are proposed for quasi-Newton methods for solving nonsmooth equations of special forms, i.e., applying the quasi-Newton techniques to the "smooth part" of the nonsmooth equations, while applying the Newton techniques to the "nonsmooth part". Along this direction, Sun and Han [86,31] proposed some modifications of Josephy's quasi-Newton methods for solving the NCP(p) and VI(C,p) with C a polyhedron, in which systems of linear equations are solved. Qi and Jiang [72] presented superlinearly convergent quasi-Newton methods for solving various semismooth KKT equations of nonlinear programming (1.1).

5. PROXIMAL NEWTON METHODS

In this section we consider the minimization problem

$$\min \{f(x), \quad x \in \mathbb{R}^n\}, \tag{5.1}$$

where $f : \mathbb{R}^n \to \mathbb{R} \cup \{+\infty\}$ is a proper closed convex function, but maybe nondifferentiable.

Bundle-type methods are the most successful ones in solving (5.1), see [36]. Since bundle-type methods are based on first-order (subgradients) models, we cannot expect to have better convergence of those methods than linear convergence. The Moreau-Yosida regularization provides a valuable tool for achieving faster convergence.

Given a convex function f, its Moreau-Yosida regularization [55,95] is defined by

$$F(x) = \min \{f(y) + \frac{1}{2}(y-x)^T M(y-x), \ \ y \in \mathbb{R}^n\}, \tag{5.2}$$

where M is an $n \times n$ symmetric positive definite matrix. Associated with the problem (5.1) is the following problem

$$\min \{F(x), \ \ x \in \mathbb{R}^n\}. \tag{5.3}$$

It is well known that problems (5.1) and (5.3) are equivalent in the sense that the solution sets of the two problems coincide with each other. One remarkable feature of F is that it is a differentiable convex function with a Lipschitz continuous gradient on $\mathbb{R}^n$. Hence the nondifferentiable convex program (5.1) is converted into a continuously differentiable convex problem (5.3). For more properties of Moreau-Yosida regularization, see [82,83,36,47,69].

Let $p(x)$ be the unique minimizer of (5.2). Then the typical proximal point algorithm may be written as follows: After having x^k, the next iterate is generated by

$$x^{k+1} \approx p(x^k) = x^k - M^{-1}\nabla F(x^k) = x^k - M^{-1}M(x^k - p(x^k)).$$

This method has been proved globally convergent if M is a positive definite matrix. Rockafellar [82] proved that the outer iteration of this algorithm is convergent superlinearly if $M = \frac{1}{\lambda}I$, $\lambda \to \infty$ and f^*, the conjugate function of f, has a Lipschitz gradient at the solution. Recently, Qian [75], Bonnans, Gilbert, Lemaréchal and Sagastizábal [2], Lemaréchal and Sagastizábal [46,47], and Mifflin [53] considered variable metric Newton proximal methods by combining the Moreau-Yosida regularization and quasi-Newton methods. Superlinear convergence of some quasi-Newton proximal methods were discussed in [2] by assuming that the gradient of f is strongly differentiable at a solution point [56].

The derivative of F is

$$G(x) := \nabla F(x) = M(x - p(x)) \in \partial f(p(x)), \tag{5.4}$$

where ∂f is the subdifferential mapping of f in the sense of convex analysis [81].

If G is semismooth at a solution of (5.3), then we may use Newton's methods for LC^1 optimization discussed in Section 3, to achieve superlinear convergence. Therefore it is crucial to derive sufficient conditions for ensuring that G is semismooth at a solution of (5.3). Qi [69] proved that F has a semismooth gradient G at x and $\partial_B G(x)$ is positive definite if and only if the conjugate function f^* of f has a semismooth gradient at $G(x)$. When f is piecewise C^2, the semismoothness of G was proved in [87,54] under some constraint qualifications.

Based upon the above analysis, Fukushima and Qi [26] presented a globally and superlinearly convergent proximal Newton algorithm by setting $M = \frac{1}{\lambda}I$. At the k-th step of their algorithm, $p(x^k)$, $F(x^k)$ and G_k are approximated by $p^a(x^k, \varepsilon_k)$, $F^a(x^k, \varepsilon_k)$ and $G^a(x^k, \varepsilon_k)$ satisfying the following relationships

$$F^a(x^k, \varepsilon_k) = f(p^a(x^k, \varepsilon_k)) + \frac{1}{\lambda}||p^a(x^k, \varepsilon_k) - x^k||^2 \leq F(x^k) + \varepsilon_k,$$

$$G^a(x^k, \varepsilon_k) = \frac{1}{\lambda}(x^k - p^a(x^k, \varepsilon_k)),$$

where ε_k is a positive number.

Given constants $\lambda > 0, \sigma \in (0,1)$ and $\rho \in (0,1)$, the k-th step of Fukushima and Qi's algorithm can be defined as follows: First find a positive definite matrix $V_k \in \mathbb{R}^n \times \mathbb{R}^n$ and define a search direction d^k by

$$d^k = -V_k^{-1} G^a(x^k, \varepsilon_k);$$

then choose a positive number $\varepsilon_{k+1} < \varepsilon_k$ and generate the next iterate by

$$x^{k+1} = x^k + \tau_k d^k,$$

where $\tau_k = \rho^{m_k}$ and m_k is the smallest nonnegative integer m such that

$$F^a(x^k + \rho^m d^k, \varepsilon_{k+1}) \leq F^a(x^k, \varepsilon_k) + \sigma \rho^m G^a(x^k, \varepsilon_k)^T d^k + \varepsilon_k.$$

Fukushima and Qi [26] established global convergence of the above algorithm under certain conditions, and local superlinear convergence of the outer iterations under the semismoothness of G at a solution and other additional conditions with λ remaining fixed. How to approximate the generalized Jacobian V_k was addressed in [71]. Further work is needed to address the inner iteration problem.

ACKNOWLEDGEMENT

The authors are grateful to Professor Masao Fukushima for his helpful comments.

REFERENCES

[1] Bonnans J.F., "Local analysis of Newton-type methods for variational inequalities and nonlinear programming". *Appl. Math. Optimization* 29, pp. 161-186, 1994.

[2] Bonnans J.F., Gilbert J., Lemaréchal C., and Sagastizábal C., "A family of variable metric proximal methods". *Math. Programming* 68, pp. 15-47, 1995.

[3] Chen X., "On the convergence of Broyden-like methods for nonlinear equations with nondifferentiable terms". *Ann. Inst. Statist. Math.* 42, pp. 387-401, 1990.

[4] Chen X., "Convergence of the BFGS method for LC^1 convex constrained optimization". *SIAM J. Control and Optimization*, forthcoming.

[5] Chen, X. and Qi L., "A parameterized Newton method and a quasi-Newton method for solving nonsmooth equations". *Computational Optimization Appl.* 3, pp. 157-179, 1994.

[6] Chen X., Qi L., and Womersley R.S., "Newton's method for quadratic stochastic programs with recourse". *J. Computational Appl. Math.* 60, pp. 29-46, 1995.

[7] Chen X., and Womersley R.S., "A parallel inexact Newton method for stochastic program with recourse". *Annals Oper. Res.*, forthcoming.

[8] Chen X., and Yamamoto T., "On the convergence of some quasi-Newton methods for nonlinear equations with nondifferentiable term". *Computing* 48, pp. 87-94, 1992.

[9] Chen X., and Yamamoto T., "Newton-like methods for solving underdetermined nonlinear equations". *J. Computational Appl. Math.* 55, pp. 311-324, 1995.

[10] Clarke F.H., "Optimization and Nonsmooth Analysis". John Wiley, New York, 1983.

[11] De Luca T., Facchinei F., and Kanzow C., "A semismooth equation approach to the solution of nonlinear complementarity problems". *Math. Programming*, forthcoming.

[12] Dennis J.E., and Moré J.J., "Quasi-Newton methods: Motivation and theory". *SIAM Review* 19, pp. 46-89, 1977.

[13] Dennis J.E., and Schnabel R.B., "Numerical Methods for Unconstrained Optimization and Nonlinear Equations". Prentice-Hall, Englewood Cliffs, New Jersey, 1983.

[14] Dirkse S.P., Ferris M.C., "MCPLIB: A collection of nonlinear mixed complementarity problems". *Optimization Methods and Software*, forthcoming.

[15] Facchinei F., "Minimization of SC^1 functions and the Maratos effect". *Oper. Res. Letter*, 17, pp. 131-137, 1995.

[16] Facchinei F., "A new recursive quadratic programming algorithm with global and superlinear convergence properties". Preprint, Revised Version, Dipartmento di Informatica e Sistemistica, Universitá di Roma "La Sapienza", Roma, Italy, 1994.

[17] Facchinei F., and Kanzow C., "A nonsmooth inexact Newton method for the solution of large-scale nonlinear complementarity problems". Tech. Rep. 95, Institute of Applied Mathematics of the University of Hamburg, Hamburg, Germany, 1995.

[18] Facchinei F., and Soares J., "A new merit function for nonlinear complementarity problems and a related algorithm". *SIAM J. on Optimization*, forthcoming.

[19] Fischer A., "A special Newton-type optimization method". *Optimization* 24, pp. 269-284, 1992.

[20] Fischer A., "On the superlinear convergence of a Newton–type method for LCP under weak conditions". *Optimization Methods and Software*, forthcoming.

[21] Fischer A., "A globally and Q-quadratically convergent Newton-type method for positive semidefinite linear complementarity problems". *J. Optimization Theory and Appl.*, forthcoming.

[22] Fischer A., "An NCP–function and its use for the solution of complementarity problems". In: D.Z. Du, L. Qi and R.S. Womersley, eds., "Recent Advances in Nonsmooth Optimization", World Scientific, New Jersey, pp. 88-105, 1995.

[23] Fischer A., "Solution of monotone complementarity problems with locally Lipschitzian functions". MATH-NM-9-1995, Technische Universität Dresden, Dresden, Germany, 1995.

[24] Fukushima M., "Equivalent differentiable optimization problems and descent methods for asymmetric variational inequality problems". *Math. Programming* 58, pp. 99-110, 1992.

[25] Fukushima M., "Merit functions for variational inequality and complementarity problems". In this proceedings.

[26] Fukushima M., and Qi L., "A globally and superlinearly convergent algorithm for nonsmooth convex minimization". *SIAM J. Optimization*, forthcoming.

[27] Gabriel S.A., and Pang J.-S., A trust region method for constrained nonsmooth equations, in: W.W.Hager, D.W. Hearn and P.M. Pardalos, eds., "Large Scale Optimization: State of the Art" Kluwer Academic Publishers, Boston, pp. 159-186, 1994.

[28] Geiger C., and Kanzow C., "On the resolution of monotone complementarity problems". *Computational Optimization and Appl.*, forthcoming.

[29] Gomes-Ruggiero M.A., Martínez J.M., and Santos S.A., "Solving nonsmooth equations by means of quasi-Newton methods with globalization". In: D. Du, L. Qi and R. S. Womersley, eds, "Recent Advances in Nonsmooth Optimization", World Scientific, New Jersey, pp. 121-140, 1995.

[30] Han S.P., Pang J.-S., and Rangaraj N., "Globally convergent Newton methods for nonsmooth equations". *Math. Oper. Res.*, 17, pp. 586-607, 1992.

[31] Han J., and Sun D., "Newton and quasi-Newton methods for normal maps with polyhedral sets". Preprint, Institute of Applied Mathematics, Academia Sinica, China, 1994 (Revised November, 1995).

[32] Han J., and Sun D., "Superlinear convergence of approximate Newton methods for LC^1 optimization problems without strict complementarity". In: D. Du, L. Qi and R. S. Womersley, eds, "Recent Advances in Nonsmooth Optimization", World Scientific, New Jersey, pp. 141-158, 1995.

[33] Harker P.T., and Pang J.-S., "Finite-dimensional variational inequality and nonlinear complementarity problem: A survey of theory, algorithms and applications". *Math. Programming* 48, pp. 161-220, 1990.

[34] Harker P.T., and Xiao B., "Newton method for the nonlinear complementarity problem: A B-differentiable equation approach". *Math. Programming (Series B)* 48, pp. 339-357, 1990.

[35] Hiriart-Urruty J.B., Strodiot J.J., and Nguyen V.H., "Generalized Hessian matrix and second-order optimality conditions for problems with $C^{1,1}$ data". *Appl. Math. and Optimization* 11, pp. 43-56, 1984.

[36] Hiriart-Urruty J., and Lemaréchal C., "Convex Analysis and Minimization Algorithms". Springer-Verlag, Bonn, 1993.

[37] Ip C.M., and Kyparisis J., "Local convergence of quasi-Newton methods for B-differentiable equations". *Math. Programming* 56, pp. 71-89, 1992.

[38] Jiang H., and Qi L., "A globally and superlinearly convergent trust region algorithm for convex SC^1 minimization problems and its application to stochastic programs". *J. Optimization Theory and Appl.*, forthcoming.

[39] Jiang H., and Qi L., "A new nonsmooth equation approach to nonlinear complementarity problems". Technical Report AMR 94/30, School of Mathematics, The University of New South Wales, Australia, 1994 (Revised July, 1995).

[40] Josephy N.H., "Newton's method for generalized equations". Technical Report 1965, Mathematics Research Center, University of Wisconsin, Madison, 1979.

[41] Josephy N.H., "Quasi-Newton methods for generalized equations". Technical Report 1966, Mathematics Research Center, University of Wisconsin, Madison, 1979.

[42] Kanzow C., "Some equation-based methods for the nonlinear complementarity problem". *Optimization Methods and Software* 3, pp. 327-340, 1994.

[43] Kojima M., and Shindo S., "Extensions of Newton and quasi-Newton methods to systems of PC^1 equations". *J. Oper. Res. Soc. Japan*, 29, pp. 352-374, 1986.

[44] Kummer B., "Newton's method for non-differentiable functions". In: J. Guddat, B. Bank, H. Hollatz, P. Kall, D. Klatte, B. Kummer, K. Lommatzsch, L. Tammer, M. Vlach and K. Zimmerman, eds., "Advances in Mathematical Optimization", Akademi-Verlag, Berlin, pp. 114-125, 1988.

[45] Kummer B., "Newton's method based on generalized derivatives for nonsmooth functions: Convergence analysis". In: W. Oettli and D. Pallaschke, eds., "Advances in Optimization", Springer-Verlag, Berlin, pp. 171-194, 1992.

[46] Lemaréchal C., and Sagastizábal C., "An approach to variable metric methods". In: "Proceedings of the 16th IFIP Conference on System Modelling and Optimization", Springer-Verlag, Berlin, 1994.

[47] Lemaréchal C., and Sagastizábal C., "Practical aspects of the Moreau-Yosida regularization: theoretical preliminaries". *SIAM J. Optimization*, forthcoming.

[48] Mangasarian O.L., "Unconstrained Lagrangians in nonlinear programming". *SIAM J. Control* 13, pp. 772-791, 1975.

[49] Mangasarian O.L., "Unconstrained methods in nonlinear programming" In: "Nonlinear Programming", SIAM-AMS Proceedings, Volume IX, American Math. Soc., Providence, Rhode Island, pp. 169-184, 1976.

[50] Martínez J.M., and Qi L., "Inexact Newton methods for solving nonsmooth equations". *J. Computational and Appl. Math.* 60, 1995.

[51] Martínez J.M., and Zambaldi M.C., "Least change update methods for nonlinear systems with nondifferentiable terms". *Numerical Functional Analysis and Optimization* 14, pp. 405-415, 1993.

[52] Mifflin R., "Semismooth and semiconvex functions in constrained optimization". *SIAM J. Control and Optimization* 15, pp. 957-972, 1977.

[53] Mifflin R., "A quasi-second-order proximal bundle algorithm". Preprint (Revised version), Department of Pure and Applied Mathematics, Washington State University, Pullman, WA, USA, 1995.

[54] Mifflin R., Qi L., and Sun D., "Properties of the Moreau-Yosida regularization of a piecewise C^2 convex function". Manuscript, School of Mathematics, The University of New South Wales, Sydney, Australia, 1995.

[55] Moreau J., "Proximité et dualité dans un espace hilbertien". *Bulletin de la Société Mathématique de France* 93, pp. 273-299, 1965.

[56] Ortega J.M., and Rheinboldt W.C., "Iterative Solution of Nonlinear Equations in Several Variables". Academic Press, New York, 1970.

[57] Outrata J.V., and Zowe J., "A Newton method for a class of quasi-variational inequalities". *Computational Optimization and Appl.* 4, pp. 5-21, 1995.

[58] Pang J.-S., "Newton's method for B-differentiable equations". *Math. Oper. Res.* 15, pp. 311-341, 1990.

[59] Pang J.-S., "A B-differentiable equation based, globally, and locally quadratically convergent algorithm for nonlinear programs, complementarity and variational inequality problems". *Math. Programming* 51, pp. 101-131, 1991.

[60] Pang J.-S., "Complementarity problems". In: R. Horst and P. Pardalos, eds., "Handbook of Global Optimization", Kluwer Academic Publishers, Boston, pp. 271-338, 1994.

[61] Pang J.-S., and Chan D., "Iterative methods for variational and complementarity problems". *Math. Programming* 24, pp. 284-313, 1982.

[62] Pang J.-S., and Gabriel S.A., "NE/SQP: A robust algorithm for the nonlinear complementarity problem". *Math. Programming* 60, pp. 295-337, 1993.

[63] Pang J.-S., and Qi L., "Nonsmooth equations: Motivation and algorithms". *SIAM J. Optimization* 3, pp. 443-465, 1993.

[64] Pang J.-S., and Qi L., "A globally convergent Newton method for convex SC^1 minimization problems". *J. Optimization Theory Appl.* 85, pp. 633-648, 1995.

[65] Polak R., Mayne D.Q., and Wardi Y., "On the extension of constrained optimization algorithms from differentiable to nondifferentiable problems". *SIAM J. Control and Optimization* 21, pp. 179-203, 1983.

[66] Qi L., "Convergence analysis of some algorithms for solving nonsmooth equations". *Math. Oper. Res.* 18, pp. 227-244, 1993.

[67] Qi L., "Superlinear convergent approximate Newton methods for LC^1 optimization problems". *Math. Programming* 64, pp. 277-294, 1994.

[68] Qi L., "Trust region algorithms for solving nonsmooth equations". *SIAM J. Optimization* 5, pp. 219-230, 1995.

[69] Qi L., "Second-order analysis of the Moreau-Yosida approximation of a convex function". AMR 94/20, Applied Mathematics Report, University of New South Wales, 1994 (Revised August, 1995).

[70] Qi L., and Chen X., "A globally convergent successive approximation method for severely nonsmooth equations". *SIAM J. Control and Optimization* 33, pp. 402-418, 1993.

[71] Qi L., and Chen X., "A superlinearly convergent proximal Newton method for nondifferentiable convex optimization". AMR 95/20, Applied Mathematics Report, University of New South Wales, Sydney, 1995.

[72] Qi L., and Jiang H., "Semismooth Karush-Kuhn-Tucker equations and convergence analysis of Newton methods and quasi-Newton methods for solving these equations". AMR 94/5, Applied Mathematics Report, University of New South Wales, 1994 (Revised August, 1995).

[73] Qi L., and Sun J., "A nonsmooth version of Newton's method". *Math. Programming* 58, pp. 353-367, 1993.

[74] Qi L., and Womersley R.S., "An SQP algorithm for extended linear-quadratic problems in stochastic programming". *Annals Oper. Res.* 56, pp. 251-285, 1995.

[75] Qian M., "The variable metric proximal point algorithm: Global and superlinear convergence". Manuscript, Dept. of Mathematics, University of Washington, Seattle, WA, 1992.

[76] Ralph D., "Global convergence of damped Newton's method for nonsmooth equations, via the path search". *Math. Oper. Res.* 19, pp. 352-389, 1994.

[77] Robinson S.M., "Generalized equations and their solutions, part II: Applications to nonlinear programming". *Math. Programming Study* 19, pp. 200-221, 1982.

[78] Robinson S.M., "Generalized equations". In: A. Bachem, M. Grötschel and B. Korte, eds., "Mathematical Programming: The State of the Art", Springer-Verlag, Berlin, pp. 346-347, 1983.

[79] Robinson S.M., "An implicit-function theorem for a class of nonsmooth equations". *Math. Oper. Res.* 16, pp. 292-309, 1991.

[80] Robinson S.M., "Newton's method for a class of nonsmooth functions". *Set-Valued Analysis* 2, pp. 291-305, 1994.

[81] Rockafellar R.T., "Convex Analysis". Princeton University Press, Princeton, 1970.

[82] Rockafellar R.T., "Monotone operators and the proximal point algorithm". *SIAM J. Control and Optimization* 14, pp. 877-898, 1976.

[83] Rockafellar R.T., "Augmented Lagrangians and applications of the proximal point algorithm in convex programming". *Math. Oper. Res.* 1, pp. 97-116, 1976.

[84] Rockafellar R.T., "Nonsmooth optimization". In: J.R. Birge and K.G. Murty, eds., "Mathematical Programming: State of the Art", The University of Michigan, Ann Arbor, pp. 248-258, 1994.

[85] Shapiro A., "On concepts of directional differentiability". *J. Optimization Theory and Appl.* 66, pp. 477-487, 1990.

[86] Sun D., and Han J., "Newton and quasi-Newton methods for a class of nonsmooth equations and related problems". *SIAM J. Optimization*, forthcoming.

[87] Sun D., and Han J., "On a conjecture in Moreau-Yosida regularization of a nonsmooth convex function". *Bulletin of Science in China* (in Chinese), forthcoming.

[88] Taji K., and Fukushima M., "A globally convergent Newton method for solving variational inequality problems with inequality constraints". In: D. Du, L. Qi and R.S. Womersley, eds., "Recent Advances in Nonsmooth Optimization", World Scientific, New Jersey, pp. 405-417, 1995.

[89] Taji K., Fukushima M., and Ibaraki T., "A globally convergent Newton method for solving strongly monotone variational inequalities". *Math. Programming* 58, pp. 369-383, 1993.

[90] Tseng P., "Growth behaviour of a class of merit functions for the nonlinear complementarity problem". *J. Optimization Theory and Appl.*, forthcoming.

[91] Uko L.U., "Remarks on the generalized Newton method". *Math. Programming* 59, pp. 405-412, 1993.

[92] Xiao B., and Harker P.T., "A nonsmooth Newton method for variational inequalities". I: Theory, *Math. Programming* 65, pp. 151-194, 1994.

[93] Xiao B., and Harker P.T., "A nonsmooth Newton method for variational inequalities". II: Numerical results, *Math. Programming* 65, pp. 195-216, 1994.

[94] Yamashita N., and Fukushima M., "Modified Newton methods for solving semismooth reformulations of monotone complementarity problems". Technical Report, Graduate School of Information Science, Nara Institute of Science and Technology, Nara, Japan, 1995.

[95] Yosida K., "Functional Analysis". Springer Verlag, Berlin, 1964.

ON THE SOLUTION OF THE MONOTONE AND NONMONOTONE LINEAR COMPLEMENTARITY PROBLEM BY AN INFEASIBLE INTERIOR-POINT ALGORITHM

Júdice J.[1], Fernandes L.[2] and Lima A.[3]

[1] Departamento de Matemática da Universidade de Coimbra,
3000 - Coimbra, Portugal.
e-mail: judice@dragao.it.uc.pt

[2] Escola Superior de Tecnologia de Tomar,
2300 - Tomar, Portugal.

[3] Faculdade de Ciências da Universidade do Porto,
4000 - Porto, Portugal.

Abstract. The use of an Infeasible Interior-Point (IIP) algorithm [14] is investigated for the solution of the Linear Complementarity Problem (LCP) . Some monotone an nonmonotone LCPs from different sources are solved by two versions of the IIP algorithm, which differ in the line-search technique that computes the stepsize. The first version, denoted by SIPP, employs the simple maximum ratio technique commonly used in IIP methods for linear programming. On the other hand the second variant (GIIP) incorporates a more sophisticated Armijo-type line-search technique , that ensures global convergence for the procedure under some hypotheses. The computational experiments indicate that both the variants process efficiently monotone LCPs and LCPs with P matrices . On the contrary, the algorithms face many difficulties for solving some LCPs with P_0 matrices and LCPs associated with zero-sum bimatrix games . However, the algorithm GIIP has succeeded in a large number of these problems than the method SIPP.

A third version that employs an iterative solver for finding the search direction is also investigated and seems to be quite efficient for the solution of the monotone LCP that arises in a spatial equilibrium network structured model.

Keywords. Linear Complementarity Problem, Interior-Point algorithms, Spatial Equilibrium Models, Bimatrix Games.

1. INTRODUCTION

The Linear Complementarity Problem (LCP) consists of finding vectors $z \in \mathbb{R}^n$ and $w \in \mathbb{R}^n$ such that

$$w = q + Mz, \quad z \geq 0, \quad w \geq 0, \quad z^T w = 0 \tag{1.1}$$

where q is an n-vector and M is a square matrix of order n. This problem has become one of the most important areas of optimization. Several applications of the LCP have appeared in many areas of science, economics and engineering and a large number of algorithms has been designed for its solution. We refer [3, 21] for two excellent books on this subject.

It is well-known [3] that the LCP is equivalent to the following Affine Variational Inequality Problem (AVI):

Find $\bar{z} \geq 0$ such that

$$(q + M\bar{z})^T(z - \bar{z}) \geq 0 \text{ for all } z \geq 0 \tag{1.2}$$

Due to this equivalence, the LCP has been classified as Monotone or Nonmonotone depending on the matrix M to be Positive Semi-Definite (PSD) or not. We recall that a matrix M is PSD if

$$x^T M x \geq 0 \text{ for all } x \in \mathbb{R}^n.$$

This class contains the Positive Definite (PD) matrices , that is, those which satisfy

$$x^T M x > 0 \text{ for all } x \in \mathbb{R}^n - \{0\}.$$

A monotone LCP can be solved in polynomial-time [15], while a nonmonotone LCP is NP-hard [21]. In the last category of LCPs, one should distinguish the cases of M being a matrix P or P_0 , where M is $P(P_0)$ if and only if all its principal minors are positive (nonnegative). It is interesting to add that a LCP with a matrix P_0 is NP-hard [15] while the complexity of a LCP with a matrix P is still an open question. The following diagram shows the relationships among these four classes of matrices

$$\begin{array}{ccc} PD & \subset & P \\ \cap & & \cap \\ PSD & \subset & P_0 \end{array} \tag{1.3}$$

where $\subset$ represents strict inclusion. Furthermore $P = PD$ and $PSD = P_0$ for symmetric matrices [21].

Interior-Point algorithms have become one of the most interesting approaches for solving the monotone LCP . As the proper name indicates, these methods work on the interior of the nonnegative orthant $\mathbb{R}^{2n}_+$, that is, $z^k > 0$ and $w^k > 0$ are forced for each iterate (z^k, w^k). Each iteration consists of a damped Newton step for the system of nonlinear equations

$$\begin{cases} Mz - w + q &= 0 \\ ZWe &= \mu_k e \end{cases} \tag{1.4}$$

where $Z = diag(z_1, \ldots, z_n)$, $W = diag(w_1, \ldots, w_n)$ and μ_k is the so-called central parameter that goes to zero as the algorithm proceeds. A stepsize is appropriately chosen so that the positivity of the variables z_i and w_i is maintained during the whole procedure. Global convergence of the algorithm is ensured by a criterious choice of the central parameter μ_k and of the stepsize that is used in these damped Newton iterations.

There are two main categories of interior-point (IP) methods, namely the Feasible (FIP) and Infeasible (IIP) algorithms. In [15], Kojima et al have discussed an algorithm that unifies all the existing FIP methods. They have been able to establish global convergence for their unified method when the LCP is monotone or M is a P-matrix. The infeasible version of this algorithm seems to work quite efficiently for monotone LCPs when the stepsize is computed by the simple maximum ratio technique commonly employed in linear programming [10, 18]. Recently, Simantiraki and Shanno [24], have discussed the use of an Armijo-type line search technique for the computation of the stepsize. They have been able to establish global convergence for the resulting method in the sense introduced by Kojima et al [14] for linear programming . They have also tested this algorithm in a number of monotone and nonmonotone LCPs. The results seem to indicate that this global version GIIP is always more efficient than the simple version SIIP of the infeasible interior-point method that simply uses the stepsize computed from the maximum ratio criterion mentioned before.

In this paper we have studied the efficiency of these two versions of the IIP method for monotone and nonmonotone LCPs . In the first experience we have solved some monotone and nonmonotone LCPs with matrices P and P_0 . The numerical results show that the methods perform in a similar way for these type of problems and there is no clear winner between them. We have also studied these two versions of the infeasible interior-point method for the solution of some NP-hard nonmonotone LCPs with a P_0 matrix and LCPs associated with two formulations of bimatrix games . The experiments have shown that both the versions face many difficulties for processing these LCPs. However, the version GIIP has succeeded in a larger number of instances than the variant SIIP. This seems to indicate that the former version should be probably a better starting point for the possible design of an interior-point method capable of dealing with nonmonotone LCPs .

As it is usual in interior-point methods, both the versions of the algorithm have been implemented using a direct solver for finding the search direction in each iteration. However, some large LCPs with a network structure [1, 11, 22] cannot be processed with this type of implementation and require an iterative solver for that propose. The use of this iterative solver needs a special stopping criterion that leads to a small number of iterations for this method. This had lead to the design of a truncated infeasible interior-point method TIIP [23] which consider in each iteration a residual vector r^k in (1.4) to get an equation of the form $Mz - w + q = r^k$. The norm of this residual vector can be controlled in such a way that the resulting procedure possesses global convergence in a sense stated in Kojima et al [14]. Some computational experience presented in this paper and elsewhere [23] shows that this TIIP algorithm is quite appropriate for the solution of these large monotone LCPs with a network structure.

The organization of this paper is as follows. In section 2 the (truncated) infeasible interior-point method is introduced, together with its variants. The section 3 reports the experiments performed with these variants for the solution of the monotone and nonmonotone LCPs mentioned before. Finally some concluding remarks are presented at the end of the paper.

2. AN INFEASIBLE INTERIOR-POINT ALGORITHM

In this section we discuss a general interior-point algorithm that can be seen as an extension of the method described in [14] for the LCP. See also [24, 26] for other papers on the same ideas. As stated in [23] we further introduce a residual vector r^k in each iteration, that is quite appropriate when the search direction is found approximately by an iterative solver.

Consider the LCP (1.1) and let $\mathcal{N}$ be the set of the points (z, w) defined by the following equations

$$\begin{array}{lll} z > 0 \quad , & w > 0 & \\ z_i w_i & \geq \gamma \frac{z^T w}{n} & i = 1, \ldots, n \\ z^T w & \geq \bar{\gamma} \|w - q - Mz\| \text{ or } \|w - q - Mz\| \leq \epsilon_2 & \end{array} \tag{2.1}$$

where $0 < \gamma < 1$, $\bar{\gamma} > 0$ are given real numbers and ϵ_2 is a small positive tolerance. Furthermore let

$$0 \leq \beta_0 < \beta_1 < \beta_2 < \beta_3 < 1$$

be given real numbers. Then the steps of the algorithm can be stated as follows

Step 0: Let $(z^0, w^0) \in \mathcal{N}$ and $k = 0$.

Step 1: If $(z^k)^T w^k \leq \epsilon_1$ and $\|w^k - q - Mz^k\| \leq \epsilon_2$ $(\epsilon_1 \leq \epsilon_2)$ stop: (z^k, w^k) is a solution of the LCP.

Step 2: Search Direction - Let $\mu_k = \sigma_k \frac{(z^k)^T w^k}{n}$, $\beta_0 \leq \sigma_k < \beta_1$ and compute $\begin{pmatrix} u^k \\ v^k \end{pmatrix}$ by

$$\begin{pmatrix} M & -I \\ W_k & Z_k \end{pmatrix} \begin{pmatrix} u^k \\ v^k \end{pmatrix} = \begin{pmatrix} -Mz^k - q + w^k \\ -Z_k W_k e \end{pmatrix} + \begin{pmatrix} r^k \\ \mu_k e \end{pmatrix} \tag{2.2}$$

where

$$\|r^k\| \leq \beta_0 \| - Mz^k - q + w^k\| \tag{2.3}$$

and

$$Z_k = diag(z_1^k, \ldots, z_n^k), \ W_k = diag(w_1^k, \ldots, w_n^k).$$

Step 3: Stepsize

(i) Let $\bar{\alpha}_k$ be the maximum $\alpha > 0$ such that

$$(z^k, w^k) + \alpha(u^k, v^k) \in \mathcal{N} \tag{2.4}$$

$$(z^k + \alpha u^k)^T (w^k + \alpha v^k) \leq (1 - \alpha(1 - \beta_2))[(z^k)^T w^k] \text{ for all } \alpha \in [0, \bar{\alpha}_k] \tag{2.5}$$

(ii) If $\bar{\alpha}_k \leq \epsilon_2$ stop. Otherwise let α_k be such that $0 < \alpha_k \leq \bar{\alpha}_k$ and

$$(z^{k+1})^T w^{k+1} \leq (1 - \bar{\alpha}_k(1 - \beta_3))(z^k)^T w^k \tag{2.6}$$

Step 4: Set

$$\begin{aligned} z^{k+1} &= z^k + \alpha_k u^k \\ w^{k+1} &= w^k + \alpha_k v^k \end{aligned} \tag{2.7}$$

$k = k + 1$ and go to Step 1.

Next we discuss some further issues of the algorithm. In step 2, the matrix Z_k is diagonal with positive diagonal elements. So its inverse Z_k^{-1} exists and is easily computable. The system (2.2) can be rewritten as follows

$$v^k = -w^k + \mu_k Z_k^{-1} e - Z_k^{-1} W_k u^k$$

$$(M + Z_k^{-1} W_k) u^k = -Mz^k - q + \mu_k Z_k^{-1} e + r^k \tag{2.8}$$

Hence the algorithm works if the matrix $M + Z_k^{-1} W_k$ is nonsingular in each iteration k. The following result holds concerning the nonsingularity of this matrix.

Theorem 1 *[15] - M is a matrix P_0 if and only if $M + Z_k^{-1} W_k$ is nonsingular for all $z^k > 0$ and $w^k > 0$.*

It is not difficult to extend the theory presented in Kojima et al.[14] to show that all the iterates belong to the set $\mathcal{N}$, provided $(z^0, w^0) \in \mathcal{N}$ and $M + Z_k^{-1} W_k$ is nonsingular for each $(z^k, w^k) \in \mathcal{N}$. This means that the algorithm may work for the nonmonotone LCP whose matrix is not P_0. Furthermore the method always stops in one of the following terminations:

(i) in Step 1 with a solution of the LCP

(ii) in Step 3 with a failure.

The drawback of this type of methodology concerns the possibility of occurence of this failure. As we see later, the termination can occur when the LCP has no solution, but the method may terminate with such form in cases where the LCP has a solution. Actually, this type of difficulty is shared by all the direct and iterative methods for the LCP [3, 21]. So, as these other linear complementarity procedures, this algorithm cannot process the LCP in all cases.

It is important at this stage to distinguish three categories of the interior-point algorithm:

(i) The Feasible Interior-Point (FIP) algorithm in which

$$r^k = 0, \quad w^k = q + Mz^k$$

for all iterations k.

(ii) The Infeasible Interior-Point (IIP) method in which

$$r^k = 0$$

for all iterations k and $w^k \neq q + Mz^k$ in general.

(iii) The Truncated Interior-Point (TIP) algorithm, where $r^k \neq 0$ in general and satisfies the inequality (2.3).

It is important to add that for the first category of method the termination in Step 1 always occurs when M is a matrix PSD or P [15].

As stated before, all those forms of the interior-point algorithm possess global convergence to one of the terminations mentioned in the Steps 1 and 3, provided the initial point belongs to the neighborhood $\mathcal{N}$ of the central path defined by (2.1). The computation of the stepsize α_k is done according to the formulas (2.4) - (2.6) of Step 3. In practice it is not easy to find a value α_k satisfying these conditions.A quite relaxed criterion for the choice of α_k has been employed in practice and simply consists of finding the largest positive value such that the variables z_i^k and w_i^k are positive in each iteration. The most common choice for α_k takes the following form

$$\alpha_k = 0.99995 \max\{\alpha > 0 : z^k + \alpha u^k > 0, w^k + \alpha v^k > 0\}$$

This choice for α_k does not guarantee global convergence, but has shown to perform quite well in practice for linear programming [20] and monotone LCPs [10]. Recently Simantiraki and Shanno [24] have adapted an idea of El-Bakry et al [8] and have proposed a line-search that guarantees global convergence for the infeasible interior-point method and seems to work well in practice. This line-search is an Armijo type technique for the merit function

$$\Phi(z, w) = \left[\|ZWe\|^2 + \|w - Mz - q\|^2\right]^{\frac{1}{2}} \tag{2.9}$$

where $e \in \mathbb{R}^n$ is a vector of ones and $\| \ \|$ denotes the l_2 norm. This function is a quite natural choice, since the algorithm pursues a point (z^k, w^k) that is a solution of the following system of nonlinear equations

$$\begin{cases} w - Mz - q &= 0 \\ ZWe &= 0 \end{cases}$$

As stated in [5] the merit function (2.9) is usualy employed for this type of nonlinear system.

The procedure starts by finding the maximum value $\bar{\alpha}_k$ such that $(z^k + \alpha u^k, w^k + \alpha v^k) \in \mathcal{N}$ for all $0 \leq \alpha \leq \bar{\alpha}_k$. This is done according to the formulas (17) - (19) in [24]. Next the stepsize α_k is computed by

$$\alpha_k = \frac{\bar{\alpha}_k}{2^t}$$

where t is the smallest nonnegative integer satisfying

$$\Phi(z^k + \alpha_k u^k, w^k + \alpha_k v^k) \leq \Phi(z^k, w^k) + \frac{\alpha_k}{10^4} \nabla\Phi(z^k, w^k)^T \begin{bmatrix} u^k \\ v^k \end{bmatrix}.$$

with $\nabla\Phi(z^k, w^k)$ the gradient of the merit function .

As is shown in Simantiraki and Shanno [24] this line-search technique is sufficient to guarantee global convergence for the infeasible interior-point method when $r^k = 0$ in all iterations k. To show that the truncated infeasible interior-point method mantains global convergence it is necessary to prove that the search direction computed by this procedure is a descent direction for the merit function (2.9). To do this we need the following result.

Theorem 2 *If $\nabla\Phi(z^k, w^k)$ denotes the gradient of the function Φ at (z^k, w^k) and (u^k, v^k) is the search direction computed by the IIP algorithm, then*

$$\nabla\Phi(z^k, w^k)^T \begin{bmatrix} u^k \\ v^k \end{bmatrix} \leq -\Phi(z^k, w^k)(1 - \sigma_k)$$

Proof. Consider the system associated to the iteration k

$$F(z, w) = \begin{bmatrix} w - Mz - q \\ ZWe \end{bmatrix} = \begin{bmatrix} r^k \\ \mu_k e \end{bmatrix}$$

If $F'(z^k, w^k)$ denotes the jacobian of this system, then by computations similar to those presented in [24] we have

$$\begin{aligned} \nabla\Phi(z^k, w^k)^T \begin{bmatrix} u^k \\ v^k \end{bmatrix} &= \frac{1}{\|F(z^k, w^k)\|} F(z^k, w^k)^T F'(z^k, w^k) F'(z^k, w^k)^{-1} \left[-F(z^k, w^k) + \begin{bmatrix} r^k \\ \mu_k e \end{bmatrix} \right] \\ &= \frac{1}{\|F(z^k, w^k)\|} \left(-\|F(z^k, w^k)\|^2 + F(z^k, w^k)^T \begin{bmatrix} r^k \\ \mu_k e \end{bmatrix} \right) \\ &= -\|F(z^k, w^k)\| + \frac{(-Mz^k + w^k - q)^T r^k + \mu_k (z^k)^T w^k}{\|F(z^k, w^k)\|} \end{aligned}$$

But

$$\|F(z^k, w^k)\| = \Phi(z^k, w^k)$$

and

$$\mu_k (z^k)^T w^k \leq \sigma_k \|Z_k W_k\|^2$$

Furthermore $\sigma_k \geq \beta_0$ and (2.3) imply

$$\frac{\|r^k\|}{\| - Mz^k + w^k - q\|} \leq \beta_0 \leq \sigma_k$$

Hence by the Cauchy-Schwarz inequality we have

$$\begin{aligned} (-Mz^k + w^k - q)^T r^k &\leq \| - Mz^k + w^k - q\| \|r^k\| \\ &\leq \sigma_k \| - Mz^k + w^k - q\|^2 \end{aligned} \tag{2.10}$$

Therefore

$$\nabla\Phi(z^k,w^k)^T\begin{bmatrix}u^k\\v^k\end{bmatrix} \leq -\Phi(z^k,w^k)+\frac{\sigma_k\left[\Phi(z^k,w^k)\right]^2}{\Phi(z^k,w^k)} = -\Phi(z^k,w^k)(1-\sigma_k)$$

□

It follows from the theorem that one of the two possible cases occur:

(i) $\Phi(z^k, w^k) = 0$ and the algorithm stops with a solution of the LCP.

(ii) $\nabla\Phi(z^k,w^k)^T\begin{bmatrix}u^k\\v^k\end{bmatrix} < 0$ and (u^k, v^k) is a descent direction.

Hence the theory presented in [24] can now be used to achieve exactly the same conclusion stated in this paper. Therefore the TIIP algorithm possesses global convergence if the Armijo-type line-search described before is used throughout the procedure.

In the next section we describe some experiments made with the three variants of the interior-point method incorporating the two line-searches discussed in this section. As stated in the first section, we denote by SIIP and GIIP the two versions of these methods that use the simple line-search and the Armijo type technique respectively. The so-called Feasible and Infeasible interior-point versions employ a direct solver for the solution of the system (2.8) that computes the search direction. Furthermore an iterative solver is used in the truncated variant of the interior-point algorithm . The choice of the direct or iterative solvers depends on the symmetry and sparsity of the matrix M of the LCP.

3. COMPUTATIONAL EXPERIENCE

In this section we describe some computational experience on a SUN SPARCstation 10 (48Mhz, 64Mb RAM) with the variants of the interior-point algorithm discussed in the previous section. Three sets of experiments have been performed, namely the solution of some LCPs with P and P_0 matrices, LCPs associated with nonzero sum bimatrix games and LCPs corresponding to a network spatial equilibrium model.

3.1. LCPs with matrices P and P_0

In this first experience we have started by considering a number of symmetric P and P_0 matrices from the literature. Since symmetric matrices $P(P_0)$ are $PD(PSD)$ then the corresponding LCPs are monotone. These matrices are presented below.

(i) Pentadiagonal PD matrices, denoted by PDL, that have been introduced in [19]. The nonzero elements are given by

$$\begin{array}{lllll} m_{ii} & = & 6 & & \\ m_{i,i-1} & = & m_{i,i+1} & = & -4 \\ m_{i,i-2} & = & m_{i,i+2} & = & 1 \end{array} \tag{3.1}$$

(ii) PD matrices that arise on the solution of the Laplace equation by finite differences [2]. We denote by LAPL this type of matrices and a description of their elements appears in [10].

(iii) PD matrices of the form $M = LL^T$, where L is a unit lower triangular matrix with all off-diagonal elements equal to 2 [9]. This type of matrix is denoted by FAT.

(iv) The well-known Hilbert matrices [21] whose elements are given by

$$m_{ij} = \frac{1}{i+j-1}$$

These matrices are PSD and are denoted by $HILB$.

(v) The so-called Pascal PSD matrices [21] that are denoted by $PASC$. Their elements satisfy

$$m_{ij} = \frac{(i+j-2)!\,\theta}{(i-1)!(j-1)!}$$

where ! denotes factorial and θ is a small positive real number (we have set $\theta = 10^{-7}$).

We have also considered in this first experience two well-known matrices P that are stated next.

(vi) The matrices L referred in (iii), which are P but not PD and are denoted by MU.

(vii) PD matrices introduced by Chandrasekaran, Pang and Stone [21], whose elements are given by

$$\begin{array}{lcllllll} m_{ii} & = & 1 & & & & & \\ m_{ij} & = & 2 & \text{if} & j > i & \text{and} & i+j & \text{is odd} \\ m_{ij} & = & -1 & \text{if} & j > i & \text{and} & i+j & \text{is even} \\ m_{ij} & = & -1 & \text{if} & j < i & \text{and} & i+j & \text{is odd} \\ m_{ij} & = & 2 & \text{if} & j < i & \text{and} & i+j & \text{is even} \end{array} \tag{3.2}$$

These matrices are denoted by $PANG$.

Finally we have constructed some matrices of the form

$$DAE$$

where A is a matrix of the categories mentioned before and D and E are diagonal matrices whose diagonal elements d_{ii} and e_{ii} are randomly generated numbers belonging to the intervals $(0, 0.1)$ and $(0, 1)$ respectively. We denote by $PSNAME$ the matrices of this form, where $NAME$ represents one of the classes of the matrices stated before.

All these matrices have been used for constructing the LCPs of our first experience. According to the suggestions of their authors, the right-hand side vectors for the LCPs with the matrices FAT, MU and $PANG$ are given by $q = -e$, where e is a vector of ones. Furthermore all the right-hand side vectors q of the remaining LCPs have been generated by the technique discussed in [10], with $scal = 0$, $|F| = \frac{n}{2}$ and $|F_d| = 0$.

The symmetric monotone LCPs have been solved by the two variants SIIP and GIIP of the infeasible interior-point method, that have been implemented according to the lines stated in [10]. The LCPs with nonsymmetric matrices have been solved by a similar implementation in which the subroutine MA27 [6] has been replaced by MA28 [6]. We note that this procedure is certainly not the best way of implementing the interior-point methods in the latter case. However, this code is suitable for comparing the efficiency of the two variants of the interior-point algorithm . These two versions SIIP and GIIP require some parameters that have been set according to the recommendations presented in [10] and [24] respectively. Our experiences have shown that $\sigma_k = \frac{1}{n}$ is usually preferable for the LCPs of smaller dimension. We have used this value for all the problems but $LAPL$ and $PSLAPL$, for which the choice presented in [24] has been followed.

As suggested in [10], we have used the initial point as follows

$$\begin{aligned} z_i^0 &= \lambda \\ w_i^0 &= \|q + Mz\|_\infty \end{aligned} \tag{3.3}$$

where λ is a small number belonging to the interval $[1,5]$. Finally the stopping criterion for both the variants of the infeasible interior-point use the tolerances $\epsilon_1 = \epsilon_2 = 10^{-8}$.

Test Problem	n	SIIP	GIIP
PDL	1000	12	12
	2000	12	12
LAPL	10000	8	8
	15000	8	8
HILB	100	17	9
	200	21	9
PASC	20	20	22
	30	29	34
FAT	100	27	21
	200	33	23
MUC	100	11	15
	200	12	15
PANG	100	20	15
	200	25	16
PSDPL	1000	32	34
	2000	39	42
PSLAPL	10000	36	40
	15000	47	53
PSHILB	100	14	10
	200	28	10
PSPASC	20	17	20
	30	21	20
PSPANG	100	15	13
	30	19	14

Table 1 - Number of iterations for LCPs with matrices P and P_0

The results displayed in Table 1 indicate that both the versions SIIP and GIIP of the infeasible interior-point algorithm are quite efficient for solving all the LCPs. We note that the matrices $PSHILB$ and $PSPASC$ are P_0 but are not P. Actually, they belong to the so-called class of sufficient matrices introduced in [4]. It is interesting to note that the Feasible IP algorithms possess global convergence for LCPs with matrices of this class, as these matrices coincide with the matrices P_* introduced in [15]. See [25] for a proof of this result.

In the second set of experiments we have considered some matrices P_0 that are neither P nor even sufficient. We start with the LCP defined by

$$q = \begin{bmatrix} 0 \\ 0 \\ -1 \\ 0 \end{bmatrix}, \quad M = \begin{bmatrix} 0 & 0 & 0 & 0 \\ 1 & 1 & 0 & 0 \\ 1 & 1 & 0 & 0 \\ 1 & 1 & 0 & 1 \end{bmatrix}$$

It is easy to see that the famous Lemke's algorithm [16] cannot process this LCP. Furthermore the LCP is feasible and possesses an infinity number of solutions of the form

$$\bar{z} = (\alpha, 0, 0, 0)\ , \alpha > 0$$

and

$$\bar{z} = (1, 0, \beta, 0) \ , \beta > 0$$

Both the variants of the infeasible interior-point methods have attained a solution of the first form in 4 iterations.

It is well-known that if M is a matrix P_0 then one of the following three possible cases must occur:

(i) The LCP has a solution.

(ii) The LCP is infeasible, that is, its linear constraints are incompatible.

(ii) The LCP is feasible but has no solution.

Actually this also holds for a general matrix M. In our next experiment we have considered some LCPs satisfying the two last cases. In all the problems the infeasible interior-point method has terminated with a failure.

After all these successful experiments we have tested both the variants of the infeasible interior-point algorithm on some NP-hard LCPs with matrices P_0 that have been introduced in [15]. Both the variants of the method have stopped with a failure in Step 3 despite the existence of a solution for these LCPs. So these experiments and several other results presented elsewhere [18] seem to indicate that both the variants are efficient for solving monotone LCPs and LCPs with P and sufficient matrices. Furthermore there is no clear winner between the two versions. The algorithm can also solve more difficult LCPs with matrices P_0 for which other traditional methods are unable to process. However, the two versions do not seem to be able to solve NP-hard LCPs with matrices P_0.

3.2. Nonzero Sum Bimatrix Games

It is well-known that any nonzero sum bimatrix game can be reduced into a nonmonotone LCP . There are two possible formulations of the bimatrix games in terms of a LCP. In the most common formulation (FORM1) the right-hand side vector q and the matrix M take the following forms

$$q = \begin{bmatrix} -e^m \\ -e^r \end{bmatrix} \ , \ M = \begin{bmatrix} 0 & A \\ B & 0 \end{bmatrix}$$

where m and r are positive integer numbers, $e^t \in \mathbb{R}^t$ is a vector of ones and $A \in \mathbb{R}^{mxr}$, $B \in \mathbb{R}^{rxm}$ are positive matrices [17]. On the other hand in the second formulation (FORM2) $-e^r$ is replaced by e^r and B is a negative matrix [7].

It is easy to find a feasible positive starting point (z^0, w^0) for both the variants of the interior-point method. Next we present a process for computing the initial point z^0 that has been used in our experiments. The vector w^0 is given by $q + Mz^0$. Let

$$k_1 = \min_{1 \leq i \leq r} \sum_{j=1}^{m} b_{ij} \quad k_2 = \min_{1 \leq i \leq m} \sum_{j=1}^{r} a_{ij}$$

Then in FORM1 we set

$$z_i^0 = \frac{\alpha_1}{k_1} \quad i = 1, \ldots, m \quad z_i^0 = \frac{\alpha_2}{k_2} \quad i = m+1, \ldots, m+r$$

where

$$\alpha_1 = \max\{r \max_{i,j} b_{ij}, \, 1.5\} \quad \alpha_2 = \max\{m \max_{i,j} a_{ij}, \, 1.5\}$$

On the other hand in FORM2 we set

$$z_i^0 = \frac{1}{\alpha_1 k_1} \qquad i = 1, \ldots, m$$

where

$$\alpha_1 = \begin{cases} \frac{1}{r} \min b_{ij} & \text{if } -1.5 < \alpha_1 < -1.0 \\ -1.5 & \text{otherwise} \end{cases}$$

Furthermore, the components $m+1, \ldots, m+r$ of z^0 are given as in FORM1.

As before, we have tested both the variants SIIP and GIIP of the interior-point algorithm . The elements of the matrices A and B of the test problems are randomly generated in the intervals $[a, b]$ and $[c, d]$ respectively. The values of these numbers a, b, c and d and the values of m and r are displayed in Table 2 together with the results of the performances of the two variants on these problems. In case of failure the value of the complementarity gap $z^T w$ at that stage is included after the letter F.

FORM	m	r	a	b	c	d	n	SIIP	GIIP
1	20	30	40	72	43	59	50	F(1E-4)	F(1E-4)
1	40	20	13	24	74	80	60	24	91
1	30	30	104	110	15	26	60	F(1E-5)	F(1E-5)
1	35	35	5	11	4	7	70	27	80
1	35	35	70	72	40	72	70	27	78
1	35	35	0.9	1.8	0.3	0.9	70	27	72
2	30	20	15	45	-35	-30	50	F(1E-3)	61
2	30	30	1	2	-2	-1	60	F(1E-1)	153
2	30	30	100	101	-300	-200	60	F(1E-5)	114
2	30	30	5E-5	4E-2	-4E-2	-5E-5	60	F(2E+0)	182

Table 2 - Nonzero Sum Bimatrix Games

The results displayed in Table 2 show that the variant GIIP seems to be more recommendable for this type of LCPs. In fact this procedure has been successful in 8 of the 10 test problems. Furthermore the value of the complementarity gap was quite small in the two failures. On the other hand the SIIP has failed in 6 problems. So these results seem to indicate that there is probably room for the use of interior-point algorithms on some nonmonotone LCPs whose matrices are not P_0 . However, there is much to be done for the design of an efficient algorithm of this type in this case.

3.3. A Spatial Equilibrium Model

This model has been discussed in [1, 11, 22] and consists of r regions that ship m commodities ($m \geq 1$) among them. Furthermore there is connection between each pair of regions and the supply/demand functions are affine. This model leads into a monotone LCP with matrix $M = JBJ^T$, where

$$B = \begin{bmatrix} B_{11} & B_{12} & \cdots & B_{1m} \\ B_{21} & B_{22} & \cdots & B_{2m} \\ \vdots & \vdots & \ddots & \vdots \\ B_{m1} & B_{m2} & \cdots & B_{mm} \end{bmatrix} \in \mathbb{R}^{mrxmr}$$

is a (symmetric or unsymmetric) PD matrix such that each submatrix B_{ij} is diagonal with positive elements. Furthermore J is a block diagonal matrix of the form

$$J = \begin{bmatrix} G & & & \\ & G & & \\ & & \ddots & \\ & & & G \end{bmatrix}$$

where $G \in \mathbb{R}^{rxr(r-1)}$ is an incidence matrix.

Therefore the dimension of the LCP is $n = mr(r-1)$, which can be quite large even for small number of regions and commodities. For instance a model of 50 commodities and 100 regions leads into a LCP of dimension 495000. Since B is a PD matrix, then $M = JBJ^T$ is a PSD matrix and the LCP is monotone. So the infeasible interior-point is recommended for solving this type of LCP. However, the large dimension of the LCP and the structure of its matrix prohibits the use of a direct solver to find the search direction in each iteration. Therefore the truncated version of the interior-point is much more suited in this application. In [23], we have developed an efficient implementation of the Truncated IIP method for the solution of this LCP. One of most important features of this implementation lies on the fact that the search direction is computed by solving a linear system of order much smaller than n. To see this, we note that each iteration of the TIIP method requires the solution of a linear system with matrix $M + E_k = JBJ^T + E_k$, where E_k is a diagonal matrix with positive diagonal elements. Since B is a PD matrix then

$$B = LDU$$

where L and U are unit lower and upper triangular matrices respectively. Furthermore no fill-in occurs during this factorization due to the special structure of B. Now suppose we wish to solve the system

$$(M + E_k)y = b \tag{3.4}$$

Hence its solution satisfies

$$y = (M + E_k)^{-1}b$$

By the Shermann-Morrison formula [5] we have

$$y = E_k^{-1}b - E_k^{-1}J^TU^{-1}(D^{-1} + UJE_k^{-1}J^TL)^{-1}L^{-1}Jb$$

Hence y can be computed by solving the systems

$$L\alpha = Jb \tag{3.5}$$

$$(D^{-1} + UJE_k^{-1}J^TL)\beta = \alpha \tag{3.6}$$

$$U\theta = \beta \tag{3.7}$$

and setting

$$y = E_k^{-1}(b - J^T\theta)$$

Since the matrices L and U are triangular, then the major computational effort for solving the system (3.4) lies on the solution of (3.6). The dimension of this latter system is mr, which is much smaller then the dimension of the LCP. To have an idea of the gap we note that for $m = 50$ and $r = 100$ the order of the system (3.6) is 5000, while the dimension of the LCP is 495000.

The system (3.6) is solved by an iterative solver, which depends on the fact of B being symmetric or not. In the first case the Preconditioned Conjugate Gradient (PCG) method should be employed. On the other hand a nonsymmetric iterative method such as the Preconditioned QMR algorithm is quite suitable for the second case. Both these iterative methods can efficiently do their job if a good preconditioner is available. An incomplete QR factorization preconditioner has shown to be recommendable in this instance and is fully described in [23].

Another important feature of the implementation is the use of a dynamic stopping criterion for the iterative solver. If we set the vector r^k of the truncated IP method as the residual of the iterative solver, then this latter procedure can stop even when the residual vector has a relatively large norm. In effect the theory of the truncated IP

number of commodities: 50

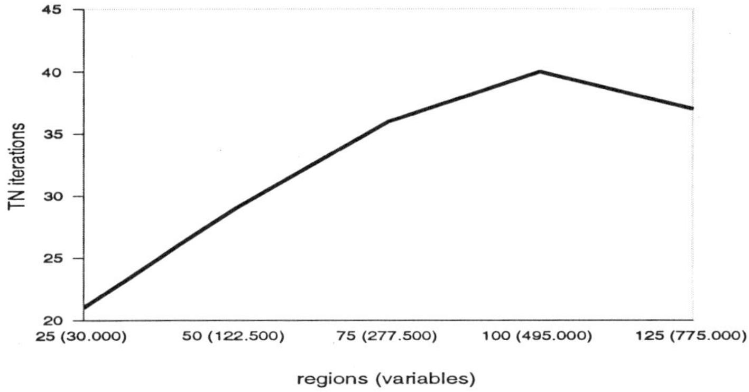

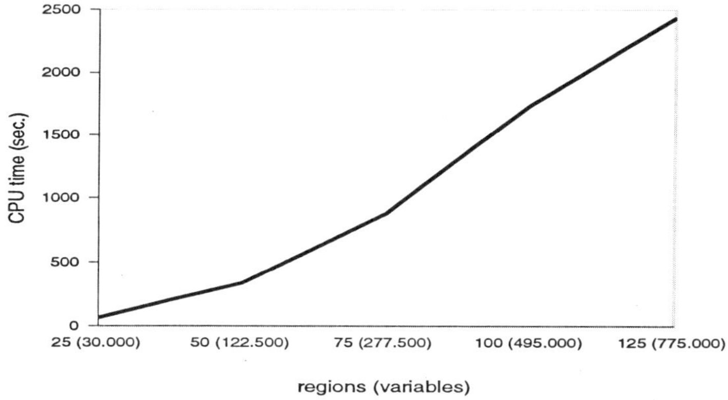

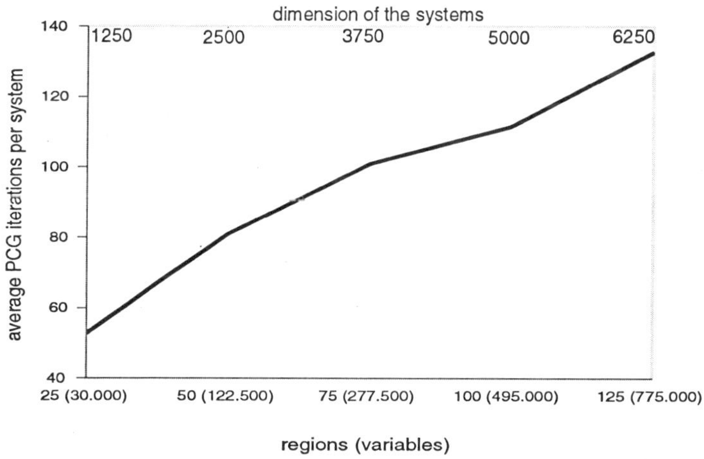

method only requires that the inequality (2.3) holds in each iteration k and this can be satisfied with $\|r^k\|$ large.

Based on the interesting features mentioned before, the implementation of the truncated IIP method (version SIIP) has shown to be quite efficient for solving this type of LCP [23]. The next figures illustrate the performance of this method for the case of some spatial equilibrium models with a symmetric PD matrix B, 50 commodities and a different number of regions. In each figure the dimension of the LCP is in brackets on the right of the corresponding number of regions. The first and the second figures represent the number of iterations and the CPU time in seconds that the truncated IP method has required for solving all these problems. The third figure contains the average number of iterations that the PCG method requires during the application of the truncated IP algorithm. We also show the dimension of the system (3.6) to be solved in each iteration, according to the procedure explained above. The results displayed in the first two figures show that the truncated IP method is quite efficient for this type of LCPs. Furthermore the performance of the algorithm does not seem to be quite affected by an increase of the dimension of the LCPs. On the other hand the average number of iterations of the PCG method grows linearly with the order of the LCP and is allways smaller than $0.05p$, where p is the order of the system to be solved. We recommend [23] for a full report of the performance of the truncated IP algorithm for this type of monotone LCP when M is symmetric or unsymmetric. These results show the great efficiency of the interior-point in this instance.

4. CONCLUDING REMARKS

In this paper we have investigated the use of an infeasible interior-point algorithm for the solution of the Linear Complementarity Problem (LCP). The results presented in this paper and elsewhere [10, 18, 23] indicate that the method is quite efficient for solving monotone LCPs and LCPs with P matrices and other subclasses of P_0 matrices. However, it has not been able to process some NP-hard LCPs with P_0 matrices. The algorithm also faces difficulties for solving LCPs associated with nonzero sum bimatrix games , despite being successful in many cases.

A robust nonenumerative method for solving a large class of nonmonotone LCPs is still to be designed. We hope that this paper will increase the interest of the research community on the nonmonotone LCP and some of its generalizations that appear quite often in the solution of nonlinear programs [21] variational inequalities [3] and global optimization [12, 13].

ACKNOWLEDGMENT

We are grateful to Luís Portugal for is suggestions on a first draft of this manuscript.

REFERENCES

[1] R. Asmuth, B. Eaves, and E. Peterson, "Computing economic equilibria on affine networks with Lemke's algorithm". *Mathematics of Operations Research*, 4:209–214, 1979.

[2] R. L. Burden, J. D. Faires, and A. Reynolds, *"Numerical Analysis"*. Prindle, Weber and Schmidt, Boston, 1981.

[3] R. Cottle, J. Pang, and R. Stone, *"The Linear Complementarity Problem"*. Academic Press, New York, 1992.

[4] R. Cottle, J. Pang, and Venkateswaran, "Sufficient matrices and the linear complementarity problem". *Linear Algebra and its Applications*, 114:231–249, 1989.

[5] J. Dennis Jr. and R. Schnabel, *"Numerical Methods for Unconstrained Optimization and Nonlinear Equations"*. Prentice-Hall, Englewood Cliffs, New York, 1983.

[6] I. Duff, A. Erisman, and J. Reid, *"Direct Methods for Sparse Matrices"*. Clarendon Press, Oxford, 1986.

[7] B. Eaves, "The linear complementarity problem". *Management Science*, 17:612–634, 1971.

[8] A. El-Bakry, R. Tapia, Y. Zhang, and T. Touchiya, "On the formulation of the primal-dual interior-point method for nonlinear programming for linear complementarity problems". TR 92-40, Rice University, USA, 1995. (revised April).

[9] Y. Fathi, "Computational complexity of linear complementarity problems associated with positive definite matrices". *Mathematical Programming*, 17:335–344, 1979.

[10] L. Fernandes, J. Júdice, and J. Patrício, "An investigation of interior-point and block pivoting algorithms for large-scale symmetric monotone linear complementarity problems". *Computational Optimization and Applications*, 5:49–77, 1996.

[11] F. Guder, J. Morris, and S. Yoon, "Parallel and serial successive overrelaxation for multicommodity spatial price equilibrium problems". *Transportation Science*, 26:48–58, 1992.

[12] J. Judice and A. Faustino, "A computational analysis of LCP methods for bilinear and concave quadratic programming". *Computers and Operations Research*, 18:645–654, 1991.

[13] J. Judice and A. Faustino, "A sequential LCP algorithm for bilinear linear programming". *Annals of Operations Research*, 34:89–106, 1992.

[14] M. Kojima, N. Meggido, and S. Mizuno, "A primal-dual infeasible interior-point algorithm for linear programming". *Mathematical Programming*, 61:263–280, 1993.

[15] M. Kojima, N. Megiddo, T. Noma, and A. Yoshise, *"A Unified Approach to Interior-Point Algorithms for Linear Complementarity Problems"*. Lecture Notes in Computer Science 538. Springer-Verlag, Berlin, 1991.

[16] C. Lemke. "On complementary pivot theory", In G. Dantzig and A. Veinott, editors, *Mathematics of Decision Sciences*, pages 95–114. American Mathematical Society, Providence, 1968.

[17] C. Lemke and J. Howson Jr, "Equilibrium points of bimatrix games". *SIAM Journal of Applied Mathematics*, 12:413–423, 1964.

[18] A. P. Lima, "A Study of the Linear Complementarity Problem". Master Thesis (in Portuguese), University of Porto, 1995.

[19] Y. Lin and J. Pang, "Iterative methods for large convex quadratic programs: a survey". *SIAM Journal on Control and Optimization*, 25:383–411, 1987.

[20] L. Lustig, R. Marsten, and D. Shanno, "Computational experience with a primal-dual method for linear programming". *Linear Algebra and its Applications*, 152:191–222, 1991.

[21] K. Murty, *"Linear Complementarity, Linear and Nonlinear Programming"*. Heldermann Verlag, Berlin, 1988.

[22] J. Pang, "A hybrid method for the solution of some multicommodity spatial equilibrium problems". *Management Science*, 27:1142–1157, 1981.

[23] L. Portugal, L.Fernandes, and J. Judice, "A truncated Newton interior-point algorithm for the solution of a multicommodity spatial equilibrium model". Technical report, Department of Mathematics, University of Coimbra, Coimbra, Portugal, 1995.

[24] E. Simantiraki and D. Shanno, "An infeasible interior-point method for linear complementarity problems". RRR 7-95, RUTCOR, New Brunswick, New Jersey, USA, 1995.

[25] H. Valiaho. "P_*-matrices are just sufficient", Technical report, Department of Mathematics, University of Helsinki, Finland, 1995.

[26] Y. Zhang. "On the convergence of a class of infeasible interior-point methods for the horizontal linear complementarity problem", *SIAM Jounal on Optimization*, 4:208–227, 1994.

ERGODIC RESULTS IN SUBGRADIENT OPTIMIZATION

Larsson Torbjörn,[1] Patriksson Michael,[2] and Strömberg Ann-Brith[3]

[1,3] Division of Optimization
Department of Mathematics
Linköping Institute of Technology
S-581 83 Linköping, Sweden
[1] e-mail: tl@math.liu.se
[3] e-mail: anstr@math.liu.se
[2] Department of Mathematics, Box 4350
University of Washington
Seattle, WA 98195-4350
e-mail: patrikss@math.washington.edu or mipat@math.liu.se

Abstract. Subgradient methods are popular tools for nonsmooth, convex minimization, especially in the context of Lagrangean relaxation; their simplicity has been a main contribution to their success. As a consequence of the nonsmoothness, it is not straightforward to monitor the progress of a subgradient method in terms of the approximate fulfilment of optimality conditions, since the subgradients used in the method will, in general, *not* accumulate to subgradients that verify optimality of a solution obtained in the limit. Further, certain supplementary information, such as convergent estimates of Lagrange multipliers, is not directly available in subgradient schemes.

As a means for overcoming these weaknesses of subgradient optimization methods, we introduce the computation of an ergodic (averaged) sequence of subgradients. Specifically, we consider a nonsmooth, convex program solved by a conditional subgradient optimization scheme (of which the traditional subgradient optimization method is a special case) with divergent series step lengths, which generates a sequence of iterates that converges to an optimal solution. We show that the elements of the ergodic sequence of subgradients in the limit fulfil the optimality conditions at this optimal solution. Further, we use the convergence properties of the ergodic sequence of subgradients to establish convergence of an ergodic sequence of Lagrange multipliers. Finally, some potential applications of these ergodic results are briefly discussed.

Key Words. Nonsmooth minimization, Conditional subgradient optimization, Ergodic sequences, Lagrange multipliers.

1. INTRODUCTION

1.1. Background and motivation

Let the function $f : \mathbb{R}^n \mapsto \mathbb{R}$ be nonsmooth and convex and thus everywhere continuous and subdifferentiable. Further, let $X \subseteq \mathbb{R}^n$ be a nonempty, closed and convex set, and assume that $\inf_{\mathbf{x}\in X} f(\mathbf{x})$ is attained at some $\mathbf{x} \in X$. These properties are assumed to hold throughout the paper. The program considered is

[P]

$$f^* = \min_{\mathbf{x}\in X} f(\mathbf{x}).$$

Defining the *lower level set* of the function f as

$$L_f(z) = \{\, \mathbf{x} \in \mathbb{R}^n \mid f(\mathbf{x}) \leq z \,\}, \quad z \in \mathbb{R},$$

the *solution set* of the program [P] is the nonempty, closed and convex set $X^* = X \cap L_f(f^*)$.

To enable the statement of optimality conditions for the program [P], we first give some basic definitions (e.g., Rockafellar, 1970a, and Hiriart-Urruty and Lemaréchal, 1993a). The *subdifferential* of f at $\mathbf{x} \in \mathbb{R}^n$ is the nonempty, convex and compact set

$$\partial f(\mathbf{x}) = \left\{ \gamma \in \mathbb{R}^n \;\middle|\; f(\mathbf{y}) \geq f(\mathbf{x}) + \gamma^T(\mathbf{y} - \mathbf{x}), \quad \forall \mathbf{y} \in \mathbb{R}^n \right\}, \tag{1.1}$$

the elements of which are called *subgradients*. The *normal cone* to the set X at some $\mathbf{x} \in X$ is (e.g., Lemaréchal and Hiriart-Urruty, 1993a, Definition III:5.2.3) the set

$$N_X(\mathbf{x}) = \left\{ \boldsymbol{\nu} \in \mathbb{R}^n \;\middle|\; \boldsymbol{\nu}^T(\mathbf{y} - \mathbf{x}) \leq 0, \quad \forall \mathbf{y} \in X \right\}, \tag{1.2}$$

which is nonempty, closed, and convex. The *tangent cone* to the set X at $\mathbf{x} \in X$ (ibid., Definition III:5.1.1) may be expressed as the *polar cone* to $N_X(\mathbf{x})$ (ibid., Corollary III:5.2.5), that is, the set

$$T_X(\mathbf{x}) = \left\{ \boldsymbol{\tau} \in \mathbb{R}^n \;\middle|\; \boldsymbol{\tau}^T\boldsymbol{\nu} \leq 0, \quad \forall \boldsymbol{\nu} \in N_X(\mathbf{x}) \right\}, \tag{1.3}$$

which is nonempty, closed and convex. Conversely, $N_X(\mathbf{x})$ is the polar cone to $T_X(\mathbf{x})$ (ibid., Proposition III:5.2.4). If $\mathbf{x} \in \operatorname{int} X$, then $N_X(\mathbf{x}) = \{\mathbf{0}\}$ and $T_X(\mathbf{x}) = \mathbb{R}^n$.

Optimality conditions for the program [P] are given in the next proposition (ibid., Theorem VII:1.1.1 and Condition VII:(1.1.3)). Here, and in the following,

$$P_S(\mathbf{v}) = \arg\min_{\mathbf{u}\in S} \| \mathbf{u} - \mathbf{v} \|$$

denotes the *Euclidean projection* of a vector $\mathbf{v} \in \mathbb{R}^n$ onto a nonempty, closed and convex set $S \subseteq \mathbb{R}^n$.

Proposition 1.1 (Optimality conditions). $\mathbf{x} \in X^*$ *if and only if there exists a* $\gamma \in \partial f(\mathbf{x})$ *such that* $P_{-T_X(\mathbf{x})}(\gamma) = \mathbf{0}$ *or, equivalently,* $\partial f(\mathbf{x}) \cap -N_X(\mathbf{x}) \neq \emptyset$. □

The next proposition states important properties of the set $\partial f(\mathbf{x}) \cap -N_X(\mathbf{x})$, $\mathbf{x} \in X^*$, and follows from the results given above and a result of Burke and Ferris (1991, Lemma 2). It will be useful later in our analysis.

Proposition 1.2 *For any* $\mathbf{x} \in X^*$, *the nonempty set* $\partial f(\mathbf{x}) \cap -N_X(\mathbf{x})$ *is convex and compact, and* $\partial f(\mathbf{y}) \cap -N_X(\mathbf{y}) = \partial f(\mathbf{x}) \cap -N_X(\mathbf{x})$ *for all* $\mathbf{y} \in X^*$. □

The following proposition, which is easily established, gives a geometric interpretation of the elements of the set $\partial f(\mathrm{x}) \cap -N_X(\mathrm{x})$, $\mathrm{x} \in X^*$.

Proposition 1.3 (Separation). *Let* $\mathrm{x} \in X^*$. *Then, the hyperplane*

$$H_\gamma(\mathrm{x}) = \left\{ \mathrm{y} \in \mathbb{R}^n \,\middle|\, \gamma^T(\mathrm{y} - \mathrm{x}) = 0 \right\}$$

separates X *and* $L_f(f^*)$ *if and only if* $\gamma \in \partial f(\mathrm{x}) \cap -N_X(\mathrm{x})$. □

As compared to the case of smooth convex minimization, the task of determining whether or not a certain $\mathrm{x} \in X$ solves the program [P] is, in general, computationally more expensive; recall in particular that even if an optimal solution $\mathrm{x} \in X^*$ is at hand, it is, in general, nontrivial to find a $\gamma \in \partial f(\mathrm{x})$ that verifies optimality (e.g., satisfies the conditions of Proposition 1.1). Both these tasks may be performed by, for example, solving the optimization problem

$$v(\mathrm{x}) = \min_{\gamma \in \partial f(\mathrm{x})} \left\| P_{-T_X(\mathrm{x})}(\gamma) \right\|^2.$$

The value $v(\mathrm{x})$ is zero if and only if $\mathrm{x} \in X^*$.[1] If, for example, the subdifferential set and the tangent cone are polyhedral with explicit representations, then the function v is evaluated by solving a quadratic program. If the function f is implicitly defined, for example when the program [P] is a Lagrangean dual problem, one may have to invoke a column (i.e., a subgradient) generation scheme to enable the evaluation of v.

We consider *subgradient optimization methods* for solving the program [P]. These methods originate in a work by N.Z. Shor from 1962; see Shor (1991) for a review of the early history of nonsmooth optimization. These methods are conceptually simple, in general easy to implement, and have shown to be practically useful. They have therefore been frequently applied, particularly in connection with Lagrangean duality (e.g., Fisher, 1981 and 1985, and Beasley, 1993).

Given an initial iterate $\mathrm{x}^0 \in X$, the traditional subgradient optimization method for solving the program [P] (e.g., Polyak, 1969, or Shor, 1985) is given by the iteration formula

$$\mathrm{x}^{t+\frac{1}{2}} = \mathrm{x}^t - \alpha_t \gamma(\mathrm{x}^t), \quad \mathrm{x}^{t+1} = P_X\left(\mathrm{x}^{t+\frac{1}{2}}\right), \quad t = 0, 1, \ldots, \tag{1.4}$$

where $\gamma(\mathrm{x}^t) \in \partial f(\mathrm{x}^t)$ and $\alpha_t > 0$ is the step length.

For the case of unconstrained optimization, Ermol'ev (1966) proved convergence of the method (1.4) using step lengths according to a divergent series, and Polyak (1967 and 1969) extended it to the case of convex constraints and gave additional convergence results.

Even though subgradient optimization methods are proven convergent, the objective value can not be guaranteed to descend in each iteration.[2] Further, in general $\{P_{-T_X(\mathrm{x}^t)}(\gamma(\mathrm{x}^t))\} \not\to \mathbf{0}$, and even if $\mathrm{x}^t \in X^*$ it is quite likely that $P_{-T_X(\mathrm{x}^t)}(\gamma(\mathrm{x}^t)) \neq \mathbf{0}$. Hence, the progress of subgradient optimization methods can *not* be monitored in terms of the approximate fulfilment of optimality conditions (e.g., those of Proposition 1.1).

In contrast, termination criteria in descent methods (for smooth or nonsmooth minimization) are often based on the approximate fulfilment of optimality conditions. Further, such methods enable the calculation of supplementary and useful information. An example of this is the convergent lower bound on the optimal value in the method of Frank

[1]The function $v : X \mapsto \mathbb{R}_+$ is an example of a *merit* function for the problem [P].

[2]However, their computational burden in each iteration is low in comparison with that of descent methods for nonsmooth minimization, and a larger number of iterations can therefore be allowed.

and Wolfe (1956) for constrained, smooth, convex minimization. Another example is the availability of convergent multiplier estimates from the solutions to the direction finding subproblems in bundle methods; their values are of particular interest in applications where the multipliers have physical interpretations (e.g., Ben-Tal and Bendsøe, 1993).

A frequently used technique for inducing properties that an original sequence is lacking is to construct an auxiliary, *ergodic*, sequence from the original one, that is, a sequence with elements that are weighted averages of those of the original sequence. Typical applications of such averaging processes are the smoothing of oscillations in a nonconvergent sequence and the filtering out of stochastic noise (see, e.g., Kornai and Lipták, 1965; Dunn, 1973; and Powell and Sheffi, 1982).

We will show that the elements of an ergodic sequence of subgradients will, in the limit, fulfil the optimality conditions of Proposition 1.1; hence, the construction of such a sequence introduces the possibility to employ a termination criterion which is related to the approximate fulfilment of these conditions and, further, it enables the calculation of convergent Lagrange multiplier estimates. Subgradient optimization methods thus have an inherent capability of utilizing information assembled during the iteration process. This observation reveals a new characteristic of subgradient optimization methods and deepens the understanding of their behaviour. The practical usefulness of the ergodic results in this work will be demonstrated in a subsequent report.

1.2. Basic concepts

The ergodic results to be presented are derived for the conditional subgradient optimization method (Larsson et al., 1992), which generalizes the method (1.4) in the sense that the feasible set X may be taken into account when determining the step direction. It was shown (ibid.) that the practical performance of the resulting method when applied to three types of structured optimization models is better than that of the method (1.4).

The basis for the development of the generalized method is a relaxation of the subdifferential definition (1.1). The *conditional subdifferential* of f with respect to the feasible set X at $\mathbf{x} \in X$ (see Dem'yanov and Šomesova, 1978 and 1980) is the nonempty, closed and convex set

$$\partial^X f(\mathbf{x}) = \left\{ \gamma \in \mathbb{R}^n \;\middle|\; f(\mathbf{y}) \geq f(\mathbf{x}) + \gamma^T(\mathbf{y} - \mathbf{x}), \quad \forall \mathbf{y} \in X \right\},$$

the elements of which will be referred to as *conditional subgradients*. Clearly, for all $\mathbf{x} \in X$, $\partial^X f(\mathbf{x}) \supseteq \partial f(\mathbf{x})$. Dem'yanov and Šomesova (1978 and 1980) show the following result.

Proposition 1.4 (Conditional subdifferential). $\partial^X f(\mathbf{x}) = \partial f(\mathbf{x}) + N_X(\mathbf{x})$, $\mathbf{x} \in X$. □

The optimality conditions in Proposition 1.1 can be stated in terms of the conditional subdifferential.

Proposition 1.5 (Optimality condition). $\mathbf{x} \in X^*$ *if and only if* $\mathbf{0} \in \partial^X f(\mathbf{x})$. □

Given an initial iterate $\mathbf{x}^0 \in X$, the *conditional subgradient optimization* method (Larsson et al., 1992) is defined by the iteration formula

$$\mathbf{x}^{t+\frac{1}{2}} = \mathbf{x}^t - \alpha_t \gamma^X(\mathbf{x}^t), \quad \mathbf{x}^{t+1} = P_X\left(\mathbf{x}^{t+\frac{1}{2}}\right), \quad t = 0, 1, \ldots, \tag{1.5}$$

where $\alpha_t > 0$ and $\gamma^X(\mathbf{x}^t)$ is a conditional subgradient of f at $\mathbf{x}^t$, that is,

$$\gamma^X(\mathbf{x}^t) = \gamma(\mathbf{x}^t) + \boldsymbol{\nu}(\mathbf{x}^t), \quad \text{with} \quad \gamma(\mathbf{x}^t) \in \partial f(\mathbf{x}^t) \quad \text{and} \quad \boldsymbol{\nu}(\mathbf{x}^t) \in N_X(\mathbf{x}^t).$$

Note that the vector $\mathbf{x}^{t+\frac{1}{2}} - \mathbf{x}^{t+1} \in N_X(\mathbf{x}^{t+1})$. Clearly, if $\{\boldsymbol{\nu}(\mathbf{x}^t)\} = \{\mathbf{0}\}$, then the method (1.5) reduces to the method (1.4). In the case where the set X is stated in terms of explicit constraints (and fulfils some constraint qualification) the normal vector $\boldsymbol{\nu}(\mathbf{x}^t)$ is a nonnegative linear combination of the outward normals of the constraints that are active at $\mathbf{x}^t$.

Choosing $\boldsymbol{\nu}(\mathbf{x}^t) = P_{N_X(\mathbf{x}^t)}(-\gamma(\mathbf{x}^t))$ results in $\gamma^X(\mathbf{x}^t) = P_{-T_X(\mathbf{x}^t)}(\gamma(\mathbf{x}^t))$ (see Larsson et al., 1992, Lemma 3.2), defining the special case of the method (1.5) that is called the *subgradient projection* method; for the case of a polyhedral feasible set, the direction $-\gamma^X(\mathbf{x}^t)$ from $\mathbf{x}^t$ is then feasible.

We consider step lengths satisfying the divergent series conditions

$$\alpha_t > 0, \ \forall t, \quad \{\alpha_t\} \to 0, \quad \text{and} \quad \sum_{t=0}^{\infty} \alpha_t = \infty, \tag{1.6}$$

and the additional condition

$$\sum_{t=0}^{\infty} \alpha_t^2 < \infty. \tag{1.7}$$

The following convergence result, where the conditions (1.6) yield $\{f(\mathbf{x}^t)\} \to f^*$ and the addition of the condition (1.7) leads to the convergence also of the sequence $\{\mathbf{x}^t\}$, is given by Larsson et al. (1992, Theorem 2.7).

Proposition 1.6 (Convergence of conditional subgradient optimization). *Let the method (1.5)–(1.7) be applied to the program* [P]. *If the sequences* $\{\gamma(\mathbf{x}^t)\}$ *and* $\{\boldsymbol{\nu}(\mathbf{x}^t)\}$ *are bounded, then* $\{\mathbf{x}^t\} \to \mathbf{x}^\infty \in X^*$. □

The boundedness of the sequence $\{\gamma(\mathbf{x}^t)\}$ is ensured if the program [P] has some suitable additional property, for example that X is bounded or that f is piecewise linear with a finite number of pieces. In contrast, boundedness of the sequence $\{\boldsymbol{\nu}(\mathbf{x}^t)\}$ may always be imposed.[3] If subgradient projection is used, then $\|\boldsymbol{\nu}(\mathbf{x}^t)\| \leq \|\gamma(\mathbf{x}^t)\|$ so that the sequence $\{\boldsymbol{\nu}(\mathbf{x}^t)\}$ is bounded whenever this is the case for the sequence $\{\gamma(\mathbf{x}^t)\}$.

To establish convergence of the method (1.5) for *adaptive* step length selection rules, based on line searches or formulas involving estimates of the optimal value, the *almost complete relaxation* strategy of Dem'yanov and Vasil'ev (1985, Sect. 3.4) can be used (Corollary 2.8 in Larsson et al., 1992). It works as follows. Define two sequences, $\{\underline{\alpha}_t\}$ and $\{\overline{\alpha}_t\}$, with $\underline{\alpha}_t \leq \overline{\alpha}_t$ for all t, and both satisfying the conditions (1.6) and (1.7).[4] Suppose that, at some iteration t, an adaptive rule has generated a (tentative) step length β_t. The step length α_t used is defined as the projection of β_t onto the interval $[\underline{\alpha}_t, \overline{\alpha}_t]$. The sequence $\{\alpha_t\}$, resulting from this safeguarding strategy, satisfies the conditions (1.6) and (1.7).[5]

Now, consider the application of the method (1.5)–(1.7) to the program [P] and suppose that during the first t iterations the subgradients $\{\gamma(\mathbf{x}^s)\}_{s=0}^{t-1}$ have been used to define the step directions. We then define the sequence $\{\mathbf{g}^t\}$ of weighted averages of subgradients used up to iteration t as

$$\mathbf{g}^t = \sum_{s=0}^{t-1} \frac{\alpha_s}{\sum_{r=0}^{t-1} \alpha_r} \gamma(\mathbf{x}^s), \quad t = 1, 2, \ldots.$$

[3]For example, by choosing the elements to satisfy $\|\boldsymbol{\nu}(\mathbf{x}^t)\| \leq c$ for some $c \geq 0$.

[4]It may, for example, be appropriate to let the sequences be given by $\underline{\alpha}_t = \underline{c}/(t+1)$ and $\overline{\alpha}_t = \overline{c}/(t+1)$, $t = 0, 1, \ldots$, where $\underline{c} > 0$ and $\overline{c} > 0$ are very small and very large constants, respectively.

[5]Traditionally, the divergent series conditions (1.6) and (1.7) are fulfilled through the choice of step lengths according to some appropriate and a priori determined formula, e.g., $\alpha_t = 1/(t+1)$, $t = 0, 1, \ldots$.

It will be shown (in Theorem 3.4) that, provided that the sequences $\{\gamma(\mathbf{x}^t)\}$ and $\{\nu(\mathbf{x}^t)\}$ are bounded, the elements of the ergodic sequence $\{\mathbf{g}^t\}$ in the limit fulfil the optimality conditions given in Proposition 1.1.

Note that the sequence $\{\mathbf{g}^t\}$ is calculated beside the sequence $\{\mathbf{x}^t\}$ and that it therefore does *not* affect the course of the method (1.5)–(1.7). The use of the sequence $\{\mathbf{g}^t\}$ for accelerating subgradient optimization algorithms is a subject for further research.

1.3 Outline

In Sect. 2 we define ergodic sequences of subgradients, normal cone elements, and objective values evaluated in the method (1.5), and of certain auxiliary functions. We also state convergence properties of the sequences of auxiliary functions. Then, in Sect. 3, we derive the main result that the elements of the ergodic sequence of subgradients tend toward the set $\partial f(\mathbf{x}) \cap -N_X(\mathbf{x})$, for some optimal solution $\mathbf{x}$. We also give ergodic results for sequences of conditional subgradients. Sect. 4 contains convergence results for an ergodic sequence of Lagrange multipliers. It is shown that, under a strong Slater constraint qualification, the elements of this ergodic sequence tend toward the set of Lagrange multipliers at an optimal solution; a corresponding result holds for multipliers associated with objective pieces. Finally, in Sect. 5, we discuss some opportunities for further research.

2. TECHNICAL PRELIMINARIES

We first define the sequences to be studied; these are ergodic sequences of the subgradients and the normal cone elements, respectively, generated in the method (1.5). The weights employed in the construction of the ergodic sequences are based on the step lengths used in this method. Then, we establish properties of sequences of certain affine functions in order to derive convergence results for the ergodic sequence of subgradients.

We denote by A_t the cumulative step length of the method (1.5) up to iteration t, that is,

$$A_t = \sum_{s=0}^{t-1} \alpha_s, \quad t = 1, 2, \dots. \tag{2.1}$$

Definition 2.1 (Ergodic sequence of subgradients). *Let the method (1.5) be applied to the program* [P]. *The ergodic sequence* $\{\mathbf{g}^t\}$ *of subgradients is defined by*

$$\mathbf{g}^t = A_t^{-1} \sum_{s=0}^{t-1} \alpha_s \gamma(\mathbf{x}^s), \quad t = 1, 2, \dots. \qquad \square$$

This ergodic sequence of subgradients is of primary interest in this work. Clearly, each vector $\mathbf{g}^t$ is a convex combination of the subgradients $\{\gamma(\mathbf{x}^s)\}_{s=0}^{t-1}$. A recursive computation of $\mathbf{g}^t$ can be implemented using the formula[6]

$$\mathbf{g}^1 = \gamma(\mathbf{x}^0), \quad \mathbf{g}^t = \frac{A_{t-1}}{A_t}\mathbf{g}^{t-1} + \frac{\alpha_{t-1}}{A_t}\gamma(\mathbf{x}^{t-1}), \quad t = 2, 3, \dots.$$

Definition 2.2 (Ergodic sequence of normal vectors). *Let the method (1.5) be applied to the program* [P]. *The ergodic sequence of normal terms of the conditional subgradients is defined by*

$$\mathbf{n}_c^t = A_t^{-1} \sum_{s=0}^{t-1} \alpha_s \nu(\mathbf{x}^s), \quad t = 1, 2, \dots,$$

[6]This remark applies to all ergodic sequences to be defined in this report.

the ergodic sequence of the projection steps is defined by

$$\mathbf{n}_p^t = A_t^{-1} \sum_{s=0}^{t-1} (\mathbf{x}^{s+\frac{1}{2}} - \mathbf{x}^{s+1}), \quad t = 1, 2, \ldots,$$

and the ergodic sequence of normal vectors is then defined by

$$\mathbf{n}^t = \mathbf{n}_c^t + \mathbf{n}_p^t, \quad t = 1, 2, \ldots. \qquad \square$$

Clearly, $\mathbf{n}_c^t$ is a convex combination of the normal terms of the conditional subgradients used up to iteration t of the method (1.5) while $\mathbf{n}_p^t$ is a positive linear combination of its projection steps. If the traditional subgradient optimization method (1.4) is employed then, for all t, $\mathbf{n}_c^t = \mathbf{0}$ and $\mathbf{n}^t = \mathbf{n}_p^t$.

The following result is the first step towards a convergence result for the ergodic sequence $\{\mathbf{g}^t\}$.

Lemma 2.3 (Boundedness of $\{\mathbf{g}^t\}$ and $\{\mathbf{n}^t\}$). *Let the sequences $\{\mathbf{g}^t\}$ and $\{\mathbf{n}^t\}$ be given by Definitions 2.1 and 2.2, respectively. If the sequences $\{\gamma(\mathbf{x}^s)\}$ and $\{\boldsymbol{\nu}(\mathbf{x}^s)\}$ are bounded, then this is also true for the sequences $\{\mathbf{g}^t\}$ and $\{\mathbf{n}^t\}$.*

Proof. Since the elements $\mathbf{g}^t$ are convex combinations of $\{\gamma(\mathbf{x}^s)\}_{s=0}^{t-1}$, the sequence $\{\mathbf{g}^t\}$ is also bounded. Analogously, since each $\mathbf{n}_c^t$ is a convex combination of $\{\boldsymbol{\nu}(\mathbf{x}^s)\}_{s=0}^{t-1}$, the sequence $\{\mathbf{n}_c^t\}$ is also bounded. By the iteration formula (1.5), $\|\mathbf{x}^{s+\frac{1}{2}} - \mathbf{x}^{s+1}\| \leq \alpha_s \|\gamma^X(\mathbf{x}^s)\| \leq \alpha_s(\|\gamma(\mathbf{x}^s)\| + \|\boldsymbol{\nu}(\mathbf{x}^s)\|)$ for all s. Using the same arguments as above, it follows that the sequence $\{\mathbf{n}_p^t\}$ is bounded. Hence, the sequence $\{\mathbf{n}^t\}$ is bounded. $\square$

The ergodic sequence of objective values obtained in the method (1.5) is defined as

$$\varphi_t = A_t^{-1} \sum_{s=0}^{t-1} \alpha_s f(\mathbf{x}^s), \quad t = 1, 2, \ldots. \tag{2.2}$$

It follows that $\varphi_t \geq f^*$ for all t, since, for all s, $\mathbf{x}^s \in X$.

Utilizing that, for all s, $\gamma(\mathbf{x}^s) \in \partial f(\mathbf{x}^s)$, and the definition (1.1), we may derive the *surrogate inequalities*

$$\sum_{s=0}^{t-1} \alpha_s f(\mathbf{x}) \geq \sum_{s=0}^{t-1} \alpha_s \left(f(\mathbf{x}^s) + \gamma(\mathbf{x}^s)^T (\mathbf{x} - \mathbf{x}^s) \right), \quad \mathbf{x} \in \mathbb{R}^n, \quad t = 1, 2, \ldots,$$

and define the *affine* functions $f_t : \mathbb{R}^n \mapsto \mathbb{R}$, $t = 1, 2, \ldots$, as

$$f_t(\mathbf{x}) = A_t^{-1} \sum_{s=0}^{t-1} \alpha_s \left(f(\mathbf{x}^s) + \gamma(\mathbf{x}^s)^T (\mathbf{x} - \mathbf{x}^s) \right), \quad \mathbf{x} \in \mathbb{R}^n. \tag{2.3}$$

Then, $f_t(\mathbf{x}) \leq f(\mathbf{x})$ holds for all $\mathbf{x} \in \mathbb{R}^n$ and all t, that is, each function f_t is an affine *underestimate* of the objective f. Clearly, the hyperplane defined by the function f_t is a weighted average of all supporting hyperplanes generated in the previous iterations of the conditional subgradient optimization method (1.5). From Definition 2.1 follows that $\nabla f_t(\mathbf{x}) = \mathbf{g}^t$ and $f_t(\mathbf{x}) = f_t(\mathbf{y}) + (\mathbf{g}^t)^T(\mathbf{x} - \mathbf{y})$ for any $\mathbf{x}, \mathbf{y} \in \mathbb{R}^n$ and all t.

We let the convex functions $\delta_t : \mathbb{R}^n \mapsto \mathbb{R}$, $t = 1, 2, \ldots$, describe the pointwise vertical distance between the functions f and f_t, that is,

$$\delta_t(\mathbf{x}) = f(\mathbf{x}) - f_t(\mathbf{x}), \quad \mathbf{x} \in \mathbb{R}^n. \tag{2.4}$$

It follows that $\delta_t(\mathbf{x}) \geq 0$ for all $\mathbf{x} \in \mathbb{R}^n$ and all t.

Further, we define the affine functions $\varrho_t : \mathbb{R}^n \mapsto \mathbb{R}$, $t = 1, 2, \ldots$, as

$$\varrho_t(\mathbf{x}) = A_t^{-1} \sum_{s=0}^{t-1} \left(\alpha_s \boldsymbol{\nu}(\mathbf{x}^s)^T (\mathbf{x}^s - \mathbf{x}) + \left(\mathbf{x}^{s+\frac{1}{2}} - \mathbf{x}^{s+1}\right)^T \left(\mathbf{x}^{s+1} - \mathbf{x}\right)\right), \quad \mathbf{x} \in \mathbb{R}^n. \quad (2.5)$$

Using that, for all s, $\boldsymbol{\nu}(\mathbf{x}^s) \in N_X(\mathbf{x}^s)$ and $\mathbf{x}^{s+\frac{1}{2}} - \mathbf{x}^{s+1} \in N_X(\mathbf{x}^{s+1})$, and the definition (1.2), it follows that $\varrho_t(\mathbf{x}) \geq 0$ for $\mathbf{x} \in X$ and all t, that is, each of these inequalities is a *valid inequality* for the set X. From Definition 2.2 follows that $\nabla \varrho_t(\mathbf{x}) = -\mathbf{n}^t$ and $\varrho_t(\mathbf{x}) = \varrho_t(\mathbf{y}) - (\mathbf{n}^t)^T(\mathbf{x} - \mathbf{y})$ for any $\mathbf{x}, \mathbf{y} \in \mathbb{R}^n$ and all t.

The following lemma is crucial for our continued analysis.

Lemma 2.4 (Convergence of $\{\delta_t(\mathbf{x}^\infty)\}$ and $\{\varrho_t(\mathbf{x}^\infty)\}$ to 0 and of $\{\varphi_t\}$ to f^*). *Let the method (1.5)–(1.7) be applied to the program* [P] *and let the sequences* $\{\delta_t(\mathbf{x})\}$, $\{\varrho_t(\mathbf{x})\}$, *and* $\{\varphi_t\}$ *be given by the definitions (2.4), (2.5), and (2.2), respectively. Suppose that the sequences* $\{\gamma(\mathbf{x}^s)\}$ *and* $\{\boldsymbol{\nu}(\mathbf{x}^s)\}$ *are bounded. Then,* $\{\delta_t(\mathbf{x}^\infty)\} \to 0$, $\{\varrho_t(\mathbf{x}^\infty)\} \to 0$, *and* $\{\varphi_t\} \to f^*$.

Proof. By Proposition 1.6, $\{\mathbf{x}^t\} \to \mathbf{x}^\infty$. From the iteration formula (1.5),

$$\begin{aligned} 0 &\leq \left\|\mathbf{x}^{t+1} - \mathbf{x}^\infty\right\|^2 \leq \left\|\mathbf{x}^{t+1} - \mathbf{x}^\infty\right\|^2 + \left\|\mathbf{x}^{t+\frac{1}{2}} - \mathbf{x}^{t+1}\right\|^2 \\ &= \left\|\mathbf{x}^{t+\frac{1}{2}} - \mathbf{x}^\infty\right\|^2 - 2\left(\mathbf{x}^{t+\frac{1}{2}} - \mathbf{x}^{t+1}\right)^T \left(\mathbf{x}^{t+1} - \mathbf{x}^\infty\right) \\ &= \left\|\mathbf{x}^t - \mathbf{x}^\infty\right\|^2 + \alpha_t^2 \left\|\gamma^X(\mathbf{x}^t)\right\|^2 - 2\left(\alpha_t \gamma^X(\mathbf{x}^t)^T(\mathbf{x}^t - \mathbf{x}^\infty) + \left(\mathbf{x}^{t+\frac{1}{2}} - \mathbf{x}^{t+1}\right)^T \left(\mathbf{x}^{t+1} - \mathbf{x}^\infty\right)\right). \end{aligned}$$

Repeated application of the second inequality and utilization of the definitions (2.4), (2.5), and (2.2), result in the inequality

$$0 \leq \left\|\mathbf{x}^0 - \mathbf{x}^\infty\right\|^2 + \sum_{s=0}^{t-1} \alpha_s^2 \left\|\gamma^X(\mathbf{x}^s)\right\|^2 - 2A_t \left(\delta_t(\mathbf{x}^\infty) + \varrho_t(\mathbf{x}^\infty) + \varphi_t - f^*\right).$$

Since $\delta_t(\mathbf{x}^\infty) \geq 0$, $\varrho_t(\mathbf{x}^\infty) \geq 0$, $\varphi_t \geq f^*$, and $A_t > 0$, it is immediate that

$$0 \leq \delta_t(\mathbf{x}^\infty) + \varrho_t(\mathbf{x}^\infty) + \varphi_t - f^* \leq \frac{1}{2A_t}\left(\left\|\mathbf{x}^0 - \mathbf{x}^\infty\right\|^2 + \sum_{s=0}^{t-1} \alpha_s^2 \left\|\gamma(\mathbf{x}^s) + \boldsymbol{\nu}(\mathbf{x}^s)\right\|^2\right).$$

Let $t \to \infty$ and invoke the conditions (1.6) and (1.7). □

Corollary 2.5 *Let the assumptions of Lemma 2.4 hold. Then,* $\{\delta_t(\mathbf{x}^t)\} \to 0$ *and* $\{\varrho_t(\mathbf{x}^t)\} \to 0$.

Proof. Definition 2.1, and the definitions (2.3) and (2.4), yield

$$0 \leq \delta_t(\mathbf{x}^t) = \delta_t(\mathbf{x}^\infty) + f(\mathbf{x}^t) - f(\mathbf{x}^\infty) - \left(\mathbf{g}^t\right)^T \left(\mathbf{x}^t - \mathbf{x}^\infty\right), \quad t = 1, 2, \ldots.$$

From Definition 2.2, and the definition (2.5), follow that

$$0 \leq \varrho_t(\mathbf{x}^t) = \varrho_t(\mathbf{x}^\infty) - \left(\mathbf{n}^t\right)^T \left(\mathbf{x}^t - \mathbf{x}^\infty\right), \quad t = 1, 2, \ldots.$$

Proposition 1.6, Lemma 2.3, the continuity of f, and Lemma 2.4 yield that the right hand sides of both the above equations tend to zero as t approaches infinity. The result follows. □

A small value of $\delta_t(\mathbf{x}^t) \geq 0$ implies that the affine underestimate f_t of the function f defines a hyperplane that is close to a support to $\mathtt{epi}\, f$ at $(\mathbf{x}^t, f(\mathbf{x}^t))$, and a small value of $\varrho_t(\mathbf{x}^t)$ yields that the hyperplane $\{\mathbf{x} \in \mathbb{R}^n \mid \varrho_t(\mathbf{x}) = 0\}$ is close to a support to X at $\mathbf{x}^t$ (while $X \subseteq \{\mathbf{x} \in \mathbb{R}^n \mid \varrho_t(\mathbf{x}) \geq 0\}$ always holds).

Lemma 2.6 ($\{\mathbf{g}^t+\mathbf{n}^t\}$ tends to $\mathbf{0}$). *Let the method (1.5)–(1.7) be applied to the program* [P] *and let the sequences* $\{\mathbf{g}^t\}$ *and* $\{\mathbf{n}^t\}$ *be given by Definitions 2.1 and 2.2, respectively. Suppose that the sequences* $\{\gamma(\mathbf{x}^s)\}$ *and* $\{\boldsymbol{\nu}(\mathbf{x}^s)\}$ *are bounded. Then,* $\{\mathbf{g}^t + \mathbf{n}^t\} \to \mathbf{0}$.

Proof. By Definitions 2.1 and 2.2, and the iteration formula (1.5),

$$\begin{aligned} \mathbf{g}^t + \mathbf{n}^t &= A_t^{-1} \sum_{s=0}^{t-1} \left(\alpha_s \gamma^X(\mathbf{x}^s) + \mathbf{x}^{s+\frac{1}{2}} - \mathbf{x}^{s+1} \right) \\ &= A_t^{-1} \sum_{s=0}^{t-1} \left(\mathbf{x}^s - \mathbf{x}^{s+\frac{1}{2}} + \mathbf{x}^{s+\frac{1}{2}} - \mathbf{x}^{s+1} \right) \;=\; A_t^{-1} \left(\mathbf{x}^0 - \mathbf{x}^t \right). \end{aligned}$$

Proposition 1.6 yields that $\{\mathbf{x}^t\} \to \mathbf{x}^\infty$. The result then follows from the definition (2.1) and the condition (1.6). □

In our continued analysis we will by $d_Y(\mathbf{x})$ denote the Euclidean distance between a point $\mathbf{x} \in \mathbb{R}^n$ and a convex set $Y \subseteq \mathbb{R}^n$, that is,

$$d_Y(\mathbf{x}) = \min_{\mathbf{y} \in Y} \|\mathbf{y} - \mathbf{x}\| .$$

The function $d_Y : \mathbb{R}^n \mapsto \mathbb{R}_+$ is convex and continuous.

3. ERGODIC RESULTS FOR SUBGRADIENTS

We first show that the distance between $\mathbf{g}^t$ and the set $\partial f(\mathbf{x}^\infty)$ tends to zero. Then, we establish that the projection of $\mathbf{g}^t$ onto the set $-T_X(\mathbf{x}^\infty)$ tends to zero. The combination of these results yields that the elements of the sequence $\{\mathbf{g}^t\}$ in the limit provide fulfilment of the optimality conditions given in Proposition 1.1. Finally, we give corresponding results for an ergodic sequence of *conditional* subgradients.

As concluded above, the affine functions f_t, $t = 1, 2, \ldots$, underestimate the convex function f everywhere, and from Lemma 2.4 follows that $\{f_t(\mathbf{x}^\infty)\} \to f(\mathbf{x}^\infty)$. Hence, in the limit, the sequence $\{f_t\}$ of affine functions provides supporting hyperplanes to $\mathtt{epi}\, f$ at $(\mathbf{x}^\infty, f(\mathbf{x}^\infty))$; this observation is further developed in the following theorem.

Theorem 3.1 (Convergence of $\{\mathbf{g}^t\}$ to $\partial f(\mathbf{x}^\infty)$). *Let the method (1.5)–(1.7) be applied to the program* [P] *and let the sequence* $\{\mathbf{g}^t\}$ *be given by Definition 2.1. Suppose that the sequences* $\{\gamma(\mathbf{x}^s)\}$ *and* $\{\boldsymbol{\nu}(\mathbf{x}^s)\}$ *are bounded. Then,*

$$\left\{ d_{\partial f(\mathbf{x}^\infty)}(\mathbf{g}^t) \right\} \to 0.$$

Proof. The sequence $\{d_{\partial f(\mathbf{x}^\infty)}(\mathbf{g}^t)\}$ is bounded since, by Lemma 2.3, the sequence $\{\mathbf{g}^t\}$ is bounded, and $\partial f(\mathbf{x}^\infty)$ is compact. Hence, we can choose a subsequence $\mathcal{T}$ such that

$$\lim_{\substack{t \in \mathcal{T} \\ t \to \infty}} d_{\partial f(\mathbf{x}^\infty)} \left(\mathbf{g}^t \right) = \limsup_{t \to \infty} d_{\partial f(\mathbf{x}^\infty)} \left(\mathbf{g}^t \right),$$

and let $\tilde{\mathbf{g}}$ be an accumulation point of the the sequence $\{\mathbf{g}^t\}_{t \in \mathcal{T}}$. From the definitions (2.3) and (2.4) follow that, for all t and any $\mathbf{y} \in \mathbb{R}^n$,

$$f(\mathbf{y}) \geq f_t(\mathbf{y}) = f_t(\mathbf{x}^\infty) + \left(\mathbf{g}^t\right)^T (\mathbf{y} - \mathbf{x}^\infty) = f(\mathbf{x}^\infty) + \left(\mathbf{g}^t\right)^T (\mathbf{y} - \mathbf{x}^\infty) - \delta_t(\mathbf{x}^\infty).$$

From Lemma 2.4 then follows that, for any $\mathbf{y} \in \mathbb{R}^n$,

$$\tilde{\mathbf{g}}^T(\mathbf{y}-\mathbf{x}^\infty) \leq \limsup_{\substack{t\in\mathcal{T}\\ t\to\infty}} \left(\mathbf{g}^t\right)^T(\mathbf{y}-\mathbf{x}^\infty) \leq f(\mathbf{y}) - f(\mathbf{x}^\infty) + \lim_{\substack{t\in\mathcal{T}\\ t\to\infty}} \delta_t(\mathbf{x}^\infty) = f(\mathbf{y}) - f(\mathbf{x}^\infty).$$

Hence, $\tilde{\mathbf{g}} \in \partial f(\mathbf{x}^\infty)$, implying that $d_{\partial f(\mathbf{x}^\infty)}(\tilde{\mathbf{g}}) = 0$. By the continuity of the function $d_{\partial f(\mathbf{x}^\infty)}$, $d_{\partial f(\mathbf{x}^\infty)}(\tilde{\mathbf{g}}) = \lim_{t\in\mathcal{T},\, t\to\infty} d_{\partial f(\mathbf{x}^\infty)}(\mathbf{g}^t)$, and the result follows. □

For any $\varepsilon \geq 0$, the *ε-subdifferential* of f at $\mathbf{x} \in \mathbb{R}^n$ is defined as (e.g., Hiriart-Urruty and Lemaréchal, 1993b, Definition XI:1.1.1)

$$\partial_\varepsilon f(\mathbf{x}) = \left\{ \boldsymbol{\gamma} \in \mathbb{R}^n \;\middle|\; f(\mathbf{y}) \geq f(\mathbf{x}) + \boldsymbol{\gamma}^T(\mathbf{y}-\mathbf{x}) - \varepsilon, \quad \forall \mathbf{y} \in \mathbb{R}^n \right\}.$$

For any $\varepsilon_2 > \varepsilon_1 \geq 0$ and any $\mathbf{x} \in \mathbb{R}^n$, $\partial_{\varepsilon_1} f(\mathbf{x}) \subseteq \partial_{\varepsilon_2} f(\mathbf{x})$, and $\partial_0 f(\mathbf{x}) = \partial f(\mathbf{x})$. Using Definition 2.1 and the definitions (2.3) and (2.4) we derive that $\mathbf{g}^t$ is a $\delta_t(\mathbf{x})$-subgradient of f at $\mathbf{x}$ for any $\mathbf{x} \in \mathbb{R}^n$; note in particular that $\mathbf{g}^t$ is a $\delta_t(\mathbf{x}^t)$-subgradient of f at $\mathbf{x}^t$ and that $\{\delta_t(\mathbf{x}^t)\} \to 0$ by Corollary 2.5.

From the definition (2.5) follows that, for every t, $\varrho_t(\mathbf{x}) \geq 0$ is a valid inequality for the convex set X, and from Lemma 2.4 that $\{\varrho_t(\mathbf{x}^\infty)\} \to 0$. Hence, in the limit, the sequence $\{\varrho_t\}$ of affine functions provides supporting hyperplanes to the set X at $\mathbf{x}^\infty$; the next theorem further develops this remark.

Theorem 3.2 (Convergence of $\{\mathbf{g}^t\}$ to $-N_X(\mathbf{x}^\infty)$). *Let the method (1.5)–(1.7) be applied to the program* [P] *and let the sequence* $\{\mathbf{g}^t\}$ *be given by Definition 2.1. Suppose that the sequences* $\{\boldsymbol{\gamma}(\mathbf{x}^s)\}$ *and* $\{\boldsymbol{\nu}(\mathbf{x}^s)\}$ *are bounded. Then,*

$$\left\{d_{-N_X(\mathbf{x}^\infty)}(\mathbf{g}^t)\right\} \to 0.$$

Proof. The sequence $\{d_{N_X(\mathbf{x}^\infty)}(\mathbf{n}^t)\}$ is bounded since, by Lemma 2.3, the sequence $\{\mathbf{n}^t\}$ is bounded, and, by the definition (1.2), $\mathbf{0} \in N_X(\mathbf{x}^\infty)$. Hence, we can choose a subsequence $\mathcal{T}$ such that

$$\lim_{\substack{t\in\mathcal{T}\\ t\to\infty}} d_{N_X(\mathbf{x}^\infty)}\left(\mathbf{n}^t\right) = \limsup_{t\to\infty} d_{N_X(\mathbf{x}^\infty)}\left(\mathbf{n}^t\right),$$

and let $\tilde{\mathbf{n}}$ be an accumulation point of the sequence $\{\mathbf{n}^t\}_{t\in\mathcal{T}}$. From the definition (2.5) follows that, for all t and any $\mathbf{y} \in X$,

$$\left(\mathbf{n}^t\right)^T(\mathbf{y}-\mathbf{x}^\infty) = -\varrho_t(\mathbf{y}) + \varrho_t(\mathbf{x}^\infty) \leq \varrho_t(\mathbf{x}^\infty).$$

From Lemma 2.4 follow that, for any $\mathbf{y} \in X$,

$$\tilde{\mathbf{n}}^T(\mathbf{y}-\mathbf{x}^\infty) \leq \limsup_{\substack{t\in\mathcal{T}\\ t\to\infty}} \left(\mathbf{n}^t\right)^T(\mathbf{y}-\mathbf{x}^\infty) \leq \lim_{\substack{t\in\mathcal{T}\\ t\to\infty}} \varrho_t(\mathbf{x}^\infty) = 0.$$

Hence, $\tilde{\mathbf{n}} \in N_X(\mathbf{x}^\infty)$, implying that $d_{N_X(\mathbf{x}^\infty)}(\tilde{\mathbf{n}}) = 0$. By the continuity of the function $d_{N_X(\mathbf{x}^\infty)}$, $d_{N_X(\mathbf{x}^\infty)}(\tilde{\mathbf{n}}) = \lim_{t\in\mathcal{T},\, t\to\infty} d_{N_X(\mathbf{x}^\infty)}(\mathbf{n}^t)$. Therefore, for any $\varepsilon > 0$ and all t that are sufficiently large, $d_{N_X(\mathbf{x}^\infty)}(\mathbf{n}^t) < \varepsilon/2$, and, by Lemma 2.6, $\|\mathbf{g}^t + \mathbf{n}^t\| < \varepsilon/2$, so that $d_{-N_X(\mathbf{x}^\infty)}(\mathbf{g}^t) \leq d_{-N_X(\mathbf{x}^\infty)}(-\mathbf{n}^t) + \|\mathbf{g}^t + \mathbf{n}^t\| < \varepsilon$. The theorem follows. □

From Definition 2.2 and the definition (2.5) follow that, for any $\mathbf{x} \in X$ and all t,

$$\mathbf{n}^t \in \left\{ \boldsymbol{\nu} \in \mathbb{R}^n \;\middle|\; \boldsymbol{\nu}^T(\mathbf{y}-\mathbf{x}) \leq \varrho_t(\mathbf{x}), \quad \forall \mathbf{y} \in X \right\}.$$

This set is nonempty, closed, and convex, and it clearly contains $N_X(\mathbf{x})$ (cf. the relationship between the ε-subdifferential and the subdifferential).

Theorem 3.3 (Projection of $\mathbf{g}^t$ onto $-T_X(\mathbf{x}^\infty)$ tends to $\mathbf{0}$). *Let the assumptions of Theorem 3.2 hold and let $T_X(\mathbf{x})$ be defined by (1.3). Then,*

$$\left\{P_{-T_X(\mathbf{x}^\infty)}(\mathbf{g}^t)\right\} \to \mathbf{0}.$$

Proof. Applying the Moreau decomposition (Zarantonello, 1971, Lemma 2.2) to $\mathbf{g}^t$, with respect to the polar cones $N_X(\mathbf{x}^\infty)$ and $T_X(\mathbf{x}^\infty)$, yields

$$\mathbf{g}^t = P_{-N_X(\mathbf{x}^\infty)}\left(\mathbf{g}^t\right) + P_{-T_X(\mathbf{x}^\infty)}\left(\mathbf{g}^t\right), \quad \forall t,$$

which implies that $\|P_{-T_X(\mathbf{x}^\infty)}(\mathbf{g}^t)\| = \|\mathbf{g}^t - P_{-N_X(\mathbf{x}^\infty)}(\mathbf{g}^t)\| = d_{-N_X(\mathbf{x}^\infty)}(\mathbf{g}^t)$. The result then follows from Theorem 3.2. □

Theorem 3.4 (Optimality fulfilment). *Let the method (1.5)–(1.7) be applied to the program* [P] *and let the sequence $\{\mathbf{g}^t\}$ be given by Definition 2.1. Suppose that the sequences $\{\gamma(\mathbf{x}^s)\}$ and $\{\boldsymbol{\nu}(\mathbf{x}^s)\}$ are bounded. Then,*

$$\left\{d_{\partial f(\mathbf{x}^\infty)\cap -N_X(\mathbf{x}^\infty)}(\mathbf{g}^t)\right\} \to 0$$

or, equivalently,

$$\left\{d_{\partial f(\mathbf{x}^\infty)}(\mathbf{g}^t)\right\} \to 0 \quad \textit{and} \quad \left\{P_{-T_X(\mathbf{x}^\infty)}(\mathbf{g}^t)\right\} \to \mathbf{0}.$$

Proof. The first statement follows from Theorems 3.1 and 3.2, and the second statement is then implied by Theorem 3.3. □

The corresponding ergodic result for the sequence $\{\mathbf{n}^t\}$ is $\{d_{N_X(\mathbf{x}^\infty)\cap -\partial f(\mathbf{x}^\infty)}(\mathbf{n}^t)\} \to 0$ or, equivalently, $\{d_{-\partial f(\mathbf{x}^\infty)}(\mathbf{n}^t)\} \to 0$ and $\{P_{T_X(\mathbf{x}^\infty)}(\mathbf{n}^t)\} \to \mathbf{0}$.

In the interesting special case of *unconstrained minimization* the optimality conditions given in Proposition 1.1 reduce to $\mathbf{0} \in \partial f(\mathbf{x})$ and Theorem 3.4 reduces to the following.

Corollary 3.5 (Convergence of $\{\mathbf{g}^t\}$ to $\mathbf{0}$). *Suppose that $X = \mathbb{R}^n$ and that the assumptions of Theorem 3.4 hold. Then,*

$$\left\{\mathbf{g}^t\right\} \to \mathbf{0}.$$

□

Finally in this section, we define the ergodic sequence of *conditional* subgradients by

$$\mathbf{g}^t + \mathbf{n}_c^t = A_t^{-1}\sum_{s=0}^{t-1} \alpha_s \gamma^X(\mathbf{x}^s), \quad t = 1, 2, \ldots, \tag{3.1}$$

where $\mathbf{g}^t$ and $\mathbf{n}_c^t$ are given by Definitions 2.1 and 2.2, respectively.

Theorem 3.6 (Optimality fulfilment). *Let the assumptions of Theorem 3.4 hold and let the sequence $\{\mathbf{g}^t + \mathbf{n}_c^t\}$ be defined by (3.1). Then,*

$$\left\{d_{\partial^X f(\mathbf{x}^\infty)\cap -N_X(\mathbf{x}^\infty)}(\mathbf{g}^t + \mathbf{n}_c^t)\right\} \to 0$$

or, equivalently,

$$\left\{d_{\partial^X f(\mathbf{x}^\infty)}(\mathbf{g}^t + \mathbf{n}_c^t)\right\} \to 0 \quad \textit{and} \quad \left\{P_{-T_X(\mathbf{x}^\infty)}(\mathbf{g}^t + \mathbf{n}_c^t)\right\} \to \mathbf{0}.$$

Proof. Analogously to the proof of Theorem 3.2 we can show that $\{d_{N_X(\mathbf{x}^\infty)}(\mathbf{n}_c^t)\} \to 0$ and $\{d_{N_X(\mathbf{x}^\infty)}(\mathbf{n}_p^t)\} \to 0$. From Lemma 2.6 and Definition 2.2, $\{\mathbf{g}^t + \mathbf{n}_c^t + \mathbf{n}_p^t\} \to \mathbf{0}$, which then implies that $\{d_{-N_X(\mathbf{x}^\infty)}(\mathbf{g}^t + \mathbf{n}_c^t)\} \to 0$. From Theorem 3.1 then follows that $\{d_{\partial f(\mathbf{x}^\infty)+N_X(\mathbf{x}^\infty)}(\mathbf{g}^t + \mathbf{n}_c^t)\} \to 0$. Utilizing the polarity of the normal and tangent cones we can, analogously to the proof of Theorem 3.3 show that the projection of $\mathbf{g}^t + \mathbf{n}_c^t$ onto $-T_X(\mathbf{x}^\infty)$ vanishes in the limit. □

The corresponding ergodic result for the sequence $\{\mathbf{n}_p^t\}$ is $\{d_{N_X(\mathbf{x}^\infty)\cap-\partial^X f(\mathbf{x}^\infty)}(\mathbf{n}_p^t)\} \to 0$ or, equivalently, $\{d_{-\partial^X f(\mathbf{x}^\infty)}(\mathbf{n}_p^t)\} \to 0$ and $\{P_{T_X(\mathbf{x}^\infty)}(\mathbf{n}_p^t)\} \to \mathbf{0}$.

4. ERGODIC RESULTS FOR LAGRANGE MULTIPLIERS

Here, we consider the cases when the feasible set and/or the objective function of the program [P] have explicit representations, and utilize the previous ergodic results to construct ergodic sequences of the Lagrange multipliers associated with these representations of the program [P]. In Subsection 4.1 we assume that the feasible set is defined by a finite number of smooth, convex inequality constraints and/or linear equality constraints, and satisfies a strong Slater constraint qualification. The ergodic sequences of multipliers associated with these constraints are shown to converge to Lagrange multipliers. In Subsection 4.2 we assume that the objective function is represented as the pointwise maximum of a finite number of smooth, convex functions. We show that the ergodic sequences of multipliers associated with the pieces of the objective function converge to Lagrange multipliers.

4.1. Multipliers of constraints

Throughout this subsection we assume that the feasible set X of the program [P] is explicitly described as

$$X = \left\{ \mathbf{x} \in \mathbb{R}^n \,\middle|\, a_i(\mathbf{x}) \le 0,\ i \in \mathcal{M};\quad \mathbf{b}_j^T\mathbf{x} = e_j,\ j \in \mathcal{P} \right\}, \tag{4.1}$$

where the functions $a_i : \mathbb{R}^n \mapsto \mathbb{R}$, $i \in \mathcal{M} = \{1, \dots, m\}$, are convex and differentiable, and $\mathbf{b}_j \in \mathbb{R}^n$ and $e_j \in \mathbb{R}$, $j \in \mathcal{P} = \{m+1, \dots, m+p\}$. The convergence analysis of this subsection relies on the assumption that the set X fulfils the following constraint qualification (e.g., Hiriart-Urruty and Lemaréchal, 1993a, Definition VII:2.3.1).

Assumption 4.1 (Strong Slater assumption). *The vectors $\mathbf{b}_j$, $j \in \mathcal{P}$, are linearly independent, and the set X has a nonempty relative interior, that is,*

$$\left\{ \mathbf{x} \in \mathbb{R}^n \,\middle|\, a_i(\mathbf{x}) < 0,\ i \in \mathcal{M};\quad \mathbf{b}_j^T\mathbf{x} = e_j,\ j \in \mathcal{P} \right\} \neq \emptyset. \qquad \square$$

We let, for $\mathbf{x} \in X$, $\mathcal{I}(\mathbf{x}) = \{i \in \mathcal{M} \mid a_i(\mathbf{x}) = 0\}$ denote the set of inequality constraints that are active at $\mathbf{x}$, and $\Pi_X(\mathbf{x})$ the set of multiplier values that satisfy the complementary slackness conditions for the program [P] at $\mathbf{x}$, that is,

$$\Pi_X(\mathbf{x}) = \left\{ \boldsymbol{\pi} \in \mathbb{R}_+^m \times \mathbb{R}^p \,\middle|\, \pi_i = 0,\ i \in \mathcal{M} \setminus \mathcal{I}(\mathbf{x}) \right\} \ni \mathbf{0}.$$

Letting $\boldsymbol{\omega} \in \mathbb{R}^{m+p}$, the normal cone of X at $\mathbf{x} \in X$ can then be expressed as

$$N_X(\mathbf{x}) = \left\{ \boldsymbol{\nu} \in \mathbb{R}^n \,\middle|\, \boldsymbol{\nu} = \sum_{i\in\mathcal{M}} \omega_i \nabla a_i(\mathbf{x}) + \sum_{j\in\mathcal{P}} \omega_j \mathbf{b}_j;\quad \boldsymbol{\omega} \in \Pi_X(\mathbf{x}) \right\}.$$

For any $\mathbf{x} \in X$, let

$$M_X(\mathbf{x}) = \left\{ \boldsymbol{\mu} \in \Pi_X(\mathbf{x}) \;\middle|\; \mathbf{0} \in \partial f(\mathbf{x}) + \sum_{i \in \mathcal{M}} \mu_i \nabla a_i(\mathbf{x}) + \sum_{j \in \mathcal{P}} \mu_j \mathbf{b}_j \right\} \tag{4.2}$$

be the (possibly empty) set of Lagrange multipliers associated with the constraints defining the set X.

The optimality conditions for the program [P] can be stated in terms of the set $M_X(\mathbf{x})$ (e.g., Hiriart-Urruty and Lemaréchal, 1993a, Theorems VII:2.1.4 and VII:2.3.2).[7]

Proposition 4.2 (Optimality conditions). *Suppose that the set X, defined by (4.1), fulfils Assumption 4.1, and let $\mathbf{x} \in X$. Then, $\mathbf{x} \in X^*$ if and only if $M_X(\mathbf{x}) \neq \emptyset$.* □

Proposition 4.3 (Properties of $M_X(\mathbf{x})$, $\mathbf{x} \in X^*$). *Suppose that the set X, defined by (4.1), fulfils Assumption 4.1. For any $\mathbf{x} \in X^*$ the nonempty set $M_X(\mathbf{x})$ is convex and compact, and $M_X(\mathbf{y}) = M_X(\mathbf{x})$ for all $\mathbf{y} \in X^*$.*

Proof. The closedness and convexity of $M_X(\mathbf{x})$ follow from its definition (4.2). The invariance over the set X^* and the boundedness follow from Proposition VII:3.1.1 and Theorem VII:2.3.2, respectively, in Hiriart-Urruty and Lemaréchal (1993a). □

Our derivation of convergence results for the ergodic sequence of Lagrange multipliers starts with the following lemma.

Lemma 4.4 *Suppose that the set X, defined by (4.1), fulfils Assumption 4.1, and let $\mathbf{x} \in X$ and $\boldsymbol{\pi} \in \mathbb{R}^{m+p}$. Then, the system*

$$\sum_{i \in \mathcal{I}(\mathbf{x})} \pi_i \nabla a_i(\mathbf{x}) + \sum_{j \in \mathcal{P}} \pi_j \mathbf{b}_j = \mathbf{0} \tag{4.3a}$$

$$\boldsymbol{\pi} \in \Pi_X(\mathbf{x}) \tag{4.3b}$$

is satisfied if and only if $\boldsymbol{\pi} = \mathbf{0}$.

Proof. Clearly, $\boldsymbol{\pi} = \mathbf{0}$ satisfies the system (4.3).

Conversely, suppose that $\boldsymbol{\pi}$ satisfies the system (4.3). From Assumption 4.1 follows that there is an $\hat{\mathbf{x}} \neq \mathbf{x}$ such that $\hat{\mathbf{x}} \in \operatorname{rint} X$. The convexity and differentiability of the functions a_i, together with the fact that $a_i(\mathbf{x}) = 0$ for all $i \in \mathcal{I}(\mathbf{x})$, then yield that $\nabla a_i(\mathbf{x})^T(\hat{\mathbf{x}} - \mathbf{x}) \leq a_i(\hat{\mathbf{x}})$ for all $i \in \mathcal{I}(\mathbf{x})$, and, since $\pi_i \geq 0$ and $a_i(\hat{\mathbf{x}}) < 0$ for all $i \in \mathcal{I}(\mathbf{x})$, that

$$\sum_{i \in \mathcal{I}(\mathbf{x})} \pi_i \nabla a_i(\mathbf{x})^T (\hat{\mathbf{x}} - \mathbf{x}) \leq \sum_{i \in \mathcal{I}(\mathbf{x})} \pi_i a_i(\hat{\mathbf{x}}) \leq 0. \tag{4.4}$$

However, since, for all $j \in \mathcal{P}$, $\mathbf{b}_j^T(\hat{\mathbf{x}} - \mathbf{x}) = 0$, it follows from the equations (4.3a) that $\sum_{i \in \mathcal{I}(\mathbf{x})} \pi_i \nabla a_i(\mathbf{x})^T (\hat{\mathbf{x}} - \mathbf{x}) = 0$, and from (4.4) we then obtain that $\sum_{i \in \mathcal{I}(\mathbf{x})} \pi_i a_i(\hat{\mathbf{x}}) = 0$. From the relations $\pi_i \geq 0$ and $a_i(\hat{\mathbf{x}}) < 0$ for all $i \in \mathcal{I}(\mathbf{x})$ it follows that $\pi_i = 0$ for all $i \in \mathcal{I}(\mathbf{x})$. The system (4.3) then reduces to the equations $\sum_{j \in \mathcal{P}} \pi_j \mathbf{b}_j = \mathbf{0}$ and $\pi_i = 0$, $i \in \mathcal{M}$. By Assumption 4.1, the vectors $\mathbf{b}_j$, $j \in \mathcal{P}$, are linearly independent, thus implying that $\pi_j = 0$ for all $j \in \mathcal{P}$. Hence, $\boldsymbol{\pi} = \mathbf{0}$. □

[7] Parts of Propositions 4.2 and 4.3 are true also under weaker constraint qualifications.

Definition 4.5 (Ergodic sequence of Lagrange multipliers). *Let the set X be defined by (4.1), and let the method (1.5) be applied to the program* [P]. *The sequences* $\{\boldsymbol{\omega}(\mathbf{x}^s)\}$ *and* $\{\boldsymbol{\lambda}(\mathbf{x}^s)\}$ *are defined by*

$$\boldsymbol{\omega}(\mathbf{x}^s) \in \left\{ \boldsymbol{\omega} \in \Pi_X(\mathbf{x}^s) \;\middle|\; \boldsymbol{\nu}(\mathbf{x}^s) = \sum_{i \in \mathcal{M}} \omega_i \nabla a_i(\mathbf{x}^s) + \sum_{j \in \mathcal{P}} \omega_j \mathbf{b}_j \right\}, \qquad s = 0, 1, \ldots,$$

and

$$\boldsymbol{\lambda}(\mathbf{x}^s) \in \left\{ \boldsymbol{\lambda} \in \Pi_X(\mathbf{x}^{s+1}) \;\middle|\; \mathbf{x}^{s+\frac{1}{2}} - \mathbf{x}^{s+1} = \sum_{i \in \mathcal{M}} \lambda_i \nabla a_i(\mathbf{x}^{s+1}) + \sum_{j \in \mathcal{P}} \lambda_j \mathbf{b}_j \right\}, \qquad s = 0, 1, \ldots,$$

where $\boldsymbol{\nu}(\mathbf{x}^s) \in N_X(\mathbf{x}^s)$ *and* $\mathbf{x}^{s+\frac{1}{2}} - \mathbf{x}^{s+1} \in N_X(\mathbf{x}^{s+1})$ *for all s. The ergodic sequence* $\{\boldsymbol{\mu}^t\}$ *of Lagrange multipliers is defined by*

$$\boldsymbol{\mu}^t = A_t^{-1} \sum_{s=0}^{t-1} \alpha_s \left(\boldsymbol{\omega}(\mathbf{x}^s) + \frac{\boldsymbol{\lambda}(\mathbf{x}^s)}{\alpha_s} \right), \qquad t = 1, 2, \ldots. \qquad \square$$

Note that $\mu_i^t \geq 0$ for all $i \in \mathcal{M}$ and all t.

Clearly, the value of the vector $\boldsymbol{\omega}(\mathbf{x}^s)$ depends of the choice of the normal term $\boldsymbol{\nu}(\mathbf{x}^s)$ in the conditional subgradients used in the method (1.5). In the special case of the traditional subgradient optimization method (1.4), $\boldsymbol{\nu}(\mathbf{x}^s) = \mathbf{0}$ for all s, and then, from Lemma 4.4, $\boldsymbol{\omega}(\mathbf{x}^s) = \mathbf{0}$ for all s. In the special case of the method (1.5) where $\gamma^X(\mathbf{x}^s)$ is a *projected subgradient*,

$$\boldsymbol{\nu}(\mathbf{x}^s) = \arg\min_{\boldsymbol{\nu}} \left\{ \|\boldsymbol{\nu}\|^2 \;\middle|\; \nabla a_i(\mathbf{x}^s)^T (\boldsymbol{\nu} + \gamma(\mathbf{x}^s)) \geq 0, i \in \mathcal{I}(\mathbf{x}^s);\; \mathbf{b}_j^T (\boldsymbol{\nu} + \gamma(\mathbf{x}^s)) = 0, j \in \mathcal{P} \right\},$$

and the elements $\omega_i(\mathbf{x}^s)$, $i \in \mathcal{I}(\mathbf{x}^s)$, and $\omega_j(\mathbf{x}^s)$, $j \in \mathcal{P}$, are Lagrange multipliers associated with the constraints of this projection problem.

The vector $\boldsymbol{\lambda}(\mathbf{x}^s)$ always contains Lagrange multipliers associated with the constraints of the projection problem

$$\mathbf{x}^{s+1} = \arg\min_{\mathbf{x}} \left\{ \left\| \mathbf{x}^{s+\frac{1}{2}} - \mathbf{x} \right\|^2 \;\middle|\; a_i(\mathbf{x}) \leq 0,\ i \in \mathcal{M};\quad \mathbf{b}_j^T \mathbf{x} = e_j,\ j \in \mathcal{P} \right\}.$$

Lemma 4.6 (Boundedness of $\{\boldsymbol{\omega}(\mathbf{x}^s)\}$ and $\{\boldsymbol{\lambda}(\mathbf{x}^s)/\alpha_s\}$). *Suppose that the set X is defined by (4.1) and fulfils Assumption 4.1, and let the method (1.5)–(1.7) be applied to the program* [P]. *If the sequences* $\{\gamma(\mathbf{x}^s)\}$ *and* $\{\boldsymbol{\nu}(\mathbf{x}^s)\}$ *are bounded, then this is also true for the sequences* $\{\boldsymbol{\omega}(\mathbf{x}^s)\}$ *and* $\{\boldsymbol{\lambda}(\mathbf{x}^s)/\alpha_s\}$, *given by Definition 4.5.*

Proof. We first prove, by contradiction, that the sequence $\{\boldsymbol{\omega}(\mathbf{x}^s)\}$ is bounded.

Suppose that this is not the case. Then there exists a subsequence $\mathcal{S}$ such that $\|\boldsymbol{\omega}(\mathbf{x}^s)\| > 0$ for all $s \in \mathcal{S}$ and $\{\|\boldsymbol{\omega}(\mathbf{x}^s)\|\}_{s \in \mathcal{S}} \to \infty$. Let $\boldsymbol{\pi}^s = \boldsymbol{\omega}(\mathbf{x}^s)/\|\boldsymbol{\omega}(\mathbf{x}^s)\|$, implying that $\boldsymbol{\pi}^s \geq \mathbf{0}$ and $\|\boldsymbol{\pi}^s\| = 1$ for all $s \in \mathcal{S}$. According to Proposition 1.6, $\{\mathbf{x}^s\} \to \mathbf{x}^\infty$. Suppose that, for some $i \in \mathcal{M}$, $a_i(\mathbf{x}^\infty) < 0$. By the continuity of the function a_i, it must then hold for all s which are sufficiently large that $a_i(\mathbf{x}^s) < 0$, $\omega_i(\mathbf{x}^s) = 0$, and $\pi_i^s = 0$; hence, for any $i \in \mathcal{M} \setminus \mathcal{I}(\mathbf{x}^\infty)$, $\pi_i^s = 0$ for all $s \in \mathcal{S}$ that are large enough. The sequence $\{\boldsymbol{\pi}^s\}_{s \in \mathcal{S}}$ is clearly bounded and thus has an accumulation point, say $\tilde{\boldsymbol{\pi}}$. Then, from the preceding, $\tilde{\boldsymbol{\pi}} \in \Pi_X(\mathbf{x}^\infty)$ and $\|\tilde{\boldsymbol{\pi}}\| = 1$.

From Definition 4.5 we have that

$$\sum_{i \in \mathcal{M}} \pi_i^s \nabla a_i(\mathbf{x}^s) + \sum_{j \in \mathcal{P}} \pi_j^s \mathbf{b}_j = \frac{\boldsymbol{\nu}(\mathbf{x}^s)}{\|\boldsymbol{\omega}(\mathbf{x}^s)\|}, \quad \forall s \in \mathcal{S}.$$

Taking the limit of the subsequence of $\mathcal{S}$ corresponding to $\tilde{\boldsymbol{\pi}}$ we obtain

$$\sum_{i \in \mathcal{I}(\mathbf{x}^\infty)} \tilde{\pi}_i \nabla a_i(\mathbf{x}^\infty) + \sum_{j \in \mathcal{P}} \tilde{\pi}_j \mathbf{b}_j = \mathbf{0},$$

since $\{\mathbf{x}^s\} \to \mathbf{x}^\infty$ and the functions ∇a_i, $i \in \mathcal{M}$, are continuous, and since the sequence $\{\boldsymbol{\nu}(\mathbf{x}^t)\}$ is bounded and the sequence $\{\|\boldsymbol{\omega}(\mathbf{x}^s)\|\}_{s \in \mathcal{S}}$ is presumed to tend to infinity. Then, by Lemma 4.4, $\tilde{\boldsymbol{\pi}} = \mathbf{0}$ must hold, and we have arrived at a contradiction. Hence, the sequence $\{\boldsymbol{\omega}(\mathbf{x}^s)\}$ is bounded.

To establish that the sequence $\{\boldsymbol{\lambda}(\mathbf{x}^s)/\alpha_s\}$ is bounded, we first observe that, from (1.5), $\|\mathbf{x}^{s+\frac{1}{2}} - \mathbf{x}^{s+1}\|/\alpha_s \leq \|\boldsymbol{\gamma}(\mathbf{x}^s) + \boldsymbol{\nu}(\mathbf{x}^s)\|$ for all s, which implies that the sequence $\{(\mathbf{x}^{s+\frac{1}{2}} - \mathbf{x}^{s+1})/\alpha_s\}$ is bounded. Assume then that the sequence $\{\boldsymbol{\lambda}(\mathbf{x}^s)/\alpha_s\}$ is not bounded and define the index set $\mathcal{S}$ analogously to the above. By, for all $s \in \mathcal{S}$, letting $\boldsymbol{\pi}^s = \boldsymbol{\lambda}(\mathbf{x}^s)/\|\boldsymbol{\lambda}(\mathbf{x}^s)\|$, Definition 4.5 yields that

$$\sum_{i \in \mathcal{M}} \pi_i^s \nabla a_i(\mathbf{x}^{s+1}) + \sum_{j \in \mathcal{P}} \pi_j^s \mathbf{b}_j = \frac{(\mathbf{x}^{s+\frac{1}{2}} - \mathbf{x}^{s+1})/\alpha_s}{\|\boldsymbol{\lambda}(\mathbf{x}^s)\|/\alpha_s}, \quad \forall s \in \mathcal{S}.$$

Clearly, the right hand side of this equality tends to the zero vector and we may analogously to the first part of the proof derive a contradiction. □

From Definition 4.5 and Lemma 4.6 follow that, under the conditions (1.6) and (1.7), the sequence $\{\boldsymbol{\mu}^t\}$ is bounded.

Theorem 4.7 (Convergence of $\{\boldsymbol{\mu}^t\}$ to $M_X(\mathbf{x}^\infty)$). *Suppose that the set X is defined by (4.1) and fulfils Assumption 4.1. Let the method (1.5)–(1.7) be applied to the program* [P] *and let the sequence $\{\boldsymbol{\mu}^t\}$ be given by Definition 4.5. Suppose that the sequences $\{\boldsymbol{\gamma}(\mathbf{x}^s)\}$ and $\{\boldsymbol{\nu}(\mathbf{x}^s)\}$ are bounded. Then,*

$$\left\{d_{M_X(\mathbf{x}^\infty)}(\boldsymbol{\mu}^t)\right\} \to 0.$$

Proof. From the continuity of the functions a_i, $i \in \mathcal{M}$, follows that $\mathcal{I}(\mathbf{x}^s) \subseteq \mathcal{I}(\mathbf{x}^\infty)$ for all s that are sufficiently large. From Definition 4.5 then follows that $\omega_i(\mathbf{x}^s) = \lambda_i(\mathbf{x}^s) = 0$ for all $i \in \mathcal{M} \backslash \mathcal{I}(\mathbf{x}^\infty)$ and all s that are sufficiently large, so that $\{\mu_i^t\} \to 0$, $i \in \mathcal{M} \backslash \mathcal{I}(\mathbf{x}^\infty)$. Hence, $\{d_{\Pi_X(\mathbf{x}^\infty)}(\boldsymbol{\mu}^t)\} \to 0$.

To simplify the notation, we define $\boldsymbol{\nu}^t = \sum_{i \in \mathcal{M}} \mu_i^t \nabla a_i(\mathbf{x}^\infty) + \sum_{j \in \mathcal{P}} \mu_j^t \mathbf{b}_j$. From Definitions 2.2 and 4.5 then follow that

$$\begin{aligned} \boldsymbol{\nu}^t &= \mathbf{n}^t - \sum_{i \in \mathcal{M}} A_t^{-1} \sum_{s=0}^{t-1} \alpha_s \omega_i(\mathbf{x}^s) \left(\nabla a_i(\mathbf{x}^s) - \nabla a_i(\mathbf{x}^\infty)\right) \\ &\quad - \sum_{i \in \mathcal{M}} A_t^{-1} \sum_{s=0}^{t-1} \lambda_i(\mathbf{x}^s) \left(\nabla a_i(\mathbf{x}^{s+1}) - \nabla a_i(\mathbf{x}^\infty)\right). \end{aligned} \tag{4.5}$$

Let $\varepsilon > 0$. Lemma 2.6 and Theorem 3.1 then yield that $d_{-\partial f(\mathbf{x}^\infty)}(\mathbf{n}^t) \leq \varepsilon/3$ for all t that are sufficiently large. We next show that the second term of the right hand side of the equation (4.5) tends to the zero vector. Lemma 4.6 implies that the sequence $\{\omega_i(\mathbf{x}^s)\}$ is bounded for each $i \in \mathcal{M}$. Hence, since the function ∇a_i is continuous, there is an iterate $t(\varepsilon)$ such that

$$\omega_i(\mathbf{x}^s) \left\|\nabla a_i(\mathbf{x}^s) - \nabla a_i(\mathbf{x}^\infty)\right\| \leq \frac{\varepsilon}{6m}, \quad \forall s \geq t(\varepsilon).$$

Therefore, for all $t > t(\varepsilon)$,

$$A_t^{-1} \sum_{s=t(\varepsilon)}^{t-1} \alpha_s \omega_i(\mathbf{x}^s) \|\nabla a_i(\mathbf{x}^s) - \nabla a_i(\mathbf{x}^\infty)\| \le A_t^{-1} \sum_{s=t(\varepsilon)}^{t-1} \alpha_s \frac{\varepsilon}{6m} \le \frac{\varepsilon}{6m}.$$

Since $\{A_t\} \to \infty$ we have, for all t that are large enough,

$$A_t^{-1} \sum_{s=0}^{t(\varepsilon)-1} \alpha_s \omega_i(\mathbf{x}^s) \|\nabla a_i(\mathbf{x}^s) - \nabla a_i(\mathbf{x}^\infty)\| \le \frac{\varepsilon}{6m}.$$

From the derivations above follow that, for all $t > t(\varepsilon)$ that are large enough,

$$\begin{aligned} 0 \ &\le \left\| \sum_{i \in \mathcal{M}} A_t^{-1} \sum_{s=0}^{t-1} \alpha_s \omega_i(\mathbf{x}^s) \left(\nabla a_i(\mathbf{x}^s) - \nabla a_i(\mathbf{x}^\infty)\right) \right\| \\ &\le \sum_{i \in \mathcal{M}} A_t^{-1} \sum_{s=0}^{t-1} \alpha_s \, \omega_i(\mathbf{x}^s) \, \| \, \nabla a_i(\mathbf{x}^s) - \nabla a_i(\mathbf{x}^\infty) \, \| \ \le \ \frac{\varepsilon}{3}. \end{aligned}$$

It can, analogously to the preceding, be shown that the same conclusion holds for the last term of the right hand side of equation (4.5). It follows that $d_{-\partial f(\mathbf{x}^\infty)}(\boldsymbol{\nu}^t) \le \varepsilon$ for all t that are sufficiently large, and hence that $\{d_{-\partial f(\mathbf{x}^\infty)}(\boldsymbol{\nu}^t)\} \to 0$. Now, let $\mathcal{T}$ be a subsequence such that

$$\lim_{\substack{t \in \mathcal{T} \\ t \to \infty}} d_{M_X(\mathbf{x}^\infty)}(\boldsymbol{\mu}^t) = \limsup_{t \to \infty} d_{M_X(\mathbf{x}^\infty)}(\boldsymbol{\mu}^t),$$

and let $\tilde{\boldsymbol{\mu}}$ be an accumulation point of the sequence $\{\boldsymbol{\mu}^t\}_{t \in \mathcal{T}}$. From the preceding and the closedness of the sets $\Pi_X(\mathbf{x}^\infty)$ and $\partial f(\mathbf{x}^\infty)$ then follow that $\tilde{\boldsymbol{\mu}} \in \Pi_X(\mathbf{x}^\infty)$ and $\sum_{i \in \mathcal{M}} \tilde{\mu}_i \nabla a_i(\mathbf{x}^\infty) + \sum_{i \in \mathcal{P}} \tilde{\mu}_j \mathbf{b}_j \in -\partial f(\mathbf{x}^\infty)$, thus implying that $\tilde{\boldsymbol{\mu}} \in M_X(\mathbf{x}^\infty)$. The theorem follows. □

4.2. Multipliers of objective pieces

Throughout this subsection we assume that the nonsmooth objective function f of the program [P] has the form

$$f(\mathbf{x}) = \max_{k \in \mathcal{L}} c_k(\mathbf{x}), \quad \mathbf{x} \in \mathbb{R}^n, \tag{4.6}$$

where the functions $c_k : \mathbb{R}^n \mapsto \mathbb{R}$, $k \in \mathcal{L} = \{1, \ldots, \ell\}$, are convex and differentiable. We let $\mathcal{K}(\mathbf{x}) = \{k \in \mathcal{L} \,|\, c_k(\mathbf{x}) = f(\mathbf{x})\}$ denote the set of objective pieces that are active at $\mathbf{x} \in \mathbb{R}^n$ and

$$\Pi_f(\mathbf{x}) = \left\{ \pi \in \mathbb{R}_+^\ell \;\middle|\; \pi_k = 0, \ k \in \mathcal{L} \setminus \mathcal{K}(\mathbf{x}); \ \ \sum_{k \in \mathcal{L}} \pi_k = 1 \right\}, \quad \mathbf{x} \in \mathbb{R}^n.$$

Letting $\boldsymbol{\xi} \in \mathbb{R}^\ell$, the subdifferential of f at $\mathbf{x}$ can then be expressed as

$$\partial f(\mathbf{x}) = \left\{ \gamma \in \mathbb{R}^n \;\middle|\; \gamma = \sum_{k \in \mathcal{L}} \xi_k \nabla c_k(\mathbf{x}); \quad \boldsymbol{\xi} \in \Pi_f(\mathbf{x}) \right\}.$$

Let the program [P] be reformulated as

$$\begin{aligned} \min \quad & z, & (4.7\text{a}) \\ \text{s.t.} \quad & c_k(\mathbf{x}) - z \le 0, \quad k \in \mathcal{L}, & (4.7\text{b}) \\ & \mathbf{x} \in X, & (4.7\text{c}) \end{aligned}$$

and introduce nonnegative multipliers η_k, $k \in \mathcal{L}$, associated with the constraints (4.7b). Optimality of an $\mathbf{x} \in X$ can then be characterized in terms of the nonemptiness of the set

$$M_f(\mathbf{x}) = \left\{ \boldsymbol{\eta} \in \Pi_f(\mathbf{x}) \;\middle|\; \mathbf{0} \in \sum_{k \in \mathcal{L}} \eta_k \nabla c_k(\mathbf{x}) + N_X(\mathbf{x}) \right\} \tag{4.8}$$

of Lagrange multipliers associated with the constraints (4.7b) (e.g., Hiriart-Urruty and Lemaréchal, 1993a, Theorem VII:2.1.4). Note that, in contrast to Propositions 4.2 and 4.3, a constraint qualification is here always fulfilled (since $\mathtt{int}(\mathtt{epi}\, f) \neq \emptyset$).

Proposition 4.8 (Optimality conditions). *Let* $\mathbf{x} \in X$. *Then,* $\mathbf{x} \in X^*$ *if and only if* $M_f(\mathbf{x}) \neq \emptyset$. □

Proposition 4.9 (Properties of $M_f(\mathbf{x})$, $\mathbf{x} \in X^*$). *For any* $\mathbf{x} \in X^*$ *the nonempty set* $M_f(\mathbf{x})$ *is convex and compact, and* $M_f(\mathbf{y}) = M_f(\mathbf{x})$ *for all* $\mathbf{y} \in X^*$.

Proof. The convexity and compactness of $M_f(\mathbf{x})$ are immediate from its definition (4.8), and its invariance over X^* follows from Proposition VII:3.1.1 in Hiriart-Urruty and Lemaréchal (1993a) applied to the reformulation (4.7) of the program [P]. □

Definition 4.10 (Ergodic sequence of multipliers). *Let the objective function f be defined by (4.6) and let the method (1.5) be applied to the program* [P]. *The sequence* $\{\boldsymbol{\xi}(\mathbf{x}^s)\}$ *is defined by*

$$\boldsymbol{\xi}(\mathbf{x}^s) \in \left\{ \boldsymbol{\xi} \in \Pi_f(\mathbf{x}^s) \;\middle|\; \gamma(\mathbf{x}^s) = \sum_{k \in \mathcal{L}} \xi_k \nabla c_k(\mathbf{x}^s) \right\}, \quad s = 0, 1, \ldots,$$

where $\gamma(\mathbf{x}^s) \in \partial f(\mathbf{x}^s)$ *for all* s. *The ergodic sequence* $\{\boldsymbol{\eta}^t\}$ *of multipliers is then defined by*

$$\boldsymbol{\eta}^t = A_t^{-1} \sum_{s=0}^{t-1} \alpha_s \boldsymbol{\xi}(\mathbf{x}^s), \quad t = 1, 2, \ldots . \qquad \square$$

Note that $\boldsymbol{\eta}^t \geq \mathbf{0}$ and $\sum_{k \in \mathcal{L}} \eta_k^t = 1$ for all t; obviously, the sequence $\{\boldsymbol{\eta}^t\}$ is bounded.

Theorem 4.11 (Convergence of $\{\boldsymbol{\eta}^t\}$ to $M_f(\mathbf{x}^\infty)$). *Let the objective function f be defined by (4.6), let the method (1.5)–(1.7) be applied to the program* [P], *and let the sequence* $\{\boldsymbol{\eta}^t\}$ *be given by Definition 4.10. Suppose that the sequences* $\{\gamma(\mathbf{x}^t)\}$ *and* $\{\boldsymbol{\nu}(\mathbf{x}^t)\}$ *are bounded. Then,*

$$\left\{ d_{M_f(\mathbf{x}^\infty)}(\boldsymbol{\eta}^t) \right\} \to 0.$$

Proof. Similarly to the proof of Theorem 4.7 we can show that $\{d_{\Pi_f(\mathbf{x}^\infty)}(\boldsymbol{\eta}^t)\} \to 0$. Defining $\gamma^t = \sum_{k \in \mathcal{L}} \eta_k^t \nabla c_k(\mathbf{x}^\infty)$, it follows from Definitions 2.1 and 4.10 that

$$\gamma^t = \mathbf{g}^t - \sum_{k \in \mathcal{L}} A_t^{-1} \sum_{s=0}^{t-1} \alpha_s \xi_k(\mathbf{x}^s) \left(\nabla c_k(\mathbf{x}^s) - \nabla c_k(\mathbf{x}^\infty) \right).$$

Let $\varepsilon > 0$. Theorem 3.1 then yields that $d_{-N_X(\mathbf{x}^\infty)}(\mathbf{g}^t) \leq \varepsilon/2$ for all t that are sufficiently large. Using the same technique as in the proof of Theorem 4.7 and utilizing the boundedness of the sequence $\{\boldsymbol{\xi}(\mathbf{x}^s)\}$, we can show that

$$0 \leq \left\| \sum_{k \in \mathcal{L}} A_t^{-1} \sum_{s=0}^{t-1} \alpha_s \xi_k(\mathbf{x}^s) \left(\nabla c_k(\mathbf{x}^s) - \nabla c_k(\mathbf{x}^\infty) \right) \right\| \leq \frac{\varepsilon}{2}.$$

It follows that $d_{-N_X(\mathbf{x}^\infty)}(\gamma^t) \leq \varepsilon$ for all t that are sufficiently large, and hence that $\{d_{-N_X(\mathbf{x}^\infty)}(\gamma^t)\} \to 0$. Using the same technique as in the proof of Theorem 4.7, we can also show that all accumulation points, $\tilde{\eta}$, of the sequence $\{\eta^t\}$ fulfil $\tilde{\eta} \in \Pi_f(\mathbf{x}^\infty)$ and $\sum_{k\in\mathcal{L}} \tilde{\eta}_k \nabla c_k(\mathbf{x}^\infty) \in -N_X(\mathbf{x}^\infty)$, thus implying $\tilde{\eta} \in M_f(\mathbf{x}^\infty)$. The theorem follows. □

5. FURTHER RESEARCH

The ergodic results of Sectan immediately be generalized to the case when both the objective function and the feasible set have explicit representations. A straightforward topic for continued studies is the derivation of ergodic results for multipliers when the constraint functions a_i, $i \in \mathcal{M}$, and/or the objective pieces c_k, $k \in \mathcal{L}$, are allowed to be nonsmooth.

We have recently constructed a convergent lower bounding procedure for the program [P] (cf. the lower bounding principle of the Frank–Wolfe method) based on the ergodic results presented in this paper, and studied its application in a subgradient scheme for a nonsmooth formulation of the linear minimum cost multicommodity network flow problem (Larsson et al., 1995). The application of the lower bounding procedure to other classes of structured, nonsmooth programs, especially Lagrangean dual problems, will be studied further.

A further application of the sequence $\{\mathbf{g}^t\}$ is the construction of an auxiliary, optimizing sequence $\{\overline{\mathbf{x}}^t\}$ of feasible solutions to the program [P], by

$$\overline{\mathbf{x}}^t = P_X\left(\mathbf{x}^t - \sigma_t \mathbf{g}^t\right), \quad t = 1, 2, \ldots,$$

where $0 < \underline{\sigma} \leq \sigma_t \leq \overline{\sigma} < +\infty$ for all t. This sequence, which is generated in a manner akin to gradient projection methods for smooth optimization, has stronger properties than the original sequence $\{\mathbf{x}^t\}$; for example, under very mild assumptions it identifies *finitely* the set of active constraints at the optimal solution $\mathbf{x}^\infty$. This is an extension of results from the differentiable case, and it illustrates well the smoothing effect of the process defining the sequence $\{\mathbf{g}^t\}$.

Consider the *general* problem of finding an $\mathbf{x} \in \mathbb{R}^n$ such that $\mathbf{0} \in F(\mathbf{x})$, where $F : \mathbb{R}^n \mapsto 2^{(\mathbb{R}\cup\{\pm\infty\})^n}$ is a maximal monotone point-to-set map. A natural method for solving this problem can be described as follows. Let

$$\mathbf{x}^0 \in \operatorname{dom} F, \quad \gamma^t \in F(\mathbf{x}^t), \quad \alpha_t > 0, \quad \text{and} \quad \mathbf{x}^{t+1} = P_{\operatorname{dom} F}(\mathbf{x}^t - \alpha_t \gamma^t), \quad t = 0, 1, \ldots. \quad (5.1)$$

For any nonempty, closed and convex set $X \subseteq \mathbb{R}^n$ and any convex function $f : \mathbb{R}^n \mapsto \mathbb{R}$, the mapping $F \equiv \partial^X f$ is maximal monotone, since it equals the sum of the two maximal monotone operators ∂f and N_X (Rockafellar, 1970b, Theorem B, and 1970c, Theorem 1); by Proposition 1.5, the problem [P] is hence a special case of the general problem and, further, the conditional subgradient method (1.5) is a special case of the method (5.1). Proposition 1.6 then establishes that $\{d_{F^{-1}(\mathbf{0})}(\mathbf{x}^t)\} \to 0$, that is, $\mathbf{0} \in F(\mathbf{x}^\infty)$. A subject for future research is to derive ergodic results for the method (5.1) applied to the general problem.

ACKNOWLEDGEMENT

The research was supported financially by grants (No. 94-292 and 282-93-1195) from the Swedish Research Council for Engineering Sciences (TFR).

REFERENCES

[1] J.E. Beasley, "Lagrangean relaxation". *In*: "Modern Heuristic Techniques for Combinatorial Problems," C.R. Reeves, ed., Blackwell Scientific Publications, Oxford, pp. 243–303, 1993.

[2] A. Ben-Tal and M.P. Bendsøe, "A new method for optimal truss topology design". *SIAM Journal on Optimization*, 3:322–358, 1993.

[3] J.V. Burke and M.C. Ferris, "Characterization of solution sets of convex programs". *Operations Research Letters*, 10:57–60, 1991.

[4] V.F. Dem'yanov and V.K. Šomesova, "Conditional subdifferentials of convex functions". *Soviet Mathematics Doklady*, 19:1181–1185, 1978.

[5] V.F. Dem'yanov and V.K. Shomesova, 1980, "Subdifferentials of functions on sets". *Cybernetics*, 16/1:24–31.

[6] V.F. Dem'yanov and L.V. Vasil'ev, "Nondifferentiable Optimization." Optimization Software, New York, NY, 1985.

[7] J.C. Dunn, "On recursive averaging processes and Hilbert space extensions of the contraction mapping principle". *Journal of the Franklin Institute*, 295:117–133, 1973.

[8] Yu.M. Ermol'ev, "Methods for solving nonlinear extremal problems". *Cybernetics*, 2/4:1–14, 1966.

[9] M.L. Fisher, "The Lagrangian relaxation method for solving integer programming problems". *Management Science*, 27:1–18, 1981.

[10] M.L. Fisher, "An applications oriented guide to Lagrangian relaxation". *Interfaces*, 15:10–21, 1985.

[11] M. Frank and P. Wolfe, An algorithm for quadratic programming. *Naval Research Logistics Quarterly*, 3:95–110, 1956.

[12] J.-B. Hiriart-Urruty and C. Lemaréchal, "Convex Analysis and Minimization Algorithms I: Fundamentals." Springer-Verlag, Berlin, 1993a.

[13] J.-B. Hiriart-Urruty and C. Lemaréchal, "Convex Analysis and Minimization Algorithms II: Advanced Theory and Bundle Methods." Springer-Verlag, Berlin, 1993b.

[14] J. Kornai and T. Lipták, "Two level planning". *Econometrica*, 33:141–169, 1965.

[15] T. Larsson, M. Patriksson, and A.-B. Strömberg, "Conditional subgradient optimization—theory and applications". Report LiTH-MAT-R-92-24, Department of Mathematics, Linköping Institute of Technology, Linköping, Sweden, 1992. Revised 1994. *European Journal of Operational Research* (to appear).

[16] T. Larsson, M. Patriksson, and A.-B. Strömberg, "Ergodic results and bounds on the optimal value in subgradient optimization". Paper presented at the Symposium on Operations Research (SOR'95), Annual Conference of the DGOR, GMÖOR, and ÖGOR, Universität Passau, September 13–15, Passau, Germany, 1995.

[17] B.T. Polyak, "A general method of solving extremum problems". *Soviet Mathematics Doklady*, 8/3:593–597, 1967.

[18] B.T. Polyak, "Minimization of unsmooth functionals". *USSR Computational Mathematics and Mathematical Physics*, 9:14–29, 1969.

[19] W.B. Powell and Y. Sheffi, "The convergence of equilibrium algorithms with predetermined step sizes". *Transportation Science*, 16:45–55, 1982.

[20] R.T. Rockafellar, "Convex Analysis." Princeton University Press, Princeton, NJ, 1970a.

[21] R.T. Rockafellar, "On the maximal monotonicity of subdifferential mappings". *Pacific Journal of Mathematics*, 33:209–216, 1970b.

[22] R.T. Rockafellar, "On the maximality of sums of nonlinear monotone operators". *Transactions of the American Mathematical Society*, 149:75–88, 1970c.

[23] N.Z. Shor, "Minimization Methods for Non-Differentiable Functions." Springer-Verlag, Berlin, 1985.

[24] N.Z. Shor, "The development of numerical methods for nonsmooth optimization in the USSR". *In*: "History of Mathematical Programming: A Collection of Personal Reminiscences," J.K. Lenstra, A.H.G. Rinnoy Kan and A. Schrijver, eds., CWI and North-Holland, Amsterdam, pp. 135–139, 1991.

[25] E.H. Zarantonello, "Projections on convex sets in Hilbert space and spectral theory". *In*: "Contributions to Nonlinear Functional Analysis," E.H. Zarantonello, ed., Proceedings of a symposium conducted by the Mathematics research center, University of Wisconsin, Madison, Academic Press, New York, NY, pp. 237–424, 1971.

PROTO-DERIVATIVES AND THE GEOMETRY OF SOLUTION MAPPINGS IN NONLINEAR PROGRAMMING

Levy Adam B.[1] and Rockafellar R.T.[2]

[1] Department of Mathematics
Bowdoin College
Brunswick, ME 04011 USA
e-mail: alevy@polar.bowdoin.edu

[2] Department of Mathematics
University of Washington
Seattle, WA 98195 USA
e-mail: rtr@math.washington.edu

Abstract. We quantify the sensitivity of KKT pairs associated with a parameterized family of nonlinear programming problems. Our approach involves proto-derivatives, which are generalized derivatives appropriate even in cases when the KKT pairs are not unique; we investigate what the theory of such derivatives yields in the special case when the KKT pairs are unique (locally). We demonstrate that the graph of the KKT multifunction is just a reoriented graph of a Lipschitz mapping, and use proto-differentiability to show that the graph of the KKT multifunction actually has the stronger property of being a reorientation of the graph of a B-differentiable mapping. Our results indicate that proto-derivatives provide the same kind of information for possibly set-valued mappings (like the KKT multifunction) that B-derivatives provide for single-valued mappings.

Key words. Proto-derivative; sensitivity analysis; nonlinear programming

1. INTRODUCTION

A nonlinear programming problem in $x \in \mathbb{R}^n$ as parameterized by $w \in \mathbb{R}^d$ can be formulated in terms of a set $X \subseteq \mathbb{R}^n$, functions f_i on $\mathbb{R}^n \times \mathbb{R}^d \to \mathbb{R}$ for $i = 0, 1, \ldots, m$, and a convex cone $K \subseteq \mathbb{R}^m$:

$$\text{minimize } f_0(w, x) \text{ over all } x \in C(w) \tag{1.1}$$

where

$$C(w) := \{x \in X : F(w, x) \in K\} \quad \text{for} \quad F(w, x) = (f_1(w, x), \ldots, f_m(w, x)). \tag{1.2}$$

It will be supposed here that X and K are closed and every f_i is of class $\mathcal{C}^2$.

An important issue in the study of such a problem is the response of solutions to changes in the parameters, but this faces the difficulty that there can be several (or many, or no) optimal solutions x associated with a particular w. The solution mapping $w \mapsto x$ is necessarily then a *multifunction* (set-valued) whose "rates of change" cannot be captured by traditional methodology. The same goes for the mapping $w \mapsto (x, y)$ that assigns to each w the pairs (x, y) in which x satisfies a first-order optimality condition with y as multiplier vector.

Because of this difficulty, much of the theory of rates of change in the dependence of solutions on parameters has centered so far on circumstances in which the multifunction under investigation happens to be single-valued in a local sense. An adequate conceptual framework is provided then by relatively elementary notions of Lipschitz continuity and one-sided differentiability. It is not necessary to appeal to the more challenging notions of such kind that have been devised for handling multifunctions in general. But by narrowing the scope to mappings that are essentially single-valued, a significant part of the overall picture could be lost.

The multifunctions that associate parameter elements w with solutions x or quasi-solution pairs (x, y) tend to be very special within the realm of multifunctions. Although not necessarily single-valued, their graphs reflect many of the geometric features of the graphs of single-valued mappings. Our aim in this paper is to develop such geometry and its consequences, with emphasis on "proto-differentiability" as providing the natural counterpart to the type of one-sided differentiability, namely B-differentiability, that has come to prevail in studies of the single-valued setting. For this purpose we direct our attention mainly to quasi-solution pairs (x, y), as follows.

The generalized Karush-Kuhn-Tucker (KKT) conditions for (1.1)–(1.2) at a feasible solution x, as established in Rockafellar [25], concern a multiplier vector $y = (y_1, \ldots, y_m)$ such that

$$y \in N_K(F(w,x)), \quad -[\nabla_x f_0(w,x)+y_1\nabla_x f_1(w,x)+\cdots+y_m\nabla_x f_m(w,x)] \in N_X(x), \quad (1.3)$$

where $N_K(F(w,x))$ is the normal cone to K at $F(w,x)$ and $N_X(x)$ is the normal cone to X at x. The first normal cone can be taken in the sense of convex analysis, but unless X happens to be convex the second should be interpreted in the broader sense of limits of proximal norm vectors (unconvexified); cf. [25]. Such a multiplier vector exists when x is locally optimal and satisfies the constraint qualification that there is no vector $y \neq 0$ such that

$$y \in N_K(F(w,x)), \qquad -[y_1\nabla_x f_1(w,x) + \cdots + y_m \nabla f_m(w,x)] \in N_X(x). \qquad (1.4)$$

(This is given in Theorem 4.2 of [25] for the case of $K = \mathbb{R}^s_- \times \mathbb{R}^{m-s}$; the version for general K can be obtained by applying Theorem 10.1 of that paper to the indicator function δ_K.)

When $X = \mathbb{R}^n$ and $K = \mathbb{R}^s_- \times \mathbb{R}^{m-s}$, which we refer to as the *conventional case* of our problem, conditions (1.3) and (1.4) reduce to the classical KKT conditions and the dual form of the Mangasarian-Fromovitz constraint qualification. Other choices of the cone K allow for coverage of positive-definite programming, for instance. In taking X to be a proper subset of $\mathbb{R}^n$ one can incorporate nonnegativity requirements, upper and lower bounds on variables, and indeed other constraints of any kind without having to introduce additional multipliers whose perturbations might need to be coped with. In the *convex case* of our problem, i.e., when X is convex, $f_0(w,x)$ is convex in x, and $F(w,x)$ is convex in x with respect to the partial ordering induced by K, the generalized KKT conditions (1.3) on (x, y) are sufficient for x to be a globally optimal solution to (1.1)–(1.2).

We are interested in the KKT multifunction that assigns to each w the pairs (x, y) satisfying (1.3). In the conventional case, at least, much is already known about the properties of this multifunction; see for example [14], [13], [16], [2], [3], [1], [4]). Robinson's property of strong regularity provides a useful criterion for localized single-valuedness and Lipschitz continuity, even B-differentiability [16]. Closely tied to such results is the analysis of parametric dependence not only on w but on certain "canonical" parameters as well. The role of these canonical elements is to ensure that the parameterization is sufficiently rich. They are also in fact the key to the graphical geometry that is the topic here.

The *primal canonical perturbation* vector $u \in \mathbb{R}^m$ shifts $F(w,x)$ to $F(w,x)+u$, while the *dual canonical perturbation* vector $v \in \mathbb{R}^n$ shifts $f_0(w,x)$ to $f_0(w,x) - \langle v, x\rangle$. In bringing these vectors in, we adopt the format of a nonlinear programming problem parameterized by $(u,v,w) \in \mathbb{R}^m \times \mathbb{R}^n \times \mathbb{R}^d$:

$$\text{minimize } f_0(w,x) - \langle v,x\rangle \text{ over all } x \in C(u,w) \tag{1.5}$$

where

$$C(u,w) := \{x \in X : F(w,x)+u \in K\} \quad \text{for} \quad F(w,x) = (f_1(w,x),\ldots,f_m(w,x)). \tag{1.6}$$

The generalized KKT conditions on x and a multiplier vector y then have the form:

$$y \in N_K(F(w,x)+u) \quad \text{and} \quad v - \nabla_x f_0(w,x) - \nabla_x F(w,x)^{\mathsf{T}} y \in N_X(x). \tag{1.7}$$

The multifunction we specifically look at in this format is

$$S_{\text{KKT}} : (u,v,w) \mapsto \{(x,y) : \text{ (1.7) holds }\}. \tag{1.8}$$

Obviously, the parameters represented by u and v could notationally be built into the specification of w, but there are advantages to making them explicit in this manner.

Our approach is to study the geometry and generalized differentiability of the multifunction S_{KKT} in (1.8) without restricting our focus only to situations where S_{KKT} exhibits single-valuedness. We identify properties that automatically turn into Lipschitz continuity or B-differentiability under single-valuedness, but have important content even without that. The issue of ascertaining single-valuedness is thus posed not as a prerequisite to the analysis of "rates of change," but as a separate matter that can be taken up subsequent to such analysis.

We begin in Sect.2 by demonstrating under mild assumptions on X (convexity would suffice but is not required) that the graph of S_{KKT} is a Lipschitz manifold in the sense introduced by Rockafellar [18] in work with subgradient multifunctions. The dimension of this manifold is $m+n+d$, the same as that of the domain space for S_{KKT}, so this shows very powerfully the "functionlike" nature of S_{KKT}, even though its values need not be singletons everywhere. The Lipschitz aspect of the manifold furnishes preliminary insight into limitations on the effects of perturbations. When S_{KKT} does happen to be single-valued locally, the graphical geometry entails that it must be locally Lipschitz continuous too. A Lagrangian version of the generalized KKT conditions is used as a stepping stone to these results.

In moving on to generalized differentiability in Sect.3, we rely on "proto-derivatives," which are distinguished by their being defined in terms of set convergence of graphs. In heuristic terms, a multifunction is proto-differentiable at a point in its graph, if the image of its graph near the point "stabilizes" as closer and closer views are examined. It was noted in [7] that a single-valued mapping is B-differentiable if and only if it is continuous and proto-differentiable with single-valued proto-derivative mapping (in which event the proto-derivatives are the same as the B-derivatives). We augment this here by showing that a single-valued locally Lipschitz continuous mapping is B-differentiable if and only if it is proto-differentiable; there is no need in this case to assume single-valuedness of the proto-derivative multifunction. As a geometric extension of this fact, we prove that the graph a proto-differentiable multifunction is locally a Lipschitz manifold if and only if it is locally a B-differentiable Lipschitz manifold.

In application to sensitivity analysis, we prove that if the sets X and K are "fully amenable" (as holds in particular if they are polyhedral convex, but allows also for cases where X is not necessarily convex), then the multifunction S_{KKT} in (1.8) is proto-differentiable. This has the consequence that the graph of S_{KKT} is a B-differentiable Lipschitz manifold, a powerful property not previously observed.

Although new insights into graphical geometry are presented in this paper, the methodology of proto-differentiability has already been found to be effective in the study of multifunctions much more general than the one treated here; cf. our earlier work in [6]. On the other hand, many questions remain unanswered, especially concerning the suppression of some of the elements involved in a setting like (1.8). In [7] we obtained proto-derivatives for quasi-solution multifunctions associated with first-order optimality conditions for similar minimization problems, but in the form $(v, w) \mapsto x$, without explicit dependence on primal perturbations u or pairing with multiplier elements y. Whether or not the graph of such a multifunction can be seen locally as a Lipschitz manifold is still unclear, however. As for the suppression of v, results of Levy [5] give approximations at least of "outer" proto-derivatives, which are weaker than true proto-derivatives, but again the graphical geometry is not yet satisfactorily understood.

Of course, when some localization of the multifunction S_{KKT} in (1.8) is single-valued and Lipschitz continuous, or B-differentiable, there is no difficulty whatever retaining these properties in suppressing some of the elements. Even the submapping $w \mapsto x$ will be single-valued and Lipschitz continuous, or B-differentiable, in that case.

2. GRAPHS AS LIPSCHITZ MANIFOLDS

The notation $S : \mathbb{R}^k \rightrightarrows \mathbb{R}^l$ indicates a multifunction, or set-valued mapping, that assigns to each $w \in \mathbb{R}^k$ a set of vectors $z \in \mathbb{R}^l$. The *graph* of S is $\operatorname{gph} S = \{(w, z) : z \in S(w)\}$. For some choices of w the set $S(w)$ may be empty, or it may just be a singleton. The *effective domain* of S is $\operatorname{dom} S = \{w : S(w) \neq \emptyset\}$. As a special case, S might be *single-valued* on a set D, meaning that $S(w)$ is a singleton for every $w \in D$ (and hence in particular that $D \subseteq \operatorname{dom} S$). A weaker concept is that of S being *locally single-valued around* $(\bar{w}, \bar{z})$, a point of $\operatorname{gph} S$, which refers to the existence of neighborhoods W of $\bar{w}$ and Z of $\bar{z}$ such that the "submapping" $w \in W \mapsto S(w) \cap Z$ is single-valued.

When such single-valuedness occurs, one can go on to ask whether the submapping in question is actually continuous or even Lipschitz continuous. If so, there are obvious consequences for the geometry of $\operatorname{gph} S$. But the same graphical geometry would persist if a smooth one-to-one mapping of $\mathbb{R}^{k+l}$ onto itself were applied to $\operatorname{gph} S$, regardless of whether the image of $\operatorname{gph} S$ were again locally the graph of a single-valued mapping. Putting this in reverse, one can have situations where $\operatorname{gph} S$ is not locally single-valued around $(\bar{w}, \bar{z})$, and yet it corresponds under a certain nonlinear local transformation, smooth in both directions, to the graph of a single-valued, Lipschitz continuous mapping. The following concept assists in formalizing the idea.

Definition 2.1. *A subset M of $\mathbb{R}^N$ is locally* a Lipschitz manifold of dimension s *around the point $\bar{u} \in M$ in the sense of [18] if, under a smooth change of coordinates around $\bar{u}$, it can be identified locally with the graph of a Lipschitz continuous mapping of dimension s, or in other words, if there is an open neighborhood U of $\bar{u}$ in $\mathbb{R}^N$ and a one-to-one mapping Φ of U onto an open set in $\mathbb{R}^s \times \mathbb{R}^{N-s}$ with both Φ and Φ^{-1} continuously differentiable, such that $\Phi(M \cap U)$ is the graph of some Lipschitz continuous mapping $H : O \to \mathbb{R}^{N-s}$ for an open set O in $\mathbb{R}^s$.*

In particular, the graph of any locally Lipschitz continuous mapping $G : \mathbb{R}^k \to \mathbb{R}^l$ (single-valued) is locally a Lipschitz manifold of dimension k in $\mathbb{R}^{k+l}$; for this one can take Φ to be the identity. The same is true then of G^{-1}, which need not be a single-valued mapping.

Indeed, if a multifunction $S : \mathbb{R}^k \rightrightarrows \mathbb{R}^l$ is such that $\operatorname{gph} S$ is a Lipschitz manifold of dimension s around $(\bar{w}, \bar{z})$, and if the multifunction $S' : \mathbb{R}^{k'} \rightrightarrows \mathbb{R}^{l'}$ is such that $\operatorname{gph} S'$ corresponds to $\operatorname{gph} S$ under a one-to-one transformation which is continuously differentiable in both directions and associates $(\bar{w}, \bar{z})$ with $(\bar{w}', \bar{z}')$ (with $k + l = k' + l'$), then $\operatorname{gph} S'$ too is locally a Lipschitz manifold of dimension s around $(\bar{z}', \bar{w}')$. (Here the transformation could merely be local.) The following elementary facts illustrate this. (By the *inverse* of S is meant the multifunction $S^{-1} : z \mapsto \{w : z \in S(w)\}$.)

Proposition 2.2. *If the graph of $S : \mathbb{R}^k \rightrightarrows \mathbb{R}^l$ is locally a Lipschitz manifold of dimension s around $(\bar{w}, \bar{z})$, then the graph of the inverse $S^{-1} : \mathbb{R}^l \rightrightarrows \mathbb{R}^k$ is likewise locally a Lipschitz manifold of dimension s around $(\bar{z}, \bar{w})$.*

Proposition 2.3. *If the graph of $S : \mathbb{R}^k \rightrightarrows \mathbb{R}^l$ is locally a Lipschitz manifold of dimension s around $(\bar{w}, \bar{z})$, and if the single-valued mapping $G : \mathbb{R}^k \to \mathbb{R}^l$ is of class $\mathcal{C}^1$, then the graph of $S + G$ is locally a Lipschitz manifold of dimension s around $(\bar{w}, \bar{z} + G(\bar{z}))$.*

Proof. The mapping $\Phi : (w, z) \mapsto (w, z + G(z))$ carries one graph onto the other. This is a $\mathcal{C}^1$ mapping for which the inverse, $\Phi^{-1} : (w, z) \mapsto (w,\ z - G(z))$, is $\mathcal{C}^1$ as well. □

Especially important is the next example, which concerns a key class of mappings which can well fail to be single-valued.

Proposition 2.4 (Minty [8]). *For any maximal monotone multifunction $T : \mathbb{R}^n \rightrightarrows \mathbb{R}^n$, the set $\operatorname{gph} T$ is locally a Lipschitz manifold of dimension n around all of its points.*

The subgradient multifunctions ∂f of proper, lower semicontinuous, convex functions $f : \mathbb{R}^n \to \overline{I\!R}$ (the extended reals) are known to be maximal monotone, cf. [17], and the same is therefore true for the graphs of normal cone multifunctions $x \mapsto N_X(x)$ when X is a closed, convex set (and $N_X(x)$ is taken to be the empty set for $x \notin X$), inasmuch as $N_X = \partial \delta_X$.

Corollary 2.5. *For any proper, lsc, convex function $f : \mathbb{R}^n \to \overline{I\!R}$ the graph of the subgradient multifunction $\partial f : \mathbb{R}^n \rightrightarrows \mathbb{R}^n$ is locally a Lipschitz manifold of dimension n around all of its points. Likewise, for any nonempty, closed, convex set $X \subseteq \mathbb{R}^n$ the graph of the normal cone multifunction $N_X : \mathbb{R}^n \rightrightarrows \mathbb{R}^n$ is locally a Lipschitz manifold of dimension n around all of its points.*

The scope of this geometric property can be greatly be extended now beyond the bounds of convex analysis on the basis of recent results of Poliquin and Rockafellar [11]. The subgradients then are "limiting proximal subgradients." Recall that a function $f : \mathbb{R}^n \to \overline{I\!R}$ is *amenable* at a point $\bar{x}$ if there is a local representation $f(x) = g(G(x))$ in which G is a $\mathcal{C}^1$ mapping into a space $\mathbb{R}^m$, the function $g : \mathbb{R}^m \to \overline{I\!R}$ is a proper, lower semicontinuous and convex, the point $G(\bar{x})$ lies in $\operatorname{dom} g$, and the constraint qualification is satisfied that there is no nonzero vector y in the normal cone $N_{\operatorname{dom} g}(G(\bar{x}))$ for which the gradient of the mapping $x \mapsto \langle y, G(x) \rangle$ vanishes. It is *strongly amenable* if there is such a representation with G of class $\mathcal{C}^2$ rather than just of class $\mathcal{C}^1$.

The category of strongly amenable functions includes all $\mathcal{C}^2$ functions f, all proper, lsc, convex functions f, and more. For instance, it includes all functions of the form $f = h + \delta_D$ in which h is the pointwise max of a finite collection of $\mathcal{C}^2$ functions on $\mathbb{R}^n$ and D is a subset of $\mathbb{R}^n$ specified by finitely many equality and inequality constraints for $\mathcal{C}^2$ functions—provided only that, at the point $\bar{x}$ where the amenability is to be tested, the Mangasarian-Fromovitz constraint qualification is fulfilled; see [19] and [10] (the terminology of amenability comes from the latter). Any function of the kind just described is in fact "fully amenable," a higher property which will be of interest in Sect.3 and will explained there when it is needed.

Theorem 2.6. *If a function $f : \mathbb{R}^n \to \overline{\mathbb{R}}$ is strongly amenable at $\bar{x}$, then for any subgradient $\bar{v} \in \partial f(\bar{x})$ the graph of the subgradient multifunction $\partial f : \mathbb{R}^n \rightrightarrows \mathbb{R}^n$ is locally a Lipschitz manifold of dimension n around $(\bar{x}, \bar{v})$.*

Proof. Proposition 2.5 of Poliquin and Rockafellar [11] establishes that f is "prox-regular" and "subdifferentially continuous" at $\bar{x}$. Theorem 3.2 of the same paper demonstrates, among other things, that when these properties hold and $\bar{v} \in \partial f(\bar{x})$ there is a neighborhood of $(\bar{x}, \bar{v})$ with respect to which the graph of ∂f is maximal submonotone, i.e., such that the multifunction $T = \partial f + \lambda I$ is maximal monotone for some $\lambda > 0$. Then the graph of T is, by a localization of Minty's theorem cited above, locally a Lipschitz manifold of dimension n around $(\bar{x}, \bar{v} + \lambda \bar{x})$. Since $\partial f = T - \lambda I$, we conclude from Proposition 2.3 that the graph of ∂f itself is locally a Lipschitz manifold of dimension n around $(\bar{x}, \bar{v})$. □

A set D is defined to be *amenable* or *strongly amenable* at a point $\bar{x}$ if its indicator function has these properties. In particular, any closed, convex set D is strongly amenable at any of its points, and so too is any set D specified locally by finitely many $\mathcal{C}^2$ constraints in such a manner that the Mangasarian-Fromovitz constraint qualification holds.

Corollary 2.7. *If a set $X \subseteq \mathbb{R}^n$ is strongly amenable at one of its points $\bar{x}$, then for any normal vector $\bar{v} \in N_X(\bar{x})$ the graph of the multifunction N_X is locally a Lipschitz manifold of dimension n around $(\bar{x}, \bar{v})$.*

Proof. This specializes Theorem 2.6 to the case of $f = \delta_X$. □

These results have laid the foundation for proving the following fact about the geometry of the general multifunction S_{KKT} introduced above.

Theorem 2.8. *Let $(\bar{x}, \bar{y}) \in S_{\text{KKT}}(\bar{u}, \bar{v}, \bar{w})$ and suppose the set X is strongly amenable at the point $\bar{x}$. Then the graph of S_{KKT} is locally a Lipschitz manifold of dimension $m + n + d$ around $(\bar{u}, \bar{v}, \bar{w}; \bar{x}, \bar{y})$.*

Proof. The relation $y \in N_K(F(w,x)+u)$ in (1.7) can be expressed in terms of the polar cone $Y = K^*$ as $F(w,x) + u \in N_Y(y)$, cf. [17], Corollary 23.5.4. Let $D = \mathbb{R}^d \times X \times Y$, so that

$$N_D(w,x,y) = \{(z', v', u') : z' = 0 \in \mathbb{R}^d,\ v' \in N_X(x),\ u' \in N_Y(y)\}. \tag{2.1}$$

Further, define the mapping $G : \mathbb{R}^d \times \mathbb{R}^n \times \mathbb{R}^m$ by

$$G(w,x,y) = (0,\ \nabla f_0(w,x) + \nabla_x F(w,x)^{\mathsf{T}} y, -F(w,x)).$$

In this notation, the KKT conditions (1.7) come down to $(0, v, u) \in N_D(w,x,y) + G(w,x,y)$. We thus have

$$(w,x,y;z,v,u) \in \operatorname{gph}(N_D + G) \iff \begin{cases} (u,v,w;x,y) \in \operatorname{gph} S_{\text{KKT}}, \\ z = 0. \end{cases} \tag{2.2}$$

Therefore, the assertion that $\operatorname{gph} S_{\text{KKT}}$ is locally a Lipschitz manifold of dimension $m + n + d$ around $(\bar{u}, \bar{v}, \bar{w}; \bar{x}, \bar{y})$ is equivalent to the assertion that $\operatorname{gph}(N_D + G)$ has this property around $(\bar{w}, \bar{x}, \bar{y}; 0, \bar{v}, \bar{u})$.

Because Y is a closed, convex cone, we know that D is strongly amenable at $(\bar{w}, \bar{x}, \bar{y})$ whenever the points $\bar{x} \in X$ and $\bar{y} \in Y$ are such that X is strongly amenable at $\bar{x}$. Then by Corollary 2.7 the graph of the normal cone multifunction N_D is locally a Lipschitz manifold of dimension $d + n + m$ around $(\bar{w}, \bar{x}, \bar{y}, 0, v', u')$ for any vectors $v' \in N_X(\bar{x})$ and $u' \in N_Y(\bar{y})$. On the other hand, the mapping G is of class $\mathcal{C}^1$ (by our blanket assumption that f_0 and the component functions f_i of F are of class $\mathcal{C}^2$). It follows then by Proposition 2.3 that the graph of $N_D + G$ is locally a Lipschitz manifold of dimension $d+n+m$ around any of its points $(\bar{w}, \bar{x}, \bar{y}, \bar{z}, \bar{v}, \bar{u})$ such that X is strongly amenable at $\bar{x}$. This gives what we need. □

3. GRAPHICAL DIFFERENTIABILITY

A mapping $G : \mathbb{R}^k \to \mathbb{R}^l$ (single-valued) is *B-differentiable* at $\bar{w}$ as defined by Robinson [15] if the difference quotient mappings

$$\Delta_t G(\bar{w}) : \omega \mapsto [G(\bar{w} + t\omega) - G(\bar{w})]/t \quad \text{for} \quad t > 0$$

converge pointwise as $t \searrow 0$ to a continuous mapping $H : \mathbb{R}^k \to \mathbb{R}^l$ and do so uniformly on bounded sets. Then $H(\omega)$ is the one-sided directional derivative of G relative to ω.

Classical differentiability is the case where $H(\omega)$ is linear in ω. The definition of B-differentiability always implies at least that $H(0) = 0$ and that H is positively homogeneous, i.e., $H(\lambda\omega) = \lambda H(\omega)$ for $\lambda > 0$ (hence also for $\lambda = 0$), but it allows $H(\omega)$ to be nonlinear in ω in other respects. We will denote H by $DG(\bar{w})$ and call this the B-derivative mapping for G at $\bar{w}$.

Of course, when G is Lipschitz continuous around $\bar{w}$ the same holds uniformly for the mappings $\Delta_t G(\bar{w})$, and mere pointwise convergence of these mappings to H as $t \searrow 0$ implies that H is globally Lipschitz continuous and that the convergence is uniform on bounded sets. In this case, therefore, B-differentiability is automatic simply from the existence of $\lim_{t \searrow 0} [G(\bar{w} + t\omega) - G(\bar{w})]/t$ for every ω.

Definition 3.1. *A subset M of $\mathbb{R}^N$ is locally* a B-differentiable Lipschitz manifold of dimension s *near the point $\bar{u} \in M$ if, under a smooth change of coordinates around $\bar{u}$, it can be identified locally with the graph of a Lipschitz continuous mapping of dimension s that happens also to be B-differentiable, or in other words, if there is an open neighborhood U of $\bar{u}$ in $\mathbb{R}^N$ and a one-to-one mapping Φ of U onto an open set in $\mathbb{R}^s \times \mathbb{R}^{N-s}$ with both Φ and Φ^{-1} continuously differentiable, such that $\Phi(M \cap U)$ is the graph of some B-differentiable, Lipschitz continuous mapping $H : O \to \mathbb{R}^{N-s}$ for an open set O in $\mathbb{R}^s$.*

The graph of any B-differentiable, locally Lipschitz continuous mapping $G : \mathbb{R}^k \to \mathbb{R}^l$ is locally a B-differentiable Lipschitz manifold in $\mathbb{R}^{k+l}$. So too is the graph of G^{-1} (generally just a multifunction). Following the same geometric patterns as in the preceding section, we arrive at the principle that if a multifunction $S : \mathbb{R}^k \rightrightarrows \mathbb{R}^l$ is such that $\operatorname{gph} S$ is locally a B-differentiable Lipschitz manifold of dimension s around $(\bar{w}, \bar{z})$, and if the multifunction $S' : \mathbb{R}^{k'} \rightrightarrows \mathbb{R}^{l'}$ is such that $\operatorname{gph} S'$ corresponds to $\operatorname{gph} S$ under a one-to-one transformation which is continuously differentiable in both directions and associates $(\bar{w}, \bar{z})$ with $(\bar{w}', \bar{z}')$, then $\operatorname{gph} S'$ too is a B-differentiable Lipschitz manifold of dimension s around $(\bar{z}', \bar{w}')$.

Facts analogous to Propositions 2.2 and 2.3 can be stated at once.

Proposition 3.2. *If the graph of $S : \mathbb{R}^k \rightrightarrows \mathbb{R}^l$ is locally a B-differentiable Lipschitz manifold of dimension s around $(\bar{w}, \bar{z})$, then the graph of the inverse $S^{-1} : \mathbb{R}^l \rightrightarrows \mathbb{R}^k$ is likewise locally a B-differentiable Lipschitz manifold of dimension s around $(\bar{z}, \bar{w})$.*

Proposition 3.3. *If the graph of $S : \mathbb{R}^k \rightrightarrows \mathbb{R}^l$ is locally a B-differentiable Lipschitz manifold of dimension s around $(\bar{w}, \bar{z})$, and if the single-valued mapping $G : \mathbb{R}^k \to \mathbb{R}^l$ is of class $\mathcal{C}^1$, then the graph of the multifunction $S + G$ is locally a B-differentiable Lipschitz manifold of dimension s around $(\bar{w}, \bar{z} + G(\bar{z}))$.*

We wish to take advantage of such geometry in treating various multifunctions that arise in optimization. For this we are led to a concept of generalized differentiability called proto-differentiability, which was introduced in [21]. Proto-differentiability is distinguished from other differentiability notions through its utilization of set convergence of graphs.

Consider any multifunction $S : \mathbb{R}^k \rightrightarrows \mathbb{R}^l$ and any pair $(\bar{w}, \bar{z}) \in \operatorname{gph} S$. For each $t > 0$ one can form the difference quotient multifunction

$$\Delta_t S(\bar{w}|\bar{z}) : \omega \mapsto [S(\bar{w} + t\omega) - \bar{z}]/t \quad \text{for} \quad t > 0.$$

(When $\bar{z}$ happens to be the sole element of $S(\bar{w})$, the notation $\Delta_t S(\bar{w})$ suffices.) Instead of asking the difference quotient multifunctions $\Delta_t S(\bar{w}|\bar{z})$ to converge in some kind of pointwise sense as $t \searrow 0$, proto-differentiability asks that they converge graphically, i.e., that their graphs converge as subsets of $\mathbb{R}^k \times \mathbb{R}^l$ to the graph of some multifunction $H : \mathbb{R}^k \rightrightarrows \mathbb{R}^l$. Then H is the *proto-derivative* multifunction at $\bar{w}$ for $\bar{z}$; the notation we will use for this multifunction H is $DS(\bar{w}|\bar{z})$. It associates with each $\omega \in \mathbb{R}^k$ some (possibly empty) subset of $\mathbb{R}^l$.

The concept of Painlevé-Kuratowski set convergence underlies the formation of these graphical limits. It refers to a kind of approximation described from two sides as follows. The *inner set limit* of a parameterized family of sets $\{C_t\}_{t>0}$ in $\mathbb{R}^N$ is the set of points η such that for *every* sequence $t_k \searrow 0$ there is a sequence of points $\eta_k \in C_{t_k}$ with $\eta_k \to \eta$. The *outer set limit* of the family is the set of points η such that for *some* sequence $t_k \searrow 0$ there is a sequence of points $\eta_k \in C_{t_k}$ with $\eta_k \to \eta$. When the inner and outer set limits coincide, the common set C is the *limit* as $t \searrow 0$.

In our framework, this is applied to sets that are the graphs of multifunctions. For a multifunction $S : \mathbb{R}^k \rightrightarrows \mathbb{R}^l$ and any pair $(\bar{w}, \bar{z})$ in $\operatorname{gph} S$, the graph of the difference quotient mapping $\Delta_t S(\bar{w}|\bar{z})$ is $t^{-1}[\operatorname{gph} S - (\bar{w}, \bar{z})]$. The multifunction $D^+S(\bar{w}|\bar{z}) : \mathbb{R}^k \rightrightarrows \mathbb{R}^l$ having as its graph the outer limit of the sets $\operatorname{gph} \Delta_t S(\bar{w}|\bar{z})$ as $t \searrow 0$ is called the *outer graphical derivative* of S at $\bar{w}$ for $\bar{z}$. In parallel, the multifunction $D^-S(\bar{w}|\bar{z}) : \mathbb{R}^k \rightrightarrows \mathbb{R}^l$ having as its graph the inner limit of these sets is the *inner graphical derivative*. Proto-differentiability of S at $\bar{w}$ for $\bar{z}$ is the case where the outer and inner derivatives agree, the common mapping being then the *proto-derivative*: $DS(\bar{w}|\bar{z}) = D^+S(\bar{w}|\bar{z}) = D^-S(\bar{w}|\bar{z})$, cf. Rockafellar [24]. (Again, if $\bar{z}$ happens to be the only element of $S(\bar{w})$, the notation can be simplified to $DS(\bar{w})$.)

The following result comes from [7] and clarifies the relationship in the single-valued case between proto-differentiability and B-differentiability.

Proposition 3.4 ([7, Proposition 2.2]). *Let O be an open neighborhood of a point $\bar{w} \in \mathbb{R}^k$ and consider a single-valued mapping $G : O \to \mathbb{R}^l$. Then G is B-differentiable at $\bar{w}$ if and only if G is continuous at $\bar{w}$ and (as a special case of a multifunction that happens to be single-valued) is proto-differentiable at $\bar{w}$ with $DG(\bar{w})$ single-valued, in which event one has the local expansion*

$$G(\bar{w} + t\omega) = G(\bar{w}) + tDG(\bar{w})(\omega) + o(t|\omega|) \quad \textit{for} \quad t > 0.$$

When G happens to be Lipschitz continuous around $\bar{w}$, an even stronger result holds.

Proposition 3.5. *In the setting of Proposition 3.4, suppose that G is also Lipschitz continuous around $\bar{w}$. Then G is B-differentiable at $\bar{w}$ if and only if G is proto-differentiable at $\bar{w}$, in which event the proto-derivative multifunction reduces to the B-derivative mapping.*

Proof. Necessity follows immediately from Proposition 3.4. For sufficiency, we apply [22, Theorem 4.3], noting that for the single-valued mapping G local Lipschitz continuity around $\bar{w}$ is equivalent to "pseudo-Lipschitz" continuity at $\bar{w}$ as a multifunction. (In [22], B-differentiability is equivalent to a property called "semi-differentiability.") □

These results mean that proto-differentiability extends to multifunctions, just in the manner that might be wished, the notion of one-sided directional differentiability deemed most appropriate in the sensitivity analysis of single-valued mappings, smooth or nonsmooth. The question of whether a certain mapping is single-valued or not can be dealt with as a separate issue, which need not be resolved before progress can be made on quantitative stability of solutions.

Theorem 3.6. *For a multifunction $S : \mathbb{R}^k \rightrightarrows \mathbb{R}^l$, let $(\bar{w}, \bar{z})$ be a point around which the graph of S is locally a Lipschitz manifold of dimension s. In order that the graph of S be locally in fact a B-differentiable Lipschitz manifold around $(\bar{w}, \bar{z})$, it is necessary and sufficient that, for all (w, z) within some neighborhood of $(\bar{w}, \bar{z})$ relative to* gph S, *the mapping S should be proto-differentiable at w for z.*

Proof. A transformation Φ as in Definitions 2.1 and 3.1 allows us to pass to the framework where S is replaced by a single-valued Lipschitz continuous mapping G. Then Proposition 3.5 can be brought into play, and the result is immediate. □

The route to applying this result to optimality conditions lies in the second-order variational analysis of subgradient multifunctions and a further form of "amenability." Recall that a convex function $g : \mathbb{R}^m \to \overline{\mathbb{R}}$ is *piecewise linear-quadratic* if the set dom g is polyhedral and can be represented as the union of finitely many polyhedral convex sets, relative to each of which g has a formula as a polynomial function of degree no more than 2. A function $f : \mathbb{R}^n \to \overline{\mathbb{R}}$ is called *fully amenable* at $\bar{x}$ if it satisfies the earlier definition of being strongly amenable at $\bar{x}$ and does so with the additional condition that the convex function g in that representation is piecewise linear-quadratic.

Functions f of this type were first studied by Rockafellar [19] for their second-order "epi-derivatives". The connection between such generalized second derivatives and proto-derivatives of the corresponding subgradient multifunctions ∂f was established for convex functions by Rockafellar [23] and for arbitrary fully amenable functions by Poliquin [9]. The following result was obtained in particular.

Theorem 3.7 ([9]). *Suppose that the function $f : \mathbb{R}^n \to \overline{\mathbb{R}}$ is fully amenable at $\bar{x}$, and let $\bar{v} \in \partial f(\bar{x})$. Then for all pairs (x, v) in some neighborhood of $(\bar{x}, \bar{v})$ relative to* gph ∂f, *the subgradient multifunction ∂f is proto-differentiable at x for v.*

We can now deduce from this an important geometric property of the graphs of subgradient multifunctions.

Theorem 3.8. *If the function $f : \mathbb{R}^n \to \overline{\mathbb{R}}$ is fully amenable at $\bar{x}$, then for any $\bar{v} \in \partial f(\bar{x})$ the graph of the subgradient multifunction $\partial f : \mathbb{R}^n \rightrightarrows \mathbb{R}^n$ is locally a B-differentiable Lipschitz manifold around $(\bar{x}, \bar{v})$.*

Proof. It is merely necessary to combine Theorem 3.7 with Theorems 2.6 and 3.6. □

There is no need here to discuss the large class of functions that are fully amenable (see [10]), because attention here is turned toward application to KKT conditions. The case of indicator functions is therefore the main one to consider.

A set $X \subseteq \mathbb{R}^n$ is called *fully amenable* at one of its points $\bar{x}$ if the indicator function δ_X is fully amenable at $\bar{x}$.

Corollary 3.9. *If a set $X \subseteq \mathbb{R}^n$ is fully amenable at $\bar{x}$, then for any $\bar{v} \in N_X(\bar{x})$ the graph of the normal cone multifunction $N_X : \mathbb{R}^n \rightrightarrows \mathbb{R}^n$ is locally a B-differentiable Lipschitz manifold around $(\bar{x}, \bar{v})$.*

The operational meaning of full amenability of a set can be elucidated as follows.

Proposition 3.10. *A set $X \subseteq \mathbb{R}^n$ is fully amenable at one of its points $\bar{x}$ if and only if there exists a local representation of X around $\bar{x}$ by a finite family of $\mathcal{C}^2$ constraints (equations, inequalities or a mixture) such that the Mangasarian-Fromovitz constraint qualification is satisfied at $\bar{x}$.*

Proof. An indicator function δ_D is convex and piecewise linear-quadratic if and only if D is a polyhedral convex set. Through its definition, therefore, full amenability of X at $\bar{x}$ means the existence of a neighborhood V of $\bar{x}$ yielding a representation $X \cap V = G^{-1}(D) \cap V$ for some polyhedral convex set D and $\mathcal{C}^2$ mapping G with the property that no nonzero vector $y \in N_D(G(\bar{x}))$ has $\nabla G(\bar{x})^{\top} y = 0$. The set D can be expressed by a system of finitely many linear equations and inequalities, and moreover this can be accomplished in such a manner that the system satisfies the Mangasarian-Fromovitz constraint qualification at $G(\bar{x})$. Namely, we can write $D = G_0^{-1}(D_0)$ for a set D_0 of the form $\mathbb{R}^q_+ \times \mathbb{R}^r$ and an affine mapping G_0 with the property that no nonzero vector $y_0 \in N_{D_0}(G_0(G(\bar{x})))$ has $\nabla G_0(G(\bar{x}))^{\top} y_0 = 0$. Then the local representation $X = [G_0 \circ G]^{-1}(D_0)$ corresponds to a standard system of $\mathcal{C}^2$ constraints around $\bar{x}$ such that the Mangasarian-Fromovitz constraint qualification is satisfied at $\bar{x}$.

Proposition 3.11. *A polyhedral convex set $X \subseteq \mathbb{R}^n$ is fully amenable at all of its points.*

Proof. This is obvious from the comment made at the beginning of the proof of the preceding proposition. The mapping G in the definition of full amenability can in this case be taken to be the identity. □

Our principle geometric result about the KKT multifunction defined in (1.8) can now be stated and proved.

Theorem 3.12. *Let $(\bar{x}, \bar{y}) \in S_{\rm KKT}(\bar{u}, \bar{v}, \bar{w})$ and suppose the set X is fully amenable at the point $\bar{x}$. Suppose also that the cone K is fully amenable at $F(\bar{w}, \bar{x}) + \bar{u}$. Then the graph of $S_{\rm KKT}$ is locally a B-differentiable Lipschitz manifold of dimension $m + n + d$ around the point $(\bar{u}, \bar{v}, \bar{w}; \bar{x}, \bar{y})$.*

Proof. The argument closely follows the lines of the one employed for the related result in Theorem 2.8. We introduce D and G in exactly the same manner as there and use them to represent the graph of $S_{\rm KKT}$ as in (2.2). The challenge becomes that of demonstrating that the graph of $N_D + G$ is locally a B-differentiable Lipschitz manifold around $(\bar{w}, \bar{x}, \bar{y}; 0, \bar{v}, \bar{u})$. By virtue of Proposition 3.3, this can be accomplished by demonstrating that the graph of N_D is locally a B-differentiable Lipschitz manifold around the point

$$(\bar{w}, \bar{x}, \bar{y}; 0, \bar{v}', \bar{u}') \quad \text{for} \quad \bar{v}' = \bar{v} - \nabla_x f_0(\bar{w}, \bar{x}) - \nabla_x F(\bar{w}, \bar{x})^{\top} \bar{y}, \quad \bar{u}' = F(\bar{w}, \bar{x}) + \bar{u}.$$

The product structure exhibited for the graph of N_D in (2.1) brings this down to verifying that the graph of N_X is a B-differentiable Lipschitz manifold around the point $(\bar{x}, \bar{v}')$, whereas the graph of N_Y is such a manifold around $(\bar{y}, \bar{u}')$.

The required property of the graph of N_X follows from Corollary 3.9 through the full amenability assumed for X. The corresponding assumption for K likewise tells us that the graph of N_K is locally a B-differentiable Lipschitz manifold around $(\bar{u}', \bar{y})$. But $N_Y = N_K^{-1}$ through the polarity between Y and K, so by Proposition 3.2 the graph of N_Y is locally a B-differentiable Lipschitz manifold around $(\bar{y}, \bar{u}')$, as desired. □

Corollary 3.13. *The conclusion of Theorem 3.12 holds in particular when both X and K are polyhedral convex sets, or when both X and K can be represented by systems of finitely many $\mathcal{C}^2$ constraints for which the Mangasarian-Fromovitz constraint qualification is fulfilled at $\bar{x}$ and $F(\bar{w}, \bar{x}) + \bar{u}$, respectively.*

Proof. This specializes to the criteria for full amenability in Propositions 3.10 and 3.11. □

REFERENCES

[1] A.L. Dontchev, "Implicit function theorems for generalized equations". To appear, Mathematical Programming, 1995.

[2] A.L. Dontchev and W.W. Hager, "On Robinson's implicit function theorem". Set-Valued Analysis and Differential Inclusions, Birkhäuser, 1991.

[3] A.L. Dontchev and W.W. Hager, "Implicit functions, Lipschitz maps, and stability in optimization". To appear, in Mathematics of Operations Research, 1995.

[4] J. Kyparisis, "Parametric variational inequalities with multivalued solution sets". Mathematics of Operations Research 17, pp. 341–364, 1992.

[5] A.B. Levy, "Implicit set-valued mapping theorems and the sensitivity analysis of variational conditions". preprint, 1995.

[6] A.B. Levy and R.T. Rockafellar, "Sensitivity analysis of solutions to generalized equations". Transactions of the American Mathematical Society 345, pp. 661–671, 1994.

[7] A.B. Levy and R.T. Rockafellar,, "Sensitivity of solutions in nonlinear programming problems with nonunique multipliers". Accepted for publication in Nonsmooth Optimization, 1995.

[8] G.J. Minty, "Monotone (nonlinear) operators in Hilbert space". Duke Math. Journal Vol. 29, pp. 341-346, 1962.

[9] R.A. Poliquin, "Proto-differentiation of subgradient set-valued mappings". Canadian Journal of Mathematics, Vol. 42, pp. 520-532, 1990.

[10] R.A. Poliquin and R.T. Rockafellar, "Amenable functions in optimization". In "Nonsmooth Optimization Methods and Applications" (F. Giannessi, ed.), Gordon and Breach, pp. 338-353, 1992.

[11] R.A. Poliquin and R.T. Rockafellar, "Prox-regular functions in variational analysis". To appear in Transactions of the American Mathematical Society.

[12] S.M. Robinson, "Generalized equations and their solutions, part I: Basic theory". Mathematical Programming Study, Vol. 10, pp. 128–141, 1979.

[13] S.M. Robinson, "Strongly regular generalized equations". Mathematics of Operations Research Vol. 5, pp. 43–62, 1980.

[14] S.M. Robinson, "Generalized equations and their solutions, part ii: Applications to nonlinear programming". Mathematical Programming Study Vol. 19, pp. 200–221, 1982.

[15] S.M. Robinson, "Local structure of feasible sets in nonlinear programming, part iii: Stability and sensitivity". Mathematical Programming Study, Vol. 30, pp. 45–66, 1987.

[16] S.M. Robinson, "An implicit-function theorem for a class of nonsmooth functions". Mathematics of Operations Research, Vol. 16, pp. 292–309, 1991.

[17] R.T. Rockafellar, "Convex Analysis". Princeton University Press, 1970.

[18] R.T. Rockafellar, "Maximal monotone relations and the second derivatives of nonsmooth functions". Annales de l'Institût Henri Poincaré - Analyse non linéaire, Vol. 2, pp. 167–184, 1985.

[19] R.T. Rockafellar, "First and second-order epi-differentiability in nonlinear programming". Transactions of the American Mathematical Society, Vol. 307, pp. 75–108, 1988.

[20] R.T. Rockafellar, "Perturbation of generalized Kuhn-Tucker points in finite-dimensional optimization". Nonsmooth Analysis and Related Topics (F.H. Clarke et al., eds.), Plenum Press, pp. 393–402, 1989.

[21] R.T. Rockafellar, "Proto-differentiability of set-valued mappings and its applications in optimization". Analyse Non Linéaire (H. Attouch, J. P. Aubin, F.H. Clarke, and I. Ekeland, eds.), Gauthier-Villars, pp. 449–482, 1989.

[22] R.T. Rockafellar, "Second-order optimality conditions in nonlinear programming obtained by way of epi-derivatives". Mathematics of Operations Research Vol. 14, pp. 462–484, 1989.

[23] R.T. Rockafellar, "Generalized second derivatives of convex functions and saddle functions". Transactions of the American Mathematical Society Vol. 322, pp. 51-77, 1990.

[24] R.T. Rockafellar, "Nonsmooth analysis and parametric optimization". In "Methods of Nonconvex Analysis" (A. Cellina, ed.), Lecture Notes in Mathematics, Vol. 1446, Springer-Verlag, pp. 137–151, 1990.

[25] R.T. Rockafellar, "Lagrange multipliers and optimality". SIAM Review Vol. 35, pp. 183–238, 1993.

NUMERICAL RESULTS FOR UNCONSTRAINED OPTIMIZATION WITHOUT DERIVATIVES

Lucidi Stefano[1] and Sciandrone Marco[1]

[1] University of Rome "La Sapienza"
via Buonarroti 12-00185 Roma,Italy
e-mail: lucidi@iasi.rm.cnr.it
e-mail: sciandro@newton.dis.uniroma1.it

Abstract. This paper is a first attempt to investigate on the computational potentialities of the line search approach proposed in [4], for solving unconstrained optimization problems without using any gradient information. The reported numerical results show that the considered approach is very promising in the field of derivative free minimization methods.

Key words. Unconstrained optimization, derivative free algorithms, line search techniques.

1. INTRODUCTION

We consider the problem of minimizing an objective function without constraints on the variables and we assume that the partial derivatives are not available. Such optimization problems arise more and more frequently in industrial and scientific applications. In fact, we more and more often have to tackle applications that require the minimization of complex functions for which is very difficult or impossible to calculate analytically their partial derivatives. Usually, in these applications the values of the objective function are obtained by measurements made on some complex system or by simulation procedure. Therefore, any evaluation of the objective function could be very expensive and affected by noise. These reasons may make the approximation of the objective function's derivatives by means of numerical differentiation formulae impracticable.

The study of derivative free minimization methods has been one of the main research argument in the field of optimization. Most of the algorithms proposed in literature come in three different classes.

In the methods of the first class, usually called *pattern search* methods, the estimate of the minimizer is made as the result of comparing the function values at points belonging to a specified geometric pattern (usually a simplex). Examples of such algorithms were proposed in [5], [7] and more recently in [12], [2]

The methods of the second class (see [8], [9], [14], [10], [4]) drawn their inspiration from the traditional gradient-based methods, in fact their common characteristic is the fact of performing one-dimensional minimizations along the search directions.

The methods of the last class are based on the idea of approximating the objective function by a model which is progressively built and updated (see [1], [11]). This approach seems to be particulary valuable in the case where the cost of evaluation of the objective function is so high as to make the use of expensive linear algebra calculations practicable.

As regard theoretical properties of the previous methods, most of them has been conceived without any convergence analisys in mind, only few theoretical results have been reported. Only recently, general global convergence analisys have been performed in the classes of the pattern search methods (see [12]) and the line search methods (see [4]). The results reported in [12] and [4] show that, even if the partial derivatives are neither calculated or approximated, it is possible to define algorithms which have convergence properties comparable to those proved for classical methods which use gradient information.

In particular, in [12] a new subclass of pattern search algorithms is proposed and their global convergence towards a stationary point is proved. Some of these algorithms have been implemented, in particular it is available a code described in [13] and called PDS, which can be considered an important tool for solving minimization problems without using any information about the derivatives of the objective function.

In [4] new stabilization criteria to force the global convergence of a linesearch-type algorithm are introduced. The distinguishing feature of these criteria is that they do not require any gradient information and, hence, they can be the basis to define new derivative free line search algorithms. However, until now, no work have been devoted to investigate the effectiveness from a computational point of view of the approach proposed in [4].

In this work we perform a first attempt to go into the computational details of an algorthm model proposed in [4] and [3]. In particular we have concentrated our attention on an algorithm model described in [3] and based on the stabilization criteria of [4], we have implemented and tested it on a large number of test problems available from CUTE collection. Then, in order to have a feel of its efficience, we have compared its numerical results with the ones obtained by the PDS code of [12], which is believed one the most efficient derivative free method.

More specifically, in section 2 we describe a linesearch-type method for the satisfaction of global convergence conditions. In section 3, we construct a derivative free algorithm employing the described method. Finally, in section 4, we show our numerical experience.

2. THE LINESEARCH APPROACH

We consider the following unconstrained minimization problem

$$\min_{x \in \mathbb{R}^n} \quad f(x),$$

where we assume that $f : \mathbb{R}^n \to \mathbb{R}$ is a continuously differentiable function and that its derivatives cannot be evaluated. For this particular optimization problem can be defined iterative algorithms which have the same structure of the minimization methods using derivatives, namely methods which have the form:

$$x_{k+1} = x_k + \alpha_k d_k, \quad k = 0, 1, \ldots \tag{2.1}$$

where $x_0 \in \mathbb{R}^n$ is a given starting point, $d_k \neq 0$ is the search direction, and α_k is the stepsize. As usual the following compactness assumption will be required.

Assumption 2.1. *The level set $\mathcal{L} = \{x \in \mathbb{R}^n : f(x) \leq f(x_0)\}$ is compact.*

In the field of unconstrained minimization many results have been proposed in literature that consider global convergence towards a stationary points of algorithms of the type of (2.1). All these results characterize the way of choosing the direction d_k and the way of computing the stepsize α_k. In order to ensure the global convergence of the algorithm considered in this paper, we recall the following proposition.

Proposition 2.2. *Let $\{x_k\}$ be the sequence generated by (2.1). Suppose that Assumption 2.1. holds and assume that, for all k, the following conditions hold.*

(i) *There exists a $\eta > 0$ such that the matrix P_k whose columns are $d_{k+j}/\|d_{k+j}\|$, for $j = 0, \ldots, n-1$ satisfies*

$$|Det(P_k)| \geq \eta; \tag{2.2}$$

(ii)

$$f(x_{k+1}) \leq f(x_k); \tag{2.3}$$

$$\lim_{k\to\infty} \frac{\nabla f(x_k)' d_k}{\|d_k\|} = 0; \tag{2.4}$$

$$\lim_{k\to\infty} \|x_{k+1} - x_k\|. \tag{2.5}$$

Then either for some index ν we have $\nabla f(x_\nu) = 0$, or the sequence $\{x_k\}$ is such that:

(a) $\{x_k\}$ remains in $\mathcal{L}$ and admits limit points;

(b) every limit point of $\{x_k\}$ belong to $\mathcal{L}$;

(c) $\lim_{k\to\infty} \|\nabla f(x_k)\| = 0$;

(d) every limit point $\bar{x}$ of $\{x_k\}$ satisfies $\nabla f(\bar{x}) = 0$.

The preceding result plays an important role in the field of derivative free optimization methods and it can be used as basis to construct globally convergent algorithms. In fact Condition (i) requires only that n consecutive directions used in (2.1) are linearly indipendent, while Condition (ii) identifies mild requirements that must be guaranted by the choice of the stepsize α_k. In [3], [4], [6] has been proposed some acceptability criteria and line search techniques for the computation of α_k which, without using any derivative information, ensure the satisfactory of Condition (ii).

In general, the fact that any gradient information is not available implies that a line search technique must tackle the following difficulties:

(a) in order to evaluate the improvement of the objective function it is not possible to use an Armjio-type rule of the form:

$$f(x_k + \alpha_k d_k) \leq f(x_k) + \gamma \alpha_k \nabla f(x_k)' d_k; \tag{2.6}$$

(b) it is not known if d_k is a descent direction, therefore it is needed to try also negative values for α_k;

(c) it is not possible to estimate the "goodness" of the search direction d_k controlling that it is not orthogonal to the gradient.

The acceptability criteria and line search techniques proposed in [4], [3], overcome the problems described in points (a), (b) and (c). In particular, in this work we use the Line Search Algorithm (LMA) proposed in [3]. This algorithm is based on the following Conditions I and II.

Let $\gamma > 0$ and $\delta \in (0,1)$ be given numbers and let $\{\xi_k\}$ be a preselected sequence of positive numbers converging to zero.

Condition I. *There exists* $\alpha \in \mathbb{R}$ *such that:*

$$f(x_k + \alpha d_k) \leq f(x_k) - \gamma\alpha^2\|d_k\|^2, \tag{2.7}$$

$$f(x_k + \frac{\alpha}{\delta}) \geq min[f(x_k + \alpha d_k), f(x_k) - \gamma(\frac{\alpha}{\delta})^2\|d_k\|^2]. \tag{2.8}$$

Condition II. *There exists* $\alpha \in \mathbb{R}$ *such that:*

$$f(x_k + \alpha d_k) > f(x_k) - \gamma\alpha^2\|d_k\|^2, \tag{2.9}$$

$$f(x_k - \alpha d_k) > f(x_k) - \gamma\alpha^2\|d_k\|^2, \tag{2.10}$$

and

$$|\alpha| \leq \xi_k.$$

Condition I tackles the points (a) and (b). In fact, condition (2.6) is replaced by (2.7) and there is no restriction on the sign of α. As usual, the role of Condition I is to determine a "sufficiently" large α_k, that provides a "sufficiently" decrease of the objective function. The need of Condition II arises from the point (c). In fact Condition II is a test for terminating the search for α_k when it is likely that d_k is not a "good" search direction.
Then we have the following algorithm, where $\Delta_k > 0$ is an initial estimate of α_k.

The Line Search Algorithm (LSA)

Data. $\Delta_k \in \mathbb{R}$, $\gamma > 0$, $\delta \in (0,1)$, $\xi_k > 0$.

Step 1. Set $\alpha = \Delta_k$.

Step 2. If there exists $u \in \{-1, 1\}$ such that

$$f(x_k + u\alpha d_k) \leq f(x_k) - \gamma\alpha^2\|d_k\|^2,$$

then determine the largest integer j such that $\alpha = u\Delta_k\delta^{-j}$ satisfies Condition I, set $\alpha_k = \alpha$ and stop.

Step 3. Otherwise, compute the smallest integer j such that either:

(i) there exists $u \in \{-1, 1\}$ such that for $\alpha = u\Delta_k\delta^j$ Condition I holds, then set $\alpha_k = \alpha$ and stop;

(ii) for $\alpha = \Delta_k\delta^j$ Condition II holds, then set $\alpha_k = 0$ and stop.

The explained algorithm has properties, summarized in the following proposition, that allows us to satisfy Condition (ii) of Proposition 2.2..

Proposition 2.3. *Suppose that Assumption 2.1. holds. The algorithm LSA is well defined and terminates with a value of α_k satisfying either Condition I or Condition II. Moreover, assume that (2.1) generates an infinite sequence $\{x_k\}$ and that $\{d_k\}$ is bounded whenever $\{x_k\}$ is bounded. Then,*

(p_1) *the sequence $\{x_k\}$ is bounded;*

(p_2) $f(x_{k+1}) \leq f(x_k)$;

(p_3) $\lim_{k\to\infty} \|x_{k+1} - x_k\|$;

(p_4) $\lim_{k\to\infty} \frac{\nabla f(x_k)' d_k}{\|d_k\|} = 0$;

(p_5) *every limit point $\bar{x}$ of $\{x_k\}$ is such that $\bar{x} \neq x_0$ satisfies $f(\bar{x}) < f(x_0)$.*

3. A COORDINATE DESCENT METHOD

On the basis of the results of Proposition 2.2.-Proposition 2.3. we can construct a global derivative free stabilization scheme which uses linearly independent search directions, and determines the stepsize along these directions by means of the LSA described in the preceding section.
Since the aim of this work is to have a first feel of the computational potentialities of the line search approach in the field of derivative free methods, we have made the simplest choice concerning the linear indipendence search directions to use in the minimization method. In fact, we consider an algorithm which use the orthonormal set $\{e_j\}, j = 1, \ldots, n$ of the coordinate directions as search directions. Therefore, the following algorithm can be considered a particular implementation of a coordinate descent method.

Algorithm 1

Data. $x_0 \in \mathbb{R}^n$, $a_j \in \mathbb{R}, j = 1, \ldots, n$, $\xi_0 \in \mathbb{R}$, $\rho \in (0,1)$.

Step 0. Set $k = 0$.

Step 1. Set $j = 1$.

Step 2. Set $d_k = e_j$ and $\Delta_k = a_j$; compute the stepsize α_k along d_k by LSA.

Step 3. If $\alpha_k \neq 0$, set $x_{k+1} = x_k + \alpha_k d_k$, $a_j = \alpha_k$ and $\xi_{k+1} = min\{|\alpha_k|, \rho\xi_k\}$.

Step 4. Set $k = k + 1$;
If $j < n$, set $j = j + 1$, go to Step 2;
If $j = n$, go to Step 1.

We remark that, from a computational point of view, a crucial role in the Algorithm 1 is played by the initial estimate Δ_k. In order to improve the efficience of the algorithm, we have tried to exploit as much as possible the information on the objective functions obtained in the past iterations of the algorithm. In particular, given the coordinate direction e_j corresponding to the search direction d_k, we set Δ_k to a value obtained in the previous line search along the same coordinate direction. The rationale behind this strategy is the fact that in general the stepsize α_k produced by a line search conveys "good" information on the behaviour of the objective function along the direction e_j. Therefore, it should be a good starting value for the next line search along the same direction e_j.

Then, using Proposition 2.2.-Proposition 2.3., we have the following convergence result.

Proposition 3.1. *Suppose that Assumption 2.1. holds and let* $\{x_k\}$ *be the sequence generated by Algorithm 1. Then either for some index* ν *we have* $\nabla f(x_\nu) = 0$*, or the sequence* $\{x_k\}$ *is such that:*

(a) $\{x_k\}$ *remains in* $\mathcal{L}$ *and admits limit points;*

(b) every limit point of $\{x_k\}$ *belong to* $\mathcal{L}$*;*

(c) $\lim_{k\to\infty} \|\nabla f(x_k)\| = 0$*;*

(d) every limit point $\bar{x}$ *of* $\{x_k\}$ *satisfies* $\nabla f(\bar{x}) = 0$.

Finally, we note that, as observed in [4], many coordinate descent methods are reported in almost all textbooks on optimization. However, their global convergence properties require that either the gradient is known, or the minimum of the objective function along any line of $\mathbb{R}^n$ is unique and an exact line search is performed along the coordinate direction. Here we point out again that the approach proposed in [4], has allowed to define the explained coordinate descent method whose global convergence property does not require exact line searches or restrictive assumptions.

4. NUMERICAL RESULTS

We have implemented the Algorithm 1 (ALG1) and we have tested it on 134 unconstrained problems of dimension n, with n ranging from 2 to 50, taken from CUTE collection. As stopping criterion we have used the following test:

$$max\{\|x_{k-j} - x_{k-i}\|, 0 \le i,j \le n-1\} \le 10^{-6}. \qquad (4.1)$$

Then we have solved the same problems with two versions of the algorithm (PDS) of Torczon described in [13]:

- the first using a grid of cardinality 256

- the second using a grid of cardinality $2n$

For both these versions we have used a stopping criterion comparable with (4.1).
We have compared our numerical results with the ones obtained by the best version (the second one) of the algorithm PDS.

Given a test problem, we say that an algorithm has *failed* if it has not been able to satisfy the stopping criterion with a number of function evaluations less or equal to 10^6. First of all we report the number of failures of the two algorithms:

- ALG1 fails in 10 problems

- PDS fails in 24 problems

As regard more specifically the 10 failures of ALG1, we note that algorithm PDS is able to solve just one of these problems. In the nine problems (HARD PROBLEMS) were both ALG1 and PDS fail, it is possible to note that, after 10^6 function evaluations, ALG1 determines a better function value. This behaviour is showed by the following table, where we report, for the HARD PROBLEMS, the values of the objective function obtained by the two algorithms and the cpu time used when they were stopped.

		ALG1		PDS	
	n	v_f	cpu	v_f	cpu
PALMER1C	8	$3.3 * 10^3$	65 sec	$3.8 * 10^4$	78 sec.
PALMER2C	8	$1.1 * 10^2$	45 sec.	$1.3 * 10^3$	59 sec.
PALMER3C	8	$1.0 * 10^1$	45 sec.	$2.5 * 10^2$	59 sec.
PALMER4C	8	$1.1 * 10^1$	45 sec.	$2.3 * 10^2$	59 sec.
NONMSQRT	9	0.75	29 sec.	0.76	44 sec.
WATSON	31	$5.1 * 10^{-6}$	180 sec.	$1.2 * 10^{-3}$	307 sec.
NCB20B	50	99	808 sec.	99	1174 sec.
PENALTY1	50	$4.3 * 10^{-2}$	88 sec.	$4.4 * 10^{-2}$	404 sec.
VARDIM	50	91	84 sec.	$1.1 * 10^2$	409 sec.

Table 1: HARD PROBLEMS

In 109 test problems (SOLVED PROBLEMS) both the algorithms have been able to satisfy the stopping criterion. For each of these problems we have compared the behaviours of the two algorithms in terms of

- number of function evaluations (n_f)
- value of the objective function (v_f)
- CPU time (cpu).

For the sake of brevity, we can not report the details of this comparison, but just few cumulative results.

First of all, in table 2 we show the total number of function evaluations and the total cpu time required by both the algorithms for solving all the SOLVED PROBLEMS.

	Total n_f	Total cpu
ALG1	4 milions	523 sec.
PDS	13 milions	7007 sec.

Table 2: cumulative results

We say that an algorithm wins, for example in terms of number of function evaluations, if the number of function evaluations required to solve a test problem is smaller or equal to the 95% of the one required by the other algorithm. We have used the same rule to calculate the number of wins in terms of obtained values of the objective function and in terms of required cpu times. In the following table we report the number of wins.

	Number of wins in terms of n_f	Number of wins in terms of v_f	Number of wins in terms of cpu
ALG1	100	51	108
PDS	8	7	1

Table 3: number of WINS

In order to characterize better the comparison between the two algorithms, we have considered the distribution of the *percentual gains* Δ_{n_f}, Δ_{v_f}, Δ_{cpu}, where these quantities are defined as follows:

- Δ_{n_f} is the ratio between the difference in absolute value of n_f in the two algorithms and the largest value of n_f;

- Δ_{v_f} is the ratio between the difference in absolute value of v_f in the two algorithms and the largest value of v_f;

- Δ_{cpu} is the ratio between the difference in absolute value of cpu in the two algorithms and the largest value of cpu.

The results about the number of wins having Δ_{n_f}, Δ_{v_f}, Δ_{cpu} belonging to fixed intervals are showed in the tables 4-6.

	$\Delta_{n_f} \in (0, 0.25]$	$\Delta_{n_f} \in (0.25, 0.5]$	$\Delta_{n_f} \in (0.5, 0.75]$	$\Delta_{n_f} \in (0.75, 1]$
ALG1	1	2	16	81
PDS	3	0	1	4

Table 4: Number of wins in terms of n_f

	$\Delta_{v_f} \in (0, 0.25]$	$\Delta_{v_f} \in (0.25, 0.5]$	$\Delta_{v_f} \in (0.5, 0.75]$	$\Delta_{v_f} \in (0.75, 1]$
ALG1	1	0	0	50
PDS	0	0	0	7

Table 5: Number of wins in terms of v_f

	$\Delta_{cpu} \in (0, 0.25]$	$\Delta_{cpu} \in (0.25, 0.5]$	$\Delta_{cpu} \in (0.5, 0.75]$	$\Delta_{cpu} \in (0.75, 1]$
ALG1	0	4	4	100
PDS	0	0	0	1

Table 6: Number of wins in terms of cpu

All the results reported in this section indicate clearly that ALG1 outperforms algorithm PDS, in fact:

- it solves a larger number of test problems
- in the HARD PROBLEMS it finds a "better" point

- in the SOLVED PROBLEMS it shows a numerical behaviour in terms of all the chosen criteria which is much better than the one of algorithm PDS.

Even if no final conclusions can be drawn, the preliminary numerical experience reported in this section seems to indicate that the approach proposed in [4], [3] is very promising for defining new efficient derivative free algorithms.

5. CONCLUSIONS

In order to have an idea on the validity of the line search approach in the field of unconstrained minimization without derivatives, in this work we have developed the computational aspects of a derivative free algorithm belonging to the class described in [4], [3]. This algorithm is a simple coordinate descent method which uses a line search technique to compute the stepsize, and it is globally convergent towards a stationary point.

The numerical results obtained by this algorithm are very encouraging and seem to indicate that the approach proposed in [4], [3] can be considered the basis to define efficient unconstrained minimization algorithms even in the case where the partial derivatives are not available.

REFERENCES

[1] A. R. Conn and P. L. Toint, "An algorithm using quadratic interpolation for unconstrained derivative free optimization". In this Volume.

[2] J. E. Dennis and V. Torczon, "Direct search methods on parallel machines". *SIAM Journal on Optimization*, 1:448–474, 1991.

[3] L. Grippo, "A class of unconstrained minimization methods for neural network training". *Optimization Methods and Software*, 4:135–150, 1994.

[4] L. Grippo, F. Lampariello, and S. Lucidi, "Global convergence and stabilization of unconstrained minimization methods without derivatives". *Journal of Optimization Theory and Applications*, 56(3):385–406, 1988.

[5] R. Hooke and T. A. Jeeves, "Direct search solution of numerical and statistical problems". *Journal of the ACM*, 8:212–229, 1961.

[6] R. D. Leone, M. Gaudioso, and L. Grippo, "Stopping criteria for linesearch methods without derivatives". *Mathematical Programming*, 30:285–300, 1984.

[7] J. Nelder and R. Mead, "A simplex method for function minimization". *Comput. J.*, 7:308–313, 1965.

[8] J. Ortega and W. Rheinboldt, *"Iterative Solution of Nonlinear Equations in Several Variables"*. Academic Press, New York, 1970.

[9] E. Polak, *"Computational Methods in Optimization"*. Academic Press, New York, 1971.

[10] M. Powell, "An efficient method for finding the minimum of a function of several variables without calculating derivatives". *Comput. J.*, 7:155–163, 1964.

[11] M. J. D. Powell, "A direct search optimization method that models the objective and constraint functions by linear interpolation". In K. A. Publishers, editor, *Advances in Optimization nad Numerical Analysis, Proceedings of the Sixth Workshop on Optimization and Numerical Analysis, Oaxaca, Mexico*, pages 51–67, Dordrecht, NL, 1994.

[12] V. Torczon, "On the convergence of the multidirectional search algorithm". *SIAM J. Optimization*, 1:123–145, 1991.

[13] V. Torczon, "Pds: Direct search methods for unconstrained optimization on either sequential or parallel machines". Technical Report 92/9, Department of Mathematical Sciences, Rice University, Houston, Texas, 1992.

[14] W. Zangwill, "Minimizing a function without calculating derivatives". *Comput. J.*, 10:293–296, 1967.

HIERARCHICAL SYSTEMS WITH WEIGHTED REACTION SET

Mallozzi Lina[1] and Morgan Jacqueline[2]

Dipartimento di Matematica e Applicazioni

Università degli Studi di Napoli Federico II

[1]Via Claudio 21, 80125 Napoli, Italia

[2]Complesso Monte S.Angelo Via Cintia, 80126 Napoli, Italia

[1]e-mail: mallozzi@vm.cised.unina.it

Abstract. A static nonzero–sum game is considered in the case where the number of alternatives available to each player is not necessarily a finite set but can be a continuum.

Key words. Nonzero-sum game; Nash equilibria; Saddle-point problem; reaction set; Stackelberg and hierarchical problems; probability measures; intermediate hierarchical equilibrium; approximate solutions.

1. INTRODUCTION

We consider a static nonzero-sum game in the case in which the number of alternatives available to each player is not necessary a finite set but can be a continuum. One of the players, called the leader, is supposed to have the leadership in playing the game, i.e. for example, he knows all the objective functions and plays first, and the other N players, called the followers, know only their objective functions and the strategy played by the leader.

Let $x \in X$ (resp. $y_j \in Y_j$, $j=1,\dots,N$) be the action variable of the leader (resp. of the j-th follower) where the action set X (resp. Y_j) is assumed to be a subset of $\mathbb{R}^n$ (resp. $\mathbb{R}^{k_j}$). l (resp. f_j, $j=1,\dots,N$) will be the extended real valued cost function of the leader (resp. of the j-th follower) defined on $X \times Y$, where $Y = Y_1 \times \dots \times Y_N$. The aim of the players, supposed to be rational, is to minimize their own objective functions, i.e. if the leader plays the strategy $x \in X$ the followers reacts by playing a strategy that solves a parametric problem, called the lower level problem which could be, for example:

Situation 1. A parametric optimization problem if $N{=}1$, $Y{=}Y_1$, $f{=}f_1$:

$$\begin{cases} \text{find } \overline{y}\in Y \text{such that} \\ \operatorname*{Inf}_{y\in Y} f(x,y)=f(x,\overline{y}) \end{cases} \qquad P(x)$$

Situation 2. A parametric Nash problem if $N > 1$:

$$\begin{cases} \text{for any } j{=}1,\ldots,N \text{ find } \overline{y}_j\in Y_j \text{ such that} \\ \operatorname*{Inf}_{y_j\in Y_j} f_j(x,\overline{y}_1,\ldots,\overline{y}_{j-1},y_j,\overline{y}_{j+1},\ldots,\overline{y}_N)= \\ \qquad f_j(x,\overline{y}_1,\ldots,\overline{y}_{j-1},\overline{y}_j,\overline{y}_{j+1},\ldots,\overline{y}_N) \end{cases} \qquad N(x)$$

Situation 3. A parametric Saddle-Point problem if $N{=}2$ and $f_2{=}-f_1{=}f$:

$$\begin{cases} \text{find } (\overline{y}_1,\overline{y}_2)\in Y_1\times Y_2 \text{ such that } \forall y_1\in Y_1,\ \forall y_2\in Y_2 \\ f(x,y_1,\overline{y}_2)\leq f(x,\overline{y}_1,\overline{y}_2)\leq f(x,\overline{y}_1,y_2) \end{cases} \qquad SP(x)$$

Generically, for any $x\in X$ let $M(x)$, called the *reaction set*, be the set of solutions to one of the previous lower level problems defined by the followers.

When $M(x)$ is a singleton $\tilde{y}(x)$ for any $x\in X$, in situation 1 (resp. 2, 3) a strategy $\overline{x}\in X$ is called a Stackelberg (resp. Stackelberg-Nash, Stackelberg-Saddle-Point) solution for the leader if:

$$\operatorname*{Inf}_{x\in X} l(x,\tilde{y}(x))=l(\overline{x},\tilde{y}(\overline{x}))$$

Such concepts, first introduced by Von Stackelberg [20] in situation 1, in the context of static economic competitions, has been presented in the framework of Control Theory by Chen and Cruz [6] and Simaan and Cruz [19], and a great deal of papers have been devoted to these problems in static and dynamic context with applications (for references see, for example, Bagchi [3], Basar and Olsder [4], Vincente and Calamai [21]).

In the case in which $M(x)$ is not a singleton for at least one $x\in X$, the behaviour of the leader depends on the choice of the strategies' followers in their reaction set. Two extremal possibilities have been already considered. The first one, called "optimistic" assumes that the leader can inforce the followers to choose the best strategies for himself. More precisely, the leader in this case solves the following problem, also called the *strong hierarchical problem* (*strong Stackelberg* in situation 1, *strong hierarchical Saddle-Point* in situation 2 and *strong hierarchical Nash* in situation 3) (Leitmann, [8]; Breton et al., [5]):

$$\begin{cases} \text{find } \overline{x}\in X \text{such that} \\ \operatorname*{Inf}_{x\in X}\ \operatorname*{Inf}_{y\in M(x)} l(x,y)=\operatorname*{Inf}_{y\in M(\overline{x})} l(\overline{x},y) \end{cases} \qquad (s-HP)$$

Let S^s be the set of solutions to the problem $(s-HP)$ and v^s be the value of the problem $(s-HP)$.

Following the second possibility, called "pessimistic", the leader is supposed to know nothing about the choice of the followers and to minimize the worst, i.e. solving the problem, called the *weak hierarchical problem* (*weak Stackelberg* in situation 1, *weak hierarchical Saddle-Point* in situation 2 and *weak hierarchical Nash* in situation 3):

$$\begin{cases} \text{find } \overline{x}{\in}X \text{ such that} \\ \\ \operatorname*{Inf}_{x\in X} \operatorname*{Sup}_{y\in M(x)} l(x,y) = \operatorname*{Sup}_{y\in M(\overline{x})} l(\overline{x},y) \end{cases} \qquad (w-HP)$$

Let S^w be the set of solutions to the problem $(w-HP)$ and v^w be the value of the problem $(w-HP)$.

In this paper we present a solution concept in the intermediate situations in which the leader cannot inforce the followers to choose the best strategies for himself, but has some informations on the followers' choice in their reaction set.

First we consider the case in which the set $M(x)$ is discrete for any $x{\in}X$ and then the case in which $M(x)$ is not discrete for any $x{\in}X$. In both situations we give examples concerning situations 1, 2 and 3, and we compare the obtained results with the weak hierarchical problem. Let us note that situation 1 has been already considered in Mallozzi and Morgan [15]. Here, in the case in which $M(x)$ is not a discrete set for any $x{\in}X$, existence results will be proved in situation 1, in line with previous results obtained for the weak and strong problems (Lignola and Morgan, [10,11]; Loridan and Morgan, [12,13]; Mallozzi and Morgan, [14,15]; Morgan, [18]; Morgan and Raucci, [17]).

2. DISCRETE REACTION SET

Suppose that the solution set to the lower level problem $M(x)$ has a finite number of elements. More precisely, let

$$M(x)=\{y^j(x),\ j=1,\ldots,m(x)\}$$

If we assume that the leader has sufficient informations on the choice of the followers in the reaction set in order to attibute for any $x{\in}X$ a discrete probability distribution on $M(x)$, $D(x) = \{p_1(x),\ldots,p_{m(x)}(x)\}$, a natural way to define an"optimal" strategy for him is to solve the following problem:

$$\text{minimize on } X \text{ the function } v(x)= \sum_{j=1}^{m(x)} p_j(x)\ l(x,y^j(x))$$

Let $D=\{D(x), x{\in}X\}$.

Definition 2.1. The problem

$$\begin{cases} \text{find } \overline{x}{\in}X \text{such that} \\ \\ \operatorname*{Inf}_{x\in X} v(x)=v(\overline{x}) \end{cases} \qquad (i-HP_D)$$

is called the *intermediate hierarchical problem* (*intermediate Stackelberg* in situation 1, *intermediate hierarchical Saddle-Point* in situation 2 and *intermediate hierarchical Nash* in situation 3) with respect to D. Let S^i_D be the set of solutions to the problem $(i-HP_D)$ and v^i_D the corresponding value; any $(\overline{x},\overline{y})\in X\times Y$ such that $\overline{x}\in S^i_D$ and $\overline{y}\in M(\overline{x})$ will be called an *intermediate hierarchical equilibrium* with respect to D.

Remark 2.1. It is easy to prove that

$$v^s \leq v^i_D \leq v^w$$

for any family of probability distributions D, then by considering the intermediate hierarchical problem one can obtain a lower value than the value corresponding to the weak hierarchical problem.

Examples of such a situation are the following:

Example 2.1 (situation 1). Let $X{=}Y{=}[-1,1]$,

$$l(x,y){=}x+y, \quad f(x,y){=}\,|y^2-x^4|$$

In this case $M(x){=}\{x^2,-x^2\}$ is the set of the solutions to the minimum problem $P(x)$:

$$\begin{cases} \text{find } \overline{y}\in Y \text{such that} \\ \\ \operatorname*{Inf}_{y\in Y} f(x,y){=}f(x,\overline{y}) \end{cases} \qquad P(x)$$

Let us consider on $M(x)$ the probability distribution $D(x){=}(\alpha,1-\alpha)$ with $\alpha\in[0,1]$, for any $x\in X$. We have

$$S^s{=}\{-1\}, \quad v^s{=}-2$$

$$S^w{=}\{-\frac{1}{2}\}, \quad v^w{=}-\frac{1}{4}$$

$$S^i_D{=}\begin{cases} \{-1\} & \text{if } \alpha\leq\frac{3}{4} \\ \\ \{\frac{-1}{2(2\alpha-1)}\} & \text{if } \alpha>\frac{3}{4} \end{cases} \qquad v^i_D{=}\begin{cases} 2\alpha-2 & \text{if } \alpha\leq\frac{3}{4} \\ \\ \frac{-1}{4(2\alpha-1)} & \text{if } \alpha>\frac{3}{4} \end{cases}$$

Let us note that for $\alpha\leq\frac{3}{4}$ the solution to the intermediate hierarchical problem coincides with the solution to the strong hierarchical problem. Moreover for $\alpha{=}1$ the intermediate hierarchical problem coincides with the weak hierarchical problem. Let us observe that $v^i<v^w$ for $\alpha\neq\frac{3}{4}$.

Example 2.2 (situation 1). Consider the same data as in example 1 and the following probability distribution on $M(x)$ depending on x:

$$D'(x){=}\begin{cases} (\frac{4}{5},\frac{1}{5}) & \text{if } -1\leq x<0 \\ 1 & \text{if } x=0 \\ (\frac{1}{2},\frac{1}{2}) & \text{if } 0<x\leq 1 \end{cases}$$

In this case we have $S^i_{D'}=\{-\frac{5}{6}\}, v^i_{D'}=-\frac{5}{12}$; again we have $v^i_{D'}<v^w$.

Example 2.3 (situation 2). Let $X=Y_1=Y_2=[0,1]$,

$$l(x,y_1,y_2)=-x-2(y_1+y_2),$$

$$f_1(x,y_1,y_2)=y_1^2(y_2^2+x+\frac{1}{4})-2y_1y_2^2, \quad f_2(x,y_1,y_2)=y_2^2-2y_1y_2+x$$

Then

$$M(x)=\begin{cases}\{(0,0),(\frac{1}{2},\frac{1}{2})\} & \text{if } x=0\\ \{(0,0)\} & \text{if } x>0\end{cases}$$

is the set of solutions to the Nash problem $N(x)$:

$$\begin{cases}\text{find } (\overline{y}_1,\overline{y}_2)\in Y_1\times Y_2 \text{ such that}\\ \operatorname*{Inf}_{y_1\in Y_1} f_1(x,y_1,\overline{y}_2)=f_1(x,\overline{y}_1,\overline{y}_2)\\ \operatorname*{Inf}_{y_2\in Y_2} f_2(x,\overline{y}_1,y_2)=f_2(x,\overline{y}_1,\overline{y}_2)\end{cases} \qquad N(x)$$

Let us consider on $M(x)$ the probability distribution

$$D(x)=\begin{cases}(\alpha,1-\alpha) & \text{if } x=0\\ 1 & \text{if } x>0\end{cases}$$

with $\alpha\in[0,1]$.

We have

$$S^s=\{0\}, \quad v^s=-2$$

$$S^w=\{1\}, \quad v^w=-1$$

$$S_D^i=\begin{cases}\{0\} & \text{if } \alpha\leq\frac{1}{2}\\ \{1\} & \text{if } \alpha>\frac{1}{2}\end{cases} \qquad v_D^i=\begin{cases}-2(1-\alpha) & \text{if } \alpha\leq\frac{1}{2}\\ -1 & \text{if } \alpha>\frac{1}{2}\end{cases}$$

Let us observe that for any $\alpha<\frac{1}{2}$ we have that $v_D^i<v^w$.

Remark 2.2. In the case in which $M(x)$, $x\in X$ is a countable set, that is $M(x)=\{(y^n(x)),\ n\in\mathcal{N}\}$, the intermediate hierarchical problem with respect to the probability distribution $D(x)=(p_n(x))$, $n\in\mathcal{N}$, is the following:

$$\text{minimize on } X \text{ the function } v(x)=\sum_{j=1}^{+\infty}p_j(x)\ l(x,y^j(x))$$

if the sum converges.

3. GENERAL REACTION SET

Suppose the solution set to the lower level problem $M(x)$ to be not discrete for at least one $x\in X$ and measurable with respect to the Lebesgue measure with non zero measure. Let us suppose that the leader attributes a probability measure $\mu_x(y)$ on $M(x)$ for any $x \in X$. In this case the intermediate hierarchical problems becomes:

$$\text{minimize on } X \text{ the function } v(x) = \int_{M(x)} l(x,y)\, d\mu_x(y)$$

if $v(x)$ is well defined for any $x \in X$.

Let $\mu = \{\mu_x, x \in X\}$.

Definition 3.1. The problem

$$\begin{cases} \text{find } \overline{x} \in X \text{such that} \\ \\ \operatorname*{Inf}_{x \in X} v(x) = v(\overline{x}) \end{cases} \qquad (i-HP_\mu)$$

is called the *intermediate hierarchical problem* with respect to μ. Again we have:

$$v^s \leq v^i_\mu \leq v^w$$

for any family of probability measures μ.

Let us give some examples:

Example 3.1 (situation 1). Let $X = Y = [-1, 1]$,

$$l(x,y) = xy, \quad f(x,y) = \max\,\{0, y(y - \frac{x}{2} + \frac{1}{2})\}$$

Then for any $x \in X$, $M(x) = [\frac{x-1}{2}, 0]$ is the set of the solutions to the minimum problem $P(x)$. Let us consider on $M(x)$ the probability measure μ associated to the uniform density for any $x \in X$.

We have

$$S^s = \{-1\}, \quad v^s = -1$$

$$S^w = \{[0,1]\}, \quad v^w = 0$$

$$S^i_\mu = \{\frac{1}{2}\}, \quad v^i_\mu = -\frac{1}{16}$$

In this case we have that $v^i_\mu < v^w$.

Example 3.2 (situation 1). Consider the same data as in example 4 but with the probability measure μ' associated to the density

$$\varphi_x(y) = \begin{cases} \frac{2y+1}{x+1} & \text{if } x \in [0,1] \\ \\ 2y + 1 - x & \text{if } x \in [-1, 0) \end{cases}$$

In this case it can be proved

$$S^i_{\mu'} = \{-\frac{1}{6}\}, \quad v^i_{\mu'} = -\frac{1}{72}$$

and again we have $v^i_{\mu'} < v^w$.

The following theorem gives sufficient conditions for existence of solutions to the intermediate Stackelberg problem in the case in which the measure μ_x is absolutely continuous with respect to the Lebesgue measure with density φ_x. In this case the intermediate hierarchical problems becomes:

$$\text{minimize on } X \text{ the function } v(x) = \int_{M(x)} l(x,y)\ \varphi_x(y)\ dy$$

Theorem 3.1. Let X and Y be compact. Assume that the following assumptions are satisfied:

i) l and f are lower semicontinuous functions on $X \times Y$;

ii) for any $(x,y) \in X \times Y$ and for any (x_n) converging to x, there exists a sequence (y_n) such that: $\limsup_{n \to +\infty} f(x_n, y_n) \leq f(x,y)$;

iii) $\varphi_x(y)$ is a continuous function on X;

iv) M is a lower semicontinuous multivalued function on X (i.e. for any $x \in X$ and for any (x_n) converging to x, $M(x) \subseteq \operatorname{Liminf}_n M(x_n)$, i.e. for any $y \in M(x)$ there exists a sequence (y_n) converging to y such that $y_n \in M(x_n)$ for n large; Aubin, [2] in a topological setting and Lignola and Morgan, [9] in a sequential setting).

Then there exists at least a solution to the intermediate Stackelberg problem.

Proof. Let us recall that from assumptions i) and ii) the function $x \mapsto \operatorname{Inf}_{y \in Y} f_1(x,y)$ is upper semicontinuous (Lignola and Morgan, [9]). Since f_1 is lower semicontinuous, we have that M is a closed graph multivalued function, i.e. for any $x \in X$ and for any (x_n) converging to x, $\operatorname{Limsup}_n M(x_n) \subseteq M(x)$, i.e. for any $y_0 \in Y$ and for any (y_n) converging to y_0 with $y_{n_k} \in M(x_{n_k})$ for a selection (n_k) we have $y_0 \in M(x)$. In fact let $y = \lim_{k \to +\infty} y_{n_k}$; then

$$f_1(x,y) \leq \liminf_{k \to +\infty} f_1(x_{n_k}, y_{n_k}) = \liminf_{k \to +\infty} \operatorname{Inf}_{y \in Y} f_1(x_{n_k}, y) \leq$$

$$\limsup_{k \to +\infty} \operatorname{Inf}_{y \in Y} f_1(x_{n_k}, y) \leq \operatorname{Inf}_{y \in Y} f_1(x,y)$$

so $f_1(x,y) = \operatorname{Inf}_{y \in Y} f_1(x,y)$ i.e. $y \in M(x)$.

By using assumption iv) we have that for any $x \in X$ and for any (x_n) converging to x, $M(x) = \operatorname{Lim}_n M(x_n)$ in the Kuratowski sense (Kuratowski, [7]), that is $\operatorname{Limsup}_n M(x_n) \subseteq M(x) \subseteq \operatorname{Liminf}_n M(x_n)$.

Now, we can prove that the function $v(x)$ is lower semicontinuous on X, i.e. for any $x \in X$ and for any (x_n) converging to x we have

$$\liminf_{n \to +\infty} \int_{M(x_n)} l(x_n, y)\ \varphi_{x_n}(y)\ dy \geq \int_{M(x)} l(x,y)\ \varphi_x(y)\ dy$$

Let ${}_A(y)$ be the characteristic function of the set A, i.e.

$${}_A(y) = \begin{cases} 1, & \text{if } y \in A, \\ 0, & \text{if } y \notin A. \end{cases}$$

It can be proved that for any $x \in X$ and for any (x_n) converging to x, $\lim_{n \to +\infty} {}_{M(x_n)}(y) = {}_{M(x)}(y)$ for any $y \in Y$. By the Fatou's lemma (Ash [1]), we have:

$$\liminf_{n \to +\infty} \int_{M(x_n)} l(x_n, y)\ \varphi_{x_n}(y)\ dy =$$

$$\liminf_{n \to +\infty} \int_Y l(x_n, y)\ {}_{M(x_n)}(y)\ \varphi_{x_n}(y)\ dy \geq$$

$$\int_Y \left(\liminf_{n \to +\infty} l(x_n, y)\ {}_{M(x_n)}(y)\ \varphi_{x_n}(y) \right) dy \geq$$

$$\int_Y l(x, y)\ {}_{M(x)}(y)\ \varphi_x(y)\ dy$$

The last inequality is satisfied if for any $y \in Y$

$$\liminf_{n \to +\infty} l(x_n, y)\ {}_{M(x_n)}(y)\ \varphi_{x_n}(y) \geq l(x, y)\ {}_{M(x)}(y)\ \varphi_x(y)$$

and this holds because of the lower semicontinuity of l, the continuity of $\varphi_x(y)$ with respect to x and the convergence of the characteristic function of $M(x_n)$ to the characteristic function of $M(x)$. □

A continuous function satisfies assumptions i) and ii), but there exist not continuous functions satisfying assumptions i) and ii) as shown in the following example.

Example 3.3. Let $X = Y = [0, 1]$,

$$f(x, y) = \begin{cases} -\frac{(y-1)y}{x} & \text{if } x \neq 0 \\ -(y-1)y & \text{if } x = 0 \end{cases}$$

The function f satisfies assumptions i) e ii) (with $y_n = 0$ for any $n \in \mathcal{N}$), but the function $x \mapsto f(x, y)$ is not upper semicontinuous in $x = 0$.

In the following example there exist no solutions to the weak hierarchical problem, but there exists a solution to the intermediate hierarchical problem.

Example 3.4 (situation 1). Let $X = [0, 1]$, $\quad Y = [-1, 1]$,

$$l(x, y) = \frac{2y - x}{2}, \quad f(x, y) = x(y - x)^2$$

Then for any $x \in X$,

$$M(x) = \begin{cases} \{x\}, & \text{if } x \neq 0, \\ [-1, 1], & \text{if } x = 0, \end{cases}$$

is the set of solutions to the problem $P(x)$. Let us consider on $M(0)$ the probability measure μ associated to the uniform density.

We have:

$$S^s = \{0\}, \quad v^s = -1$$

$$S^w = \emptyset, \quad v^w = 0$$

$$S^i_\mu = \{0\}, \quad v^i_\mu = 0$$

In this case $v^i_\mu = v^w$.

If we consider on $M(0)$, the measure μ' associated to the following density:

$$\varphi_x(y) = \begin{cases} y^2 + 1, & \text{if } y \in [-1, 0], \\ (y-1)^2, & \text{if } y \in (0, 1], \end{cases}$$

we have that

$$S^i_{\mu'} = \{0\}, \quad v^i_{\mu'} = -\frac{2}{3}$$

then $v^i_{\mu'} < v^w$.

When M is not a lower semicontinuous multivalued function, for any $\varepsilon > 0$, the set of approximated solutions to the problem $P(x)$, i.e. the set

$$M(x, \varepsilon) = \{y \in Y : f(x, y) \leq \varepsilon + \operatorname*{Inf}_{y \in Y} f(x, y)\}$$

can be considered. In Loridan and Morgan [12,13] and Lignola and Morgan [11] it has been shown that under suitable assumptions on the data, the multivalued function $x \rightrightarrows M(x, \varepsilon)$ is continuous on X (i.e. lower semicontinuous and closed graph) for any $\varepsilon > 0$.

So we define a regularization for the intermediate hierarchical problem in the following way. Let us suppose that the leader can attribute a probability measure μ^ε_x on $M(x, \varepsilon)$ for any $x \in X$ with density φ^ε_x. In this case one can consider for any $\varepsilon > 0$, the regularized intermediate hierarchical problem:

$$\text{minimize on } X \text{ the function } v(x, \varepsilon) = \int_{M(x,\varepsilon)} l(x, y)\ \varphi^\varepsilon_x(y)\ dy$$

Let $\mu^\varepsilon = \{\mu^\varepsilon_x, x \in X\}$.

Definition 3.2. The problem

$$\begin{cases} \text{find } \overline{x}_\varepsilon \in X \text{such that} \\ \operatorname*{Inf}_{x \in X} v(x, \varepsilon) = v(\overline{x}_\varepsilon, \varepsilon) \end{cases} \qquad (i\text{–}HP_{\mu^\varepsilon})(\varepsilon)$$

is called the *regularized intermediate hierarchical problem* with respect to μ^ε and $v^i_{\mu^\varepsilon}(\varepsilon)$ is its value.

In this case we have:

$$v^i_{\mu^\varepsilon}(\varepsilon) \leq v^w(\varepsilon)$$

for any family of probability measures μ^ε and for any $\varepsilon > 0$.

The following theorem gives sufficient conditions in order to have the existence of solutions to the regularized intermediate Stackelberg problem; the result is given without assuming the lower semicontinuity of the multivalued function $x \rightrightarrows M(x)$ as in Theorem 3.1.

Theorem 3.2. Let X and Y be compact. Suppose that the following assumptions hold:
i) l and f are lower semicontinuous functions on $X \times Y$
ii) for any $(x,y) \in X \times Y$ and for any (x_n) converging to x, there exists a sequence (y_n) such that: $\limsup_{n \to +\infty} f(x_n, y_n) \leq f(x,y)$
iii) $\varphi_x^\varepsilon(y)$ is a continuous function on X
iv) Y is convex and $f(x,y)$ is strictly quasiconvex on Y for any $x \in X$ (i.e. $f(x, \theta y + (1-\theta)z) < \max\{f(x,y),\ f(x,z)\}$ for any $\theta \in]0,1[$ and for any $y, z \in Y$ verifying $f(x,y) \neq f(x,z)$; Mangasarian [16]) then there exists at least a solution to the regularized intermediate Stackelberg problem for any $\varepsilon > 0$.

Proof. From assumption iv), ii) and from the lower semicontinuity of f, we have that for any $x \in X$ and for any (x_n) converging to x, $M(x,\varepsilon) \subseteq \operatorname{Liminf}_n M(x_n, \varepsilon)$ for any $\varepsilon > 0$ (Loridan and Morgan, 1992), i.e. for any $y \in M(x,\varepsilon)$ there exists a sequence (y_n) converging to y such that $y_n \in M(x_n, \varepsilon)$ for n large; since the multivalued function $x \rightrightarrows M(x,\varepsilon)$ is closed graph too, then for any $x \in X$ and for any (x_n) converging to x, $M(x_n, \varepsilon)$ converges to $M(x,\varepsilon)$ in the Kuratowski sense and it can be proved that the function

$$v(x,\varepsilon) = \int_{M(x,\varepsilon)} l(x,y)\ \varphi_x^\varepsilon(y)\ dy$$

is lower semicontinuous on X for any $\varepsilon > 0$. Therefore there exists at least a solution to the regularized intermediate Stackelberg problem. □

Example 3.5. Let $X = Y = [0,1]$,

$$l(x,y) = -xy, \quad f(x,y) = y\left(x - \frac{1}{2}\right).$$

Then, for any $x \in X$,

$$M(x) = \begin{cases} \{1\}, & \text{if } x \in [0, \frac{1}{2}), \\ [0,1], & \text{if } x = \frac{1}{2}, \\ \{0\}, & \text{if } x \in (\frac{1}{2}, 1], \end{cases}$$

is the set of solutions to the minimum problem $P(x)$.

Let $w(x) = \operatorname{Sup}_{y \in M(x)} l(x,y)$; in the considered example we have:

$$w(x) = \begin{cases} -x, & \text{if } x \in [0, \frac{1}{2}), \\ 0, & \text{if } x \in [\frac{1}{2}, 1], \end{cases}$$

and there exist no solution to the weak Stackelberg problem; the value is $v^w = -\frac{1}{2}$.

Let us consider the the probability measure μ associated to the uniform density $M(x)$ for any $x \in X$. We have:

$$v(x)=\begin{cases} -x & \text{if } x\in[0,\frac{1}{2}) \\ -\frac{1}{4} & \text{if } x=\frac{1}{2} \\ 0 & \text{if } x\in(\frac{1}{2},1] \end{cases}$$

and again there exists no intermediate Stackelberg equilibrium to this problem; the value is $v^i_\mu = -\frac{1}{2} = v^w$. Observe that the multivalued function $x \rightrightarrows M(x)$ is not lower semicontinuous in $x=\frac{1}{2}$.

Let us compute for $\varepsilon > 0$, the ε-solutions to the intermediate Stackelberg problem. For any $\varepsilon > 0$, one can compute the approximated solutions to the problem $P(x)$:

$$M(x,\varepsilon)=\begin{cases} [1-\frac{2\varepsilon}{(1-2x)},1], & \text{if } x\in[0,\frac{1}{2}-\varepsilon), \\ [0,1], & \text{if } x\in[\frac{1}{2}-\varepsilon,\frac{1}{2}+\varepsilon], \\ [0,\frac{2\varepsilon}{(2x-1)}], & \text{if } x\in(\frac{1}{2}+\varepsilon,1]. \end{cases}$$

If we consider the family of measures μ^ε associated to uniform density on $M(x,\varepsilon)$ for any $x \in X$, we have:

$$v(x,\varepsilon)=\begin{cases} \frac{x\varepsilon}{(1-2x)}-x & \text{if } x\in[0,\frac{1}{2}-\varepsilon) \\ -\frac{x}{2} & \text{if } x\in[\frac{1}{2}-\varepsilon,\frac{1}{2}+\varepsilon] \\ -\frac{x\varepsilon}{(2x-1)} & \text{if } x\in(\frac{1}{2}+\varepsilon,1] \end{cases}$$

and $\overline{x}_\varepsilon = \frac{1}{2} - \frac{\sqrt{\varepsilon}}{2}$ is the solution to the approximate intermediate Stackelberg problem and the corresponding value is $v^i_{\mu^\varepsilon}(\varepsilon) = -\frac{1}{2} - \frac{\varepsilon}{2}$ for any $\varepsilon > 0$.

ACKNOWLEDGEMENT

This research has been partially supported by the Italian Ministry for University (M.U.R.S.T.), which is here gratefully acknowledged.

REFERENCES

[1] Ash R. B., "Real Analysis and Probability" Academic Press, New York, 1972.

[2] Aubin J. P., "Mathematical Methods of Game and Economic Theory". North-Holland Publishing Company, Amsterdam, Holland, 1982.

[3] Bagchi A., "Stackelberg Differential Games in Economic Model". Lecture Notes in Control and Information Sciences 64, Springer Verlag, Berlin, Germany, 1984.

[4] Basar T. and Olsder G. J, "Dynamic Noncooperative Games". Academic Press, New York 1982.

[5] Breton M., Alj A. and Haurie A., Sequential Stackelberg equilibria in two-person games. *J. Opt. Th. and Appl.* 59:71, 1988.

[6] Chen C. I., and Cruz J. B., "Stackelberg solution for two-person games with biased information patterns". *IEEE Trans. Aut. Control* AC-17:791, 1972.

[7] Kuratowski C., "Topology". Academic Press, New York, 1966.

[8] Leitmann G., "On generalized Stackelberg strategies". *J. Opt. Th. and Appl.* 26:637, 1978.

[9] Lignola M. B. and Morgan J., "Semicontinuity of marginal functions in a sequential setting". *Optimization* 24:241, 1992.

[10] Lignola M. B., and Morgan J., "Regularized bilevel programming problem". Preprint n.22, Dip. di Mat. e Appl., Università di Napoli Federico II, 1993.

[11] Lignola M. B. and Morgan J., "Topological existence and stability for Stackelberg problems". *J. Opt. Th. and Appl.* 151:164, 1995.

[12] Loridan P. and Morgan J., "New results on approximate solutions in two-level optimization". *Optimization* 20:819, 1989.

[13] Loridan P. and Morgan J., "On strict ε-solutions for a two-level optimization problem". In Proc. Inter. Conf. Oper. Res. 90 Vienna, Edited by G. Feichtinger, W. Buhler, F.J. Radermacher and P. Staly, Springer Verlag, 1992.

[14] Mallozzi L. and Morgan J., "ε-mixed strategies for static continuous Stackelberg problem". *J. Opt. Th. and Appl.* 78:303, 1993.

[15] Mallozzi L. and Morgan J., "Problemi di Stackelberg con risposte pesate". Atti del XIX Convegno A.M.A.S.E.S., Bari, Italia, 1995.

[16] Mangasarian O. L., "Non Linear Programming". Mc Graw-Hill, New York, 1969.

[17] Morgan J. and Raucci, R., "Approximate solutions for strong hierarchical saddle-point problems". Preprint n.23, Dip. di Mat. e Appl., Università degli Studi di Napoli, 1995.

[18] Morgan J., "Constrained well-posed two-level optimization problems". In "Non-smooth Optimization and Related Topics", Edited by F. H. Clarke, V. F. Demyanov and F. Giannessi, Plenum Press, New York and London, 1989.

[19] Simaan H. and Cruz, J. B., "On the Stackelberg strategy in nonzero-sum games". *J. Opt. Th. and Appl.* 11:533, 1973.

[20] von Stackelberg H., "Marktform und Gleichgewicht". Springer Verlag, Vienna, 1934.

[21] Vincente L. N., and Calamai P. H., "Bilevel and multilevel programming: a bibliografy review". *J. Global Opt.* 5:291, 1994.

MATHEMATICAL PROGRAMMING IN MACHINE LEARNING

Mangasarian Olvi L.

Computer Sciences Department
University of Wisconsin
Madison, WI 53705, USA

Abstract. We describe in this work a number of central problems of machine learning and show how they can be modeled and solved as mathematical programs of various complexity.

Key words. Machine Learning, Neural Networks, Generalization.

1. INTRODUCTION

Machine learning can be thought of as generalizing information gleaned from given data to new unseen data. As such it can be considered as determining a mapping between an input set and an output set in a robust manner that is amenable to generalization. In this work we shall concentrate on a number of fundamental problems of machine learning, and show how mathematical programming plays a significant role in their formulation and solution. In Section 2 we consider the classical problem of discriminating between two point sets in the n-dimensional real space $\mathbb{R}^n$, and show that its complexity ranges from polynomial-time to NP-complete, depending on the measure of error employed. When the traditional distance of a misclassified point to a separating plane is used as an error, a single linear program [6, 15, 16, 4] usually solves the problem. Recently [10, 18, 2, 7] a more complex, and for certain applications more realistic, error measure has been considered, namely the number of misclassified points by a separating plane. This problem, even though shown to be NP-complete [7], can be effectively solved by a parametric [2] or a hybrid method [7]. In Section 3 we describe a central problem of machine learning, that of improving generalization [27]. We give a very simple model which justifies the often accepted rule-of-thumb of machine learning and approximation theory, that overfitting leads to poor generalization. In fact we go the opposite direction, and show that inexact fitting can lead to improved generalization. In Section 4 we use an equivalence between the step function and the complementarity problem to show that the problem of training a neural network can be represented as mathematical program with equilibrium constraints (MPEC) which has been studied recently in the literature [14].

A word about our notation now. For a vector x in the n-dimensional real space $\mathbb{R}^n$, x_+ will denote the vector in $\mathbb{R}^n$ with components $(x_+)_i := \max\{x_i, 0\},\ i = 1,\ldots,n$. Similarly x_* will denote the vector in $\mathbb{R}^n$ with components $(x_*)_i := (x_i)_*,\ i = 1,\ldots,n$, where $()_*$ is the step function that maps a nonpositive number into zero and a positive number into one. The p-norm will be denoted by $\|\cdot\|_p$ for $p = 1, 2, \ldots, \infty$, while $\|\cdot\|$ will denote an arbitrary norm. We will also make use of the function x_- which will denote the vector in $\mathbb{R}^n$ with components $(x_-)_i := \min\{x_i, 1\}$, $i = 1, \ldots, n$. The notation $A \in \mathbb{R}^{m \times n}$ will signify a real $m \times n$ matrix. For such a matrix, A^T will denote the transpose while A_i will denote row i. For two vectors x and y in $\mathbb{R}^n$, $x^T y$ will denote the scalar product, while $x \perp y$ will denote $x^T y = 0$. A vector of ones in a real space of arbitrary dimension will be denoted by e. The symbols ":=" and "=:" will denote a definition of a term adjacent to the colon by a term adjacent to the equality.

2. LINEAR DISCRIMINATION

We begin with the fundamental problem of constructing a linear discriminator between two given point sets $\mathcal{A}$ and $\mathcal{B}$ in $\mathbb{R}^n$. That is we look for a plane

$$x^T w = \theta, \tag{2.1}$$

such that

$$\begin{aligned} x^T w > \theta \quad & \text{for } x \in \mathcal{A} \\ x^T w < \theta \quad & \text{for } x \in \mathcal{B}. \end{aligned} \tag{2.2}$$

Here w is the normal to the plane and $\dfrac{|\theta|}{\|w\|_2}$ is the Euclidean distance from the origin to the plane. In general it is not possible to satisfy (2.2) except in the special case when the convex hulls of $\mathcal{A}$ and $\mathcal{B}$ do not intersect. Thus, one resorts in the general case to optimizing some error criterion in the satisfaction of (2.2). The simplest such criterion is to use linear programming in order to construct a plane (2.1) that maximizes a weighted sum of the distances of correctly classified points to the plane [16, 4] as follows:

$$\max_{w,\theta,y,z} \left\{ e^T y + e^T z \mid Aw \geq e\theta + y,\ Bw \leq e\theta - z,\ y \leq e,\ z \leq e \right\} \tag{2.3}$$

Here the rows of the matrices $A \in \mathbb{R}^{m \times n}$ and $B \in \mathbb{R}^{k \times n}$ represent the m points in $\mathcal{A}$ and the k points in $\mathcal{B}$ respectively, while e is a vector of ones of appropriate dimension. The objective function of (2.3) represents the sum of distances, positive and truncated to one for correctly classified points and nonpositive for incorrectly classified points, to the plane $x^T w = \theta$, multiplied by $\|w\|_2$. Although, apparently different from the robust linear program of [4, Proposition 2.4], it is equivalent to it if we set the weights $\delta_1 = \delta_2 = 1$ in the latter and make a simple change of variables. A principal advantage of the formulation (2.3), is that it ties more easily with the classification maximization formulation (2.7) below, once we make use of the function

$$(\zeta)_- := \min\{\zeta, 1\} = 1 - (1 - \zeta)_+. \tag{2.4}$$

By using this nondecreasing piecewise-linear concave function, the linear program (2.3) can be written as the following unconstrained concave maximization problem:

$$\max_{w,\theta} e^T (Aw - e\theta)_- + e^T (-Bw + e\theta)_- \tag{2.5}$$

Another simplifying feature of the linear programming formulation (2.3) is the absence of a normalization vector e from both terms of (2.5), and from the first two constraints of (2.3). The linear program (2.3) also maintains the non-nullity properties of w [4, Theorems 2.5 & 2.6], which can be summarized as follows here. The point $(w = 0, \gamma, y, z)$ is a solution of (2.3) if and only if $e^T A = e^T B$ and $m = k$, in which case the solution is never unique in $w = 0$. If the convex hulls of $\mathcal{A}$ and $\mathcal{B}$ are disjoint, then all points in $\mathcal{A}$ and $\mathcal{B}$ are correctly classified, and the equivalent programs (2.3) and (2.5) yield a maximum value of $m+k$, equal to the total number of points in $\mathcal{A}$ and $\mathcal{B}$ that have been completely separated by the plane $x^Tw = \theta$. However, in the general case of intersecting convex hulls, the linear program (2.3) obtains an approximate separating plane that maximizes a weighted sum of distances of points as described above. For this case the number of correctly classified points is given by:

$$e^T y_* + e^T z_* \quad \text{or equivalently} \quad e^T((Aw - e\theta)_-)_* + e^T((-Bw + e\theta)_-)_*, \tag{2.6}$$

where (w, θ, y, z) is a solution of (2.3) and, as indicated earlier, $(\cdot)_*$ is the step function. Although the linear programming formulation (2.3) is very effective for practical problems [21] and can be used in the construction of neural networks [3] as well multi-surface discriminators [1, 4], it does not minimize the number of misclassified points by the plane (2.1), which may be an important consideration in certain applications. In order to minimize the number of misclassified points, we need to *maximize* the number of satisfied components of the inequalities (2.2). This corresponds to solving the following problem:

$$\max_{w,\theta} e^T(Aw - e\theta)_* + e^T(-Bw + e\theta)_* \tag{2.7}$$

We refer to this problem as the *classification maximization problem*, that is the problem of maximizing the number of correctly classified points. Although this is an NP-complete problem [7, Proposition 2], effective methods for its solution have been proposed [18, 7] and successfully tested on real world problems [2, 7]. We outline two of these approaches briefly now.

We first describe a hybrid approach, recently proposed and tested successfully on ten publicly available databases [7]. The idea of this approach is to combine the two criteria described above as follows. For a fixed orientation w, translate the plane (2.1) by varying θ, so as to minimize the number of misclassified points, that is solve (2.7) for a fixed w, which is then a one dimensional problem in θ with a finite number of objective function values. Then for a fixed θ, rotate the plane (2.1) by varying w, so as to minimize the weighted average of the sum of the distances of the misclassified points to the plane (2.1), that is solve the linear program (2.3) in w for a fixed value of θ. The algorithm stops, when successive line searches in θ fail to decrease the number of misclassified points. Needless to say, there is no guarantee that this approach will give a global solution to the NP-complete problem (2.7). However it seems to have the best generalization [7] as determined by tenfold cross-validation [26] on the ten data sets employed.

We describe now another approach for solving the classification maximization problem (2.7), by reducing it to an LPEC, a linear program with equilibrium constraints [18]. We begin with a variation of a lemma of [18, Lemma 2.1] which ensures that the backward implication of the lemma holds also for zero components of a.

Lemma 1 Equilibrium characterization of step function $()_*$. *For $r \in \mathbb{R}^m$, $u \in \mathbb{R}^m$, $a \in \mathbb{R}^m$ and e, a vector of ones in $\mathbb{R}^m$:*

$$r = (a)_*,\ u = (a - \epsilon e)_+ \iff \begin{pmatrix} 0 \le r \perp u - a + \epsilon e \ge 0 \\ 0 \le u \perp -r + e \ge 0 \end{pmatrix}, \tag{2.8}$$

where ϵ is a sufficiently small positive number, that is

$$0 < \epsilon < \inf_{a_i \neq 0} |a_i|. \tag{2.9}$$

Proof. The points $r = (a)_*$ and $u = (a - \epsilon e)_+$ uniquely solve the dual linear programs

$$\max_r \ \{(a - \epsilon e)^T r | 0 \le r \le e\} \ \text{and} \ \min_u \ \{e^T u | u \ge a - \epsilon e,\ u \ge 0\}. \tag{2.10}$$

The right hand side of the equivalence (2.8) is merely the Karush-Kuhn-Tucker necessary and sufficient optimality conditions for $r = (a)_*$ and $u = (a - \epsilon e)_+$ to solve (2.10). □

We note that the use of ϵ is unnecessary in [18], because of the following equivalence:

$$r = (a)_*,\ u = (a)_+ \iff (r,u) \in \arg\min_{r,u} \{e^T r | 0 \le r \perp u - a \ge 0,\ 0 \le u \perp -r + e \ge 0\}, \tag{2.11}$$

and because the term $e^T r$ is minimized in [18], but is being maximized here (see (2.12) for example). With Lemma 1, we can reformulate the classification maximization problem (2.7) as the following LPEC with ϵ sufficiently small and positive:

$$\begin{array}{ll} \underset{w,\theta,r,u,s,v}{\text{maximize}} & e^T r + e^T s \\ \text{subject to} & \begin{array}{l} 0 \le r \perp u - Aw + e\theta + \epsilon e \ge 0 \\ 0 \le u \perp -r + e \ge 0 \\ 0 \le s \perp v + Bw - e\theta + \epsilon e \ge 0 \\ 0 \le v \perp -s + e \ge 0 \end{array} \end{array} \tag{2.12}$$

Note that with the exception of the "*perp*" condition, all constraints and the objective function are linear. To overcome the nonlinear effect of the $\perp$-condition, an implicitly exact penalty function formulation has been proposed as well as a parametric approach [18]. The parametric approach is preferable, because (2.12) has infinitely many stationary points as was pointed out in [18]. The reason for this anomaly is that any (w, θ) determining a plane $x^T w = \theta$ that does not contain any points from either the sets $\mathcal{A}$ or $\mathcal{B}$, is a stationary solution for problem (2.12). This is so because a slight perturbation of the plane does not change the number of misclassified points. To overcome this difficulty a parametric reformulation was proposed in [18] and implemented in [2]. For the classification maximization problem (2.7), the parametric reformulation of (2.12) is the following:

$$\begin{array}{ll} \underset{w,\theta,r,u,s,v}{\text{minimize}} & [r^T(-Aw + e\theta + \epsilon e) + e^T u] + [s^T(Bw - e\theta + \epsilon e) + e^T v] =: f(\mu) \\ \text{subject to} & \begin{array}{l} 0 \le r,\quad u - Aw + e\theta + \epsilon e \ge 0 \\ 0 \le u,\quad -r + e \ge 0 \\ 0 \le s,\quad v + Bw - e\theta + \epsilon e \ge 0 \\ 0 \le v,\quad -s + e \ge 0 \\ e^T r + e^T s \ge \mu \\ \mu \in [0, \infty) \end{array} \end{array} \tag{2.13}$$

Here μ is a parameter that represents the number of points correctly classified. The largest value of μ, such that the objective function has a minimum of zero, is the maximum number of points that can be correctly classified by a plane $x^T w = \theta$. Note that $f(\mu)$ is a nondecreasing function of μ, and the largest value $\bar{\mu}$ for which $f(\bar{\mu}) = 0$, constitutes a maximum to the NP-complete classification maximization problem (2.7). The parametric approach consists of starting at some large $\mu > \bar{\mu}$, solving (2.13) by a Frank-Wolfe algorithm [8, 5], for decreasing values of μ until $\bar{\mu}$ is reached. Efficient estimation of successive values of μ can be achieved by a secant method applied to $f(\mu)$. The method seems to work quite well as evidenced by computational results given in [2, 7].

3. IMPROVING GENERALIZATION

In this section we shall consider a fundamental problem of machine learning: How to train a system on a given training set so as to improve generalization on a new unseen testing set [13, 24, 28]. We shall concentrate on some very recent results [27] obtained for a simple linear model and which make critical use of mathematical programming ideas. These ideas, although rigorously established for a simple linear model only, seem to extend to much more complex systems, including neural networks [27].

The model that we shall consider here consists of the *training set* $\{A, a\}$ where A is a given $m \times n$ real matrix and a is a given $m \times 1$ real vector. A vector x in $\mathbb{R}^n$ is to be "learnt" such that the linear system

$$Ax = a, \tag{3.1}$$

which does not have an exact solution, is satisfied in some approximate fashion, and such that the error in satisfying

$$Cx = c, \tag{3.2}$$

for some unseen *testing set* $(C, c) \in \mathbb{R}^{k \times n} \times \mathbb{R}^k$, is minimized. Of course, if we disregard the testing set error (3.2), the problem becomes the standard least-norm problem:

$$\min_{x \in \mathbb{R}^n} \|Ax - a\|, \tag{3.3}$$

where $\|\cdot\|$ is some norm on $\mathbb{R}^m$. However with an eye to possible perturbations in the given training set $\{A, a\}$, we pose the following motivational question: If the vector a of the training set is known only to an accuracy of τ, where τ is some positive number, does it make sense to attempt to drive the error to zero as is done in (3.3), or is it not better to tolerate errors in the satisfaction of $Ax = a$ up to a magnitude of τ? In other words, instead of (3.1), we should try to satisfy the following system of inequalities, in some best sense:

$$-e\tau \leq Ax - a \leq e\tau \tag{3.4}$$

To do that, we solve the following regularized quadratic program for some nonnegative τ and a small positive ϵ:

$$\begin{array}{ll} \underset{x,y,z}{\text{minimize}} & \frac{1}{2}\|y\|_2^2 + \frac{1}{2}\|z\|_2^2 + \frac{\epsilon}{2}\|x\|_2^2 \\ \text{subject to} & -z - e\tau \leq Ax - a \leq e\tau + y \\ & y, z \geq 0. \end{array} \tag{3.5}$$

Here ϵ is a small fixed positive regularization constant that ensures the uniqueness of the x component of the solution. We note immediately, that if $\tau = 0$, problem (3.5) degenerates to the regularized classical least squares problem:

$$\min_{x \in \mathbb{R}^n} \frac{1}{2}\|Ax - a\|_2^2 + \frac{\epsilon}{2}\|x\|_2^2. \tag{3.6}$$

The key question to ask here, is this: Under what conditions does a solution $x(\tau)$ of (3.5), for some $\tau > 0$, give a smaller error on a testing set? We are able to give an answer to this question and corroborate it computationally [27], by considering a general testing set $(C, c) \in \mathbb{R}^{k \times n} \times \mathbb{R}^k$ as well as a simpler testing set, where only the right side of (3.1) is perturbed. We begin with the latter and simpler perturbation, that is:

$$Ax = a + t, \tag{3.7}$$

where t is some arbitrary perturbation in $\mathbb{R}^m$, and consider the following associated error function:

$$f(\tau) := \frac{1}{2} \|Ax(\tau) - a - t\|_2^2. \tag{3.8}$$

In particular we would like to know when is $f(0)$ *not* a local minimum of $f(\tau)$ on the set $\{\tau \mid \tau \geq 0\}$. In fact we are only interested in the τ-interval $[0, \hat{\tau}]$, where $\hat{\tau}$ is defined by

$$\hat{\tau} := \min_x \|Ax - a\|_\infty, \tag{3.9}$$

because the minimum value of (3.5) approaches zero, for $\tau \geq \hat{\tau}$, as ϵ approaches zero. Since $x(\tau)$ is continuous and piecewise-linear on $\tau \geq 0$ it follows that $f(\tau)$ defined by (3.8) is continuous piecewise-quadratic on $[0, \hat{\tau}]$, and hence attains a minimum at some τ in $[0, \hat{\tau}]$. Since $f(\tau)$ is directionally differentiable, it follows that *if* the directional derivative $f'(\tau; 1)$ at $\tau = 0$ in the positive direction is negative, then $\tau = 0$ is a strict local maximum of $f(\tau)$. Hence, as measured by the error criterion (3.8), $x(\tau)$ for some positive τ provides a better point. The following theorem gives a sufficient condition for $f'(0; 1) < 0$ and thus ensuring that solving (3.5) for some positive τ produces an $x(\tau)$ that generalizes better on the system (3.7) than that obtained by solving a plain regularized least squares problem (3.6), that is $f(\bar{\tau}) < f(0)$ for some $\bar{\tau} \in (0, \hat{\tau}]$.

Theorem 2 *[27]* **Improved generalization on $Ax = a + t$ with positive training tolerance.** *The testing set error function $f(\tau)$ of (3.8) has a strict local maximum at* 0 *and a global minimum on $[0, \hat{\tau}]$, where $\hat{\tau}$ is defined by (3.9), at some $\bar{\tau} > 0$, whenever*

$$(\epsilon x(0) + A^T t)^T (x(\tau) - x(0)) > 0 \tag{3.10}$$

for some $\tau \in (0, \tilde{\tau}]$, for some sufficiently small $\tilde{\tau}$.

For the more general testing model given by $Cx = c$ of (3.2), we have the following result for improved generalization.

Theorem 3 *[27]* **Improved generalization on $Cx = c$ with positive training tolerance.** *Let $x(\tau)$ be defined by the tolerant training of $Ax = a$ by the quadratic program (3.5) with tolerance $\tau \geq 0$. Let $g(\tau)$ denote the error generated by $x(\tau)$ in the testing model $Cx = c$, defined by:*

$$g(\tau) := \frac{1}{2} \|Cx(\tau) - c\|_2^2. \tag{3.11}$$

The zero-tolerance error $g(0)$ generated by $x(0)$ is a strict local maximum over $\tau \geq 0$ whenever

$$\|r(0)\|_2^2 > r(\tau)^T r(0) \textit{ for some } \tau \in (0, \tilde{\tau}] \tag{3.12}$$

for some sufficiently small $\tilde{\tau}$, where $r(\tau)$ is defined by

$$r(\tau) := Cx(\tau) - c. \tag{3.13}$$

Computational results carried out in [27] have corroborated the improved generalization results of Theorem 2 above, as well as for more complex models such as neural networks, where a threshold tolerance in measuring the error in the backpropagation algorithm [23, 11, 20] is allowed.

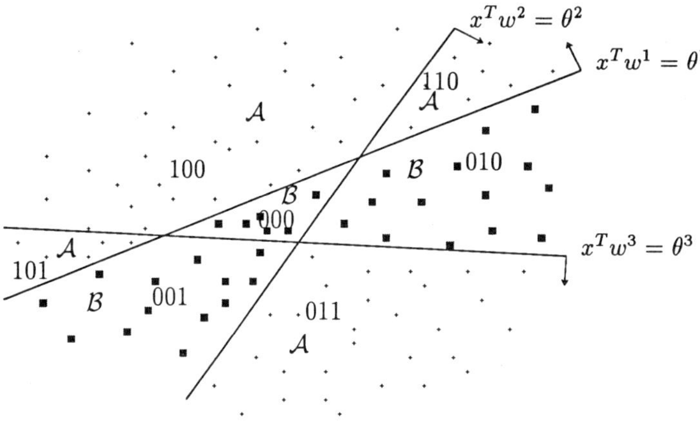

Figure 1: **Seven polyhedral regions in $\mathbb{R}^2$ generated by three planes: $x^Tw^1 = \theta^1$, $x^Tw^2 = \theta^2$ and $x^Tw^3 = \theta^3$. Each region contains elements of only one set $\mathcal{A}$ or $\mathcal{B}$ and is tagged by a binary number, the ith digit of which denotes whether the region is on the 1-side, $x^Tw^i > \theta^i$, or 0-side, $x^Tw^i < \theta^i$, of the ith plane.**

4. NEURAL NETWORKS AS MATHEMATICAL PROGRAMS WITH EQUILIBRIUM CONSTRAINTS

A neural network, which is a generalization of a separating plane in $\mathbb{R}^n$, can be defined as a nonlinear map from $\mathbb{R}^n$ into some set, typically {0,1}. One intuitive way to generate such a map is to divide $\mathbb{R}^n$ into various polyhedral regions, each of which containing elements of only one of two given disjoint point sets $\mathcal{A}$ and $\mathcal{B}$. (See Figure 1.) In its general form, this problem is again an extremely difficult and nonconvex problem. However, various greedy sequential constructions of the planes determining the various polyhedral regions [16, 19, 1] have been quite successful in obtaining very effective algorithms for training neural networks. These algorithms are much faster than the classical online backpropagation (BP) gradient algorithm [23, 11, 20], where the training is done on one point at a time. Often online BP is erroneously referred to as a descent algorithm, which it is not.

In this section of the paper we relate the polyhedral regions into which $\mathbb{R}^n$ is partitioned, to a classical neural network with one hidden layer of linear threshold units (LTUs) and one output LTU. (See Figure 2.) An LTU is an abstraction of a human neuron which fires if its input exceeds its threshold value. Thus the LTU, depicted by its threshold value of θ^1 in Figure 2, will have the output $(x^Tw^1 - \theta^1)_*$, where $(\cdot)_*$ is the step function defined earlier. An obvious representation of such an LTU would be by the plane $x^Tw^1 = \theta^1$. It turns out that every neural network mapping $\mathbb{R}^n$ into the set {0,1} can be related to a partitioning of $\mathbb{R}^n$ into polyhedral regions, but not conversely. However, any two disjoint point sets in $\mathbb{R}^n$ can be discriminated between by *some* polyhedral partition that corresponds to a neural network with one hidden layer with a sufficient number of hidden units [12, 19].

We describe now precisely when a specific partition of $\mathbb{R}^n$ by h separating planes

$$x^Tw^i = \theta^i,\ i = 1, \ldots, h, \tag{4.1}$$

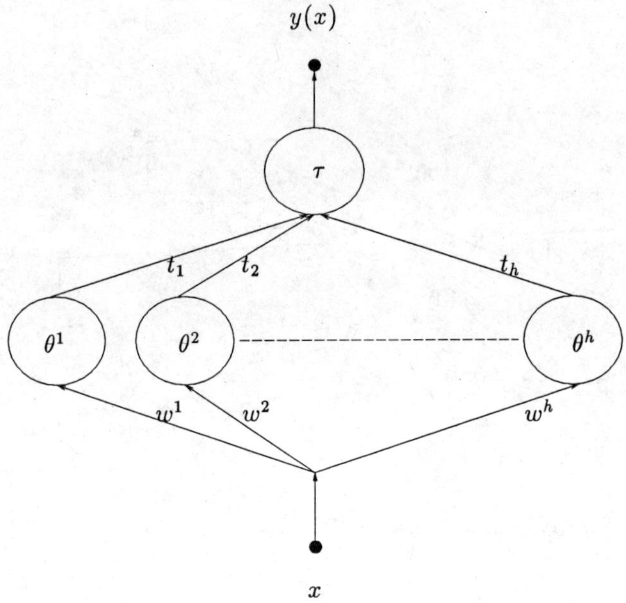

Figure 2: **A typical feedforward neural network with a single layer of h hidden linear threshold units (LTUs), input $x \in \mathbb{R}^n$, and output $y(x) \in \{0, 1\}$. The output of hidden unit i is $(x^T w^i - \theta^i)_*$, $i = 1, \ldots, h$. The output $y(x)$ of the output LTU is $(\sum_{i=1}^h (x^T w^i - \theta^i)_* t_i - \tau)_*$.**

corresponds to a neural network with h hidden units. (See Figures 1 and 2.) The h separating planes (4.1) divide $\mathbb{R}^n$ into at most p polyhedral regions, where [9]

$$p := \sum_{i=0}^{n} \binom{h}{i}. \tag{4.2}$$

We shall assume that $\mathcal{A}$ and $\mathcal{B}$ are contained in the interiors of two mutually exclusive subsets of these regions. (See Figure 1.) Each of these polyhedral regions can be mapped uniquely into a vertex of the unit cube in $\mathbb{R}^h$,

$$\{z | z \in \mathbb{R}^h,\ 0 \leq z \leq e\} \tag{4.3}$$

by using the map:

$$(x^T w^i - \theta^i)_*,\ i = 1, \ldots, h, \tag{4.4}$$

where x is a point in $\mathbb{R}^n$ belonging to some polyhedral region. If the p polyhedral regions of $\mathbb{R}^n$ constructed by the h planes (4.1) are such that vertices of the cube (4.3) corresponding to points in $\mathcal{A}$, are linearly separable in $\mathbb{R}^h$ from the vertices of (4.3) corresponding to points in $\mathcal{B}$, by a plane

$$r^T t = \tau, \tag{4.5}$$

as in the example of Figure 3, then the polyhedral partition of $\mathbb{R}^n$ corresponds to a neural network with h hidden linear threshold units (with thresholds θ^i, incoming arc weights w^i, $i = 1, \ldots, h$) and output linear threshold unit (with threshold τ and incoming arc weights t_i, $i = 1, \ldots, h$) [17]. This condition is necessary and sufficient for the polyhedral

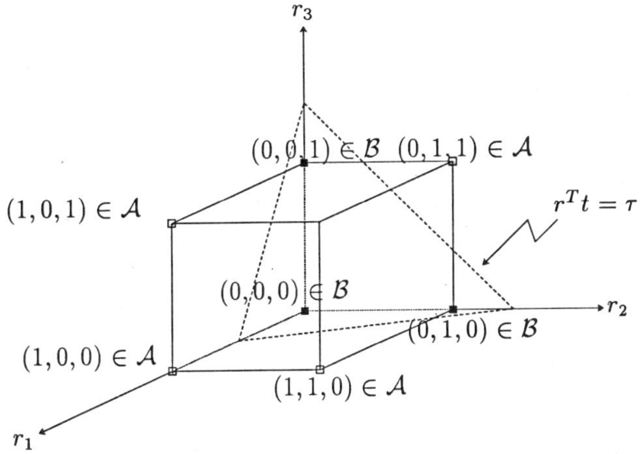

Figure 3: **The vertices of the unit cube into which the sets $\mathcal{A}$ and $\mathcal{B}$ of Figure 1 are mapped by the three planes shown in that figure, or equivalently by three hidden LTUs of a neural network. A plane, $r^T t = \tau$, separates the vertices associated with $\mathcal{A}$ from those associated with $\mathcal{B}$. This plane corresponds to the output LTU of a neural network and the weights of its incoming arcs.**

partition of $\mathbb{R}^n$ in order for it to correspond to a neural network with one layer of hidden units. For more details, see [17].

"Training" a neural network consists of determining $(w^i, \theta^i) \in \mathbb{R}^{n+1}$, $i = 1, \ldots, h$, $(t, \tau) \in \mathbb{R}^{h+1}$, such that the following nonlinear inequalities are satisfied as best as possible:

$$
\begin{aligned}
\sum_{i=1}^{h} (Aw^i - e\theta^i)_* t_i > e\tau \\
\sum_{i=1}^{h} (Bw^i - e\theta^i)_* t_i < e\tau
\end{aligned}
\tag{4.6}
$$

This can be achieved by maximizing a weighted sum of correctly classified points in $\mathbb{R}^h$ by solving the following unconstrained maximization problem (as in the equivalent programs (2.3) and (2.5)):

$$
\begin{aligned}
\max_{w^i, \theta^i, t_i, \tau} \; & e^T \left(\sum_{i=1}^{h} (Aw^i - e\theta^i)_* t_i - e\tau\right)_- \\
+ & e^T \left(\sum_{i=1}^{h} -(Bw^i - e\theta^i)_* t_i + e\tau\right)_-,
\end{aligned}
\tag{4.7}
$$

where the function $(\cdot)_-$ is defined in (2.4). If instead of the step function $(\zeta)_*$ the sigmoid function $\sigma(\zeta)$ is used in (4.7), where $\sigma(\zeta) := \frac{1}{1+e^{-\alpha\zeta}}$, $\alpha > 0$, we obtain an error function similar to the error function that backpropagation attempts to find a stationary point for, and for which a convergence proof is given in [20], and stability analysis in [25]. We note that the classical exclusive-or (XOR) example [22] for which $A = \begin{bmatrix} 1 & 0 \\ 0 & 1 \end{bmatrix}$, $B = \begin{bmatrix} 0 & 0 \\ 1 & 1 \end{bmatrix}$, gives a maximum value of four for (4.7) with the following solution:

$$
\begin{array}{l}
(w^1,\theta^1) = ((2\ \ -2),\ 1),\ (w^2,\theta^2) = ((-2\ \ 2),\ 1) \\
(v,\tau) = ((2\ \ 2),\ 1)
\end{array}
\tag{4.8}
$$

This corresponds to correctly separating the two points in $\mathcal{A}$ from the two points in $\mathcal{B}$.

It is interesting to note that the same solution for the XOR example is given by the greedy multisurface method tree (MSMT) [1]. MSMT attempts to separate as many points of $\mathcal{A}$ and $\mathcal{B}$ as possible by a first plane obtained by solving (2.3), and then repeats the process for each of the ensuing halfspaces, until adequate separation is obtained. For this example, the first plane obtained [4] is $(w^1,\theta^1) = ((2\ \ -2),\ 1)$, which separates $\{(1,0)\}$ from $\{(0,0),(0,1),(1,1)\}$. The second plane obtained is $(w^2,\theta^2) = ((-2\ \ 2),\ 1)$, separates $\{(0,1)\}$ from $\{(0,0),(1,1)\}$, and the separation is complete between $\mathcal{A}$ and $\mathcal{B}$. These planes correspond to the same neural network obtained by solving (4.7), which of course is not always the case when using the greedy MSMT method. However MSMT frequently gives better solutions than those generated by BP and is much faster than BP.

We now set up the problem (4.7) as an MPEC. We first use the equivalence between the step function $(\cdot)_*$ and an equilibrium condition given by Lemma 1 and obtain the following problem, where ϵ is a sufficiently small positive number:

$$
\begin{array}{ll}
\underset{w^i,\theta^i,r^i,u^i,s^i,v^i,t_i,\tau}{\text{maximize}} & e^T(\sum_{i=1}^{h} r^i t_i - e\tau)_- + e^T(\sum_{i=1}^{h} -s^i t_i + e\tau)_- \\
 & 0 \le r^i \perp u^i - Aw^i + e\theta^i + \epsilon e \ge 0 \\
 & 0 \le u^i \perp -r^i + e \ge 0 \\
\text{subject to} & 0 \le s^i \perp v^i - Bw^i + e\theta^i + \epsilon e \ge 0 \\
 & 0 \le v^i \perp -s^i + e \ge 0 \\
 & i = 1,\ldots,h.
\end{array}
\tag{4.9}
$$

By using the equivalence between the formulations (2.3) and (2.5) we can formulate (4.9) as the following MPEC:

$$
\begin{array}{ll}
\underset{w^i,\theta^i,r^i,u^i,s^i,v^i,t_i,\tau,y^i,z^i}{\text{maximize}} & e^T\sum_{i=1}^{h} y^i + e^T\sum_{i=1}^{h} z^i \\
 & \sum_{i=1}^{h} r^i t_i - e\tau \ge y^i,\ \ y^i \le e \\
 & \sum_{i=1}^{h} -s^i t_i + e\tau \ge z^i,\ \ z^i \le e \\
 & 0 \le r^i \perp u^i - Aw^i + e\theta^i + \epsilon e \ge 0 \\
\text{subject to} & 0 \le u^i \perp -r^i + e \ge 0 \\
 & 0 \le s^i \perp v^i - Bw^i + e\theta^i + \epsilon e \ge 0 \\
 & 0 \le v^i \perp -s^i + e \ge 0 \\
 & i = 1,\ldots,h
\end{array}
\tag{4.10}
$$

In a manner similar to the parametric reformulation (2.13) of the LPEC (2.12) associated with the classification maximization problem (2.7), the above MPEC can be reformulated as the following parametric bilinear program:

$$
\begin{array}{ll}
\underset{w^i,\theta^i,r^i,u^i,s^i,v^i,t_i,\tau,y^i,z^i}{\text{minimize}} & \sum_{i=1}^h (r^i)^T(-Aw^i + e\theta^i + \epsilon e) + e^T u^i \\
& \quad + \sum_{i=1}^h (s^i)^T(-Bw^i + e\theta^i + \epsilon e) + e^T v^i =: g(\mu) \\
\text{subject to} & \sum_{i=1}^h r^i t_i - e\tau \geq y^i, \quad y^i \leq e \\
& \sum_{i=1}^h -s^i t_i + e\tau \geq z^i, \quad z^i \leq e \\
& 0 \leq r^i, u^i - Aw^i + e\theta^i + \epsilon e \geq 0 \\
& 0 \leq u^i, -r^i + e \geq 0 \\
& 0 \leq s^i, v^i - Bw^i + e\theta^i + \epsilon e \geq 0 \\
& 0 \leq v^i, -s^i + e \geq 0 \\
& i = 1, \ldots, h \\
& e^T \sum_{i=1}^h y^i + e^T \sum_{i=1}^h z^i \geq \mu \\
& \mu \in [0, \infty)
\end{array}
\tag{4.11}
$$

Here μ is a parameter that represents the number of points correctly classified, and will equal $m + k$ if complete separation is achieved by the neural network. The largest value of μ for which the objective function has a minimum of zero is the maximum value of the MPEC (4.10), which corresponds to training a neural network on the sets $\mathcal{A}$ and $\mathcal{B}$. Note that $g(\mu)$ is a nondecreasing function of μ, and the largest value $\bar{\mu}$ for which $g(\bar{\mu}) = 0$, constitutes a maximum to problem (4.9). The parametric approach consists of starting at some large μ, say $\mu = m + k$, solving (4.11), for decreasing values of μ for which $g(\mu) > 0$, until $\bar{\mu}$ such that $g(\bar{\mu}) = 0$ is reached. Efficient estimation of successive values of μ can be achieved by a secant method applied to the nondecreasing function $g(\mu)$. Note that the nonconvex problem (4.11) has a bilinear objective and two sets of bilinear constraints. Although no computation has been done with this model of a neural network, it is felt that the Frank-Wolfe approach utilized to solve efficiently numerous NP-complete problems in [5] could also be effective here as well. Briefly the approach would consist of fixing t_i, $i = 1, \ldots, h$ and solving (4.11) by the bilinear approach of [5] which involves successive linear programs and line searches. Then (t_i, y^i, z^i) $i = 1, \ldots, h$ and τ are updated by solving a single linear program. The bilinear approach corresponds to adjusting the thresholds and incoming arc weights for the hidden units of the neural network as well as adjusting the threshold of the output unit, while holding the weights of the incoming arcs to the output unit fixed. The linear program then attempts to get a best linear separation between vertices of the unit cube in $\mathbb{R}^h$ that represent $\mathcal{A}$ and $\mathcal{B}$, by readjusting the threshold of the output unit of the neural network as well as the weights of its incoming arcs.

5. CONCLUSION

Significant problems associated with machine learning have been cast as a variety of mathematical programs, ranging in complexity from polynomial-time-solvable linear programs to NP-complete problems. Effective methods for solving some of these problems have been outlined. Modeling and efficiently solving many of these problems constitute an important and challenging field of research for mathematical programming.

ACKNOWLEDGEMENTS

This paper is based on research supported by Air Force Office of Scientific Research Grant F49620-94-1-000036 and National Science Foundation Grant CCR-9322479.

REFERENCES

[1] K. P. Bennett, "Decision tree construction via linear programming". In M. Evans, editor, *Proceedings of the 4th Midwest Artificial Intelligence and Cognitive Science Society Conference*, pages 97–101, Utica, Illinois, 1992.

[2] K. P. Bennett and E. J. Bredensteiner, "A parametric optimization method for machine learning". Department of Mathematical Sciences Report No. 217, Rensselaer Polytechnic Institute, Troy, NY 12180, 1994.

[3] K. P. Bennett and O. L. Mangasarian, "Neural network training via linear programming". In P. M. Pardalos, editor, *Advances in Optimization and Parallel Computing*, pages 56–67, Amsterdam, 1992. North Holland.

[4] K. P. Bennett and O. L. Mangasarian, "Robust linear programming discrimination of two linearly inseparable sets". *Optimization Methods and Software*, 1:23–34, 1992.

[5] K. P. Bennett and O. L. Mangasarian, "Bilinear separation of two sets in n-space". *Computational Optimization & Applications*, 2:207–227, 1993.

[6] A. Charnes, "Some fundamental theorems of perceptron theory and their geometry". In J. T. Lou and R. H. Wilcox, editors, *Computer and Information Sciences*, pages 67–74, Washington, D.C., 1964. Spartan Books.

[7] Chunhui Chen and O. L. Mangasarian, "Hybrid misclassification minimization". Technical Report 95-05, Computer Sciences Department, University of Wisconsin, Madison, Wisconsin, February 1995. Advances in Computational Mathematics, submitted. Available from ftp://ftp.cs.wisc.edu/math-prog/tech-reports/95-05.ps.Z.

[8] M. Frank and P. Wolfe, "An algorithm for quadratic programming". *Naval Research Logistics Quarterly*, 3:95–110, 1956.

[9] G. M. Georgiou, "Comments on hidden nodes in neural nets". *IEEE Transactions on Circuits and Systems*, 38:1410, 1991.

[10] David Heath, *"A geometric Framework for Machine Learning"*. PhD thesis, Department of Computer Science, Johns Hopkins University–Baltimore, Maryland, 1992.

[11] J. Hertz, A. Krogh, and R. G. Palmer, *"Introduction to the Theory of Neural Computation"*. Addison-Wesley, Redwood City, California, 1991.

[12] K. Hornik, M. Stinchcombe, and H. White, "Multilayer feedforward networks are universal approximators". *Neural Networks*, 2:359–366, 1989.

[13] Y. le Cun, J. S. Denker, and S. A. Solla, "Optimal brain damage". In D. S. Touretzky, editor, *Advances in Neural Information Processing Systems II (Denver 1989)*, pages 598–605, San Mateo, California, 1990. Morgan Kaufmann.

[14] Z.-Q. Luo, J.-S. Pang, D. Ralph, and S.-Q. Wu, "Exact penalization and stationarity conditions of mathematical programs with equilibrium constraints". Technical Report 275, Communications Research Laboratory, McMaster University, Hamilton, Ontario, Hamilton, Ontario L8S 4K1, Canada, 1993. Mathematical Programming, to appear.

[15] O. L. Mangasarian, "Linear and nonlinear separation of patterns by linear programming". *Operations Research*, 13:444–452, 1965.

[16] O. L. Mangasarian, "Multi-surface method of pattern separation". *IEEE Transactions on Information Theory*, IT-14:801–807, 1968.

[17] O. L. Mangasarian, "Mathematical programming in neural networks". *ORSA Journal on Computing*, 5(4):349–360, 1993.

[18] O. L. Mangasarian, "Misclassification minimization". *Journal of Global Optimization*, 5:309–323, 1994.

[19] O. L. Mangasarian, R. Setiono, and W. H. Wolberg, "Pattern recognition via linear programming: Theory and application to medical diagnosis". In T. F. Coleman and Y. Li, editors, *Large-Scale Numerical Optimization*, pages 22–31, Philadelphia, Pennsylvania, 1990. SIAM. Proceedings of the Workshop on Large-Scale Numerical Optimization, Cornell University, Ithaca, New York, October 19-20, 1989.

[20] O. L. Mangasarian and M. V. Solodov, "Serial and parallel backpropagation convergence via nonmonotone perturbed minimization". *Optimization Methods and Software*, 4(2):103–116, 1994.

[21] O. L. Mangasarian, W. Nick Street, and W. H. Wolberg, "Breast cancer diagnosis and prognosis via linear programming". Technical Report 94-10, Computer Sciences Department, University of Wisconsin, Madison, Wisconsin 53706, 1994. Operations Research 43(4) 1995, to appear. Available from ftp://ftp.cs.wisc.edu/math-prog/tech-reports/94-10.ps.Z.

[22] M. Minsky and S. Papert, *"Perceptrons: An Introduction to Computational Geometry"*. MIT Press, Cambridge, Massachusetts, 1969.

[23] D. E. Rumelhart, G. E. Hinton, and R. J. Williams, "Learning internal representations by error propagation". In D. E. Rumelhart and J. L. McClelland, editors, *Parallel Distributed Processing*, pages 318–362, Cambridge, Massachusetts, 1986. MIT Press.

[24] C. Schaffer, "Overfitting avoidance as bias". *Machine Learning*, 10:153–178, 1993.

[25] M. V. Solodov and S. K. Zavriev, "Stability properties of the gradient projection method with applications to the backpropagation algorithm". Computer Sciences Department, Mathematical Programming Technical Report 94-05, University of Wisconsin, Madison, Wisconsin, June 1994. SIAM Journal on Optimization, submitted.

[26] M. Stone, "Cross-validatory choice and assessment of statistical predictions". *Journal of the Royal Statistical Society*, 36:111–147, 1974.

[27] W. Nick Street and O. L. Mangasarian, "Improved generalization via tolerant training". Technical report, Computer Sciences Department, University of Wisconsin, Madison, Wisconsin, 1995. To appear.

[28] D. H. Wolpert, editor, *"The Mathematics of Generalization"*, Reading, MA, 1995. Addison-Wesley.

SMOOTHING TECHNIQUES FOR MACROMOLECULAR GLOBAL OPTIMIZATION

Moré Jorge J. and Wu Zhijun

Mathematics and Computer Science Division
Argonne National Laboratory
9700 South Cass Avenue
Argonne, Illinois 60439

Abstract. We study global optimization problems that arise in macromolecular modeling, and the solution of these problems via continuation and smoothing. Our results unify and extend the theory associated with the use of the Gaussian transform for smoothing. We show that the Gaussian transform can be viewed as a special case of a generalized transform and that these generalized transforms share many of the properties of the Gaussian transform. We also show that the smoothing behavior of the generalized transform can be studied in terms of the Fourier transform and that these results indicate that the Gaussian transform has superior smoothing properties.

Key words. Smoothing, continuation, global optimization, macromolecular modeling, Gaussian transform, Fourier transform

1. INTRODUCTION TO MACROMOLECULAR OPTIMIZATION

Macromolecular modeling gives rise to a wide variety of global optimization problems with interesting features. These problems usually require the determination of the global minimum of a function with $n = 3m$ variables, where m is the number of atoms in the molecule. A typical problem requires the determination of a molecular structure such that the positions of the atoms $x_1, \ldots, x_m$ in $\mathbb{R}^3$ satisfy certain constraints on the structure or such that a given potential energy function is minimized. There is also interest in determining structures that nearly satisfy these constraints or that have nearly minimal potential value.

The problem of determining a structure that satisfies geometric constraints on the structure falls under the general area of distance geometry. Distance data between atoms and other geometric constraints (for example, angle constraints) can be obtained from nuclear magnetic resonance (NMR) data, but in most cases only partial information is available. In particular, for distance data only a fraction of the distances $\delta_{i,j}$ between the (i, j) pair of atoms is available for a subset $\mathcal{S}$ of the atom pairs. Given this information, we seek positions $x_1, \ldots, x_m$ in $\mathbb{R}^3$ of the atoms in the molecule such that

$$\|x_i - x_j\| = \delta_{i,j}, \qquad (i, j) \in \mathcal{S}. \tag{1.1}$$

This work was supported by the Mathematical, Information, and Computational Sciences Division subprogram of the Office of Computational and Technology Research, U.S. Department of Energy, under Contract W-31-109-Eng-38 and by the Argonne Director's Individual Investigator Program.

Since the data obtained from NMR is usually inaccurate, distance geometry problems that arise in the determination of protein structure are usually associated with the more general problem of finding positions $x_1, \ldots, x_m$ in $\mathbb{R}^3$ such that

$$l_{i,j} \le \|x_i - x_j\| \le u_{i,j}, \qquad (i,j) \in \mathcal{S}, \tag{1.2}$$

where $l_{i,j}$ and $u_{i,j}$ are lower and upper bounds on the distance constraints, respectively.

Problems (1.1) and (1.2) can be formulated in terms of finding the global minimum of a function that measures the deviation of the structure from the constraints. We formulate problem (1.1) in terms of finding the global minimum of the function

$$f(x) = \sum_{i,j \in \mathcal{S}} p_{i,j}(x_i - x_j), \tag{1.3}$$

where the pairwise potential $p_{i,j} : \mathbb{R}^3 \mapsto \mathbb{R}$ is defined by

$$p_{i,j}(x) = \left(\|x\|^2 - \delta_{i,j}^2\right)^2. \tag{1.4}$$

Similarly, we formulate problem (1.2) in terms of finding the global minimum of (1.3) where

$$p_{i,j}(x) = \min{}^2 \left\{ \frac{\|x\|^2 - l_{i,j}^2}{l_{i,j}^2}, 0 \right\} + \max{}^2 \left\{ \frac{\|x\|^2 - u_{i,j}^2}{u_{i,j}^2}, 0 \right\}. \tag{1.5}$$

For both of these problems, $x = \{x_1, \ldots, x_m\}$ solves problem (1.2) if and only if x is a global minimizer of f and $f(x) = 0$ (see Crippen and Havel [4]).

In general, distance geometry calculations impose additional constraints on the structure. For additional information on distance geometry, see Crippen and Havel [4], Havel [11], Kuntz, Thomason, and Oshiro [16], and Brünger and Nilges [1].

In macromolecular modeling we are also interested in determining structures with minimal or nearly minimal potential energy. Various potential functions arise in applications. We are interested, in particular, in the determination of positions $x_1, \ldots, x_m$ in $\mathbb{R}^3$ of the atoms so that the potential energy

$$f(x) = \sum_{i \ne j} \frac{1}{\|x_i - x_j\|} + \sum_{i=1}^{m} \|x_i\|^2$$

is minimized. For information and references to this problem, see Hasse and Schiffer [10], Rafac, Schiffer, Hangst, Dubin, and Wales [21], and Schiffer [25].

The problems that we have introduced can be described in terms of finding the global minimum of a function of the form

$$f(x) = \sum_{i,j \in \mathcal{S}} p_{i,j}(x_i - x_j) + \alpha \sum_{i=1}^{m} \|x_i\|^2, \tag{1.6}$$

where the pairwise potential $p_{i,j} : \mathbb{R}^3 \mapsto \mathbb{R}$ is defined by

$$p_{i,j}(x) = h_{i,j}(\|x\|), \tag{1.7}$$

for some functions $h_{i,j} : \mathbb{R} \mapsto \mathbb{R}$ and $\alpha \ge 0$. Our approach for determining the global minimum of functions of the general form (1.6,1.7) is based on smoothing and continuation. The smoothing process transforms a function $f : \mathbb{R}^n \mapsto \mathbb{R}$ into a smoother function $\langle f \rangle_\lambda$ with fewer local minimizers. The parameter λ controls the degree of smoothing; the original function is obtained if $\lambda = 0$, while smoother functions are obtained as λ increases. Given the smooth function $\langle f \rangle_\lambda$, we use continuation on λ to trace the minimizers of $\langle f \rangle_\lambda$ back to the original function.

In this paper we unify and extend the work of Wu [31] and Moré and Wu [18, 19] on using the Gaussian transform $\langle f\rangle_\lambda$ for smoothing. Our results in Sections 2. and 3. show that the Gaussian transform can be viewed as a special case of a generalized transform and that these generalized transforms share many of the properties of the Gaussian transform. Of interest are the results in Section 3. on the smoothing behavior of the generalized transform in terms of the Fourier transform, since these results indicate that the Gaussian transform has superior smoothing properties.

Sections 5. and 6. complete this paper with a discussion of techniques for computing the generalized transform. Section 5. introduces the decomposable functions and shows how the generalized transform of a decomposable function can be expressed in terms of one-dimensional integrals, which can be either evaluated analytically or, at worst, approximated by Gaussian quadratures. Section 6. outlines several techniques that can be used to compute the Gaussian transform of any function defined by (1.6,1.7). We show that the Gaussian transform can be expressed in terms of special functions or one-dimensional integrals. These techniques are only applicable to the Gaussian transform, and thus show that the Gaussian transform plays a special role in problems that arise in macromolecular modeling.

2. SMOOTHING

The value $\langle f\rangle_\lambda$ of the transformed function at a point $x \in \mathbb{R}^n$ is generally obtained by computing weighted averages of f in a neighborhood centered at x. The weights are determined by a probability density function, that is, a nonnegative function $\rho : \mathbb{R}^n \mapsto \mathbb{R}$ such that

$$\int_{\mathbb{R}^n} \rho(x)\, dx = 1.$$

We could use any density function, but the use of the Gaussian density function

$$\rho(y) = \frac{1}{\pi^{n/2}} \exp\left(-\|y\|^2\right), \tag{2.1}$$

is natural. We also want the weight of the density function concentrated on the origin, so we use the Gaussian density function

$$\rho_\lambda(y) = \frac{1}{\lambda^n}\rho\left(\frac{y}{\lambda}\right) = \frac{1}{\pi^{n/2}\lambda^n} \exp\left(-\frac{\|y\|^2}{\lambda^2}\right) \tag{2.2}$$

with mean zero and variance $\frac{1}{2}\lambda^2$.

Definition 2.1 *The Gaussian transform $\langle f\rangle_\lambda$ of a function $f : \mathbb{R}^n \mapsto \mathbb{R}^m$ is*

$$\langle f\rangle_\lambda(x) = \frac{1}{\pi^{n/2}\lambda^n} \int_{\mathbb{R}^n} f(y) \exp\left(-\frac{\|y-x\|^2}{\lambda^2}\right) dy. \tag{2.3}$$

This definition shows that the value of the transformed function $\langle f\rangle_\lambda$ at a point $x \in \mathbb{R}^n$ is a weighted average of the values of f in a neighborhood of x. This observation can also be made by noting that the change of variables $y = x + \lambda u$ in (2.3) shows that the Gaussian transform can be written in the form

$$\langle f\rangle_\lambda(x) = \frac{1}{\pi^{n/2}} \int_{\mathbb{R}^n} f(x+\lambda u) \exp(-\|u\|^2)\, du. \tag{2.4}$$

Finally, one can also think of the Gaussian transform as the convolution of f with the Gaussian density function ρ_λ. This last observation comes into play when we analyze the smoothing properties of the Gaussian transform.

We have defined the Gaussian transform for a vector-valued mapping $f : \mathbb{R}^n \mapsto \mathbb{R}^m$ because we are also interested in the more general case. In particular, this extended definition will come up when discussing the smoothing of gradients and Hessian matrices.

The idea of transforming a function into a smoother function has been used extensively in macromolecular modeling. Straub [29] reviews work in this area. The diffusion equation method of Piela, Kostrowicki, and Scheraga [20] is central to our work since the idea of the Gaussian transform underlies the diffusion equation method. Other approaches include the packet annealing method of Shalloway [27, 26] and the effective energy simulated annealing method of Coleman, Shalloway, and Wu [2, 3].

We generalize the definition of the Gaussian transform by replacing the Gaussian density function with another density function $\rho : \mathbb{R}^n \mapsto \mathbb{R}$.

Definition 2.2 *Given a density function $\rho : \mathbb{R}^n \mapsto \mathbb{R}$, the generalized transform $\langle\langle f \rangle\rangle_\lambda$ of a function $f : \mathbb{R}^n \mapsto \mathbb{R}^m$ is*

$$\langle\langle f \rangle\rangle_\lambda(x) = \frac{1}{\lambda^n} \int_{\mathbb{R}^n} f(y)\, \rho\left(\frac{x-y}{\lambda}\right) dy. \tag{2.5}$$

Transformations of the general form (2.5) have been used in stochastic optimization, often in the equivalent formulation

$$\langle\langle f \rangle\rangle_\lambda(x) = \int_{\mathbb{R}^n} f(x - \lambda v)\, \rho(v)\, dv. \tag{2.6}$$

Any density function can be used in (2.5) but from a computational viewpoint, the Gaussian density function (2.1) and the uniform density function

$$\rho(x) = \tfrac{1}{2}, \qquad \|x\|_\infty \leq 1, \tag{2.7}$$

are of special interest.

The earliest reference to transformations of the form (2.5) in stochastic optimization seems to be due to Katkovnik (see, for example, Katkovnik and Kulchitskii [12]). References and generalizations of this work can be found in the work of Rubinstein [22, 23, 24], where the generalized transformation (2.5) is called a *smoothed functional.* Related work and other references can be found in Kreimer and Rubinstein [15] and Ermoliev, Norkin, and Wets [6]. Transformations of the form (2.5) also arise in the theory of distributions, but in this work the density ρ is usually infinitely differentiable and of compact support.

The emphasis of the work on stochastic optimization has been on the approximation of non-smooth functions. There has been little emphasis on computational issues. In particular, the transformation (2.5) is never computed explicitly; instead, a Monte Carlo method is used to approximate the transformation. As we shall see, the emphasis of the work on global optimization is quite different.

3. PROPERTIES OF THE GENERALIZED TRANSFORM

The Gaussian transform has many interesting properties, with most of these properties being shared by the generalized transform $\langle\langle f \rangle\rangle_\lambda$. We first consider the questions of existence of the transformations.

The Gaussian transform is defined if f is continuous almost everywhere, and if

$$|f(x)| \leq \beta_1 \exp(\beta_2 \|x\|) \tag{3.1}$$

for positive constants β_1 and β_2. Existence follows from (2.4), since (3.1) implies that

$$\left| f(x + \lambda v) \exp\left(-\|v\|^2\right) \right| \leq \beta_1 \exp\left(\beta_2 \|x\|\right) \exp\left(\beta_2 \lambda \|v\| - \|v\|^2\right).$$

The situation for the generalized transform is more delicate. The generalized transform $\langle\langle f \rangle\rangle_\lambda$ is defined if we assume that the density function ρ in (2.5) is continuous almost everywhere and has compact support. We can also admit density functions with infinite support if we restrict the class of functions f and density functions ρ. For example, if we assume that

$$|f(x)| \le \beta_1 + \beta_2 \|x\|^p$$

holds for positive constants β_1 and β_2 and exponent p, then the generalized transform $\langle\langle f \rangle\rangle_\lambda$ is defined if

$$\int_{\mathbb{R}^n} \|v\|^k \rho(v)\, dv, \qquad k = 0, 1, \ldots, p$$

exists. In this section we assume that f satisfies assumption (3.1) whenever we are discussing the Gaussian transform $\langle f \rangle_\lambda$ or the generalized transform $\langle\langle f \rangle\rangle_\lambda$ for a density function with compact support. The modifications needed to handle arbitrary density functions should be clear from this discussion.

The generalized transform is a linear operator in the vector space of functions that are continuous almost everywhere and satisfy (3.1), since

$$\langle\langle \alpha f \rangle\rangle_\lambda = \alpha \langle\langle f \rangle\rangle_\lambda, \qquad \langle\langle f_1 + f_2 \rangle\rangle_\lambda = \langle\langle f_1 \rangle\rangle_\lambda + \langle\langle f_2 \rangle\rangle_\lambda$$

for any scalar α and functions f_1 and f_2. The generalized transform $\langle\langle f \rangle\rangle_\lambda$ is also an isotone operator, since it preserves the standard order relation between functions, that is,

$$f_1 \le f_2 \qquad \text{implies that} \qquad \langle\langle f_1 \rangle\rangle_\lambda \le \langle\langle f_2 \rangle\rangle_\lambda.$$

In particular,

$$\min\{f(y) : \mathbb{R}^n\} \le \langle\langle f \rangle\rangle_\lambda(x) \le \max\{f(y) : \mathbb{R}^n\}.$$

These properties are direct consequences of the definition of the generalized transform $\langle\langle f \rangle\rangle_\lambda$.

The differentiability properties of the generalized transform $\langle\langle f \rangle\rangle_\lambda$ follow from general results (see, for example, Lang [17, Chapter 13]) on the differentiability of functions of the form

$$g(x) = \int_{\mathbb{R}^n} h(x, y)\, dy,$$

where the mapping $h : \mathbb{R}^n \mapsto \mathbb{R}$ is integrable in y. If $\partial_x h$ is continuous almost everywhere in an open set of the form $B \times \mathbb{R}^n$, and

$$|\partial_x h(x, y)| \le h_B(y), \qquad (x, y) \in B \times \mathbb{R}^n, \tag{3.2}$$

for some integrable function $h_B : \mathbb{R}^n \mapsto \mathbb{R}$, then g is differentiable and

$$g'(x) = \int_{\mathbb{R}^n} \partial_x h(x, y)\, dy.$$

This result can be applied, in particular, to the Gaussian transform.

Theorem 3.1 *The Gaussian transform $\langle f \rangle_\lambda$ is infinitely differentiable.*

Proof. Define

$$h(x, y) = f(y) \rho\left(\frac{x - y}{\lambda}\right),$$

and note that

$$\partial_x h(x, y) = \frac{1}{\lambda} f(y) \rho'\left(\frac{x - y}{\lambda}\right).$$

A calculation based on (3.1) shows that (3.2) holds, and thus $\langle f \rangle_\lambda$ is differentiable. The argument can be repeated to show that $\langle f \rangle_\lambda$ is infinitely differentiable. □

Theorem 3.1 does not hold for the generalized transform $\langle\langle f \rangle\rangle_\lambda$ unless we make additional assumptions on f or on the density function ρ. Consider, for example, the uniform density function

$$\rho(s) = \tfrac{1}{2}, \qquad |s| \leq 1.$$

In this case the generalized density function reduces to

$$\langle\langle f \rangle\rangle_\lambda(x) = \tfrac{1}{2} \int_{-1}^{1} f(x - \lambda s)\, ds,$$

and thus

$$\langle\langle f \rangle\rangle_\lambda{}'(x) = \frac{1}{2\lambda}\Big(f(x+\lambda) - f(x-\lambda)\Big).$$

This calculation shows that $\langle\langle f \rangle\rangle_\lambda$ is differentiable but that if f is not continuous, then $\langle\langle f \rangle\rangle_\lambda$ is not continuously differentiable.

The functions that arise in our work on macromolecular modeling are usually twice differentiable, and in this case we can show that the generalized transform $\langle\langle f \rangle\rangle_\lambda$ is also twice differentiable.

Theorem 3.2 *If $f : \mathbb{R}^n \mapsto \mathbb{R}$ is twice continuously differentiable almost everywhere on $\mathbb{R}^n$ and*

$$\|\nabla^2 f(x)\| \leq \gamma_1 \exp(\gamma_2 \|x\|) \tag{3.3}$$

for some positive constants γ_1 and γ_2, then

$$\nabla \langle\langle f \rangle\rangle_\lambda(x) = \langle\langle \nabla f \rangle\rangle_\lambda(x), \qquad \nabla^2 \langle\langle f \rangle\rangle_\lambda(x) = \langle\langle \nabla^2 f \rangle\rangle_\lambda(x).$$

An informal proof of this result can be obtained by differentiating under the integral sign in (2.4) to obtain that

$$\nabla \langle\langle f \rangle\rangle_\lambda(x) = \int_{\mathbb{R}^n} \nabla f(x - \lambda u)\rho(u)\, du = \langle\langle \nabla f \rangle\rangle_\lambda(x),$$

which is the desired result for the gradient. If we repeat the process, we obtain that

$$\nabla^2 \langle\langle f \rangle\rangle_\lambda(x) = \int_{\mathbb{R}^n} \nabla^2 f(x - \lambda u)\rho(u)\, du = \langle\langle \nabla^2 f \rangle\rangle_\lambda(x),$$

so that the generalized transform of the Hessian matrix is the Hessian of $\langle\langle f \rangle\rangle_\lambda$.

Theorem 3.2 was stated informally by Wu [31] for the Gaussian density function; a formal proof under assumption (3.3) appears in Moré and Wu [18]. The proof for the generalized transform follows the arguments used by Moré and Wu [18]; the only tricky part in the proof is to show that assumption (3.3) guarantees that we can differentiate under the integral sign.

Theorem 3.2 is of interest from a computational viewpoint because optimization algorithms require the gradient and Hessian of $\langle\langle f \rangle\rangle_\lambda$. This result shows that the gradient and Hessian of $\langle\langle f \rangle\rangle_\lambda$ are also smooth functions in the sense that they are obtained by transforming the gradient and Hessian matrix, respectively.

4. SMOOTHING PROPERTIES OF THE GENERALIZED TRANSFORM

One of the main attractions of the generalized transform is the ability to reduce the high-frequency components of the function. We quantify this statement in terms of the Fourier transform

$$\hat{f}(w) = \int_{\mathbb{R}^n} f(x) \exp\left(-iw^T x\right) dx$$

of f, since $\widehat{f}(w)$ is the component of f associated with the frequency w.

The key to estimating $\widehat{f}(w)$ is to express the generalized transform in terms of the convolution of f with the density function. Since

$$\langle\langle f\rangle\rangle_\lambda(x) = \frac{1}{\lambda^n}\int_{\mathbb{R}^n} f(y)\,\rho\left(\frac{x-y}{\lambda}\right)\,dy = \int_{\mathbb{R}^n} f(x-v)\,\rho_\lambda(v)\,dv,$$

where

$$\rho_\lambda(v) = \frac{1}{\lambda^n}\rho\left(\frac{v}{\lambda}\right),$$

we can write $\langle\langle f\rangle\rangle_\lambda$ as the convolution

$$\langle\langle f\rangle\rangle_\lambda(x) = (f * \rho_\lambda)(x)$$

of f with ρ_λ. If f is integrable, then the Fourier transform of the convolution of f with any integrable function is the product of the convolutions (see, for example, Folland [7, Chapter 8]), thus

$$\widehat{\langle\langle f\rangle\rangle}_\lambda(w) = \widehat{(f * \rho_\lambda)}(w) = \widehat{f}(w)\widehat{\rho_\lambda}(w). \tag{4.1}$$

The following result expresses (4.1) in terms of the original density function ρ.

Theorem 4.1 *If $f : \mathbb{R}^n \mapsto \mathbb{R}$ is integrable, then*

$$\widehat{\langle\langle f\rangle\rangle}_\lambda(w) = \widehat{f}(w)\widehat{\rho}(\lambda w). \tag{4.2}$$

Proof. Since $\langle\langle f\rangle\rangle_\lambda$ is integrable whenever f is integrable, the result follows from (4.1) by verifying that $\widehat{\rho_\lambda}(w) = \widehat{\rho}(\lambda w)$. □

We use Theorem 4.1 to estimate the rate of decay of the Fourier transform for various density functions. The estimates are simplified when the density function ρ satisfies

$$\rho(x) = \prod_{j=1}^{n} \rho_j(x_j),$$

where each $\rho_j : \mathbb{R} \mapsto \mathbb{R}$ is a one-dimensional density function, since then we have

$$\widehat{\rho}(w) = \prod_{j=1}^{n} \widehat{\rho}_j(w_j).$$

This result shows that we can compute $\widehat{\rho}$ once we know the one-dimensional $\widehat{\rho}_j$. The Fourier transform of the one-dimensional uniform density function is obtained directly from the definition; the Fourier transform of the one-dimensional Gaussian density function is a standard calculation. In this manner we obtain that the Fourier transform of the Gaussian density function (2.1) is

$$\widehat{\rho}(w) = \exp\left(-\tfrac{1}{4}\|w\|^2\right) \tag{4.3}$$

while for the uniform density function (2.7),

$$\widehat{\rho}(w) = \prod_{j=1}^{n} \frac{\sin w_j}{w_j}. \tag{4.4}$$

Note that $|\widehat{\rho}(w)| \le 1$ for both density functions, as must be true for any density function.

We use Theorem 4.1 to estimate the rate of decay of the Fourier transform in terms of the largest frequency component. If $|w_{\max}|$ is the largest component of w, then $\|w\| \ge |w_{\max}|$, and thus (4.3) shows that

$$\left|\widehat{\langle\langle f\rangle\rangle}_\lambda(\omega)\right| \le \exp\left(-\tfrac{1}{4}\lambda^2|w_{\max}|^2\right)\left|\widehat{f}(\omega)\right|$$

for the Gaussian density function. In contrast, for the uniform density function, (4.4) shows that

$$\left| \widehat{\langle\!\langle f \rangle\!\rangle_\lambda}(\omega) \right| \le \frac{\sin(\lambda |w_{\max}|)}{\lambda |w_{\max}|} \left| \hat{f}(\omega) \right| .$$

Clearly, the estimate for the Gaussian density function is more favorable.

Theorem 4.1 and the analysis of the rate of decay of the Fourier transform are an extension of the results obtained by Wu [31] for the Gaussian transform. Although we have considered only the Gaussian and uniform density functions, it is clear that this analysis can be carried out provided we are able to estimate the Fourier transform of the density function.

Figures 4..1 and 4..2 illustrate the smoothing transform as applied to the two dimensional ($n = 2$) version of the Griewank function,

$$f(x) = 1 + \sum_{i=1}^{n} \left(\frac{x_i^2}{200} \right) - \prod_{i=1}^{n} \cos\left(\frac{x_i}{\sqrt{i}} \right).$$

This function was constructed by Griewank [9] as a global optimization test function. The Gaussian and uniform transforms for the Griewank function are

$$\langle f \rangle_\lambda(x) = 1 + \sum_{i=1}^{n} \left(\frac{x_i^2}{200} + \frac{\lambda^2}{400} \right) - \prod_{i=1}^{n} \exp\left(-\frac{\lambda^2}{4i} \right) \cos\left(\frac{x_i}{\sqrt{i}} \right)$$

and

$$\langle\!\langle f \rangle\!\rangle_\lambda(x) = 1 + \sum_{i=1}^{n} \left(\frac{x_i^2}{200} + \frac{\lambda^2}{600} \right) - \prod_{i=1}^{n} \left(\sin\left(\frac{\lambda}{\sqrt{i}} \right) / \left(\frac{\lambda}{\sqrt{i}} \right) \right) \cos\left(\frac{x_i}{\sqrt{i}} \right),$$

respectively.

Figures 4..1 and 4..2 show the transformed functions for the Griewank function using the Gaussian and uniform transforms, respectively. The graphs in the left column are the transformed functions, and those in the right are the corresponding contours. The values of λ increase as we go from the graph on the top to the graph in the bottom. For these graphs the values of λ are $0, 1, 2$, and 4, with $\lambda = 0$ for the graph on the top. From these pictures, we see that both transforms smooth the function well for increasing λ values. However, for a given λ value, the Gaussian transform is slightly smoother than the uniform transform. In other words, in order to obtain a sufficiently smooth function, a larger λ value may be required for the uniform transform than for the Gaussian transform.

We have applied a simple Matlab minimization procedure to the transformed functions in the figures, first the bottom one, and then the one next to the bottom, and so on. The symbol $*$ in the contours marks the solutions obtained. The pictures show that after smoothing the function, by either the Gaussian or the uniform transform, the global minimizer of the function was found with a few continuation steps.

5. COMPUTING THE GENERALIZED TRANSFORM

A serious drawback to the general use of the generalized transform is that computing $\langle\!\langle f \rangle\!\rangle_\lambda$ for a function defined on $\mathbb{R}^n$ requires the computation of n-dimensional integrals. For functions that arise in applications we can usually avoid this difficulty by reducing the computation of the n-dimensional integrals to k-dimensional integrals where $k \ll n$. In this section we consider a class of functions for which the computation of the generalized transform requires the computation of only 1-dimensional generalized transforms.

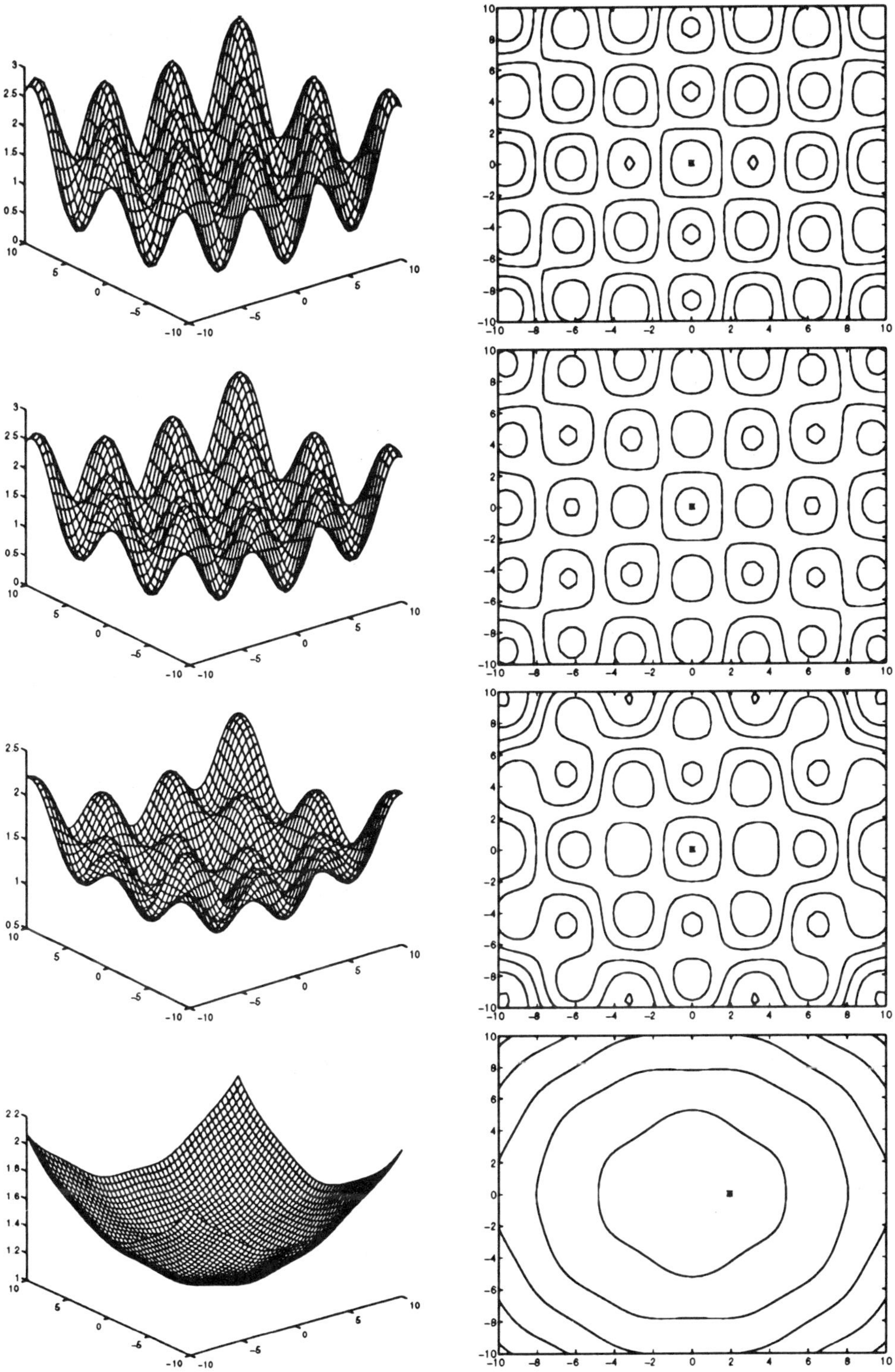

Figure 4..1: The Gaussian transform for the Griewank function

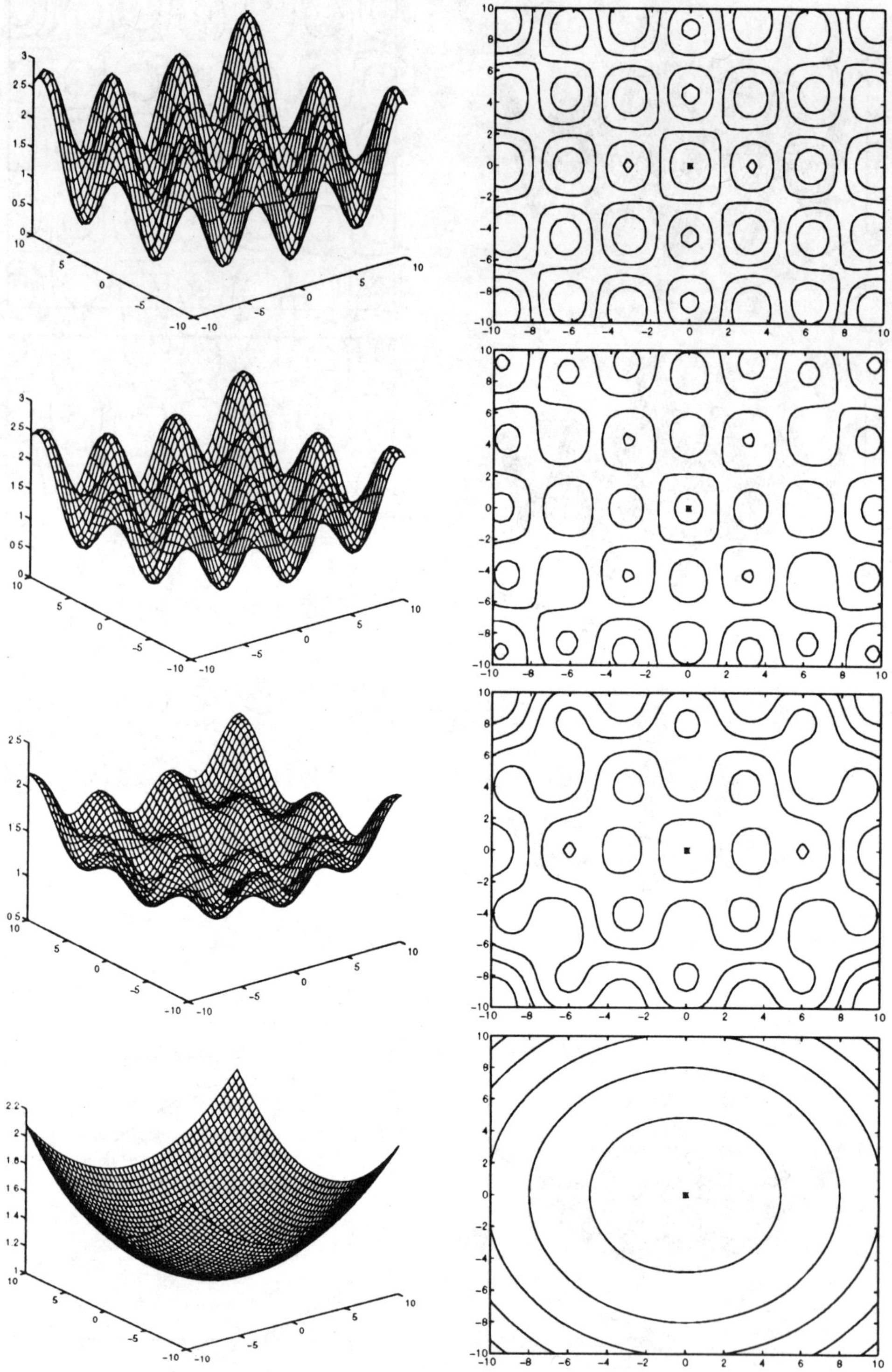

Figure 4..2: The uniform transform for the Griewank function

Definition 5.1 *A function $f : \mathbb{R}^n \mapsto \mathbb{R}$ is decomposable if f can be written in the form*

$$f(x) = \sum_{k=1}^{m} \left(\prod_{j=1}^{n} f_{k,j}(x_j) \right), \tag{5.1}$$

for some set of functions $\{f_{k,j}\}$, where $f_{k,j} : \mathbb{R} \mapsto \mathbb{R}$.

This class of functions was introduced by Wu [31] under the term generalized multilinear functions; Moré and Wu [18] introduced the term decomposable to avoid confusion with the use of multilinear for a function that is linear in each argument.

Decomposable functions form an algebra, that is, a vector space that is closed under multiplication. Linear and quadratic functions in $\mathbb{R}^n$ are decomposable. Polynomial functions, that is, functions that are linear combinations of terms of the form

$$x_1^{p_1} x_2^{p_2} \cdots x_n^{p_n},$$

for arbitrary integer powers $p_i \geq 0$, are also decomposable. An interesting example of decomposable functions arises in the fitting of data by minimizing the l_2 norm

$$\Phi(\alpha, x) = \sum_{i=1}^{m} \left(\sum_{j=1}^{n} \alpha_j \exp(-t_i x_j) - y_i \right)^2$$

of the difference between an exponential model and data $y_1, \ldots, y_m$ at times $t_1, \ldots, t_m$. Clearly, Φ is a decomposable function of the variables $(\alpha_1, x_1), \ldots, (\alpha_n, x_n)$.

The decomposable functions are of interest with respect to the Gaussian transform because computing the Gaussian transform of a decomposable function requires the computation of only a one-dimensional Gaussian transform. Indeed, if f is defined by (5.1), then

$$\langle f \rangle_\lambda(x) = \sum_{k=1}^{m} \left(\prod_{j=1}^{n} \langle f_{k,j} \rangle_\lambda(x_j) \right).$$

Thus, computing $\langle f \rangle_\lambda$ for a decomposable function requires the computation of only the one-dimensional integrals for each $\langle f_{k,j} \rangle_\lambda$.

We extend the above result to the generalized transform $\langle\langle f \rangle\rangle_\lambda$ by imposing an additional assumption on the density function ρ, that is, we assume that

$$\rho(x) = \prod_{j=1}^{n} \rho_j(x_j), \tag{5.2}$$

where each $\rho_j : \mathbb{R} \mapsto \mathbb{R}$ is a one-dimensional density function. This assumption on ρ means that the random variables associated with each density functions ρ_j are independent. The Gaussian (2.1) and uniform (2.7) density functions, as well as most other n-dimensional density functions, satisfy this assumption.

Theorem 5.2 *If $f : \mathbb{R}^n \mapsto \mathbb{R}$ is decomposable and the density function $\rho : \mathbb{R}^n \mapsto \mathbb{R}$ satisfies* (5.2), *then*

$$\langle\langle f \rangle\rangle_\lambda(x) = \sum_{k=1}^{m} \left(\prod_{j=1}^{n} \langle\langle f_{k,j} \rangle\rangle_\lambda(x_j) \right).$$

Proof. The result follows from (5.2) and the definition of the generalized transform. □

Theorem 5.2 can be used to compute the generalized transform of a decomposable function provided we are able to compute the generalized transform of the component functions $f_{k,j}$. In the remainder of this section we explore the case where the component function is analytic.

We obtain an expression for the generalized transform of an analytic function f by noting that in this case the Taylor series converges uniformly, and thus

$$\langle\langle f\rangle\rangle_\lambda(x) = \sum_{l=0}^{+\infty} f^{(l)}(x)\frac{(-1)^l\lambda^l}{l!}\int_{-\infty}^{+\infty} u^l \rho(u)\, du.$$

This expression simplifies when the density function is even.

Theorem 5.3 *If $f : \mathbb{R} \mapsto \mathbb{R}$ is analytic and $\rho : \mathbb{R} \mapsto \mathbb{R}$ is even, then the generalized transform $\langle\langle f\rangle\rangle_\lambda$ satisfies*

$$\langle\langle f\rangle\rangle_\lambda(x) = \sum_{l=0}^{+\infty} f^{(2l)}(x)\frac{\lambda^{2l}}{(2l)!}\int_{-\infty}^{+\infty} u^{2l} \rho(u)\, du.$$

Proof. Just note that the integrals with odd powers vanish by symmetry for any density function that is symmetric around the origin. □

Theorem 5.3 shows that if ρ is the uniform density function then

$$\langle\langle f\rangle\rangle_\lambda(x) = \sum_{l=0}^{+\infty} f^{(2l)}(x)\frac{\lambda^{(2l)}}{(2l+1)!} \tag{5.3}$$

is the generalized transform of f. In particular, for the sin, cos, and exp functions, expression (5.3) shows that

$$\sin(x)\frac{\sin\lambda}{\lambda}, \qquad \cos(x)\frac{\sin\lambda}{\lambda}, \qquad \exp(x)\frac{\sinh\lambda}{\lambda}$$

are, respectively, the generalized transforms for these functions. Theorem 5.3 also yields an expression for the Gaussian transform if we make use of the identity

$$\frac{1}{\pi^{1/2}}\int_{-\infty}^{+\infty} u^{2l}\exp\left(-u^2\right)\, du = \frac{(2l)!}{4^l l!},$$

which can be verified from standard tables or by integration by parts. Hence, Theorem 5.3 shows that

$$\langle f\rangle_\lambda(x) = \sum_{l=0}^{+\infty} \frac{1}{l!} f^{(2l)}(x)\left(\frac{\lambda}{2}\right)^{(2l)} \tag{5.4}$$

is the Gaussian transform of an analytic function f. As a special case of (5.4),

$$\sin(x)\exp(-\tfrac{1}{4}\lambda^2), \qquad \cos(x)\exp(-\tfrac{1}{4}\lambda^2), \qquad \exp(x)\exp(\tfrac{1}{4}\lambda^2)$$

are, respectively, the generalized transforms for the sin, cos, and exp functions.

Piela, Kostrowski, and Scheraga [20] used (5.4) as the motivation for the diffusion equation method by noting that the mapping

$$h(x,t) = \langle f\rangle_{2\sqrt{t}}(x) = \sum_{l=0}^{+\infty} \frac{1}{l!} f^{(2l)}(x) t^l$$

satisfies the diffusion equation

$$\frac{\partial^2 h}{\partial x^2}(x,t) = \frac{\partial h}{\partial t}(x,t), \qquad h(x,0) = f(x).$$

Thus, in this approach, the transformation is defined as a solution to the n-dimensional diffusion equation. In later work ([13, 14]) it was shown that (2.4) could also be used to define this transformation in $\mathbb{R}^n$. In our work we have used (2.4) as the definition of the Gaussian transform and derived all results from this definition.

Theorem 5.4 *If $f : \mathbb{R} \mapsto \mathbb{R}$ is the monic polynomial $f(x) = x^k$, then*

$$\langle f \rangle_\lambda(x) = \sum_{l=0}^{\lfloor k/2 \rfloor} \left(\frac{k!}{(k-2l)!\; l!} \right) \left(\frac{\lambda}{2} \right)^{2l} x^{k-2l}$$

is the Gaussian transform, while if ρ is the uniform density function, then

$$\langle\langle f \rangle\rangle_\lambda(x) = \sum_{l=0}^{\lfloor k/2 \rfloor} \left(\frac{k!}{(k-2l)!} \right) \frac{\lambda^{2l}}{(2l+1)!} x^{k-2l}$$

is the generalized transform.

Proof. The result follows directly from (5.3) and (5.4). □

As an application of Theorem 5.4 to an n-dimensional function, consider the general quadratic

$$f(x) = \tfrac{1}{2} x^T Q x + c^T x,$$

where $Q \in \mathbb{R}^{n \times n}$ and $c \in \mathbb{R}^n$. A computation shows that

$$\langle\langle f \rangle\rangle_\lambda(x) = \tfrac{1}{2} x^T Q x + c^T x + \frac{\lambda^2}{4} \left(\sum_{i=1}^{n} q_{i,i} \right)$$

for the Gaussian density function, while

$$\langle\langle f \rangle\rangle_\lambda(x) = \tfrac{1}{2} x^T Q x + c^T x + \frac{\lambda^2}{6} \left(\sum_{i=1}^{n} q_{i,i} \right)$$

for the uniform density function. Thus, for quadratic functions, both transforms only differ by a re-scaling of the parameter λ. This result holds for any even density function, since (5.3) shows that

$$\langle\langle f \rangle\rangle_\lambda(x) = x^2 + \lambda^2 \int_{-\infty}^{+\infty} u^2 \rho(u)\, du$$

when $f(x) = x^2$.

6. COMPUTING THE GAUSSIAN TRANSFORM

For macromolecular modeling problems we are interested in transforming a class of functions in terms of the distances between pairs of atoms, where $x_i \in \mathbb{R}^p$ is the position of the i-th atom. In general we are concerned with three-dimensional problems where $p = 3$, but values of $p > 3$ are also of interest. Given functions $p_{i,j} : \mathbb{R}^p \mapsto \mathbb{R}$ of the distance between atoms x_i and x_j, we outline several techniques that can be used to compute the Gaussian transform of the potential function

$$f(x) = \sum_{i,j \in \mathcal{S}} p_{i,j}(x_i - x_j), \tag{6.1}$$

where $\mathcal{S}$ is some subset of all pairs of atoms, and the mappings $p_{i,j}$ are of the form

$$p_{i,j}(x) = h_{i,j}\left(\|x\| \right), \tag{6.2}$$

for some mapping $h_{i,j} : \mathbb{R} \mapsto \mathbb{R}$.

The example $x \mapsto \|x\|$ points out that functions of the general form (6.1) and (6.2) are not usually decomposable. On the other hand, we now show that we can still reduce the computation of the Gaussian transform to the computation of one-dimensional integrals in terms of $h_{i,j}$. Note that in these problems f is defined on $\mathbb{R}^n$, where $n = mp$ and m is the number of atoms, but that $h_{i,j}$ is defined in $\mathbb{R}$.

The following result of Moré and Wu [19] is needed to prove that computing the Gaussian transform of (6.1) requires only the Gaussian transform of $p_{i,j}$.

Theorem 6.1 *If $f : \mathbb{R}^n \mapsto \mathbb{R}$ and $h : \mathbb{R}^p \mapsto \mathbb{R}$ are related by*

$$f(x) = h(P^T x),$$

for some matrix $P \in \mathbb{R}^{n \times p}$ such that $P^T P = \sigma^2 I$, then

$$\langle f \rangle_\lambda(x) = \langle h \rangle_{\sigma\lambda}(P^T x).$$

As an application of Theorem 6.1 consider the mapping $f : \mathbb{R}^n \mapsto \mathbb{R}$ defined by (6.1). Computing the Gaussian transform of this mapping is immediate if we are able to compute the Gaussian transform of $f_0 : \mathbb{R}^n \mapsto \mathbb{R}$ defined by

$$f_0(x) = p_{i,j}(x_i - x_j).$$

Clearly, we can find a matrix $P \in \mathbb{R}^{n \times p}$ of the form

$$P = \left(e_{i_1} - e_{j_1}, \ldots, e_{i_p} - e_{j_p} \right)$$

such that $P^T x = x_i - x_j$. Since $P^T P = \sigma^2 I$, where $\sigma^2 = 2$, Theorem 6.1 shows that

$$\langle f_0 \rangle_\lambda(x) = \langle p_{i,j} \rangle_{\sqrt{2}\lambda}(x_i - x_j).$$

An immediate consequence of this result is that

$$\langle f \rangle_\lambda(x) = \sum_{i,j \in \mathcal{S}} \langle p_{i,j} \rangle_{\sqrt{2}\lambda}(x_i - x_j)$$

is the Gaussian transform of the potential function defined by (6.1). In this case f is defined on $\mathbb{R}^n$, but $p_{i,j}$ is defined on $\mathbb{R}^p$.

The other ingredient needed for computing the Gaussian transform of functions defined by (6.1) and (6.2) is the Gaussian transform of the function $p_{i,j}$ defined by (6.2). Note that, unlike Theorem 6.1, the following result requires that f be defined on $\mathbb{R}^3$.

Theorem 6.2 *If $f : \mathbb{R}^3 \mapsto \mathbb{R}$ is of the form $f(x) = h(\|x\|)$ for some function $h : \mathbb{R} \mapsto \mathbb{R}$, then*

$$\langle f \rangle_\lambda(x) = \frac{1}{\lambda\sqrt{\pi} r} \int_0^\infty s h(s) \left[\exp\left(-\frac{(r-s)^2}{\lambda^2} \right) - \exp\left(-\frac{(r+s)^2}{\lambda^2} \right) \right] ds, \qquad (6.3)$$

where $r = \|x\|$. If the mapping h is an even function, then

$$\langle f \rangle_\lambda(x) = \frac{1}{\lambda\sqrt{\pi} r} \int_{-\infty}^{+\infty} s h(s) \exp\left(-\frac{(r-s)^2}{\lambda^2} \right) ds. \qquad (6.4)$$

Moré and Wu [19] proved Theorem 6.2 and used (6.4) to approximate the Gaussian transform of (1.5). More generally, note that the integrals that appear in (6.3) and (6.4) can be expressed explicitly or in terms of special functions. For example, if $h(s) = s$, then (6.3) can be expressed in terms of the complementary error function

$$x \mapsto \frac{2}{\sqrt{\pi}} \int_x^{+\infty} \exp\left(-s^2\right) ds.$$

However, in most cases we need to approximate the integrals with a quadrature. An adaptive quadrature could be used, but this is quite likely to be expensive in terms of function evaluations. Moré and Wu [19] recommended the use of Gaussian quadratures because this leads to a discrete transformation that shares many of the properties of the standard Gaussian transform. For the one-dimensional Gauss transform

$$\langle f \rangle_\lambda(x) = \frac{1}{\sqrt{\pi}} \int_{-\infty}^{+\infty} f(x + \lambda s) \exp\left(-s^2\right) ds,$$

the use of Gaussian quadratures on this integral yields the Gauss-Hermite transform

$$\langle f \rangle_{\lambda,q} = \frac{1}{\sqrt{\pi}} \sum_{i=1}^{q} w_i f(x + \lambda s_i), \tag{6.5}$$

which is exact for all polynomials of degree less than $2q$, where q is the number of nodes in the quadrature. The weights w_i and nodes s_i in (6.5) are independent of f and can be found in the literature (for example, Stroud and Secrest [30]) or can be computed with some of the subroutines in ORTHOPOL (Gautschi [8]). The computation of the Gauss-Hermite transform requires q function evaluations, but even for modest values of q we obtain a good approximation to the Gauss transform. For additional information on Gaussian quadrature, see Stroud and Secrest [30] and Davis and Rabinowitz [5]. Steen, Byrne, and Gelbard [28] have Gaussian quadratures for the integrals in (6.3).

ACKNOWLEDGMENTS

Our research on macromolecular global optimization problems has been influenced by John Schiffer's work on ionic systems and Julius Jellinek's work on clusters. Steve Pieper deserves special mention for bringing the work on ionic systems to our attention and for sharing his insights on this problem.

REFERENCES

[1] A. T. Brünger and M. Nilges, "Computational challenges for macromolecular structure determination by X-ray crystallography and solution NMR-spectroscopy". *Q. Rev. Biophys.*, 26:49–125, 1993.

[2] T. F. Coleman, D. Shalloway, and Z. Wu, "Isotropic effective energy simulated annealing searches for low energy molecular cluster states". *Comp. Optim. Applications*, 2:145–170, 1993.

[3] T. F. Coleman, D. Shalloway, and Z. Wu, "A parallel build-up algorithm for global energy minimizations of molecular clusters using effective energy simulated annealing". *J. Global Optim.*, 4:171–185, 1994.

[4] G. M. Crippen and T. F. Havel, *"Distance Geometry and Molecular Conformation"*. John Wiley & Sons, 1988.

[5] P. J. Davis and P. Rabinowitz, *"Methods of Numerical Integration"*. Academic Press, second edition, 1984.

[6] Y. M. Ermoliev, V. I. Norkin, and R. J.-B. Wets, "The minimization of discontinuous functions: Mollifier subgradients". *SIAM J. Control Optim.*, 33:149–167, 1995.

[7] G. B. Folland, *"Real Analysis"*. John Wiley & Sons, 1984.

[8] W. Gautschi, "Algorithm 726: ORTHOPOL – A package of routines for generating orthogonal polynomials and Gauss-type quadrature rules". *ACM Trans. Math. Software*, 20:21–62, 1994.

[9] A. Griewank, "Generalized descent for global optimization". *J. Optim. Theory Appl.*, 34:11–39, 1981.

[10] R. W. Hasse and J. P. Schiffer, "The structure of the cylindrically confined coulomb lattice". *Ann. Physics*, 203:419–448, 1990.

[11] T. F. Havel, "An evaluation of computational strategies for use in the determination of protein structure from distance geometry constraints obtained by nuclear magnetic resonance". *Prog. Biophys. Mol. Biol.*, 56:43–78, 1991.

[12] V. Y. Katkovnik and O. Y. Kulchitskii, "Convergence of a class of random search algorithms". *Automat. Remote Control*, 8:81–87, 1972.

[13] J. Kostrowicki and L. Piela, "Diffusion equation method of global minimization: Performance for standard functions". *J. Optim. Theory Appl.*, 69:269–284, 1991.

[14] J. Kostrowicki, L. Piela, B. J. Cherayil, and H. A. Scheraga, "Performance of the diffusion equation method in searches for optimum structures of clusters of Lennard-Jones atoms". *J. Phys. Chem.*, 95:4113–4119, 1991.

[15] J. Kreimer and R. Y. Rubinstein, "Nondifferentiable optimization via smooth approximation: General analytical approach". *Math. Oper. Res.*, 39:97–119, 1992.

[16] I. D. Kuntz, J. F. Thomason, and C. M. Oshiro, "Distance geometry". In N. J. Oppenheimer and T. L. James, editors, *Methods in Enzymology*, volume 177, pages 159–204. Academic Press, 1993.

[17] S. Lang, *"Real Analysis"* Addison-Wesley, second edition, 1983.

[18] J. J. Moré and Z. Wu, "Global continuation for distance geometry problems". Preprint MCS-P505-0395, Argonne National Laboratory, Argonne, Illinois, 1995.

[19] J. J. Moré and Z. Wu, "ε-optimal solutions to distance geometry problems via global continuation". Preprint MCS-P520-0595, Argonne National Laboratory, Argonne, Illinois, 1995.

[20] L. Piela, J. Kostrowicki, and H. A. Scheraga, "The multiple-minima problem in the conformational analysis of molecules: Deformation of the protein energy hypersurface by the diffusion equation method". *J. Phys. Chem.*, 93:3339–3346, 1989.

[21] R. Rafac, J. P. Schiffer, J. S. Hangst, D. H. E. Dubin, and D. J. Wales, "Stable configurations of confined cold ionic systems". *Proc. Natl. Acad. Sci. U.S.A.*, 88:483–486, 1991.

[22] R. Y. Rubinstein, *"Simulation and the Monte Carlo Method"*. John Wiley & Sons, 1981.

[23] R. Y. Rubinstein, "Smoothed functionals in stochastic optimization". *Math. Oper. Res.*, 8:26–33, 1983.

[24] R. Y. Rubinstein, *"Monte Carlo Optimization, Simulation and Sensitivity of Queueing Networks"*. John Wiley & Sons, 1986.

[25] J. P. Schiffer, "Phase transitions in anisotropically confined ionic crystals". *Phys. Rev. Lett.*, 70:818–821, 1993.

[26] D. Shalloway, "Application of the renormalization group to deterministic global minimization of molecular conformation energy functions". *J. Global Optim.*, 2:281–311, 1992.

[27] D. Shalloway, "Packet annealing: A deterministic method for global minimization, application to molecular conformation". In C. Floudas and P. Pardalos, editors, *Recent Advances in Global Optimization*, pages 433–477. Princeton University Press, 1992.

[28] N. M. Steen, G. D. Byrne, and E. M. Gelbard, "Gaussian quadratures for the integrals $\int_0^\infty \exp(-x^2)f(x)\,dx$ and $\int_0^b \exp(-x^2)f(x)\,dx$". *Math. Comp.*, 23:661–674, 1969.

[29] J. E. Straub, "Optimization techniques with applications to proteins". Preprint, Boston University, Department of Chemistry, Boston, Massachusetts, 1994.

[30] A. H. Stroud and D. Secrest, *"Gaussian Quadrature Formulas"*. Prentice-Hall, Inc., 1966.

[31] Z. Wu, "The effective energy transformation scheme as a special continuation approach to global optimization with application to molecular conformation". Preprint MCS-P442-0694, Argonne National Laboratory, Argonne, Illinois, 1994.

CONTINUOUS APPROACHES TO DISCRETE OPTIMIZATION PROBLEMS

Pardalos Panos M.

Center for Applied Optimization
Department of Industrial and Systems Engineering
Weil Hall 303
University of Florida, Gainesville FL 32611
e-mail: pardalos@math.ufl.edu

Abstract. This paper contains expository notes about continuous approaches to several discrete optimization problems. There are many ways to formulate discrete problems as equivalent continuous problems or to embed the discrete feasible domain in a larger continuous space (relaxation). The surprising variety of continuous approaches reveal interesting theoretical properties which can be explored to develop new algorithms for computing (sub)optimal solutions to discrete optimization problems.

Key Words. Discrete optimization, global optimization, graph problems, interior point algorithms, semidefinite programming, max clique, satisfiability, minimax theory, quadratic assignment, Steiner problem.

1. INTRODUCTION

Discrete (or combinatorial) optimization problems, that is, problems with a discrete feasible domain and/or a discrete domain objective function, model a large spectrum of applications in computer science, operations research and engineering.

Solution methods for discrete optimization problems can be classified into combinatorial and continuous approaches. A typical combinatorial approach generates a sequence of states, which represent a partial solution, drawn from a discrete finite set. Continuous approaches for solving discrete optimization problems are based on different equivalent characterizations in a continuous space. These characterizations include equivalent continuous formulations, or continuous relaxations, that is, embeddings of the discrete domain in a larger continuous space. The topological and geometric structure of the continuous space play an essential role in studying properties of discrete problems and developing efficient computational algorithms.

In this paper we give a brief overview of some continuous approaches to discrete optimization problems. The problems we discuss include the satisfiability problem, integer programming, the maximum clique problems, the Steiner problem in graphs and minimax approaches.

2. THE MAXIMUM CLIQUE PROBLEM

A well studied example of a discrete problem is the maximum clique problem and its variants. Consider a graph $G = G(V, E)$, where $V = \{1, \ldots, n\}$ denotes the set of vertices (nodes), and E denotes the set of edges. Denote by (i, j) an edge joining vertex i and vertex j. A clique of G is a subset C of vertices with the property that every pair of vertices in C is joined by an edge. In other words, C is a clique if the subgraph $G(C)$ induced by C is complete. The maximum clique problem is the problem of finding a clique set C of maximal cardinality. In principle one can solve this problem by enumerating all cliques C. However, due to the possible exponential number of cliques in a graph, such an approach will be hopeless even for moderate size problems.

Consider the continuous indefinite quadratic programming problem

$$\begin{aligned} \max \quad & f_G(x) = \sum_{(i,j)\in E} x_i x_j = \tfrac{1}{2} x^T A_G x \\ \text{s.t.} \quad & x \in S = \{x = (x_1, \ldots, x_n)^T : \sum_{i=1}^{n} x_i = 1, \\ & x_i \geq 0 \quad (i = 1, \ldots, n)\}, \end{aligned} \tag{2.1}$$

where A_G is the adjacency matrix of the graph G. If $\alpha = \max\{f_G(x) : x \in S\}$, then G has a maximum clique C of size $\omega(G) = 1/(1 - 2\alpha)$. This maximum can be attained by setting $x_i = 1/k$ if $i \in C$ and $x_i = 0$ if $i \notin C$.

The maximum clique problem can be modeled using an indefinite quadratic program subject to a spherical constraint and non negativity requirements. Given a graph G, the complement graph is denoted by $\bar{G}$ and $A_{\bar{G}}$ represents the node-node adjacency matrix of $\bar{G}$. If there is a k-clique in G the optimization problem

$$V(k) = \min_{x \in D} \ \frac{1}{2} x^T A_{\bar{G}} x + (\sum_{i=1}^{n} x_i - 1)^2,$$

where $D = \{\sum_{i=1}^{n} x_i^2 \leq \frac{1}{k},\ x_i \geq 0\ i = 1, \ldots, n\}$, yields a solution x^* with optimal objective value zero, the positive components of which identify a -k-clique. A search is performed to find the maximum value of k, that allows the optimal objective value of the above program to be equal zero [9,10].

Although it is NP-complete (see also [2,3]) to compute the size of the maximum clique $\omega(G)$ and the minimum number of colors $\chi(G)$ needed to color the vertices of G, Lovász [23] proposed an ingenious method for computing in polynomial time a real number $\theta(G)$ such that

$$\omega(G) \leq \theta(\bar{G}) \leq \chi(G)$$

where $\bar{G}$ is the complement of the graph G. The Lovász number $\theta(\bar{G})$ is computed by solving a continuous problem (semidefinite program) which is a relaxation of the maximum clique problem. Knuth [22] gives a very nice expository of several ways of characterizing the Lovász number θ. The various continuous formulations of θ can motivate efficient polynomial time algorithms for its computation.

3. SEMIDEFINITE PROGRAMMING APPROACHES

Semidefinite programming is a special case of convex programming. More precisely, a semidefinite program is the problem of optimizing a linear function of a symmetric matrix subject to linear equality constraints and the constraint that the matrix is positive semidefinite. Since the set of positive semidefinite matrices constitutes a convex cone, semidefinite programming is also called linear programming over cones or cone-LP [1]. Given any $\epsilon > 0$, semidefinite programming can be solved within an additive error of ϵ in polynomial time, using the ellipsoid algorithm or interior point methods.

The significance of semidefinite programming is that it leads to tighter continuous relaxations than linear programming relaxations for many discrete optimization problems. For example, as we mentioned before, the work on the Lovász number of a graph is one of the first applications of semidefinite programming.

Recently there is a lot of research activity in semidefinite programming regarding its application to discrete optimization. Goemans and Williamson [12] gave an improved approximation algorithm for maximum cut and satisfiability problems using semidefinite programming relaxations. Their algorithm for the max cut problem depends on techniques on randomly rounding a solution to a continuous nonlinear relaxation that can be seen as a semidefinite program or an eigenvalue minimization problem. In fact, eigenvalue minimization problems can be formulated as semidefinite programs. For the use of eigenvalues in combinatorial optimization see the survey in [26]. Following the techniques described in [12], Karger etal [18] have developed an approximate algorithm for the problem of coloring k-colorable graphs with the fewest possible colors. A general approach to relaxation of 0-1 integer programming problems using semidefinite programming problems was presented in [24]. Since most graph problems can be formulated as 0-1 integer programming problems, in principle, semidefinite programming can be used in obtaining good bounds and approximate algorithms in many of these problems. Although semidefinite programming provides tighter relaxations to many discrete problems and it can be solved (in theory) in polynomial time, up-to-date there are no efficient computational algorithms for solving large-scale semidefinite problems.

4. GLOBAL OPTIMIZATION FOR SATISFIABILITY

The satisfiability (SAT) problem is central in mathematical logic, computing theory, artificial intelligence, and many industrial application problems. Traditional methods treat the SAT problem as a constrained decision problem. In recent years, researchers have applied global optimization methods to solve the SAT problem. In this approach, universal satisfiability models (*UniSAT*), are formulated that transform a discrete SAT problem on Boolean space $\{0,1\}^m$ into a continuous SAT problem on real space R^m. Thus, this decision problem is transformed into a global optimization problem which can be solved by global optimization methods [13,14].

In the typical *UniSAT* models using the universal DeMorgan laws, Boolean $\vee$ and $\wedge$ connectors in conjunctive normal form (CNF) formulas are transformed into $\times$ and $+$ of ordinary addition and multiplication operations, respectively. The *true* value of the CNF formula is thus converted to the global minimum value, i.e., 0, of the objective function. The objective function can be chosen to be a multivariable polynomial, in exponential or logarithmic form.

Given a CNF formula $F(\mathbf{x})$ from $\{0,1\}^m$ to $\{0,1\}$ with n clauses $C_1, \ldots, C_n$, we define a real function $f(\mathbf{y})$ from E^m to E that transforms the SAT problem into an unconstrained global optimization problem

$$\min_{\mathbf{y} \in \mathbf{E}^m} f(\mathbf{y}) \tag{4.1}$$

where

$$f(\mathbf{y}) = \sum_{i=1}^{n} c_i(\mathbf{y}). \tag{4.2}$$

A clause function $c_i(\mathbf{y})$ is a product of m *literal functions* $q_{ij}(y_j)$ $(1 \leq j \leq m)$:

$$c_i = \prod_{j=1}^{m} q_{ij}(y_j), \tag{4.3}$$

where

$$q_{ij}(y_j) = \begin{cases} |y_j - 1| & \text{, if literal } x_j \text{ is in clause } C_i, \\ |y_j + 1| & \text{, if literal } \bar{x}_j \text{ is in clause } C_i, \\ 1 & \text{, if neither } x_j \text{ nor } \bar{x}_j \text{ is in } C_i. \end{cases} \tag{4.4}$$

The correspondence between $\mathbf{x}$ and $\mathbf{y}$ is defined as follows (for $1 \leq i \leq m$):

$$x_i = \begin{cases} 1 & \text{, if } y_i = 1, \\ 0 & \text{, if } y_i = -1, \\ undefined & \text{otherwise.} \end{cases}$$

Clearly, $F(\mathbf{x})$ is true iff $f(\mathbf{y})$=0 on the corresponding $\mathbf{y} \in \{-1, 1\}^m$.

In another approach the global optimization problem is defined to be:

$$\min_{\mathbf{y} \in E^m} f(\mathbf{y}), \tag{4.5}$$

where

$$f(\mathbf{y}) = \sum_{i=1}^{n} c_i(\mathbf{y}). \tag{4.6}$$

A clause function $c_i(\mathbf{y})$ is a product of m *literal functions* $q_{ij}(y_j)$ $(1 \leq j \leq m)$:

$$c_i = \prod_{j=1}^{m} q_{ij}(y_j), \tag{4.7}$$

where

$$q_{ij}(y_j) = \begin{cases} (y_j - 1)^{2p} & \text{, if } x_j \text{ is in clause } C_i, \\ (y_j + 1)^{2p} & \text{, if } \bar{x}_j \text{ is in clause } C_i, \\ 1 & \text{, if neither } x_j \text{ nor } \bar{x}_j \text{ is in } C_i. \end{cases} \tag{4.8}$$

where p is a positive integer.

Again the correspondence between $\mathbf{x}$ and $\mathbf{y}$ is defined as follows (for $1 \leq i \leq m$):

$$x_i = \begin{cases} 1 & \text{, if } y_i = 1, \\ 0 & \text{, if } y_i = -1, \\ undefined, & \text{otherwise.} \end{cases}$$

Clearly, $F(\mathbf{x})$ is true iff $f(\mathbf{y})$=0 on the corresponding $\mathbf{y} \in \{-1, 1\}^m$.

Hence, the *UniSAT* models transform the SAT problem from a discrete, constrained decision problem into an unconstrained global optimization problem. Many powerful solution techniques have been developed to solve unconstrained optimization problems, which are based primarily upon calculus, rather than upon algebra and pivoting, as in the simplex method. *UniSAT* models establish a correspondence between the global minimum points of the objective function and the solutions of the original SAT problem. A *CNF* $F(\mathbf{x})$ is true *if and only if* $f(\mathbf{y})$ takes the global minimum value 0 on the corresponding **y**. Extensive computational testing and comparisons with classical techniques such as the Davis-Putnam algorithm and interior point methods indicate the significance of the global optimization approaches [13,14,15].

5. THE LINEAR COMPLEMENTARITY PROBLEM

Given matrices $A_{n\times n}$, $B_{n\times l}$ and a vector $b \in R^n$ with rational entries, the mixed integer feasibility problem is to find (x, z), such that $x \in R^n$, $x \geq 0$, $z \in \{0,1\}^l$ that satisfy $Ax + Bz = b$.

This problem can be formulated as a concave quadratic minimization problem with linear constraints of the form:

$$\begin{aligned} \min \quad & f(z) = z^T(e - z) \\ \text{s.t.} \quad & Ax + Bz = b,\ 0 \leq z \leq e,\ x \geq 0, \end{aligned}$$

where e is a l-dimensional vector of 1's. The function $f(z^*) = 0$ iff the mixed integer feasibility problem has a solution z^*. Other non-obvious formulations can be used to obtain different characterizations in a continuous space.

Theorem 5.1. The mixed integer feasibility problem can be reduced to the solution of a linear complementarity problem.

Proof. The condition $z_i \in \{0,1\}$ is equivalent to:

$$z_i + w_i = 1, z_i \geq 0, w_i \geq 0, z_i w_i = 0.$$

With this transformation z_i is a continuous variable and for each z_i a new continuous variable w_i is introduced. In addition, let $s, t \in R^n$ be such that

$$s = Ax + Bz - b \geq 0, t = -Ax - Bz + b \geq 0.$$

The only way for these two inequalities to be satisfied is to have $s = t = 0$, which implies that $Ax + Bz = b$. Then, the mixed integer feasibility problem can be reduced to solution of the following linear complementarity problem $LCP(M, q)$: Find v, y such that

$$v \geq 0, y \geq 0, v^T y = 0, v = My + q,$$

where

$$y = \begin{pmatrix} z \\ x \\ \theta \end{pmatrix}, v = \begin{pmatrix} w \\ s \\ t \end{pmatrix}, M = \begin{pmatrix} -I & 0 & 0 \\ B & A & 0 \\ -B & -A & 0 \end{pmatrix}, q = \begin{pmatrix} e \\ b \\ -b \end{pmatrix},$$

where $\theta \in R^n$ and $e \in R^l$ is the vector of all 1's. □

Associated with the $LCP(M, q)$ is the quadratic problem

$$\min_{y \in S} f(y) = q^T y + y^T My,$$

where $S = \{y : My + q \geq 0, y \geq 0\}$. The $LCP(M, q)$ has a solution iff the global minimum y^* of the above quadratic program exists and $f(y^*) = 0$. Therefore, the mixed integer feasibility problem can be formulated as a continuous quadratic optimization problem with known optimal objective function value.

Furthermore, it can be shown that any linear complementarity problem can be solved by solving a corresponding mixed integer feasibility problem. Given the $LCP(M,q)$ consider the following mixed zero-one integer problem:

$$
\begin{aligned}
0 \leq My + \alpha q &\leq e - z, \\
\alpha \geq 0, 0 \leq y &\leq z, \\
z &\in \{0,1\}^n.
\end{aligned}
\tag{5.1}
$$

Let (α^*, y^*, z^*) with $\alpha^* > 0$ be a feasible point of the above mixed integer problem (which is always feasible). Then $x^* = y^*/\alpha^*$ solves the LCP. If $\alpha^* = 0$ is the only feasible value of α, then the LCP has no solution [28].

The equivalence of mixed integer programming and linear complementarity strongly supports the point of view that discrete problems can be viewed as continuous global optimization problems. Complementarity is a fundamental tool in continuous optimization since it expresses optimality conditions [16,17].

6. LINEAR AND QUADRATIC 0-1 PROGRAMMING

Consider the following linear zero-one programming problem:

$$
\begin{aligned}
&\min \quad c^T x \\
&\text{s.t.} \quad Ax \leq b,\ x_i \in \{0,1\} \quad (i = 1, \ldots, n)
\end{aligned}
$$

where A is a real $(m \times n)$–matrix, $c \in R^n$ and $b \in R^m$. Let $e^T = (1, \ldots, 1) \in R^n$ denote the vector whose components are all equal to 1. As with the integer feasibility problem, the zero-one integer linear programming problem is equivalent to the following concave minimization problem:

$$
\begin{aligned}
&\min \quad f(x) = c^T x + \mu x^T (e - x) \\
&\text{s.t.} \quad Ax \leq b,\ 0 \leq x \leq e
\end{aligned}
$$

where μ is a sufficiently large positive number. The function $f(x)$ is concave since $-x^T x$ is concave. The equivalence of the two problems is based on the facts that a concave function attains its minimum at a vertex, and that $x^T(x - e) = 0, 0 \leq x \leq e$, implies $x_i = 0$ or 1 for $i = 1, \ldots, n$. Note that a vertex of the feasible domain is not necessarily a vertex of the unit hypercube $0 \leq x \leq e$, but the global minimum is attained only when $x^T(e - x) = 0$, provided that μ is large enough.

Similar techniques can be applied to reduce nonlinear zero-one problems to equivalent concave minimization problems. For example, consider the quadratic zero-one problem of the following form:

$$
\begin{aligned}
&\min \quad f(x) = c^T x + x^T Q x \\
&\text{s.t.} \quad x_i \in \{0,1\} \quad (i = 1, \ldots, n)
\end{aligned}
$$

where Q is a real symmetric $(n \times n)$ matrix. Given any real number μ, let $\overline{Q} = Q + \mu I$ where I is the $(n \times n)$ unit matrix, and $\overline{c} = c - \mu e$. Then the above zero-one quadratic problem is equivalent to the problem:

$$
\begin{aligned}
&\min \quad \overline{f}(x) = \overline{c}^T x + x^T \overline{Q} x \\
&\text{s.t.} \quad x_i \in \{0,1\} \quad (i = 1, \ldots, n)
\end{aligned}
$$

because of $\overline{f}(x) = f(x)$.

If we choose μ such that $\overline{Q} = Q + \mu I$ becomes a negative semidefinite matrix (e.g., $\mu = -\lambda$, where λ, is the largest eigenvalue of Q), then the objective function $\overline{f}(x)$ becomes concave and the constraints can be replaced by $0 \leq x \leq e$. Therefore, the problem is equivalent to the minimization of a quadratic concave function over the unit hypercube. For similar relations between general integer programming problems and continuous global optimization see [8].

6.1. The quadratic assignment problem

Given a set $\mathcal{N} = \{1, 2, \ldots, n\}$ and $n \times n$ matrices $F = (f_{ij})$ and $D = (d_{kl})$, the quadratic assignment problem (QAP) can be stated as follows:

$$\min_{p \in \Pi_{\mathcal{N}}} \sum_{i=1}^{n} \sum_{j=1}^{n} f_{ij} d_{p(i)p(j)},$$

where $\Pi_{\mathcal{N}}$ is the set of all permutations of $\mathcal{N}$. One of the major applications of the QAP is in location theory where the matrix $F = (f_{ij})$ is the flow matrix, i.e. f_{ij} is the flow of materials from facility i to facility j, and $D = (d_{kl})$ is the distance matrix, i.e. d_{kl} represents the distance from location k to location l. The cost of simultaneously assigning facility i to location k and facility j to location l is $f_{ij} d_{kl}$. The objective is to find an assignment of all facilities to all locations (i.e. a permutation $p \in \Pi_{\mathcal{N}}$), such that the total cost of the assignment is minimized.

In addition to its application in facility location problems, the QAP has been found useful in such applications as scheduling the backboard wiring problem in electronics, and statistical data analysis [33].

The QAP is, computationally, one of the most difficult combinatorial optimization problems. This problem, of which the traveling salesman problem, graph isomorphism, graph partitioning, and the band-width reduction problem are special cases, is NP-hard. Moreover, unless P=NP, there exists no polynomial-time algorithm to find an ϵ-approximate solution to the QAP. Furthermore, the quadratic assignment problem is PLS-Complete with respect to a Kernighan-Lin like neighborhood. Computationally, general problems of size $n > 20$ are not, to this date, practically solvable to optimality.

The QAP can be formulated as a quadratic minimization problem with a concave objective function. To see this, first formulate the problem as a quadratic 0-1 programming problem:

$$\min \quad \sum_{i=1}^{n} \sum_{j=1}^{n} \sum_{k=1}^{n} \sum_{l=1}^{n} f_{ij} d_{kl} x_{ik} x_{jl} \tag{6.1}$$

$$\text{s.t.} \quad \sum_{i=1}^{n} x_{ij} = 1 \ (j = 1, \ldots, n) \ \sum_{j=1}^{n} x_{ij} = 1 \ (i = 1, \ldots, n), \tag{6.2}$$

$$x_{ij} \in \{0, 1\} \ (i, j = 1, \ldots, n).$$

If we denote the feasible domain of the above problem by D, then the problem can be written as

$$\begin{aligned} \min \quad & x^T S x \\ \text{s.t.} \quad & x \in D. \end{aligned} \tag{6.3}$$

where the $(n^2 \times n^2)$–matrix S has nonnegative entries.

Let $m = n^2$, and $Q = S - \alpha I$, where I is the $(m \times m)$ unit matrix (all entries are 1), and $\alpha > ||S||_\infty$. Moreover, let $x = (x_{11}, x_{12}, \ldots, x_{nn})^T = (x_1, x_2, \ldots, x_m)^T$, and consider the quadratic form $x^T Q x$. Assume, without loss of generality, that Q is symmetric. Then, it can be shown that the matrix Q is negative definite. Finally, the QAP is equivalent to the following concave programming problem:

$$\begin{aligned} \min \quad & x^T Q x \\ \text{s.t.} \quad & x \in \Omega, \end{aligned}$$

where $Q = S - \alpha I, \alpha > ||S||_\infty$, and Ω is the set of all $x = (x_{11}, x_{12}, \ldots, x_{nn})^T \in R^{n^2}$ satisfying

$$\begin{aligned} \textstyle\sum_{j=1}^n x_{ij} &= 1 \; (i = 1, \ldots, n), \\ \textstyle\sum_{i=1}^n x_{ij} &= 1 \; (j = 1, \ldots, n), \\ x_{ij} \geq 0, & \qquad (i, j = 1, \ldots, n). \end{aligned}$$

Global concave minimization has been the subject of extensive research the last decade [16,17]. Many algorithms developed for concave optimization can be adapted to take advantage of the special structure of the above formulation of QAP.

7. THE HAMILTONIAN CYCLE PROBLEM

On the well-known studied problems in combinatorial optimization is the Hamiltonian Cycle Problem (HCP), which can be described as follows: Given a directed graph, find a path that enters every node exactly once before returning to the starting node, or determine that no such path exists.

The HCP is considered to be a very difficult problem from an algorithmic point of view. We can solve the HCP using the results of a new approach to the HCP described in [7]. This approach involves an embedding of the problem in a controlled Markov Chain and the exploration of the induced ergodic structure. Eventually the problem is reduced to that of solving a specially structured continuous indefinite quadratic program.

Consider a moving object tracing out a directed path on the graph G with its movement "controlled" by a function π mapping the nodes N into the arcs A. This function induces a "zero-one" $N \times N$ Markov matrix $P(\pi)$ whose positive entries correspond to the arcs "selected" by π at the respective nodes. Suppose that this motion continues forever, regard $P(\pi)$ as a Markov Chain, and consider its "stationary distribution", contained in its *limit Cesaro-sum matrix:*

$$P^*(\pi) := \lim_{T \to \infty} \frac{1}{T} \sum_{t=1}^{T} P^{t-1}(\pi),$$

where $P^0(\pi) := I_N$.

The next step in this approach is to use the relationship which exists between the ergodic class/transient state structure of these Markov Chains, and the possible cycles in the graph. The limiting behavior of the chains is then inducing a new characterization of Hamiltonian cycles.

This step, together with the fact that the functions π can be regarded *as controls,* embeds the HCP in a class of models called either *Controlled Markov Chains*, or *Markov Decision Processes* (MDPs, for short).

structure which considerably complicates the analysis of the model. Therefore a perturbation of probability transitions is used to simplify the structure of the Markov chains.

Let node 1 denote the "home" node; for each pair of nodes i, j (not equal to 1) corresponding to a (deterministic) arc (i, j), replace that arc by a pair of "stochastic arcs" $(i, 1)$ and (i, j):

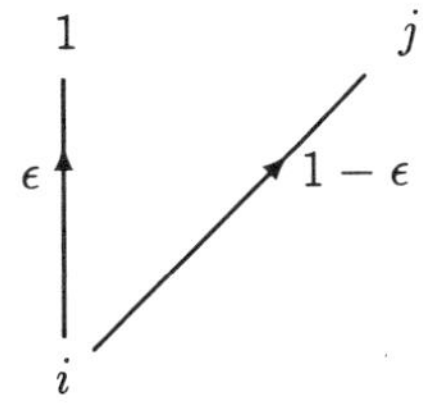

with weights ϵ and $(1-\epsilon)$ respectively ($\epsilon \in [0, 1)$). This stochastic perturbation has the interpretation that a decision to move along arc (i, j) results in movement along (i, j) only with probability of $(1-\epsilon)$, and with probability ϵ it results in return to the home node 1.

The Markov control problem Γ thus changes to an ϵ-perturbed problem $\Gamma(\epsilon)$ that is *unichain*, that is, every Markov Chain induced by a stationary policy possesses only a single ergodic class plus a set (possibly empty) of transient states. For instance, the policy π_2 now has the Markov chain matrix

$$P_\epsilon(\pi_2) = \begin{pmatrix} 0 & 1 & 0 & 0 \\ 1 & 0 & 0 & 0 \\ \epsilon & 0 & 0 & 1-\epsilon \\ \epsilon & 0 & 1-\epsilon & 0 \end{pmatrix}.$$

Let $X(\epsilon)$ be the frequency space of the perturbed MDP model $\Gamma(\epsilon)$. In [7] the following polyhedral subset $X(\epsilon)$ is constructed:

$$X'(\epsilon) = \{x \in X(\epsilon) | \sum_a x_{1a} = \frac{1}{d(\epsilon)}\} = \{x | Ax = \mathbf{b}, x \geq \mathbf{0}\},$$

where $d(\epsilon) = 1 + \sum_{i=2}^{N}(1-\epsilon)^{i-2}$, and $\epsilon \in (0, 1)$, and where A ($\mathbf{b}$) is an appropriate matrix (vector). This polyhedron possesses the following important property: For $\epsilon \in (0, 1)$ the Hamiltonian cycles of G are precisely the extreme points of $X'(\epsilon)$ that are induced by the deterministic stationary policies of the MDP $\Gamma(\epsilon)$. Also, the dimension of A is no more than $(N+2) \times N(N-1)$.

This leads to a new characterization of the Hamiltonian cycles of a graph; namely, as global solutions (with objective function value equal to 0) of an indefinite quadratic program of the form:

$$\text{(HQP)} \qquad min\{x^T Q x : \ x \in X'(\epsilon)\},$$

where Q is a block-diagonal matrix with each block $Q_i = J_i - I_i$, and where J_i and I_i are the matrix of all ones and the identity matrix respectively. The size of a given block is equal to the number n_i of arcs emanating from the given node in the graph.

This new characterization of the Hamiltonian cycles of a graph, is the basis for our approach using the techniques we have developed for indefinite quadratic problems. Preliminary computational results can be found in [7].

8. CONTINUOUS MINIMAX APPROACHES

Techniques and principles of minimax theory play a key role in many areas of research, including game theory, optimization, scheduling, location, allocation, packing, and computational complexity. In general, a minimax problem can be formulated as

$$\min_{x \in X} \max_{y \in Y} f(x, y) \tag{8.1}$$

where $f(x, y)$ is a function defined on the product of X and Y spaces [6].

A long-standing open problem on minimum networks, the Gilbert-Pollak conjecture on the Steiner ratio, was solved by using a continuous minimax approach [4]. The Steiner tree problem is a classic intractable problem with many applications in the design of computer circuits and telecommunication networks. Given a set of points in a metric space, the problem is to find a shortest network (called the Steiner minimum tree) interconnecting the points in the set. A minimum spanning tree on a set of points is the shortest network interconnecting the given points with all edges between the points. While the Steiner tree problem is intractable, the minimum spanning tree can be computed very fast. The Steiner ratio in a metric space is the largest lower bound for the ratio between lengths of a minimum Steiner tree and a minimum spanning tree for the same set of points in the metric space, which is a measure of performance for the minimum spanning tree as a polynomial-time approximation of the minimum Steiner tree. Determining the Steiner ratio in the Euclidean plane was a hard problem. Gilbert and Pollak [11] conjectured that it equals $\sqrt{3}/2$. This conjecture was proved by Du and Hwang [4]. The center part of this approach is a new theorem about the following continuous minimax problem.

$$\min_{x \in X} \max_{i \in I} f_i(x)$$

where X is a convex region X in n-dimensional Euclidean space R^n, I is a finite index set, and $f_i(x)$'s are continuous functions over X.

Minimax Theorem (Du & Hwang). Let $g(x) = \max_{i \in I} f_i(x)$. If every $f_i(x)$ is a concave function, then the minimum value of $g(x)$ over the polytope X is achieved at some point x^* satisfying the following condition:

There exists an extreme subset Z of X such that $x^* \in Z$ and the $I(x^*)$ ($= \{i \mid g(x^*) = f_i(x^*)\}$) is maximal over Z. A subset Z of X is called an *extreme subset* of X if

$$\left. \begin{array}{l} x, y \in X \\ \lambda x + (1-\lambda) y \in Z \text{ for some } 0 < \lambda < 1 \end{array} \right\} \Rightarrow x, y \in Y.$$

□

The Gilbert-Pollak conjecture was proved by first transforming the Steiner ratio problem to a continuous minimax problem. Then by the minimax theorem, the problem is reduced to the corresponding problem at critical points. The transformation between the Steiner ratio problem and the minimax problem is based on a mapping between sets of n points in the Euclidean plane and points in the $(2n-3)$-dimensional space. Furthermore, each critical point is transferred to a set of n points with a nice geometric structure, called a critical structure. Finally, the conjecture is verified on the point set with the critical structure (for details see [4]).

There are many other problems than those on Steiner trees where a continuous minimax approach can be used. Consider the problems of packing circles in a square. What is the maximum radius of n equal circles that can be packed into a unit square?

Observe that $\mathcal{F}$ is the fixed charge formulation of the SPG.

Next, we give an ϵ-error bound on the approximation of the solution of the SPDG by the solution of one sequence $\mathcal{F}_\epsilon$ of the series of concave optimization problems given above.

Theorem 9.1. The error on approximating f^* by $f^*(\epsilon_0)$ is less than $\frac{\epsilon_0}{1+\epsilon_0} \sum_{(i,j)\in\mathcal{A}^d} c_{ij}$.

Proof. Since $f(\epsilon, Y)$ is a convex and decreasing function of ϵ, we have

$$\begin{aligned}
0 \le f^* - f^*(\epsilon_0) &< \lim_{\epsilon\to 0} f(\epsilon, Y^*(\epsilon_0)) - f^*(\epsilon_0) \\
&\le (1 - \frac{1}{1+\epsilon_0}) \sum_{(i,j)\in\mathcal{A}^d} c_{ij} \\
&\le \frac{\epsilon_0}{1+\epsilon_0} \sum_{(i,j)\in\mathcal{A}^d} c_{ij}. \square
\end{aligned}$$

This new formulation allows the SPG to be treated with continuous combinatorial techniques. In particular, minimum concave cost network flow problems have been extensively studied and many computational algorithms are available for their solution.

10. CONCLUDING REMARKS

A large class of discrete optimization problems can be formulated as continuous nonconvex optimization problems. New properties and efficient algorithms have resulted from these continuous formulations. Furthermore, a variety of approaches exist to embed the discrete feasible domain in a larger continuous space (relaxation). The efficient solution of the relaxed problem can be used to compute sub-optimal solutions, or to obtain new approximate algorithms to many of these problems. It is evident that techniques from continuous optimization can be effectively used in the development of new algorithms for combinatorial optimization problems. However, further research and computational experimentation is needed for the efficient implementation of these algorithms.

REFERENCES

[1] F. Alizadeh, *"Optimization Over Positive Semi-Definite Cone; Interior-Point Methods and Combinatorial Applications"*. In "Advances in Optimization and Parallel Computing", P.M. Pardalos, editor, North–Holland, 1992.

[2] S. Arora and S. Safra, *"Probabilistic checking of proofs: A new characterization of NP"*. Proceedings of the 33rd IEEE Symp. on Foundations of Computer Science, pp. 2-13, 1992.

[3] S. Arora , C. Lund, R. Motwani, M. Sudan and M. Szegedy, *"Proof verification and hardness of approximation problems"*. Proceedings of the 33rd IEEE Symp. on Foundations of Computer Science, pp. 14-23, 1992.

[4] D.-Z. Du and F.K. Hwang, *"An approach for proving lower bounds: solution of Gilbert-Pollak's conjecture on the Steiner ratio"*. Proceedings of 31th FOCS Conference, pp. 76-85, 1990.

[5] D.-Z. Du, D.F. Hsu and K.-J Xu, *"Bounds on guillotine ratio"*. Congressus Numerantium 58, pp. 313-318, 1987.

[6] D.-Z. Du and P. M. Pardalos (Editors), *"Minimax and Applications"*. Kluwer Academic Publishers, 1995.

[7] J.A. Filar, M.G.M. Oberije and P.M. Pardalos, *"Hamiltonian cycle problem, controlled markov chains and quadratic programming"*. In The Proceedings of The 12th National Conference of The Australian Society For Operations Research, Adelaide July 7-9, pp. 263–281, 1993.

[8] F. Giannessi and F. Niccolucci, *"Connections between nonlinear and integer programming problem"*. In "Symposia Mathematica Vol. XIX, Istituto Nazionale di Alta Matematica", Acad. Press N.Y. pp. 161-176, 1976.

[9] L. Gibbons, D. Hearn and P.M. Pardalos, *"A continuous based heuristic for the maximum clique problem"*. In *"Clique, Graph Coloring, and Satisfiability: Second DIMACS Implementation Challenge"* (D.S. Johnson and M.A. Trick, Editors), DIMACS Series, American Mathematical Society, 1995.

[10] L. Gibbons, D. Hearn, P.M. Pardalos and M. Ramana, *"Continuous characterizations of the maximum clique problem"*. Research Report 95-7, ISE Department, University of Florida.

[11] E.N. Gilbert and H.O. Pollak, *"Steiner minimal trees"*. SIAM J. Appl. Math., 16, pp. 1-29, 1968.

[12] M. X. Goemans and D. P. Williamson, *"Improved Approximation Algorithms for Maximum Cut and Satisfiability Problems Using Semidefinite Programming"*. *J. ACM*; to appear. A preliminary version appeared under ".878-Approximation Algorithms for MAXCUT and MAX2SAT", *Proc. of the 26th Symposium on the Theory of Computing*, 422–431, Montréal, Canada, 1994.

[13] Jun Gu, *"Optimization algorithms for the satisfiability (SAT) problem"*. In "Advances in Optimization and Approximation" (D.-Z. Du and J. Sun, Editors), "Nonconvex Optimization and its Applications", Vol. 1, Kluwer Academic Publishers, pp. 72-154, 1994.

[14] Jun Gu, *"Global Optimization for Satisfiability (SAT) Problem"*. *IEEE Trans. on Knowledge and Data Engineering*, Vol. 6, No. 3, pp. 361-381, 1994.

[15] Jun Gu, *"Parallel Algorithms for Satisfiability (SAT) Problem"*. In *"Parallel Processing on Discrete Optimization Problems"* (Editors, P.M. Pardalos et al), American Math. Society, DIMACS Series Vol. 22, pp. 105-161, 1995.

[16] R. Horst and P. M. Pardalos (Editors), *"Handbook of Global Optimization"*. Nonconvex Optimization and its Applications Vol. 2, Kluwer Academic Publishers, 1995.

[17] R. Horst, P.M. Pardalos and N.V. Thoai, *"Introduction to Global Optimization"*. Kluwer Academic Publishers, 1995.

[18] D. Karger, R. Motwani and M. Sudan, *"Approximate graph coloring by semidefinite programming"*. Manuscript, 1995.

[19] N. Karmarkar, *"An interior-point approach for NP-complete problems"*. Contemporary Mathematics 114, pp. 297–308, pp. 1026–1047, 1990.

[20] B. Khoury, P.M. Pardalos and D. Hearn, *"Equivalent formulations of the Steiner tree problem in graphs"*. In "Network Optimization Problems" (Editors: D.-Z. Du and P.M. Pardalos), World Scientific, pp. 53-62, 1993.

[21] J. Kleinberg and M. X. Goemans, *"The Lovász Theta Function and a Semidefinite Programming Relaxation of Vertex Cover"*. Submitted to *SIAM J. Disc. Math.*, 1995.

[22] D. E. Knuth, *"The Sandwich Theorem"*. The Electronic Journal of Combinatorics, Vol. 1, #A1, 1994.

[23] L. Lovász, *"On the Shannon capacity of a graph"*. IEEE Transactions on Information Theory IT-25, pp. 1-7, 1979.

[24] L. Lovász and A. Schrjiver, *"Cones of matrices and set functions and 0-1 optimization"*. SIAM J. Opt. 1, pp. 166-190, 1991.

[25] C. Maranas, C. Floudas and P.M. Pardalos, *"New results in the packing of equal circles in a square"*. Discrete Mathematics 142, pp. 287-293, 1995.

[26] B. Mohar and S. Poljak, *"Eigenvalues in Combinatorial Optimization"*. In "Combinatorial Graph-Theoretical Problems in Linear Algebra" (Editors: R. Brualdi, S. Friedland and V. Klee), IMA Vo. 50, Springer-Verlag, pp. 107-151.

[27] P. M. Pardalos, *"On the passage from local to global in optimization"*. In "Mathematical Programming: State of the Art 1994" (J.R. Birge & K.G. Murty, Editors), The University of Michigan, pp. 220-247.

[28] P. M. Pardalos, *"The Linear Complementarity Problem"*. In "Advances in Optimization and Numerical Analysis", (Ed. S. Gomez & J.P. Hennart), Kluwer Academic Publishers, pp. 39-49, 1994.

[29] P. M. Pardalos (Editor), *"Complexity in Numerical Optimization"*. World Scientific, 1993.

[30] P. M. Pardalos and Y. Li, *"Integer programming"*. In "Handbook of Statistics", (Edited by C.R. Rao) Vol. 9, pp. 279-302, 1993.

[31] P. M. Pardalos and J. B. Rosen, *"Constrained Global Optimization: Algorithms and Applications"*. Lecture Notes in Computer Science, Vol. 268, Springer-Verlag, Berlin, 1987.

[32] P. M. Pardalos and J. Xue, *"The maximum clique problem"*. *Journal of Global Optimization*, 4, pp. 301-328, 1994.

[33] P.M. Pardalos and H. Wolkowicz (Editors), *"Quadratic Assignment and Related Problems"*. DIMACS Series, Vol. 16, American Math. Society, 1994.

[34] N. Z. Shor, *"Dual estimates in multiextremal problems"*. Journal of Global Optimization Vol. 2, No. 4, pp. 411–418, 1992.

[35] L. Tuncel, H. Wolkowicz and Q. Zhao, *"Semidefinite Programming: Theory, Applications, Algorithm"*. Kluwer Academic Publishers (to appear 1995).

HOMOTOPIES BASED ON NONSMOOTH EQUATIONS FOR SOLVING NONLINEAR VARIATIONAL INEQUALITIES

Sellami Hichem[1] and Robinson Stephen M.[2]

[1]Faculté des Sciences Economiques et de Gestion de Sfax
Université du Sud
Sfax, Tunisie
e-mail: fsegs!sellamih@tunisia.eu.net

[2]Department of Industrial Engineering
University of Wisconsin-Madison
1513 University Avenue
Madison, WI 53706-1572, USA
e-mail: smr@cs.wisc.edu

Abstract. This paper describes a globally convergent path-following method for solving nonlinear equations containing particular kinds of nonsmooth functions called *normal maps*. These normal maps express nonlinear variational inequalities over polyhedral convex sets in a form convenient for analysis and computational solution. The algorithm is based on the well known predictor-corrector method for smooth functions, but it operates in the piecewise linear normal manifold induced by the convex set, and thus extends and implements earlier ideas of Alexander, Kellogg, Li, and Yorke. We discuss how the implementation works, and present some preliminary computational results.

Key words. Variational inequality, generalized equation, normal map, continuation method, homotopy method, path-following

1. INTRODUCTION

This paper is part of a program of research attempting to develop implementable, effective homotopy methods of predictor-corrector type for the numerical solution of certain systems of nonsmooth nonlinear equations derived from variational inequalities. Such inequalities, the study of which was initiated by Stampacchia [43, 44], are ubiquitous in the description and analysis of equilibrium problems. An excellent survey of theory, algorithms, and applications as of 1990 is given in [19]. The methods we discuss are intended to help in the numerical solution of such problems, particularly in cases in which good starting points (approximate solutions) are not available.

Our purpose here is primarily expository; we explain the class of methods with which we are concerned, indicate briefly how they can be mathematically justified, and then present example calculations on some sample problems. Proofs of existence and continuity are presented in [39] and in the dissertation [38], and more detail on the numerical

The classical MDP is a process observed at discrete time points $t = 1, 2, 3, \ldots$. If at time t the system is in a *state* i and the controller chooses an *action* a, a transition to a new state j occurs with a probability p_{iaj}. Such a *policy* is a rule that tells the controller what action (or a probability distribution on the set of actions) to select under every possible circumstance. If this rule depends only on the current state, and does not depend on the way that state is reached, then such a policy is said to be a *stationary* policy. A stationary policy which never randomizes is called *deterministic or pure.*

To make the link to the HCP, the *limiting state-action frequencies* have to be considered as well. Given a policy π (satisfying certain regularity conditions), and a fixed initial state distribution, the *limiting frequency of the state-action pair (j,a)* is defined by

$$x_{ja}(\pi) = \lim_{T\to\infty} \frac{1}{T}\sum_{t=1}^{T} Pr_\pi(X_t = j, A_t = a)$$

where the random variables X_t, and A_t denote the state and action at time t. Now, the vector $x(\pi)$ of state-action frequencies is the vector whose entries are $x_{ja}(\pi)'s$ ordered in the natural fashion. The set X of all such frequency vectors is called *the frequency space* of the limiting average MDP, and is known to be an explicitly characterized polytope. Note that if π is a stationary policy, then π is the vector whose typical entry is $\pi(i,a) := Pr$ (controller chooses action a in state i, whenever i is visited). Thus each π determines a Markov Chain characterized by the probability transition matrix $P(\pi) = (p_{ij}(\pi))_{i,j=1}^N$ (where N is the number of states) whose entries are given by $p_{ij}(\pi) := \sum_a p_{iaj}\pi(i,a)$. The corresponding *limit Cesaro-sum matrix*, (also called the *stationary distribution matrix*) was defined above. We illustrate this embedding with the following simple example.

Consider the complete graph G on four nodes (with no self-loops) and think of the nodes as the states of an MDP, denoted by Γ, and of the arcs emanating from a given node as of actions available at that state. With this the set of possible actions $A(i)$ at node i is defined as: $A(i) = \{a|(i,a) \text{ is an arc in G}\}$. In a natural way, the Hamiltonian cycle $c_1 : 1 \to 2 \to 3 \to 4 \to 1$ corresponds to the non-randomized stationary policy $\pi_1 : \{1,2,3,4\} \to \{2,3,4,1\}$, where $\pi_1(2) = 3$ corresponds to the controller choosing arc (2,3) in state 2 with probability 1. The Markov chain induced by π_1 is given by the transition matrix

$$P(\pi_1) = \begin{pmatrix} 0 & 1 & 0 & 0 \\ 0 & 0 & 1 & 0 \\ 0 & 0 & 0 & 1 \\ 1 & 0 & 0 & 0 \end{pmatrix}$$

which is *irreducible*, that is, all the states belong to one ergodic class. On the other hand, the union of two sub-cycles: $1 \to 2 \to 1$ and $3 \to 4 \to 3$ corresponds to the policy $\pi_2 : \{1,2,3,4\} \to \{2,1,4,3\}$ which identifies the Markov chain transition matrix

$$P(\pi_2) = \begin{pmatrix} 0 & 1 & 0 & 0 \\ 1 & 0 & 0 & 0 \\ 0 & 0 & 0 & 1 \\ 0 & 0 & 1 & 0 \end{pmatrix}$$

containing two distinct ergodic classes. How these observations lead to an embedding of the Hamiltonian cycle problem in a Markov control problem Γ is described in [7]. That Markov control problem has deterministic transitions and a multi-chain ergodic with respect to $\mathcal{R}$ on G, to be a directed tree where all nodes in $\mathcal{R}\backslash\{r\}$ can be reached from r via directed paths in T, and all leaf nodes are in $\mathcal{R}$. Next, we define the SPDG.

Instance: A graph $G^d = (\mathcal{N}, \mathcal{A}^d, \mathcal{C}^d)$, a node subset $\mathcal{R}$ and a root node r in $\mathcal{R}$.
Question: Find the minimum cost directed Steiner tree, on G with respect to $\mathcal{R}$, that has node r as root node.

In the light of the SPG and SPDG definitions given above, it is not hard for one to see that the SPDG is a generalization of the SPG. In fact, the SPG can be transformed to a special case of the SPDG where the arc cost structure is symmetric. This transformation can be done as follows. Given a graph $G = (\mathcal{N}, \mathcal{A}, \mathcal{C})$, create the corresponding directed graph $G^d = (\mathcal{N}, \mathcal{A}^d, \mathcal{C}^d)$, where every undirected arc $(i,j) \in \mathcal{A}$ corresponds to two oppositely directed arcs (i,j) and $(j,i) \in \mathcal{A}^d$ associated with the same cost ($c^d_{ij} = c^d_{ji} = c_{ij}$). Potentially, any mathematical formulation for the SPDG can be used to formulate the SPG.

9.1. Continuous formulations of the SPG

There are many ways to formulated the SPG as an optimization problem. These formulations are of three different types: mixed integer formulations, integer formulations and continuous formulations [20]. Here, we present a new continuous formulation in the context of continuous approaches to the SPG. However, in this formulation, the source of concavity is still attached to the objective function. Indeed, we present a series of continuous ϵ-concave optimization problems, and we prove that the limit of the series when ϵ goes to zero is the fixed charge formulation of the SPDG. Also, we give an error bound on approximating the solution of the SPDG by the solution of a sequence in the series.

Assume we are given a directed graph $G = (\mathcal{N}, \mathcal{A}^d, \mathcal{C}^d)$. Let $\mathcal{R}$ be a subset of nodes on the graph, set of regular nodes. Furthermore, let node r, the root node, be an arbitrary node in $\mathcal{R}$. In addition, let y_{ij} be a flow variable that is associated with arc (i,j). Moreover, let ϵ be a strictly positive real number. Now, define $\mathcal{F}_\epsilon$, a concave minimization problem, as follows.

$$(\mathcal{F}_\epsilon) \qquad \text{minimize } f(\epsilon, Y) = \sum_{(i,j)\in\mathcal{A}^d} c_{ij}\frac{y_{ij}}{y_{ij}+\epsilon}$$
$$\text{subject to:} \quad AY = B$$
$$Y \geq 0.$$

where A is an $|\mathcal{N}| \times |\mathcal{A}^d|$ node-arc incidence matrix, and Y is an $|\mathcal{A}^d| \times 1$ vector whose entries are the y_{ij}s, and B is an $|\mathcal{N}| \times 1$ vector whose entries b_i correspond to the set of nodes; furthermore, $b_i = 0$ if $i \in \mathcal{N}\backslash\mathcal{R}$, and $b_i = -1$ if $i \in \mathcal{R}\backslash\{r\}$, and $b_r = |\mathcal{R}|-1$. This is basically a minimum cost flow problem where the objective function is a separable concave function. Note that due to the strict concavity of $f_\epsilon(Y)$, the unimodularity of A and the integrality of B, the optimal solution to $\mathcal{F}_\epsilon$ is always integer.

Now, if we let $x_{ij} = \dfrac{y_{ij}}{y_{ij}+\epsilon}$, the limit of $\mathcal{F}_\epsilon$ when ϵ goes to zero becomes

$$(\mathcal{F}) \qquad \text{minimize } f(Y) = \sum_{(i,j)\in\mathcal{A}^d} c_{ij}x_{ij}$$
$$\text{subject to:} \quad AY = B$$
$$x_{ij} = 0(1) \text{ for } y_{ij} = 0(>0) \ (\forall (i,j) \in \mathcal{A}^d)$$
$$Y \geq 0.$$

This problem is equivalent to the following: How should n points be arranged into a unit square such that the minimum distance between them is greatest? In fact, let r_n denote the maximum radius in the first problem and d_n the max-min distance in the second problem. It is easy to show that

$$r_n = \frac{d_n}{2(1+d_n)}.$$

We can write the second problem in the following form:

$$\min_{x_i \in [0,1]\times[0,1]} \max_{1 \leq i < j \leq n} -\|x_i - x_j\|.$$

For fixed i and j, $\|x_i - x_j\|$ is clearly a convex function. Using this formulation new results on this problem have been obtained in [25].

Another interesting problem which can be approached with minimax techniques is the problem of rectangular partitions. Consider a rectangle R which is partitioned into several smaller rectangles with different colors. Let $L_i(R)$ denote the total length of segments in the inside of R. We want to cut R into monochromatic pieces by a sequence of guillotine cuts. Each guillotine cut divides one piece into two pieces. Let $L_g(R)$ be the minimum total length of guillotine cuts. One is interested in the following *guillotine ratio*:

$$\rho = \inf_R \frac{L_g(R)}{L_i(R)}.$$

It was conjectured in [5] that $\rho = 1.5$. The best known result is $1.5 \leq \rho \leq 1.75$. Choose the segment-lengths of R as parameters. Then $L_i(R)$ is a linear function. For a fixed sequence s of guillotine cuts, the total length $L(R,s)$ of the cuts is also a linear function. Thus, $f_s(R) = L_i(R) - \frac{2}{3}L(R,s)$ is a linear function. Note that

$$\rho = \min_R \max_s f_s(R).$$

Restrict the total length of R by setting it equal to one. Then we obtain a minimax problem satisfying the above minimax theorem.

9. THE STEINER PROBLEM

The next example we discuss is the Steiner problem on graphs. This a very difficult problem to solve and many nonconvex network flow problems can be reduced to it.

Let $G = (\mathcal{N}, \mathcal{A}, \mathcal{C})$ be an undirected graph, where $\mathcal{N} = \{1, \ldots, n\}$ is a set of nodes, $\mathcal{A}$ is a set of undirected arcs (i,j) with each arc incident to two nodes, and $\mathcal{C}$ is a set of nonnegative costs c_{ij} associated with undirected arcs (i,j). Then, the Steiner problem in Graphs (SPG) is defined as follows:

Instance: A graph $G = (\mathcal{N}, \mathcal{A}, \mathcal{C})$, a node subset $\mathcal{R} \in \mathcal{N}$.
Question: Find the minimum cost tree, on G, that would connect all the vertices in $\mathcal{R}$.

The Steiner Problem in Directed Graphs (SPDG) is the directed graph version of the Steiner problem. Consider $G^d = (\mathcal{N}, \mathcal{A}^d, \mathcal{C}^d)$ where $\mathcal{N} = \{1, \ldots, n\}$ is a set of nodes, and $\mathcal{A}^d$ is a set of directed arcs (i,j) incident to nodes in $\mathcal{N}$. $\mathcal{C}^d = \{c^d_{ij}\}$ is a set of costs associated with $\mathcal{A}^d$.

Furthermore, let $\mathcal{R}$ (a set of regular nodes) be a subset of $\mathcal{N}$, and let r (a root node) be an arbitrary node in $\mathcal{R}$. Also, define a directed Steiner tree T rooted at node r, implementation and additional computational results are in [41]. Much of the descriptive material was initially presented in the unpublished technical report [40].

This paper is organized in five sections. In the rest of this section, we define several terms that we shall employ in what follows. In Sect.2 we show how to formulate variational inequalities in the equivalent forms of generalized equations and of normal maps, and we give some examples of particular generalized equations that we shall later use as computational examples. In that section we also briefly explain the Newton method for generalized equations, and use its properties to indicate why we are interested in implementing homotopy methods. Sect.3 gives a brief development of the classical predictor-corrector method for smooth functions, with references, and then explains ways of adapting this method to the nonsmooth functions that we employ. Sect.4 presents some preliminary computational results from application of our method. Finally we present in the concluding Sect.5 a few observations and comments on areas that remain to be investigated.

In the rest of this paper we use terms and conventions from convexity. One idea that we shall use repeatedly is that of the *normal cone* to a convex set C in $\mathbb{R}^n$, which at a point x of $\mathbb{R}^n$ is defined to be

$$N_C(x) = \begin{cases} \{x^* \in \mathbb{R}^n \mid \text{for each } c \in C,\ \langle x^*, c - x\rangle \leq 0\} & \text{if } x \in C, \\ \emptyset & \text{otherwise.} \end{cases}$$

For other terms from convex analysis we refer to Rockafellar [37].

We also introduce two ideas that we shall need in the computational work. The first is that of the tangent vector induced by a matrix: if A is an $n \times (n+1)$ matrix with rank $(A) = n$, we call the unique vector $t \in \mathbb{R}^{n+1}$ satisfying the three conditions $At = 0$, $\|t\| = 1$, and

$$\det \begin{pmatrix} A \\ t^T \end{pmatrix} > 0,$$

the *tangent vector induced by* A, and denote it by $t(A)$. We shall also need the *Moore-Penrose generalized inverse* of this same A, namely the matrix A^+ defined by $A^+ = A^T(AA^T)^{-1}$.

2. GENERALIZED EQUATIONS

Generalized equations are devices, equivalent to variational inequalities, that are very useful in expressing solution conditions for optimization and equilibrium problems. A typical generalized equation is

$$0 \in F(x) + N_C(x), \tag{2.1}$$

where F is in general a multifunction from a Hilbert space L to itself, but is often an ordinary single-valued function, and $N_C(x)$ is the normal cone to a closed convex set C in L. Sometimes the multifunction $N_C(x)$ is replaced by a more general maximal monotone operator [7], but in this paper we only treat problems of the form (2.1) with F single-valued (and, for purposes of the theory, smooth). The inclusion (2.1) expresses the geometric idea that $F(x)$ is an inward normal to C at x, or equivalently that $-F(x)$ belongs to $N_C(x)$. It is clear that (2.1) is equivalent to the *variational inequality* problem of finding x such that:

$$\text{For each } c \in C, \quad \langle F(x), c - x\rangle \geq 0. \tag{2.2}$$

Before proceeding to detail methods to deal with this generalized equation, we present a few examples illustrating how it can be used to formulate various problems. Consider first the mathematical programming problem

$$\begin{array}{rll} \min & f(y) & \\ \text{subject to} & g(y) & \in K^\circ \\ & y & \in L, \end{array} \tag{2.3}$$

where f and g are Fréchet (F-)differentiable functions from an open set $U \subset \mathbb{R}^p$ to $\mathbb{R}$ and $\mathbb{R}^m$ respectively, L is a closed convex set in $\mathbb{R}^p$, K is a closed convex cone in $\mathbb{R}^m$ and K° denotes the polar cone of K. This formulation is general enough to express a wide range of specific optimization problems found in practice. For example if the constraints of the problem we are dealing with are of the form

$$\begin{array}{lll} g_1(y) \leq 0, & \ldots, & g_k(y) \leq 0, \\ g_{k+1}(y) = 0, & \ldots, & g_{k+l}(y) = 0, \end{array}$$

then we only have to set $L = \mathbb{R}^p$, $K = \mathbb{R}^k_+ \times \mathbb{R}^l$, so that $K_1^\circ = \mathbb{R}^k_- \times \{0\}^l$. In this case if we define the standard Lagrangian associated with (2.3) by

$$\mathcal{L}(y,u) = f(y) + \langle u, g(y) \rangle, \ \ y \in U, \ u \in \mathbb{R}^k_+ \times \mathbb{R}^l,$$

then if y is a local optimizer of (2.3) at which an appropriate constraint qualification holds there will exist $u \in \mathbb{R}^k_+ \times \mathbb{R}^l$ which, with y, satisfies the necessary optimality conditions

$$\begin{array}{ll} 0 \in & \partial/\partial y \mathcal{L}(y,u) + N_L(y) \\ 0 \in & -\partial/\partial u \mathcal{L}(y,u) + N_K(u). \end{array} \tag{2.4}$$

So if we set $x = (y,u)$, $F(x) = [\partial/\partial y \mathcal{L}(y,u), -\partial/\partial u \mathcal{L}(y,u)]$ and $C = L \times K$, noting that $N_{L \times K} = N_L \times N_K$, we see that (2.4) is equivalent to (2.1).

For a second example, consider a function F from an open subset of $\mathbb{R}^m$ to itself, and a closed convex cone $K \subset \mathbb{R}^n$. The *generalized complementarity problem* for F and K is that of finding a point x such that

$$x \in K, \ F(x) \in K^*, \ \langle x, F(x) \rangle = 0, \tag{2.5}$$

where $K^* = -K^\circ$ is the dual cone of K. Since K is a cone we have:

$$N_K(x) = \begin{cases} \{y \in K^\circ \mid \langle y, x \rangle = 0\} & \text{if } x \in K, \\ \emptyset & \text{if } x \notin K. \end{cases}$$

Therefore (2.5) is just the statement that $0 \in F(x) + N_K(x)$. When $K = \mathbb{R}^n_+$ this generalized complementarity problem reduces to the well known *nonlinear complementarity problem*.

Our third example involves equilibrium conditions in economics. Hansen and Koopmans describe in [17] a mathematical model of a capital stock invariant under optimization. This model furnishes a good example of the utility of the generalized-equation formalism. Robinson presents in [34] a brief summary of the model with a derivation of the equivalent generalized equation. The technology in the model has constant returns to scale, and the model contains three goods (capital goods, resources and consumption goods). The authors ask how to find an operating path over time that maximizes the discounted utility of consumption and in which the stock of capital goods is invariant. They reduce this to a single-period problem involving the variables $z \in \mathbb{R}^L$ (capital goods), $w \in \mathbb{R}^M$ (resources), and $x \in \mathbb{R}^l$ (activity levels). The variables are related by inequalities, shown next, in which A, B, and C are linear transformations and in which the letter in front of each inequality denotes the corresponding dual variable:

$$\begin{array}{rrcl} (q_A) & Ax & \leq & z \\ (q_B) & z & \leq & Bx \\ (r) & Cx & \leq & w \\ & x & \geq & 0. \end{array} \tag{2.6}$$

Hansen and Koopmans then introduce a differentiable, concave objective function $v(x)$ and use this together with (2.6) (including the dual variables, so it is not a conventional optimization problem) to express the conditions. The analysis in [34] shows that solving this problem is equivalent to solving the generalized equation

$$0 \in \begin{bmatrix} 0 & 0 & A^T & -B^T & C^T \\ 0 & 0 & -\alpha I & I & 0 \\ -A & I & 0 & 0 & 0 \\ B & -I & 0 & 0 & 0 \\ -C & 0 & 0 & 0 & 0 \end{bmatrix} \begin{bmatrix} x \\ z \\ q_A \\ q_B \\ r \end{bmatrix} + \begin{bmatrix} -v'(x) \\ 0 \\ 0 \\ 0 \\ w \end{bmatrix} + N_{C'} \begin{bmatrix} x \\ z \\ q_A \\ q_B \\ r \end{bmatrix}, \tag{2.7}$$

where $C' = \mathbb{R}^l_+ \times \mathbb{R}^L \times \mathbb{R}^L_+ \times \mathbb{R}^L_+ \times \mathbb{R}^M_+$ and α is a parameter.

Since a substantial number of useful and interesting problems can be formulated as generalized equations, it is important to have effective methods for solving them. It is well known that Newton methods are among the most effective tools for numerically solving smooth nonlinear equations, and therefore a number of authors have suggested Newton-type methods for solving generalized equations. Early (unpublished) proposals include those of Wilson [48, 49], Robinson [33], Eaves [13], and Josephy [20, 21]. A summary of Josephy's results was published in [34], and the method in its basic form is now often referred to as the Newton-Josephy method. Later papers giving Newton-type methods of interest in connection with this problem include [15, 24, 28, 30, 32, 36, 50, 51], among others.

In general, methods of this class proceed by considering the following linearization of the generalized equation (2.1):

$$0 \in F(x_k) + dF(x_k)(x - x_k) + N_C(x). \tag{2.8}$$

A rough statement of the applicable convergence result is that if the first linearized equation (that is, (2.8) for $k = 0$) has a solution x_1 satisfying a certain niceness property and if certain other inequalities hold, then (2.8) has solutions for each value of k and the sequence of solutions converges R-quadratically to a solution of (2.1).

Like most computational procedures using the Newton technique, Josephy's method required a sufficiently good starting point, which means that one must have at least a rough idea about the location of the solution before attempting to solve the problem. Some of the later proposals cited above have successfully incorporated strategies such as damping and/or the use of merit functions (such as the so-called Fischer function; see *e.g.* [15]), which appear to increase the size of the region in which convergence can be expected.

By contrast, we shall present here an algorithm that can solve problem (2.1) without necessarily requiring knowledge of a good starting point. This algorithm extends to certain nonsmooth equations the well known path-following, or homotopy, method for solving smooth equations, and therefore to use it on (2.1) we must begin by transforming the generalized equation into an equivalent equation. We do this using the idea of a *normal map*, which we shall explain next. We then present in Sect.3 a path-following method especially adapted for normal maps, and in Sect.4 the results of some computational tests of that method.

To explain the idea of a normal map, we begin with the generalized equation (2.1) and define Π_C to be the Euclidean projector on the closed convex set C. Then we write

$$F_C = F \circ \Pi_C + I - \Pi_C.$$

This function F_C is the *normal map* induced by the function F and the set C. We shall show that finding a solution of (2.1) is equivalent to finding a zero of F_C.

Suppose $x_0 \in \mathbb{R}^n$ is such that $F_C(x_0) = 0$. Let $c_0 = \Pi_C(x_0)$, so that $-F(c_0) = x_0 - c_0$. Now $c_0 = \Pi_C(x_0)$ if and only if $\langle x_0 - c_0, c - c_0 \rangle \le 0$ for each $c \in C$; hence $\langle F(c_0), c - c_0 \rangle \ge 0$ for each $c \in C$, which means that c_0 satisfies (2.2), hence also (2.1). Conversely, suppose c_0 solves (2.1). Let $x_0 = c_0 - F(c_0)$ so that $x_0 - c_0 = -F(c_0)$. We have $-F(c_0) \in N_C(c_0)$ if and only if $\langle -F(c_0), c - c_0 \rangle \le 0$ for each $c \in C$; hence $\Pi_C(x_0) = c_0$ and $F_C(x_0) = F(\Pi_C(x_0)) + x_0 - \Pi_C(x_0) = 0$. Therefore having a solution of

$$F_C(x_0) = 0 \tag{2.9}$$

gives us a solution of (2.1) and vice-versa. The first use of this device known to us is in [11]; it was applied there to general variational inequalities. Later applications were given to the nonlinear complementarity problem in [27] and [3], and to an economic problem in [45], while Kojima showed in [22] how to use this device in connection with nonlinear programming problems.

We have just seen that solving a generalized equation is equivalent to solving an ordinary single-valued equation containing a normal map. However, in general the normal map appearing in this equation will not be F-differentiable even if the function f in (2.1) was F-differentiable, because of the presence of the projector Π_C. This means that the classical theory for smooth continuation methods has to be extended to accommodate this nonsmoothness. In the next section we indicate some ways in which this extension can be done, after first briefly reviewing the classical theory.

3. CONTINUATION METHODS: SMOOTH AND NONSMOOTH

In the preceding section we demonstrated how to transform the generalized equation (2.1) into an equation involving a normal map. In order to be able to find a solution of (2.1) with no requirement for a good starting point, we propose to adapt a classical technique called continuation. In this section we describe the type of continuation method that we shall apply to the normal map. First we give a brief overview of the general class of continuation methods, with references for further information. We then explain the particular kind of nonsmooth continuation method that we employed, and we give a brief summary of the steps in the numerical implementation. More detail is in [41].

In the continuation approach we embed the equation $F_C(x) = 0$ in a one-parameter family of equations $H(x, \lambda) = 0$, where $H(x, 0) = g(x)$ is a simple function with a known zero and $H(x, 1) = F_C(x)$ for every x. Then we try to follow a path of zeros of H from $(x_0, 0)$, where x_0 is a zero of g, to $(x^*, 1)$, where x^* is a zero of F_C. This path may contain *turning points,* at which the direction of movement of the parameter λ changes. In particular, except under strong assumptions we cannot expect that λ will increase monotonically.

Two examples of such homotopies are the *convex homotopy* given by

$$H(x, \lambda) = (1 - \lambda)(x - a) + \lambda F_C(x), \tag{3.1}$$

and the *global homotopy* defined by

$$H(x, \lambda) = F_C(x) - (1 - \lambda) F_C(x_0). \tag{3.2}$$

For the case of differentiable functions, this technique is well known and has been extensively studied. Excellent references are the book of Allgower and Georg [4], which contains a large bibliography, the supplemental bibliography in [5], and their more recent survey paper [6].

The methods that have been investigated and used can be generally divided into two classes: predictor-corrector methods and piecewise linear methods. Methods of the latter class proceed by approximating the homotopy H by a piecewise linear map $H_{\mathcal{T}}$ which affinely interpolates H at the nodes of a triangulation (or other subdivision) $\mathcal{T}$ of $\mathbb{R}^{N+1}$. They then trace the piecewise linear 1-manifold $H_{\mathcal{T}}^{-1}(0)$. We do not consider such methods here.

By contrast, the predictor-corrector method does not require the approximating map $H_{\mathcal{T}}$. Here the idea is to trace a curve $c(s)$ in the 1-manifold $H^{-1}(0)$ starting from a point x_0 satisfying $H(x_0, 0) = 0$. At each iteration one calculates the tangent to the curve c at that point, makes a "predictor" step along the tangent direction, then makes a "corrector" step to return to $H^{-1}(0)$, using a variant of Newton's method applicable to functions from $\mathbb{R}^{N+1}$ to $\mathbb{R}^N$ whose derivatives have maximal rank. The predictor step is very commonly carried out by using the *Euler predictor:*

$$v = u + ht\left(dH(u)\right),$$

where $t\left(dH(u)\right)$ is the tangent vector induced by $dH(u)$ and h is the stepsize of the predictor step. The corrector step can be performed by computing the sequence

$$\begin{aligned} \mathcal{N}^0(v) &= v \\ \mathcal{N}^{i+1}(v) &= \mathcal{N}^i(v) - dH(\mathcal{N}^i(v))^+ H(\mathcal{N}^i(v)), \end{aligned}$$

where $dH(u)^+$ is the Moore-Penrose generalized inverse of the matrix $dH(u)$. Justification for this Newton corrector can be found in, *e.g.,* Theorem 3.4.1 of [4].

Unfortunately, for normal maps the function F_C is rarely smooth, as we have seen in Sect.2. Therefore, we cannot directly apply the predictor-corrector method just described unless we resort to devices like the Mangasarian function for the nonlinear complementarity problem (see [26] and [47]), which is not always available and which may cause numerical difficulties even when it is available. Some more recent proposals for smoothing methods are in [9] and [42]; there is not yet an extensive base of numerical experience with these.

Various authors have considered ways to deal directly with nonsmoothness in implementing continuation methods. Many of these employ the piecewise-linear approach, using grids or triangularizations on which continuous functions are approximated by piecewise affine ones. For example Eaves and Scarf [14] describe a path-following method to find a solution to $F(x) = c$, where F is a piecewise affine function defined on the union of a finite number of compact convex polyhedra. The approach of that paper is expanded in the excellent tutorial paper [12]. A very large number of papers dealing with this general approach appeared in the years following the mid-1970s, many of which are referenced in the bibliographies of [4] and [5].

A few authors, however, have dealt directly with nonsmoothness in the context of the predictor-corrector method, and we discuss some of these papers because they relate more directly to our numerical implementation. The earliest approaches of this kind that we know of are those of Alexander *et al.* [1, 2, 3]. In particular, the authors of [3] present a path-following method for piecewise smooth functions on a piecewise smooth domain. They study the case of the nonlinear complementarity problem and constrained optimization, along lines similar to those in [2] and in the appendix of [1]. They prove that under suitable conditions, for almost all starting points lying in a specified open set, $H^{-1}(0)$ will be a piecewise smooth curve that eventually remains arbitrarily close to the set $\{f^{-1}(0)\} \times \{1\}$, where $f = H(\cdot, 1)$ (for example, see Theorem 2 of [3]).

The papers [1, 2, 3] do not contain any numerical results, and we are not aware of any subsequent numerical implementations of their framework. Extensions of their theoretical results are given in [39], where it is shown that if the generalized equation (2.1) is strongly regular at each solution lying in a certain ball and if a standard "outwardness" condition holds, then for almost every starting point in a certain open set, the homotopy curve actually has finite arc length and converges to a solution.

Although the method we explain below solves normal-map equations over arbitrary polyhedral convex sets C, when specialized to the nonlinear complementarity problem (that is, to the special case $C = \mathbb{R}^N_+$) it yields the method of [3]. Thus, the work we discuss here amounts to *(i)* showing how to apply the theoretical framework of [3] to variational inequalities over general polyhedral convex sets, and *(ii)* exhibiting preliminary numerical results from a computational implementation of that framework.

Although Alexander *et al.* did not explicitly phrase their work in terms of normal maps, some more recent work has been cast in that framework. Three references in this class are [29, 31, 46]. In the first of these, Pang establishes the feasibility of a predictor-corrector continuation method using the zero vector as predictor. He considered a parametric variational inequality of the form

$$\text{For each } c \in C, \quad \langle H(x, \lambda), c - x \rangle \geq 0, \tag{3.3}$$

and for a given point (x_k, λ_k) on the homotopy path he proposed to increase λ slightly to λ_{k+1} and then find x_{k+1} such that (x_{k+1}, λ_{k+1}) solved (3.3). One would expect that strong conditions would be necessary for such method to work (for example, if this method is to succeed the curve should not have any turning points), and Pang identified such conditions.

The zero-predictor scheme was further investigated in [31], and a method using a tangential predictor based on a directional derivative was considered in [46]. Computational results for both methods on five problems are presented in [31]: one is described as a nonlinear programming problem with 15 variables, two as spatial equilibrium problems from [25], and two as Nash equilibrium problems from [18].

We note here that the methods of these three papers differ fairly substantially from ours in that at each step a nonlinear variational inequality has to be solved by Newton's method. By contrast, as we show below, in our method only a system of smooth nonlinear equations has to be solved, and a convenient modified Newton method is available.

Another recent example is the paper of Cao and Ferris [8], who present a path-following algorithm for solving the problem 2.1 when f is an affine function ($f(x) = Ax - a$) and C is a polyhedral convex set. They prove that if A_C is coherently oriented (see Robinson [35]) for definition) their method terminates at the unique solution of $A_C(x) = a$. In case A_C is not coherently oriented, the program still can process the problem: that is, it will find a solution unless the problem is infeasible in a well specified sense.

Some authors have given methods designed specifically for nonlinear programming problems, which are a subclass of the general problem category that we consider here. In this area we mention as examples the paper of Kojima and Hirabayashi [23] and the recent book of Guddat *et al.* [16].

In the final portion of this section we give a brief description of how our numerical implementation proceeds. For implementation of the method we assume that a homotopy

$H(x,\lambda)$ has been chosen; for the purpose of our numerical illustrations in Sect.4, this will be the convex homotopy (3.1). We write z for the pair (x,λ).

Step 1. The first step is to select a starting point x_0 and to project x_0 onto C to obtain $\Pi_C(x_0)$, using a standard quadratic-programming procedure (we used the subroutine `E04NAF` from the NAG library). From this projection we obtain the set of active linear inequalities and equalities among those defining C, and therefore also the cell σ of the normal manifold in which x_0 is located. From this point on we do not use the quadratic programming code again; rather, as we cross cell boundaries we add and delete active constraints to obtain the affine function representing the projection operator Π_C in any given cell of the normal manifold. We set $\lambda_0 = 0$ and perform a QR factorization to obtain the form

$$dH(x_0,\lambda_0) = \begin{pmatrix} R^T & 0 \end{pmatrix} Q,$$

where Q is an $(n+1)\times(n+1)$ orthogonal matrix and R is an $n\times n$ lower triangular matrix, which will be nonsingular provided that $dH(x_0,\lambda_0)$ has rank n as predicted by the theory. We take the initial tangent vector to be the last row of Q, together with an orientation scalar that is $+1$ if $Q_{n+1,n+1}\geq 0$ and -1 otherwise. This scalar serves the purpose of beginning the path in a direction that increases λ from its initial value of 0. Some additional initialization is then performed.

Step 2. We calculate the distance $\bar{\theta}$ from the current point z along the current tangent vector t to the next cell boundary (possibly $+\infty$). If the current stepsize h is less than $\bar{\theta}$ then we skip to *Step 3* below to perform a predictor-corrector step in the current cell. Otherwise, we jump along the current tangent vector to the cell boundary, identify the new active constraint and proceed to update the affine representation of the projector as well as the tangent vector. We then perform a Newton correction *in the cell boundary* (a hyperplane of dimension $n-1$) to identify precisely the point at which the homotopy path intersects the boundary. If this correction is successful, we set the stepsize at the minimum of *(i)* 75% of the distance to the next cell boundary and *(ii)* the maximum of the current stepsize and the initial stepsize, and we proceed into the interior of the next cell. If it is not successful, we go back to the current point (in the old cell), restore the representation information, set the stepsize at 75% of the distance from the current point to the cell boundary, and perform a predictor-corrector step (as described in Step 3 below) in the old cell. The motivation for this is that after such a step we should be closer to the cell boundary and therefore the likelihood of success in the boundary correction step should be greater.

Step 3. We compute a trial (predicted) point v using the current point z, the tangent vector, and the stepsize. We then proceed to execute an ordinary predictor-corrector step in the current cell; this is done generally as in Program 1 on pp. 266–272 of [4], except that if at any point the Newton iterate leaves the current cell we cut the stepsize and reattempt the predictor-corrector procedure. We use the Jacobian at the predicted point during the Newton iteration, rather than recomputing it at each point: that is, we have a modified Newton corrector.

Step 4. We check to see if the new point $z=(x,\lambda)$ has $\lambda > 1 - ptol$, where *ptol* is a specified tolerance. If so, we switch to a Newton steplength as is done in [4]. If a Newton steplength is being used, we also check to see if the stepsize h satisfies a termination criterion using a prescribed value *hmin1*; if so, we stop the procedure and report the current values of x, λ, and related quantities as the solution. Otherwise we update the stepsize as described in Program 1 of [4] and return to *Step 2.*

If at any time the steplength h becomes less in absolute value than a prescribed value *hmin2*, then we terminate the procedure and report failure. We also make abnormal stops in case the number of evaluations of the Jacobian exceeds a prescribed number, or if the conditioning of the Jacobian becomes too bad.

In this section we have given a brief review of predictor-corrector methods and have explained at a general level the numerical procedure we used for the implementation. In the next section we exhibit some preliminary numerical results from this work.

4. PRELIMINARY NUMERICAL RESULTS

In this section we give some preliminary computations illustrating the numerical performance of the nonsmooth continuation method on numerical examples of two types. The first type consists of variational inequalities derived from the first-order necessary optimality conditions for nonlinear programming, as described in Sect.2. Within this type we give results for two problems. The first of these is a constrained Rosenbrock function that we used as a convenient source of small test examples by formulating the first-order optimality condition for the problem of minimizing the objective ϕ over the polyhedral convex set C as the normal-map equation $(d\phi)_C(x) = 0$. The second problem is the well known Shell dual problem originally given as Test Problem No. 2 on p. 22 of [10] and frequently employed since then as a test for nonlinear-programming algorithms. This problem has 15 variables, five quadratic inequality constraints, and 15 nonnegativity conditions on the variables, for a total of 20 constraints in all. The other problem type consisted of the parametric invariant capital-stock problem of Hansen and Koopmans, explained in Sect.2 above. The values used for the parameter α were those employed in [17], namely 0.7, 0.8, and 0.9. The other data were as in Sect.8 of [17]. The resulting problem has 18 variables and the convex set C is $\mathbb{R}^{16}_{+} \times \mathbb{R}^2$.

The information we provide here for each problem includes the starting points x_0 used for the various runs reported. Next are the numbers of successful interior and boundary predictor-corrector steps taken by the method, designated by s_I and s_B respectively. After this come the number n_J of Jacobian evaluations (reported since evaluation and QR factorization of the Jacobian is generally the most time-consuming part of the algorithm) and the number n_F of function evaluations required in addition to any function evaluations carried out as part of the Jacobian calculation. Finally we report the running time t in seconds on a DECstation 5000/125, and a number labeled "Residual," which is an upper bound on the absolute values of the components of $f_C(\bar{x})$ at the reported solution $\bar{x}$. The running time measurements are user times measured with the `time` routine of Ultrix 4.3A, which is not always consistent: we observed up to 5% variation on runs of the same problem from the same starting point. Times are reported to the nearest tenth of a second.

For all problems reported here we set the initial stepsize h to 10^{-3} and the parameters $hmin1$ and $hmin2$ to 10^{-5} and 10^{-20} respectively. $hmax$ was 20.0 except for the Rosenbrock problem, whose ill conditioning caused us to use the smaller value of 0.1 for $hmax$. The criterion *ptol* for switching to the Newton steplength was set in all problems to 10^{-3}.

We note that, as is discussed in Sect.5, our numerical implementation is currently not very sophisticated. There are several places in the program where improvements might be made to shorten the running time, and definitive time comparisons with other methods should probably await those improvements.

The constrained Rosenbrock problem was

$$\text{minimize } 100(x_2 - x_1^2)^2 + (1 - x_1)^2$$

$$\text{subject to } \begin{pmatrix} 1 & 0 \\ -3 & 1 \end{pmatrix} x \geq \begin{pmatrix} 1.05 \\ -3 \end{pmatrix}.$$

The constraints were designed to cut off the unconstrained minimizer at $(1,1)$; the constrained problem has an isolated minimizer at $(1.05, 1.1025)$, at which the spectral condition number of the objective Hessian is approximately 2,936. Using the convex homotopy (3.1) we obtained convergence to this minimizer with the results shown in Table 1.

Starting point	s_I	s_B	n_F	n_J	t	Residual
(0,0)	64	1	203	68	0.0	3.31E-07
(2,2)	33	2	131	40	0.0	2.44E-06
(-10,20)	367	2	997	373	0.1	3.94E-07
(-20,10)	239	0	554	240	0.1	2.33E-06

Table 1: Results for constrained Rosenbrock problem.

Table 2 presents results for the Shell dual problem of Colville. This problem was reformulated with Lagrange multipliers, in the way explained in Sect.2, yielding a variational inequality over the polyhedral set $\mathbb{R}^{20}_{+}$. The starting points are indicated in the table.

Starting point	s_I	s_B	n_F	n_J	t	Residual
$(-1,-1,\ldots,-1)$	83	24	414	135	2.5	2.05E-06
$(-10,-10,\ldots,-10)$	66	16	289	100	1.9	5.93E-07
$(+15,-15,+15,\ldots,-15)$	123	12	532	149	2.7	2.02E-07
$(20,20,\ldots,20)$	115	10	479	138	2.3	1.91E-09

Table 2: Results for Shell dual problem.

Table 3 presents results for the parametric invariant capital stock problem of [17]. It is important to note that this problem does *not* satisfy the general assumptions about differentiability, etc. that are normally made in applying continuation methods, because the function $v(x)$ whose derivative appears in (2.7) is

$$v(x) = (x_1 + 2.5x_2)^{0.2}(2.5x_3 + x_4)^{0.2}(2x_5 + 3x_6)^{0.2}, \tag{4.1}$$

and therefore the function F appearing in the resulting generalized equation contains singularities whenever any pair of variables appearing in parentheses in (4.1) has both of its variables equal to zero.

The first two columns in Table 3 give the value of α and a number β; the starting point was then taken to be $(\beta,\ldots,\beta)$. The remaining columns are the same as in Tables 1 and 2.

5. CONCLUSION

In this paper we have exhibited some preliminary results from the implementation of a path-following method for normal maps. This method solves generalized equations of the form (2.1) in which the set C is a polyhedral convex set, by converting the generalized equation to an equivalent normal-map equation. The algorithm is based on the well known predictor-corrector method for smooth functions in a version described

α	β	s_I	s_B	n_F	n_J	t	Residual
0.7	0.1	42	14	234	74	3.9	1.22E-08
0.7	10.0	135	22	630	184	4.1	2.25E-07
0.8	2.0	67	16	358	103	3.9	3.74E-08
0.8	8.0	127	20	568	169	4.2	2.53E-07
0.8	10.0	138	22	614	185	4.5	2.10E-07
0.9	10.0	128	16	567	162	4.2	9.41E-09
0.9	100.0	279	22	1220	330	11.7	1.51E-06

Table 3: Results for Hansen-Koopmans capital stock problem.

by Allgower and Georg in [4], but it operates in the piecewise linear normal manifold induced by the convex set, and thus extends and implements earlier ideas of Alexander *et al.* [1, 2, 3]. However, the results here are preliminary both because we have not yet tried the method on a wide variety of problems and because further implementation work still needs to be done to improve the speed of the algorithm as well as other aspects of its performance.

We thank E. L. Allgower and K. Georg for making available their implementation of the predictor-corrector algorithm for smooth functions. The existence of this implementation saved a great deal of work in developing the code for our implementation.

ACKNOWLEDGMENT

The research reported here was sponsored by the Air Force Office of Scientific Research, Air Force Materiel Command, USAF, under grant number F49620-95-1-0222, and by the National Science Foundation under grant number CCR-9109345. The U. S. Government has certain rights in this material, and is authorized to reproduce and distribute reprints for Governmental purposes notwithstanding any copyright notation thereon. The views and conclusions contained herein are those of the authors and should not be interpreted as necessarily representing the official policies or endorsements, either expressed or implied, of the sponsoring agencies or the U. S. Government.

REFERENCES

[1] J. C. Alexander, "The topological theory of an embedding method". In: "Continuation methods," H. Wacker, ed., Academic Press, New York, pp. 37–68, 1978.

[2] J. C. Alexander, R. B. Kellogg, T.–Y. Li, and J. A. Yorke, "Piecewise smooth continuation". Manuscript, 1979.

[3] J. C. Alexander, T.–Y. Li and J. A. Yorke, "Piecewise smooth homotopies". In: "Homotopy methods and global convergence," B. C. Eaves, F. J. Gould, H.–O. Peitgen, and M. J. Todd, eds., Plenum Press, New York, pp. 1–14, 1983.

[4] E. L. Allgower and K. Georg, "Numerical Continuation Methods: An Introduction". Springer-Verlag, Berlin 1990.

[5] E. L. Allgower and K. Georg, "Continuation and path following". *Acta Numerica* **2** (1993) 1–64.

[6] E. L. Allgower and K. Georg, "Numerical path following". In: "Handbook of Numerical Analysis," P. G. Ciarlet and J. L. Lions, eds., North-Holland, Amsterdam (forthcoming).

[7] S. C. Billups, "Algorithms for Complementarity Problems and Generalized Equations." Dissertation, Department of Computer Sciences, University of Wisconsin–Madison, 1995.

[8] M. Cao and M. C. Ferris, "A pivotal method for affine variational inequalities." *Technical Report No. 1114,* Computer Sciences Department, University of Wisconsin–Madison, Madison, WI, October 1992; forthcoming in *Mathematics of Operations Research.*

[9] C. Chen, "Smoothing Methods in Mathematical Programming." Dissertation, Computer Sciences Department, University of Wisconsin–Madison, Madison, WI 1995.

[10] A. R. Colville, "A comparative study on nonlinear programming codes". *IBM New York Scientific Center Report No. 320-2949,* IBM Corporation, Philadelphia Scientific Center [*sic*], Philadelphia, PA 1968.

[11] B. C. Eaves, "On the basic theorem of complementarity". *Mathematical Programming* **1**, pp. 68–75, 1971.

[12] B. C. Eaves," A short course in solving equations with PL homotopies". *in:* "Nonlinear Programming, SIAM-AMS Proceedings Vol 9," R. W. Cottle and C. E. Lemke, eds., American Mathematical Society, Providence, RI, pp. 73–143, 1976.

[13] B. C. Eaves, "A locally quadratically convergent algorithm for computing stationary points". *Technical Report,* Department of Operations Research, Stanford University, Stanford, CA 1978.

[14] B. C. Eaves and H. Scarf, "The solution of systems of piecewise linear equations." *Mathematics of Operations Research* **1**, pp. 1–27, 1976.

[15] F. Facchinei and J. Soares, "Testing a new class of algorithms for nonlinear complementarity problems". In: "Variational Inequalities and Network Equilibrium Problems," F. Giannessi and A. Maugeri, eds., Plenum Press, New York and London, pp. 69–83, 1995.

[16] J. Guddat, F. Guerra Vasquez and H. Th. Jongen, "Parametric Optimization: Singularities, Pathfollowing and Jumps." Teubner, Stuttgart; Wiley, Chichester 1990.

[17] T. Hansen and T. C. Koopmans, "On the definition and computation of a capital stock invariant under optimization". *Journal of Economic Theory* **5**, pp. 487–583, 1972.

[18] P. T. Harker, "Accelerating the convergence of the diagonalization and projection algorithms for finite-dimensional variational inequalities". *Mathematical Programming* **41**, pp. 29–59, 1988.

[19] P. T. Harker and J. S. Pang, "Finite-dimensional variational inequality and nonlinear complementarity problems: A survey of theory, algorithms and applications". *Mathematical Programming* **48**, pp. 161–220, 1990.

[20] N. H. Josephy, "Newton's Method for Generalized Equations and the PIES Energy Model." Dissertation, Department of Industrial Engineering, University of Wisconsin–Madison, Madison, WI 1979.

[21] N. H. Josephy, "Newton's method for generalized equations." *Technical Summary Report No. 1965,* Mathematics Research Center, University of Wisconsin-Madison, Madison, WI 1979.

[22] M. Kojima, "Strongly stable stationary solutions in nonlinear programs". In: "Analysis and Computation of Fixed Points," S. M. Robinson, ed., Academic Press, New York, pp. 93–138, 1980.

[23] M. Kojima and R. Hirabayashi, "Continuous deformation of nonlinear programs". In: "Sensitivity, Stability, and Parametric Analysis," A. V. Fiacco, ed., *Mathematical Programming Study 21,* North-Holland, Amsterdam 1984.

[24] T. De Luca, F. Facchinei, and C. Kanzow, "A semismooth equation approach to the solution of nonlinear complementarity problems". Preprint, 1995.

[25] J. G. MacKinnon, "A technique for the solution of spatial equilibrium models". *Journal of Regional Science* **16**, pp. 293–307, 1976.

[26] O. L. Mangasarian, "Equivalence of the complementarity problem to a system of nonlinear equations". *SIAM Journal on Applied Mathematics* **31**, pp. 89–92, 1976.

[27] N. Megiddo and M. Kojima, "On the existence and uniqueness of solutions in nonlinear complementarity theory". *Mathematical Programming* **12**, pp. 110–130, 1977.

[28] J. Outrata and J. Zowe, "A Newton method for a class of quasi-variational inequalities". *Computational Optimization and Applications* **4**, pp. 5–21, 1995.

[29] J.-S. Pang, "Solution differentiability and continuation of Newton's method for variational inequality problems over polyhedral sets". *Journal of Optimization Theory and Applications* **66**, pp. 121–135, 1990.

[30] J. S. Pang, "Newton's method for B-differentiable equations". *Mathematics of Operations Research* **15**, pp. 311–341, 1990.

[31] J.-S. Pang and Z. Wang, "Embedding methods for variational inequality and nonlinear complementarity problems". *Technical Report 5–26,* Department of Mathematical Sciences, The Johns Hopkins University, Baltimore, MD 1991.

[32] L. Qi and J. Sun, "A nonsmooth version of Newton's method". *Mathematical Programming* **58**, pp. 353–367, 1993.

[33] S. M. Robinson, "An implicit-function theorem for generalized variational inequalities". *Technical Summary Report No. 1672,* Mathematics Research Center, University of Wisconsin–Madison 1976.

[34] S. M. Robinson, "Generalized equations". In: "Mathematical Programming: The State of the Art, Bonn 1982," A. Bachem, M. Grötschel, and B. Korte, eds., Springer-Verlag, Berlin 1983.

[35] S. M. Robinson, "Normal maps induced by linear transformations". *Mathematics of Operations Research* **17**, pp. 691–714, 1992.

[36] S. M. Robinson, "Newton's method for a class of nonsmooth functions". *Set-Valued Analysis* **2**, pp. 291–306, 1994.

[37] R. T. Rockafellar, "Convex Analysis". Princeton University Press, Princeton, NJ 1970.

[38] H. Sellami, "A Continuation Method for Normal Maps." Dissertation, Departments of Mathematics and Industrial Engineering, University of Wisconsin–Madison, Madison, WI 1994.

[39] H. Sellami, "A homotopy continuation method for solving generalized equations." Preprint, 1995.

[40] H. Sellami and S. M. Robinson, "A continuation method for normal maps". *Technical Report 93-8,* Department of Industrial Engineering, University of Wisconsin–Madison, 1993.

[41] H. Sellami and S. M. Robinson, "Implementation of a continuation method for normal maps." Preprint, 1995.

[42] M. Solodov, "Nonmonotone and Perturbed Optimization". Dissertation, Computer Sciences Department, University of Wisconsin–Madison, Madison, WI 1995.

[43] G. Stampacchia, "Formes bilinéaires coercitives sur les ensembles convexes". *Comptes Rendus de l'Académie des Sciences de Paris* **258**, pp. 4413–4416, 1964.

[44] G. Stampacchia, "Variational inequalities". In: "Theory and Applications of Monotone Operators" (Proceedings of a NATO Advanced Study Institute held in Venice, Italy June 17–30, 1968), Edizioni "Oderisi," Gubbio, Italy 1969.

[45] M. J. Todd, "A note on computing equilibria in economies with activity analysis models of production". *Journal of Mathematical Economics* **6**, pp. 135–144, 1976.

[46] Z. Wang, "Continuation Methods for Solving the Variational Inequality and Nonlinear Complementarity Problems." Dissertation, Department of Mathematical Sciences, The Johns Hopkins University, Baltimore, MD 1990.

[47] L. T. Watson, "Globally convergent homotopy methods." A tutorial, *Technical Report 86-22*, Department of Industrial and Operations Engineering, University of Michigan, Ann Arbor, MI 1986.

[48] R. B. Wilson, "A Simplicial Method for Concave Programming". Dissertation, Graduate School of Business Administration, Harvard University, Boston, MA 1963.

[49] R. B. Wilson, "Subroutine SOLVER description". Mimeographed manuscript 1964.

[50] B. Xiao and P. T. Harker, "A nonsmooth Newton method for variational inequalities, I: Theory". *Mathematical Programming* **65**, pp. 151–194, 1994.

[51] B. Xiao and P. T. Harker, "A nonsmooth Newton method for variational inequalities, II: Numerical results". *Mathematical Programming* **65**, pp. 195–216, 1994.

ABS METHODS FOR KT EQUATIONS

Spedicato Emilio[1], Chen Z.[2], Bodon E.[1]

[1]Department of Mathematics
University of Bergamo
Bergamo 24129, Italy
e–mail: emilio@unibg.it
[2]Department of Basic Sciences
Agricultural Engineering University
Beijing 100083, China

Abstract. In this paper we present a class of ABS methods for solving the KT equations. We compare several methods in this class with the classical methods of Aasen and the method based upon the QR factorization with Householder rotations. When the number of degrees of freedom is small two of the considered ABS methods are faster than the Aasen and the QR based methods by a factor respectively about 2 and 4. Moreover when the first block in the KT equations is diagonal and a sequence of problems have to be solved where only such a block changes, for small number of degrees of freedom the solution can be updated in order two operations by the ABS methods, while order three operations are required by the other methods. Finally, numerical testing over 300 problems has shown that the ABS methods give more accurate results in about 80

Key Words. Indefinite symmetric equations, KT equations, Aasen method, QR factorization, ABS methods.

1. INTRODUCTION

The KT (Kuhn-Tucker) equations, which should be more appropriately be named KKKT (Kantorovich-Karush-Kuhn-Tucker) equations, are the special linear system $Ax = b$, where the matrix A has the form

$$A = \begin{bmatrix} G & C^T \\ C & 0 \end{bmatrix} \tag{1.1}$$

where $G \in \mathbf{R}^{n,n}$, $C \in \mathbf{R}^{m,n}$. If G is nonsingular, then matrix A is nonsingular iff $CG^{-1}C^T$ is nonsingular. Usually G is nonsingular, symmetric and positive definite. In several cases arising in interior point methods G is diagonal with positive elements. Here we will assume that G is symmetric positive definite, even if some of the methods considered later can be applied under less restrictive assumptions. Notice that in such a case A is symmetric, but indefinite. KT equations are important because they char-

Nonlinear Optimization and Applications
Edited by G. Di Pillo and F. Giannessi, Plenum Press, New York, 1996

acterize the stationarity conditions of first order in linearly constrained optimization and they appear in several other problems.

We will consider some classical and some ABS methods that take into account the structure of A. We will compare the storage and complexity of these methods, showing that for the case when m is close to n, i.e. when there are few degrees of freedom in the context of a constrained optimization problem, the ABS methods are faster, up to a factor more than four. Numerical experiments will moreover show that ABS methods are more accurate than both the Aasen method, considered to be the marginally most accurate of the methods for indefinite systemsand the special QR based method usually utilized for this problem.

1.1. Classical methods for KT equations

KT equations can be solved by general methods for indefinite symmetric systems. Among such methods, see Broyden (1994) for a review, we quote the method of Bunch and Parlett (1971), see also Bunch and Kaufmann (1977), and several conjugate gradient type methods, see for instance Stoer and Freund (1982) or Broyden (1993). A method shown in numerical experiments by Barwell and George (1976) to be the marginally most accurate is the method of Aasen (1971), based upon the following factorization of A

$$PAP^T = LTL^T \tag{1.2}$$

where P is a permutation matrix that expresses the row-column interchanges which are a feature of this method, L is lower triangular and T is symmetric tridiagonal.

Another method commonly used, see Gill, Murray and Wright (1991), is based upon a QR factorization of C and use of the structure of A, that we now describe. Let us write the system $Ax = b$ in the following form, with obvious partitioning of x and b

$$\begin{bmatrix} G & C^T \\ C & 0 \end{bmatrix} \begin{bmatrix} p \\ z \end{bmatrix} = \begin{bmatrix} g \\ c \end{bmatrix} \tag{1.3}$$

Let Q be the orthogonal matrix, usually a product of Householder matrices, such that

$$QC^T = \begin{bmatrix} U \\ 0 \end{bmatrix} \tag{1.4}$$

where U is upper triangular and nonsingular. From the first block row of equation (1.3), premultiplying by Q, we get

$$QGQ^TQp + QC^Tz = Qg \tag{1.5}$$

Define now $W = QGQ^T$, $s = Qp$, $h = Qg$. From (1.4), with obvious partitioning, (1.5) implies

$$\begin{bmatrix} W_{11} & W_{12} \\ W_{21} & W_{22} \end{bmatrix} \begin{bmatrix} s_1 \\ s_2 \end{bmatrix} + \begin{bmatrix} U \\ 0 \end{bmatrix} z = \begin{bmatrix} h_1 \\ h_2 \end{bmatrix} \tag{1.6}$$

or

$$G_{11}s_1 + G_{12}s_2 + Uz = h \tag{1.7}$$

$$G_{21}s_1 + G_{22}s_2 = h \tag{1.8}$$

Now from (1.3) we have $CQ^TQp = c$, which, using (1.4) and the above partitioning of p, gives

$$U^T s_1 = c \tag{1.9}$$

Hence s_1 is obtained by (1.9) and s_2 by (1.8) since G_{22} is easily shown to be nonsingular if A is nonsingular. From (1.7) we obtain z. Finally p is obtained from s by $p = Q^T s$.

If Q is obtained via Householder rotation matrices, then this method has a storage need of $nm + n^2/2 + (n-m)^2/2$. The number of multiplications is $5n^2m/2 + (n-m)^3 - m^3/3$ (lower order terms being neglected).

2. THE ABS METHODS

The basic ABS class has been introduced by Abaffy, Broyden and Spedicato (1984) for solving the system $Ax = b$, $x \in \mathbf{R}^n$, $b \in \mathbf{R}^m$, $m < n$, $A = (a_1, \dots, a_m)^T$. It is based upon the following recursions, where e_i the i–th unit vector.

The basic ABS class

(A) Give an arbitrary vector $x_1 \in \mathbf{R}^n$, an arbitrary nonsingular matrix $H \in \mathbf{R}^{n,n}$, set $i = 1$.

(B) Compute $s_i = H_i a_i$ and $\tau_i = a_i^T x_i - b^T e_i$. If $s_i \neq 0$ go to (C). If $s_i = 0$ and $\tau_i = 0$ set $x_{i+1} = x_i$, $H_{i+1} = H_i$ and go to (F), the i-th equation is a linear combination of the previous equations. Otherwise stop, the system has no solution.

(C) Define the search vector $p_i \in \mathbf{R}^n$ by

$$p_i = H_i^T z_i \tag{2.1}$$

where z_i, the parameter of Broyden, is arbitrary save that

$$z_i^T H_i a_i \neq 0 \tag{2.2}$$

(D) Update the estimate of the solution x_i by

$$x_{i+1} = x_i - \alpha_i p_i \tag{2.3}$$

where the stepsize α_i is given by

$$\alpha_i = (a_i^T x_i - b^T e_i)/a_i^T p_i \tag{2.4}$$

(E) Update the (Abaffian) matrix H_i by

$$H_{i+1} = H_i - H_i a_i w_i^T H_i / w_i^T H_i a_i \tag{2.5}$$

where $w_i \in \mathbf{R}^n$, the parameter of Abaffy, is arbitrary save that

$$w_i^T H_i a_i \neq 0 \tag{2.6}$$

(F) Increment the index i by one and go to (B).

The above recursions define a class of algorithms, whose parameters are z_i, w_i and H_1, the parameter of Spedicato. The algorithm is well defined iff the inequalities (2.2) and (2.6) are satisfied. These conditions are called the well definiteness conditions. The matrices H_i are called, since the first ABS International Conference, held in Luoyang in 1991, Abaffian matrices or simply Abaffians.

In the following we will assume for simplicity that rank$(A) = m$, which implies $H_i a_i \neq 0$ for all i's. We define the n by i matrices A_i, W_i, P_i by

$$A_i = (a_1, \ldots, a_i) \tag{2.7}$$

$$W_i = (w_1, \ldots, w_i) \tag{2.8}$$

$$P_i = (p_1, \ldots, p_i) \tag{2.9}$$

Then some properties of the above recursions, see for proofs Abaffy and Spedicato (1989), are the following:

(a) Implicit factorization property

$$A_i^T P_i = L_i \tag{2.10}$$

with L_i nonsingular lower triangular. Notice that, if $m = n$, then P_n is a factor of the inverse of A.

(b) Null space characterizations

$$\text{Null}(H_{i+1}) = \text{Range}(A_i) \tag{2.11}$$

$$\text{Null}(H_{i+1}^T) = \text{Range}(W_i) \tag{2.12}$$

(c) Generalized idempotency

$$H_i H_1^{-1} H_i = H_i \tag{2.13}$$

(d) Compact formula for the Abaffian (formula of Broyden)

$$H_{i+1} = H_1 - H_1 A_i (W_i^T H_1 A_i)^{-1} W_i^T H_1 \tag{2.14}$$

(e) Representation of the Abaffian in terms of $2i$ vectors

$$H_{i+1} = H_1 - S_i U_i^T \tag{2.15}$$

where the matrices $S_i = (s_1, \ldots, s_i)$ and $U_i = (u_1, \ldots, u_i)$ are defined by $s_1 = H_1 a_1$, $u_1 = H_1^T w_1$ and

$$s_i = H_1 a_i - \sum_{k<i} s_k u_k^T a_i \tag{2.16}$$

$$u_i = H_1^T w_i - \sum_{k<i} u_k s_k^T w_i \tag{2.17}$$

(f) Representation of all solutions of the first i equations

$$x = x_{i+1} + H_{i+1}^T q \tag{2.18}$$

where $q \in \mathbf{R}^n$ is arbitrary.

Particular algorithms are obtained by specific choices of the parameters H_1, z_i, w_i. The important implicit LU algorithm of the basic ABS class corresponds to the following choices

$$H_1 = I, \quad z_i = w_i = e_i \tag{2.19}$$

where e_i is the i–th unit vector. Among the properties of this algorithm the following are recalled:

(a') For $m = n$ the algorithm is well defined iff A is regular (strongly nonsingular), i.e. if the principal submatrices $A^{(i)}$ are nonsingular for $i = 1, \dots, n$. Otherwise column pivoting is necessary (or, if $m = n$, row pivoting).

(b') Only the first i components of p_i can be nonzero and $p_i^T e_i = 1$. Hence if $m = n$, (2.10) implies the factorization $A = LU$, with $L = L_n$, $U = P_n^{-1}$ unit upper triangular.

(c') The Abaffian H_{i+1} has the following form, with $S_i \in \mathbf{R}^{n-i,i}$, I_{n-i} the identity matrix in $\mathbf{R}^{n-i,n-i}$:

$$H_{i+1} = \begin{bmatrix} 0 & \dots & 0 \\ \dots & \dots & \dots \\ 0 & \dots & 0 \\ S_i & & I_{n-i} \end{bmatrix} \tag{2.20}$$

(d') The number of multiplications required by the algorithm is $nm^2 - 2m^3/3$ plus lower terms, hence, for $m = n$, $n^3/3 + O(n^2)$.

(e') If x_1 is the zero vector, then x_{i+1} has the i–th component equal to one and the last $n - i$ components identically zero, therefore it is a (special) basic type vector.

Another important algorithm is the Huang or implicit Gram-Schmidt algorithm, obtained by the choices

$$H_1 = I, \quad z_i = w_i = a_i \tag{2.21}$$

This algorithm is well defined without conditions on A, the search vectors p_i are orthogonal and coincide with the vectors obtained by applying the Gram-Schmidt orthogonalization procedure on the rows of A. It requires $3mn^2/2$ multiplications in the standard implementation. A fundamental property of the Huang algorithm is that, if x_1 is the zero vector, then x_{i+1} is the solution of least Euclidean norm of the first i equations and moreover the solution x^+ of least Euclidean norm of the whole system is approached by the sequence x_i monotonically and from below (in the Euclidean norm).

A variation of the Huang algorithm, which is more stable in non exact arithmetic, is obtained by taking $z_i = H_i a_i$ and updating the Abaffian by the following formula, defining the so called modified Huang algorithm

$$H_{i+1} = H_i - p_i p_i^T / p_i^T p_i \tag{2.22}$$

A generalization of the Huang algorithm is the symmetric algorithm, where H_1 is arbitrary symmetric positive definite and the parameters z_i, w_i are chosen as in the Huang algorithm.

3. THE METHOD OF XIA FOR KT EQUATIONS

Let us write the inverse of the KT matrix in the following form

$$\begin{bmatrix} G & C^T \\ C & 0 \end{bmatrix}^{-1} = \begin{bmatrix} V & S^T \\ S & U \end{bmatrix} \tag{3.1}$$

then, if G and $CG^{-1}C^T$ are nonsingular, the following formulas are true, see for instance Fletcher (1987)

$$V = G^{-1} - G^{-1}C^T(CG^{-1}C^T)^{-1}CG^{-1} \tag{3.2}$$

$$U = -(CG^{-1}C^T)^{-1} \tag{3.3}$$

$$S = (CG^{-1}C^T)^{-1}CG^{-1} \tag{3.4}$$

Assume that G is symmetric and positive definite. Then the following special case of the symmetric algorithm in the basic ABS class is well defined

$$H_1 = G^{-1} \tag{3.5}$$

This algorithm is known as the algorithm of Xia, having been introduced by Xia (1991) in the context of quadratic programming problems with equality constraints (whose optimality conditions are just the considered KT equations).

Let us first apply the algorithm of Xia for m steps on the matrix C and let H_{m+1} be the obtained Abaffian. From Broyden's compact formula for the Abaffian we obtain the identity

$$H_{m+1} = V \tag{3.6}$$

Let us now apply the algorithm of Xia on a nonsingular matrix C' whose first m rows consist of the matrix C and let P' be the m by m nonsingular matrix, whose columns are the search vectors generated by the algorithm of Xia. Write P' as

$$P' = (P, P'') \tag{3.7}$$

where P is generated from C and P'' from the remaining $n - m$ rows of C'. From a general property of the symmetric algorithm, see Abaffy and Spedicato (1989), the search vectors are H^{-1}–conjugate, hence in this case

$$P'^T G P' = D \tag{3.8}$$

with D diagonal nonsingular. By scaling the columns of P' we can assume without loss of generality that

$$P'^T G P' = I \tag{3.9}$$

From (40) we have $P'^T G = P'^{-1}$, or $P'^T = P'^{-1}G$ or

$$G^{-1} = P'P'^T = PP^T + P''P''^T \tag{3.10}$$

Now to determine S and U we only need the product CG^{-1}. From (2.10) it follows, with $L = L_n$

$$CP'' = 0, \quad CP = L \tag{3.11}$$

implying

$$CG^{-1} = LP^T \tag{3.12}$$

$$CG^{-1}C^T = LL^T \tag{3.13}$$

and

$$S = L^{-T}P^T \tag{3.14}$$

$$U = L^{-T}L^{-1} \tag{3.15}$$

Therefore only the matrix C is used and there is no need to know P''. The method of Xia is particularly convenient when a sequence of problems has to be solved where G is unchanged and changes are made only in C, as is the case typically when dealing with linear inequalities constraints in an active set framework. In such a case S and U can be updated in only order two multiplications. Also the case when G^{-1} is actually a Quasi-Newton approximation of the inverse Hessian and is updated by a rank-two correction can be dealt with in order two multiplications. For details see Xia, Liu and Zhang (1992), Xia (1994) and Xia, Zhang and Liu (1995).

4. A CLASS OF ABS METHODS FOR KT EQUATIONS

Here we develop a class of methods for solving the KT equations, whose parameters are the parameters H_1, z_i and w_i used in dealing with the subsystem of the KT equations containing m equations. Let us consider the KT equations in the form (3), namely

$$Gp + C^T z = g \tag{4.1}$$

$$Cp = c \tag{4.2}$$

We first deal with the subsystem (4.2) obtaining the following expression for its general solution

$$p = p_{m+1} + H_{m+1}^T q \tag{4.3}$$

with q arbitrary. The parameter choices made to construct p_{m+1} and H_{m+1} are arbitrary and define a class of algorithms.

Since the KT equations have a unique solution, there must be a choice of q in (4.3) which makes p be the unique first $n-$dimensional subvector of the solution x. Notice that since H_{m+1} is singular, q is not uniquely defined.

By multiplying equation (4.1) on the left by H_{m+1} and noticing that by properties of the Abaffians $H_{m+1}C^T = 0$, we obtain

$$H_{m+1}Gp = H_{m+1}g \tag{4.4}$$

an equation not containing z. Now there are two possibilities to determine p:

Case A

Consider the system formed by equations (4.2) and (4.4). Such a system is solvable but overdetermined. Since $\text{rank}(H_{m+1}) = n - m$, m equations will be eliminated in step (B) of any ABS algorithm applied to solve it. If we construct H_{m+1} with a generalized implicit LU algorithm, then H_{m+1} has the form

$$H_{m+1} = \begin{bmatrix} 0 \\ \\ S \end{bmatrix} \tag{4.5}$$

and the first m equations of (4.4) are trivially satisfied, leaving a determined system consisting of (4.2) and the equation

$$SGp = Sg \tag{4.6}$$

Case B

In equation (4.5) we substitute p with the expression of the general solution (4.4) obtaining

$$H_{m+1}GH_{m+1}^T q = H_{m+1}g - H_{m+1}Gp_{m+1} \tag{4.7}$$

The above system can again be solved by any ABS method for a particular solution q, m equations being again removed at step (B) of the ABS algorithm as linearly dependent. If again we use the generalized implicit LU algorithm, then we have

$$H_{m+1}GH_{m+1}^T = \begin{bmatrix} 0 & 0 \\ 0 & SGS^T \end{bmatrix} \tag{4.8}$$

and, letting $q = \begin{bmatrix} q_1 \\ q_2 \end{bmatrix}$, equation (4.7) is solved by taking q_1 arbitrary and q_2 satisfying the equation

$$SGS^T q_2 = Sg - SGp_{m+1} \tag{4.9}$$

Once p is determined, there are again two possible ways to determine z, namely:

Case C

Solve by any ABS method the overdetermined compatible system

$$C^T z = g - Gp \tag{4.10}$$

by eliminating at step (B) of the ABS algorithm $n - m$ dependent equations.

Case D

Let $P = (p_1, \ldots, p_m)$ be the matrix whose columns are the search vectors generated on the system $Cp = c$. Now $CP = L$, with L nonsingular lower diagonal. Multiplying equation (4.10) on the left by P^T we obtain the following triangular system, defining z uniquely

$$L^T z = P^T g - P^T Gp \tag{4.11}$$

It can be shown that the presented class of methods contains also the first ABS method for KT equations, proposed by Yang (1989). A first formulation of the proposed class was given by Chen and Spedicato (1993), in the framework of ABS methods for computing the null space of a matrix. Finally, another ABS method for KT equations has been proposed by Deng (1995).

5. NUMERICAL EXPERIMENTS

Some of the above described algorithms have been numerically tested to compare their accuracy on a set of test problems. The results that we will present indicate that the tested versions of the algorithms in the new ABS class are not only preferable in terms of overhead for the case when m is close to n, but they are usually more accurate, up to three orders in single precision, than both the Aasen and the QR method.

The tested methods are the following:

AASEN – the method of Aasen

QRGAU – the QR based method, using Householder rotation matrices with column pivoting to enhance stability. The indefinite system with the matrix G_{22} is solved with Gaussian elimination with partial pivoting

QRAAS – same method as QRGAU, except that the system with the matrix G is solved using Aasen method

IMLU1 – the new ABS method where H_{m+1} is built using the implicit LU algorithm and steps (A), (C) are used

IMLU2 – as above, but using steps (B) and (C)

MHUA1 – the new ABS method where H_{m+1} is built using the modified Huang algorithm in its standard form and steps (A) and (C). In solving the overdetermined system $H_{m+1}Ax = H_{m+1}c$ at the i–th step that equation is chosen having the largest residual

MHUA2 – the new ABS method where H_{m+1} is built using the modified Huang algorithm in the form using $2i$ vectors

XIA – the method of Xia, where $H_1 = G^{-1}$ is computed using Gaussian elimination with total pivoting

The following Tables give the storage requirement and the number of multiplications (only higher order terms) required by the above methods.

Table 1. Storage requirement

Method	general m	$m = 0$	$m = n$
AASEN	$(n+m)^2/2$	$n^2/2$	$2n^2$
QRGAU	$nm + n^2/2 + (n-m)^2/2$	n^2	$3n^2/2$
QRAAS	$nm + n^2/2$	$n^2/2$	$3n^2/2$
IMLU1	$3n^2/2$	$3n^2/2$	$3n^2/2$
IMLU2	$n^2 + m^2/2$	n^2	$3n^2/2$
MHUA1	$2n^2 + m^2/2 + nm$	$2n^2$	$7n^2/2$
MHUA2	$3n^2/2 + m^2/2 + nm$	$3n^2/2$	$3n^2$
XIA	$2n^2 + nm$	$2n^2$	$3n^2$

From Table 1 we see that for small m the ABS methods require more memory than AASEN or QRAAS. For large m methods IMLU1 and IMLU2 have similar requirement as the QR methods.

Table 2. Overhead

Method	general m	$m = 0$	$m = n$
AASEN	$(n+m)^3/6$	$n^3/6$	$4n^3/3$
QRGAU	$5n^2m/2 - m^3/3 + (n-m)^3/3$	$n^3/3$	$13n^3/6$
QRAAS	$5n^2m/2 - m^3/3 + (n-m)^3/6$	$n^3/3$	$13n^3/6$
IMLU1	$n^3/3 + n^2m - nm^2 + m^3/6$	$n^3/3$	$n^3/2$
IMLU2	$n^3/3 + n^2m/2 - m^3/3$	$n^3/3$	$n^3/2$
MHUA1	$4n^3$	$4n^3$	$4n^3$
MHUA2	$5n^3/2 + n^2m - nm/2$	$5n^3/2$	$3n^3$
XIA	$7n^3/6 + 3n^2m/2$	$7n^3/6$	$19n^3/6$

From Table 2 we see that for small m the cheapest method is AASEN, while the QR methods and IMLU1, IMLU2 have the same cost. For large m, IMLU1 and IMLU2 are more than twice cheaper than AASEN and more than four times faster than both versions of the QR algorithm, a fact of remarkable interest.

The algorithms have been tested on 300 problems, having the following features:

- G is nonsingular, symmetric, but not always positive definite (in this case method XIA may not be well defined)
- C is full rank
- the entries of G, C and of the solution are integers and the right hand side is computed exactly

The following families of problems were considered:

1– G with random elements, but diagonally dominant

2– G with random elements

3– G diagonally dominant ill conditioned, having the i–th and j–th rows and columns equal save for the first element, where $G_{ik} = G_{ki} = 0$, $G_{jk} = G_{kj} = 2^{-p}$, i, j, k arbitrary. In such a case we set $x_i = x_j = 2^p$, p an integer.

4– G random ill conditioned, with the above procedure

5– G the Micchelli ill conditioned matrix, where $G_{ij} = |i-j|$

6– G a generalized Micchelli ill conditioned matrix, where $G_{ij} = |v_i - v_j|$, $v_1, \ldots, v_n$ arbitrary integers

7– C formed by m arbitrary rows of an m by m diagonally dominant matrix with random entries

8– C having random entries

9– C formed as in case 7 from a diagonally dominant but ill conditioned matrix, generated as in case 3

10– C formed from a random ill conditioned matrix, generated as in case 3

In the experiments some combinations of the above families were considered. All combinations were considered of the following values for n and m: $n = 10, 20, 30, 40, 50$; $m = n/5, n/2, n-1$. Overall 300 problems were tried.

The codes were written in FORTRAN 77 and run in single and double precision.

The performance of the methods was measured by the relative errors ES and ER, defined as follows, with x the computed solution and x^+ the exact solution:

$$ES = \|x - x^+\| / \|x^+\| \tag{5.1}$$

$$ER = \|Ax - b\| / \|b\| \tag{5.2}$$

For a full presentation of the results, see Bodon and Spedicato (1995). The following Table 3 and 4 give a synthetic presentation of the results, in single precision, by comparing the different methods in this way: at the intersection of the i–th row with the j–th column there are two numbers α, β. α indicates how many times the method

at the heading of the i–th row gives a smaller error than the method at the heading of the j–th column. β indicates how many times the errors differ from only the third digit on.
From an inspection of Table 3 and Table 4 below, the following observations can be made:

- there is little difference in the performance of methods AASEN, QRGAU and QRAAS. QRGAU seems to be marginally more accurate than QRAAS
- there is little difference between methods IMLU1 and IMLU2. IMLU1 is marginally more accurate than IMLU2, which however has less storage for small m
- method MHUA2 is marginally more accurate than MHUA1 and moreover it has less storage and less overhead
- method MHUA2 is marginally more accurate than both IMLU1 and IMLU2 in ER, but is marginally less accurate in ES
- method XIA gives the worst performance. Closer examination of the result indicates that this is due to its sensitivity to errors in the initial $H_1 = G^{-1}$, when G is ill conditioned. It can solve accurately problems where G is well conditioned but C is ill conditioned
- IMLU1, IMLU2, MHUA1, MHUA2 are all more accurate than AASEN and the QR codes, especially in terms of residual error

The above observations lead to the important conclusion, which certainly needs to be confirmed by further experiments, that in the new class of ABS methods for KT equations there are methods that are superior to the classical methods not only in terms of overhead, for large m, but also in terms of accuracy.

Table 3. Comparison of methods in solution error ES

	AASEN	QRGAU	QRAAS	IMLU1	IMLU2	MHUA1	MHUA2	XIA
AASEN		141 4	151 5	106 3	109 3	125 4	119 4	237 0
QRGAU	155 4		91 165	105 3	114 3	141 10	116 2	235 0
QRAAS	144 5	44 65		99 3	109 6	134 8	105 3	230 0
IMLU1	191 3	192 3	198 3		107 108	192 2	170 4	250 0
IMLU2	188 3	183 3	185 6	85 108		180 3	159 5	252 1
MHUA1	171 4	149 10	158 8	106 2	117 3		113 9	229 0
MHUA2	177 4	182 2	192 3	126 4	136 5	178 9		249 1
XIA	63 0	65 0	70 0	50 0	47 1	71 0	50 1	

Table 4. Comparison of methods in residual error ER

	AASEN	QRGAU	QRAAS	IMLU1	IMLU2	MHUA1	MHUA2	XIA
AASEN		142 3	155 1	56 2	78 6	62 8	41 2	264 1
QRGAU	155 3		107 119	49 8	81 0	55 3	35 5	262 0
QRAAS	144 1	74 19		45 3	75 3	40 2	22 5	261 0
IMLU1	242 2	243 8	252 3		145 66	163 11	135 8	281 2
IMLU2	216 6	219 0	222 3	89 66		150 4	126 6	281 0
MHUA1	230 8	242 3	258 2	126 11	146 4		82 19	279 1
MHUA2	257 2	260 5	273 5	157 8	168 6	199 19		279 0
XIA	35 1	38 0	39 0	17 2	19 0	29 1	21 0	

Some detailed results are given in the following Table.

Table 5. Results of some problems (A diagonally dominant)

n	m	method	B diag.dom.		B random		B ill cond.	
			ES	ER	ES	ER	ES	ER
50	25							
		AASEN	$3E-7$	$4E-7$	$9E-6$	$5E-7$	$2E-1$	$2E-6$
		QRGAU	$4E-7$	$4E-7$	$5E-6$	$4E-7$	$7E-2$	$4E-7$
		IMLU1	$1E-7$	$9E-8$	$5E-6$	$8E-7$	$6E-2$	$6E-7$
		MHUA2	$2E-7$	$2E-7$	$1E-6$	$2E-7$	$6E-1$	$4E-7$
		XIA	$3E-7$	$2E-7$	$3E-6$	$2E-7$	$3E-1$	$1E-6$
50	49							
		AASEN	$6E-7$	$4E-7$	$3E-4$	$4E-7$	$9E-1$	$7E-7$
		QRGAU	$5E-7$	$4E-7$	$8E-5$	$4E-7$	$4E-0$	$6E-7$
		IMLU1	$1E-7$	$9E-8$	$5E-5$	$3E-7$	$2E-0$	$4E-7$
		MHUA2	$2E-7$	$2E-7$	$1E-5$	$1E-7$	$2E-0$	$2E-7$
		XIA	$3E-7$	$2E-7$	$4E-5$	$2E-7$	$3E-0$	$2E-4$

6. THE CASE OF G DIAGONAL

An important case, arising in particular in the interior point methods for linear programming, is when G is diagonal, with positive diagonal elements. In such a case storage and overhead are reduced, as shown in the following Tables, where only the order three terms are considered.

Table 6. Storage when G is diagonal

Method	general m	$m=0$	$m=n$
AASEN	$nm+m^2/2$	0	$3n^2/2$
QRAAS	n^2+nm	n^2	$2n^2$
IMLU1	$3nm-2m^2+(n-3m)^2/4$	$n^2/4$	n^2
IMLU2	$nm+(n-m)^2$	n^2	n^2
MHUA2	$3n^2/2+nm+m^2/2$	$3n^2/2$	$3n^2$
XIA	$nm+n^2/2$	0	$3n^2/2$

Table 7. Overhead when G is diagonal

Method	general m	$m=0$	$m=n$
AASEN	$nm^2/2+m^3/2$	0	$2n^3/3$
QRAAS	$2n^3/3+3n^2m/2$	$2n^3/3$	$13n^3/6$
IMLU1	$n^3/6+n^2m/2-nm^2/2+m^3/3$	$n^3/6$	$n^3/2$
IMLU2	$n^3/3-n^2m/2+nm^2-m^3/3$	$n^3/3$	$n^3/2$
MHUA2	$5n^3/2+n^2m/2-nm^2/2$	$5n^3/2$	$5n^3/2$
XIA	$2nm^2$	0	$2n^3$

From the above Tables we can observe:

- for small m, AASEN and XIA have the least storage, while for large m the least storage pertains to methods IMLU1 and IMLU2
- for small m, AASEN and XIA are again the cheapest methods while for large m IMLU1 and IMLU2 are the fastest methods, more than twice faster than the QR based method

An important problem in interior point methods is to compute the solution for a sequence of problems where only G, which is diagonal, changes. In such a case the ABS methods, which initially work on the matrix C, which is unchanged, are advantaged, particularly for the case of large m, where the dominant cubic term decreases with m and disappears for $m=n$, so that the overhead is dominated by the second order terms. For such a case the overheads are given by the following Table (second order terms are omitted).

Table 8. Overhead when only (diagonal) G changes

Method	general m	$m=0$	$m=n$
AASEN	$nm^2/2+m^3/2$	0	$2n^3/3$
QRAAS	$n^3/2+(n-m)^3/6$	$2n^3/3$	$n^3/2$
IMLU1	$n^3/6+n^2m/2-3nm^2/2+5m^3/3$	$n^3/6$	0
IMLU2	$n^3/3-n^2m/2+m^3/6$	$n^3/3$	0
MHUA2	$5n^3/2-5nm^2/2$	$5n^3/2$	0
XIA	$2nm^2$	0	$2n^3$

We notice that the fact that the KT equations can be solved for large m with quadratic overhead was also observed by Spedicato and Yang (1993), using a modification of the QR factorization, and Li (1995). This result is certainly of great interest for the practical implementation of interior point methods. Preliminary numerical experiments indicate that method MHUA2, closely followed by IMLU1, is the most accurate for this type of problems.

REFERENCES

[1] J.O. Aasen, *"On the reduction of a symmetric matrix to tridiagonal form"*. BIT 11:233, 1971.

[2] J. Abaffy and E.Spedicato, *"ABS Projection Algorithms, Mathematical Techniques for Solving Linear and Nonlinear Equations"*. Ellis Horwood, Chichester 1989.

[3] J. Abaffy, C.G. Broyden and E. Spedicato, *"A class of direct methods for linear systems"*. Numerische Mathematik 45:361, 1984.

[4] W. Barwell and J.A.George, *"A comparison of algorithms for solving the symmetric indefinite systems"*. ACM Trans.Math. Soft. 2:242, 1976.

[5] E. Bodon, *"Numerical results on the ABS algorithms for linear systems of equations"*. Report DMSIA 93/9, 1993.

[6] E. Bodon and E. Spedicato, *"Some numerical experiments on KT type indefinite systems of linear equations"*. Preprint, University of Bergamo 1995.

[7] C.G.Broyden, *"Block conjugate gradient methods"*. Optimization Methods and Software 2:1, 1993.

[8] C.G.Broyden, *"Linear equations in optimisation"*. In "Algorithms for Continuous Optimization, the State of the Art", Kluwer, E. Spedicato editor 1994.

[9] J.R.Bunch and B.N. Parlett, *"Direct methods for solving symmetric indefinite systems of linear equations"*. SIAM J. Numer. Anal. 8:639, 1971.

[10] J.R.Bunch and L. Kaufmann, *"Some stable methods for calculating inertia and solving symmetric linear systems"*. Math. Comp. 31:162, 1977.

[11] Z. Chen and N. Deng, *"A general algorithm for underdetermined linear systems"*. Proceedings of the First ABS Conference, Bergamo University 1992.

[12] Z. Chen and E. Spedicato, *"Computing the null space of a matrix via the ABS approach"*. preprint, University of Bergamo 1993.

[13] N. Deng, *"An algorithm for KT equations"*. Communication, Second ABS Conference, Beijing 1995.

[14] P.E. Gill, W. Murray and M.H. Wright, *"Numerical linear algebra and optimization"*. Vol. 2, Addison-Wesley 1991.

[15] S. Li, *"Fast computation of Karmarkar's projections"*. Communication, Second ABS Conference, Beijing 1995.

[16] E. Spedicato, *"Ten years of ABS methods: a review of theoretical results and computational achievements"*. Survey on Mathematicsfor Industry, 3:217, 1993.

[17] E. Spedicato and E. Bodon, *"Solution of linear least squares via the ABS algorithm"*. Mathematical Programming 58:111, 1993.

[18] E. Spedicato and Z. Yang, *"A faster method for computing Karmarkar's projection when $m > n/2$"*. Report DMSIA 93/15, 1993.

[19] E. Spedicato and M.T. Vespucci, *"Variations on the Gram-Schmidt and the Huang algorithms for linear systems: a numerical study"*. Applications of Mathematics 2, 81-100, 1992.

[20] J. Stoer and R.Freund, *"On the solution of large indefinite systems of linear equations by conjugate gradients"*. in Computing Methods in Applied Sciences and Engineering V, North Holland, Amsterdam (R. Glowinski and J.L.Lions editors).

[21] Z. Xia, *"Quadratic programming via the ABS algorithm I:linear constraints"*. Report DMSIA 91/5, University of Bergamo 1991.

[22] Z. Xia, *"An efficient ABS method for linearly constrained optimization"*. Report DMSIA 94/15, University of Bergamo 1994.

[23] Z. Xia, Y. Liu and L. Zhang, *"Application of a representation of ABS updating matrices to linearly constrained optimization I"*. Northeast Operational Research 7, pp. 1-9, 1992.

[24] Z. Xia, L. Zhang and Y. Liu, *"An ABS algorithm for minimizing a nonlinear function subject to linear inequalities"*. To appear.

[25] Z. Yang, *"ABS algorithms for solving certain systems of indefinite equations"*. Report DMSIA 89/6, 1989.

CONTRIBUTORS

Bodon E.	Department of Mathematics, University of Bergamo, Bergamo 24129, ITALY
Buti Sara	Dipartimento di Matematica e Fisica, Università di Camerino, Camerino, ITALY.
Byrd Richard H.	Computer Science Department, University of Colorado at Boulder, Boulder, Colorado 80309, USA
Castellani Marco	Dept. of Mathematics, Univ. Pisa, Via Buonarroti 2, 56127 Pisa, ITALY. e–mail: castella@gauss.dm.unipi.it
Chen Z.	Department of Basic Sciences, Agricultural Engineering University, Beijing 100083, CHINA
Chen Xiaojung	School of Mathematics, The University of New South Wales, Sydney, NSW 2052, AUSTRALIA.
Conn Andrew R.	IBM T.J. Watson Research Center, P.O. Box 218, Yorktown Heights, N.Y.10598, USA
De Leone Renato	Dipartimento di Matematica e Fisica, Università di Camerino, Camerino, ITALY. e–mail: deleone@camcic.unicam.it
Dem'yanov Vladimir F.	St. Petersburg State University, Bibliotechnaya pl. 2 , Staryi Peterhof, 198904 St. Petersburg, RUSSIA. e–mail: demyanov@robot.apmath.lgu.spb.su
Diniz-Ehrhardt Maria A.	Department of Applied Mathematics, IMECC-UNICAMP, University of Campinas, CP 6065, CEP 13081-970 Campinas - SP, BRAZIL. e–mail: cheti@ime.unicamp.br
Di Pillo Gianni	Dipartimento di Informatica e Sistemistica, Università di Roma "La Sapienza". Via Buonarroti 12, 00185 - Roma, ITALY. e–mail: dipillo@itcaspur.caspur.it
Evtushenko Yuri G.	Computing Center of Russian Academy of Sciences, Vavilov Str.40, Moscow 117967 GSP-1, RUSSIA. fax: +7-095-938-20-80 e–mail: evtushenko@ccas.ru
Facchinei Francisco	Dipartimento di Informatica e Sistemistica, Università di Roma "La Sapienza", Via Buonarroti 12, 00185 - Roma, ITALY. e–mail: soler@peano.dis.uniroma1.it
Fernandes L.	Escola Superior de Tecnologia de Tomar, 2300 – Tomar, PORTUGAL.
Ferris Michael C.	Computer Sciences Department, University of Wisconsin, Madison, WI, USA. e–mail: ferris@cs.wisc.edu
Fischer Andreas	Institute of Numerical Mathematics, Technical University of Dresden, D-01062 Dresden, GERMANY. e–mail: fischer@math.tu-dresden.de
Fukushima Masao	Graduate School of Information Science, Nara Institute of Science and Technology, Ikoma, Nara 630-01, JAPAN

	e–mail: fuku@is.aist-nara.ac.jp
Giannessi Franco	Dipartimento di Matematica, Università di Pisa, Via Buonarroti 2, 56100-Pisa, ITALY. e–mail: gianness@gauss.dm.unipi.it
Grippo Luigi	Dipartimento di Informatica e Sistemistica, Università di Roma "La Sapienza", Via Buonarroti 12, 00185 - Roma, ITALY. e–mail: grippo@newton.dis.uniroma1.it
Jiang Houyuan	School of Mathematics, The University of New South Wales, Sydney, NSW 2052, AUSTRALIA. e-0-mail: jiang@solution.maths.unsw.edu.au
Júdice Joaquim	Departamento de Matematica, Universidade de Coimbra, 3000 Coimbra, PORTUGAL. e–mail: judice@lince.mat.uc.pt
Kanzow Christian	Institute of Applied Mathematics, University of Hamburg, Bundesstrasse 55 D-20146 Hamburg, GERMANY. e–mail: kanzow@math.uni-hamburg.de
Larsson Torbjörn	Division of Optimization, Department of Mathematics, Linköping Institute of Technology, S-581 83 Linköping, SWEDEN. e-mail: tl@math.liu.se
Levy Adam B.	Department of Mathematics, Bowdoin College, Brunswick, ME 04011, USA. e–mail: alevy@polar.bowdoin.edu
Lima A.	Faculdade de Ciências da Universidade do Porto, 4000 – Porto, PORTUGAL.
Lucidi Stefano	Dipartimento di Informatica e Sistemistica, Università di Roma "La Sapienza", Via Buonarroti 12, 00185 - Roma, ITALY. e–mail: lucidi@iasi.rm.cnr.it
Mallozzi Lina	Dipartimento di Matematica ed Applicazioni, Università di Napoli "Federico II", Complesso Monte S. Angelo, Via Cintia, 80126 Napoli, ITALY. e–mail: mallozzi@vm.cised.unina.it
Mangasarian Olvi L.	Computer Science Department, University of Wisconsin, Madison, WI, 53706 USA. e–mail: olvi@cs.wisc.edu
Mastroeni Giandomenico	Dipartimento di Matematica, Università di Pisa, Via Buonarroti 2, 56100-Pisa, ITALY. e–mail: giando@karush.dm.unipi.it
Martinez J.M.	Department of Applied Mathematics, IMECC-UNICAMP University of Campinas, CP 6065, 13081-970 Campinas SP, BRAZIL. e-mail: martinez@ccvax.unicamp.br
Merelli Emanuela	Scuola di Informatica, Università di Camerino, Via Roma 20, 62032 Camerino, ITALY. e–mail: merelli@camcie.unicam.it
Moré Jorge, J.	Mathematics and Computer Science Division, Argonne National Laboratory, 9700 South Cass Avenue, Argonne, IL 60439-4803, USA. e–mail: more@mcs.anl.gov

Morgan Jaqueline	Dipartimento di Matematica e Applicazioni, Università degli Studi di Napoli "Federico II", Complesso Monte S.Angelo, Via Cintia, 80126 Napoli, Italia.
Nocedal Jorge	Electrical Engineering and Computer Science, Northwestern University, Evaston, IL 60208-0001, USA. e–mail: nocedal@eecs.nwu.edu
Pappalardo Massimo	Dipartimento di Matematica, Università di Pisa, Via Buonarroti 2, 56100-Pisa, ITALY. e–mail: pappalar@dm.unipi.it
Pardalos Panos M.	Center for Applied Optimization and ISE Department, University of Florida, 303 Weil Hall, Gainesville, FL 32611-6595, USA e–mail:pardalos@math.ufl.edu
Patriksson Michael	Department of Mathematics, Box 4350, University of Washington Seattle, WA 98195-4350, USA e-mail: patrikss@math.washington.edu or mipat@math.liu.se
Qi Liqun	School of Mathematics, University of New South Wales, Sydney, NSW 2052, AUSTRALIA. e–mail: L.Qi@unsw.edu.au
Robinson Stephen M.	Department of Industrial Engineering, University of Wisconsin at Madison, 1513 University Avenue, Madison, WI, 53706-1572 USA. e-mail: smr@cs.wisc.edu
Rockafellar R.T.	Department of Mathematics, University of Washington GN-50, Seattle, WA 98195, USA. e–mail: rtr@math.washington.edu
Rutherford Thomas F.	Department of Economics, The University of Colorado, Boulder, Colorado 80309, USA
Sciandrone Marco	Dipartimento di Informatica e Sistemistica, Università di Roma "La Sapienza", Via Buonarroti 12, 00185 - Roma, ITALY. e–mail: sciandro@newton.dis.uniroma1.it
Sellami Hichem	Faculté des Sciences Economiques et de Gestion de Sfax, Université du Sud, Sfax, TUNISIE. e-mail: fsegs!sellamih@tunisia.eu.net
Spedicato Emilio	Dipartimento di Matematica, University of Bergamo, 24129 Bergamo, ITALY. e–mail: emilio@ibguniv.bitnet
Strömberg Ann-Brith	Department of Mathematics, Linköping Institute of Technology, S-581 83 Linköping, SWEDEN. e–mail: anstr@math.liu.se
Sun Defeng	School of Mathematics, The University of New South Wales, Sydney, NSW 2052, AUSTRALIA.
Toint Philippe L.	Department of Mathematics, Facultés Universitaires ND de la Paix 61, rue de Bruxelles, B-5000 Namur, BELGIUM. e–mail: pht@math.fundp.ac.be

Vitali G.Zadan	Computing Center of Russian Academy of Sciences, Vavilov Str.40, Moscow 117967 GSP-1, RUSSIA.
Wu Zhijun	Mathematics and Computer Science Division, Argonne National Laboratory, 9700 South Cass Avenue, Argonne, IL 60439-4803, USA.
Zhu Ciyou	Department of Electrical Engineering and Computer Science, Northwestern, University, Evanston Il 60208, USA

INDEX